Structural studies on molecules of biological interest

Dorothy Hodgkin: a portrait taken at the time of her award of the Nobel prize for chemistry, 1964. (Photograph by Lotte Meitner-Graf, London.)

Structural studies on molecules of biological interest

**A VOLUME IN HONOUR OF
PROFESSOR DOROTHY HODGKIN**

EDITED BY
GUY DODSON, JENNY P. GLUSKER,
AND DAVID SAYRE

CLARENDON PRESS · OXFORD
1981

Oxford University Press, Walton Street, Oxford OX2 6DP

OXFORD LONDON GLASGOW
NEW YORK TORONTO MELBOURNE WELLINGTON
KUALA LUMPUR SINGAPORE JAKARTA HONG KONG TOKYO
DELHI BOMBAY CALCUTTA MADRAS KARACHI
NAIROBI DAR ES SALAAM CAPE TOWN

© The several contributors listed on pp. ix–xiv 1981
Published in the United States by Oxford University Press, New York

British Library Cataloguing in Publication Data

Structural studies on molecules of biological interest.
  1. Molecular biology—Addresses, essays, lectures
  I. Dodson, Guy   II. Glusker, Jenny Pickworth
  III. Sayre, David   IV. Hodgkin, Dorothy
  574.8'8   QH506   80–40398

    ISBN 0–19–855362–5

Printed in Great Britain by
Butler & Tanner Ltd
Frome and London

Contents

III METHODS IN CRYSTALLOGRAPHY

IV PROTEIN STRUCTURE

V INSULIN: BIOLOGY, CHEMISTRY, AND STRUCTURE

List of contributors

MARGARET J. ADAMS
Laboratory of Molecular Biophysics, Department of Zoology, University of Oxford, England

IAN G. ARCHIBALD
Laboratory of Molecular Biophysics, Department of Zoology, University of Oxford, England

E. N. BAKER
Department of Chemistry, Biochemistry, and Biophysics, Massey University, New Zealand

THE BEIJING (PEKING) INSULIN RESEARCH GROUP
Institute of Biophysics, Academia Sinica, Peking, People's Republic of China

G. A. BENTLEY
Institut Laue-Langevin, Grenoble, France

M. BERGER
Department of Medicine, University of Dusseldorf, W. Germany

J. M. BIJVOET
Winterwijk, Meddo, The Netherlands

TOM BLUNDELL
Department of Crystallography, Birkbeck College, London, England

DIETRICH BRANDENBURG
Deutches Wollforschungsinstitut an der RWTH, Aachen, W. Germany

E. E. BÜLLESBACH
Deutsches Wollforschungsinstitut an der RWTH, Aachen, W. Germany

C. H. CARLISLE
Emeritus Professor, Department of Crystallography Birkbeck College, London, England

H. J. CUPPERS
University of Düsseldorf, W. Germany

J. F. CUTFIELD
Department of Biochemistry, University of Otago, New Zealand

S. M. CUTFIELD
Department of Biochemistry, University of Otago, New Zealand

J. G. DAVIES
Laboratory of Molecular Biophysics, Department of Zoology, University of Oxford, England

E. J. DODSON
Department of Chemistry, University of York, England

G. G. DODSON
Department of Chemistry, University of York, England

J. D. DUNITZ
Laboratorium für Organische Chemie, Eidgenossische Technische Hochschule, Zurich, Switzerland

STEFAN O. EMDIN
Department of Pathology, University of Umeå, Sweden

STURE FALKMER
Department of Pathology, University of Lund, Sweden

C. R. GANELLIN
The Research Institute, Smith, Kline, and French Ltd, Welwyn Garden City, England

H.-G. GATTNER
Deutsches Wollforschungsinstitut an der RWTH, Aachen, W. Germany

JENNY P. GLUSKER
Institute for Cancer Research, The Fox Chase Cancer Centre, Philadelphia, Pennsylvania, USA

R. A. G. DE GRAAFF
Department of Chemistry, University of Leiden, The Netherlands

DRAGO GRDENIĆ
Laboratory of General and Inorganic Chemistry, University of Zagreb, Yugoslavia

P. A. HALBAN
Institut de Biochemie Clinique, University of Geneva, Switzerland

MARJORIE HARDING
Chemistry Department, University of Edinburgh, Scotland

PAULINE M. HARRISON
Department of Chemistry, University of Leiden, The Netherlands

JOHN R. HELLIWELL
Daresbury Laboratory, Science Research Council, Daresbury, England

S. M. HOARE
Laboratory of Molecular Biophysics, Department of Zoology, University of Oxford, England

THEO HOFMANN
Department of Biochemistry, University of Toronto, Canada

M. V. HOSUR
Department of Physics and ICMR Centre of Genetics and Cell Biology, Indian Institute of Science, Bangalore, India

I-NAN HSU
Department of Chemistry, California State University, Northridge, California, USA

NEIL ISAACS
St Vincent's School of Medical Research, Victoria Parade, Melbourne, Australia

MICHAEL N. G. JAMES
MRC Group in Protein Structure and Function, Department of Biochemistry, University of Alberta, Canada

SUSAN E. JENKINS
Daresbury Laboratory, Science Research Council, Daresbury, England

B. KAITNER
Laboratory of General and Inorganic Chemistry, University of Zagreb, Yugoslavia

B. KAMENAR
Laboratory of General and Inorganic Chemistry, University of Zagreb, Yugoslavia

PANAYOTIS G. KATSOYANNIS
Department of Biochemistry, Mount Sinai School of Medicine of the City University of New York, New York, USA

S. K. KATTI
Department of Physics and ICMR Centre of Genetics and Cell Biology, Indian Institute of Science, Bangalore, India

A. LEWIT-BENTLEY
Institut Laue-Langevin, Grenoble, France

S. A. MASON
Institut Laue-Langevin, Grenoble, France

D. A. MERCOLA
Agricultural Research Council Unit for Research on Muscle Mechanisms and Insect Physiology, Department of Zoology, University of Oxford, England

V. K. NAITHANI
Deutsches Wollforschungsinstitut an der RWTH, Aachen, W. Germany

RAMESH NARAYAN
Raman Research Institute, Bangalore, India

R. E. OFFORD
Laboratory of Molecular Biophysics, Department of Zoology, University of Oxford, England

LINUS PAULING
Linus Pauling Institute of Science and Medicine, Menlo Park, California, USA

M. F. PERUTZ
MRC Laboratory of Molecular Biology, University Postgraduate Medical School, Cambridge, England

SIR DAVID C. PHILLIPS
Laboratory of Molecular Biophysics, Department of Zoology, University of Oxford, England

K. PROUT
Chemical Crystallography Laboratory, University of Oxford, England

S. RAMAKUMAR
Department of Physics, Indian Institute of Science, Bangalore, India

S. RAMASESHAN
Raman Research Institute, Bangalore, India

C. D. REYNOLDS
Department of Chemistry, University of York, England

D. P. RILEY
IBM Research Center, Yorktown Heights, New York, USA

JOHN H. ROBERTSON
School of Chemistry, The University, Leeds, England

J. M. ROBERTSON
42 Bailie Drive, Bearsden, Glasgow, Scotland

KIWAKO SAKABE
Department of Chemistry, Nagoya University, Japan

NORIYOSHI SAKABE
Department of Chemistry, Nagoya University, Japan

KYOYU SASAKI
Department of Chemistry, Nagoya University, Japan

D. SAYRE
IBM Research Center, Yorktown Heights, New York, USA

TREVOR SEWELL
Department of Crystallography, Birkbeck College, London, England

ELI SHEFTER
School of Pharmacy, State University of New York at Buffalo, New York, USA

ANITA R. SIELECKI
MRC Group in Protein Structure and Function, Department of Biochemistry, University of Alberta, Canada

ROBERT A. SPARKS, *California Scientific Systems, Sunnyvale, California, USA*

DONALD F. STEINER, *Department of Biochemistry, University of Chicago, Illinois, USA*

TANG YOU-CHI
Institute of Physical Chemistry, Peking University, People's Republic of China

S. P. TOLLEY
Department of Chemistry, University of York, England

K. N. TRUEBLOOD
Department of Chemistry, University of California at Los Angeles, Los Angeles, California, USA

BILL TURNELL
Department of Crystallography, Birkbeck College, London, England

B. K. VAINSHTEIN
Institute of Crystallography, Academy of Sciences of the USSR, Moscow, USSR

D. VALLELY
Department of Chemistry, University of York, England

K. VENKATESAN
Department of Organic Chemistry, Indian Institute of Science, Bangalore, India

M. VIJAYAN
Molecular Biophysics Unit, Indian Institute of Science, Bangalore, India

M. A. VISWAMITRA
Department of Physics and ICMR Centre of Genetics and Cell Biology, Indian Institute of Science, Bangalore, India

F. WELCH
62 Botley Road, Littlemore, Oxford, England

J. G. WHITE
Chemistry Department, Fordham University, New York, New York, USA

STEPHEN W. WHITE
Daresbury Laboratory, Science Research Council, Daresbury, England

AXEL WOLLMER
Abteilung Physiologische Chemie, RWTH, Aachen, W. Germany

List of contributors

H. ZAHN
Deutsches Wollforschungsinstitut an der RWTH, Aachen, W. Germany

ZHANG YOU-SHANG
Shanghai Institute of Biochemistry, Academia Sinica, Shanghai, People's Republic of China

Introduction

THROUGH her many years of work in the study of the structures of complex molecules by the X-ray method, Dorothy Hodgkin has played a major role in establishing one of the most characteristic and successful features of contemporary science: the approach to understanding biological function through the understanding of molecular structure. The present volume is affectionately dedicated to Dorothy by her colleagues on her seventieth birthday. It contains papers on historical aspects of X-ray crystallography, on structural chemistry and biochemistry, crystallographic methods, protein structure, and insulin. These are the subjects to which Dorothy's own work contributed so outstandingly, and it is our hope that she and others will find the papers presented here of interest and value. In many cases the work described was either begun in her Oxford laboratory or was strongly influenced by work carried out there.

Dorothy Mary Crowfoot was born in Cairo, Egypt, on 12 May 1910, the daughter of John Winter Crowfoot and Grace Mary (Hood) Crowfoot. Her father, an archaeologist and historian, was serving with the Egyptian Ministry of Education, and later became Principal of Gordon College at Khartoum and Director of Education and Antiquities in the Sudan. Her mother was recognized in her later years as a leading authority on ancient textiles. Dorothy was educated at the Sir John Leman School, Beccles, Suffolk, and at Somerville College, Oxford (1928–31). She studied for a Ph.D. (1936) at Cambridge from 1932–4, working with J. D. Bernal, and was appointed a research fellow of Somerville College, Oxford, in 1933, becoming tutor and fellow in 1936. She became a University lecturer in 1946, University Reader in 1955, and Wolfson Research Professor of the Royal Society in 1960. She married Thomas Lionel Hodgkin, a tutor in adult education and later an African historian, in 1937 and they had three children, Luke (1938), Elizabeth (1941), and Toby (1946).

Dorothy's first X-ray studies were carried out in Oxford with H. M. Powell on thallium dialkyl halides. In 1932 she joined J. D. Bernal at Cambridge; shortly before this Bernal had taken X-ray photographs of a number of sterols and had pointed out that his photographs were incompatible with the then accepted Windaus–Wieland formula for the sterols, a view which led to rapid revision of the formula and acceptance of the present structure. Dorothy made similar measurements on over 100 steroids, distinguishing different molecular weight groups and other complexities. The steroid work culminated in her determination with C. H. Carlisle of the partial three-dimensional electron-density distribution in cholesteryl iodide which demonstrated both the chemical and stereochemical form of the molecule.

During the Second World War Dorothy worked on penicillin and was responsible, together with Barbara Rogers-Low, A. Turner-Jones, and Charles Bunn, for the elucidation of its structure. Penicillin was discovered by Fleming at St Mary Hospital, London, in 1929, and isolated by Florey and Chain and others in Oxford during the Second World War. The chemical evidence for the structure was conflicting, and the correct solution appeared through the X-ray analysis of crystals of sodium, potassium, and rubidium benzylpenicillin. In solving the structure a variety of aids to structure analysis were investigated, including isomorphous replacement, optical analogues, and difference maps; also for the first time the conclusions of a structure analysis were shown as a full three-dimensional electron-density map. This study showed, perhaps more clearly than any previous one, the power of the X-ray method to find chemical structures.

This power was demonstrated still more clearly in the structure analysis of vitamin B_{12} in the early 1950s. Here X-ray analysis led to knowledge of the arrangement of the atoms and hence the chemical bonding in more than half of a very much larger molecule. The structure included a number of unexpected features, in particular a hitherto undiscovered ring system, now called the corrin ring. As a result of this work the vitamin has been synthesized by R. B. Woodward and A. E. Eschenmoser and their collaborators. One group of papers in this volume is devoted to reminiscences of this X-ray study.

In 1934 J. D. Bernal took the first X-ray diffraction photographs of a crystalline protein (pepsin) immersed in its mother liquor. The photographs indicated that the protein molecules were packed with sufficient regularity in the crystals to permit, at least in principle, the solution of their structure by X-ray analysis. Dorothy helped Bernal with the pepsin photographs and a year later was able to take X-ray photographs of a second protein, insulin, in Oxford. Insulin was then too large a problem for her to tackle, but through the years of work on other structures she continued to work on insulin, applying new crystallographic techniques to its analysis as they became available. Thus insulin in her hands received a considerable share of the early exploratory work on isomorphous replacement, analysis of Patterson functions, and model building. Finally she directed all her attention to the insulin problem, and with the help of her co-workers the chemical and crystallographic difficulties were overcome and the structure of the 2Zn insulin crystal was determined in August 1969. A major part of this volume is devoted to the reporting of current studies on insulin structure and related topics.

It has been a characteristic of Dorothy's work that, in addition to helping elucidate biological function, it has influenced chemistry greatly. For this we cannot do better than refer the reader to the concluding words of J. D. Dunitz's paper in this volume.

Dorothy has received numerous awards, the most notable of which are a Fellowship of the Royal Society (1947), Nobel Prize for Chemistry (1964), and Order of Merit (1965). She has been Chancellor of Bristol University since 1970, and served as President of the International Union of Crystallography (1972–5) and President of the British Association for the Advancement of Science (1977–8). She was the third woman to win the Nobel Prize for Chemistry, the others being Mme Marie Curie in 1911 and her daughter, Irene Joliot-Curie, in 1935. She was the second woman to be honoured by the British sovereign with the Order of Merit, the highest civilian honour in the United Kingdom, the first being Florence Nightingale in 1907.

Other honours include foreign membership in the Royal Netherlands Academy of Science and Letters, the American Academy of Arts and Sciences, the Yugoslav Academy of Sciences, the Ghana Academy of Sciences, the Puerto Rico Academy of Sciences, the Leopoldina, the USSR Academy of Sciences, the Australian Academy of Sciences, and the National Academy of Sciences of the USA. She received a Royal Medal of the Royal Society in 1956 and its Copley Medal in 1976. She has been awarded honorary degrees by the following universities: Leeds, Manchester, Cambridge, Sussex, Warwick, Hull, East Anglia, London, Bristol, Bath, Exeter, Kent, York, Modena, Zagreb, Delhi, Ghana, Harvard, Brown, Chicago, and Mount Sinai (New York).

An introduction to a book for Dorothy would be incomplete if it made no mention of her personality. In a world in which science is often performed in large impersonal groups, Dorothy has shown that absolutely first-class science may still be joined with a loving attention to individual values. Hundreds of students and guests have come to her laboratory from all over the world; it is safe to say that each one of these remembers some evidence of Dorothy's concern for their comfort and circumstances in Oxford. Many more have enjoyed the hospitality of the open and friendly house kept by the Hodgkins, and have realized how much the devotion, balance, and humour of their family life have helped in the busy lives Thomas and Dorothy lead. The friendships formed through and around Dorothy, which link crystallographers, chemists, and biochemists in ways which transcend nationalities, are a token of her constant effort to break down the barriers between people and countries and, by personal action, to counteract some of the destructive tendencies of our time.

Dorothy retired in 1977, but there is some doubt as to whether she has heard the news. She travels widely in connection with science; and, housed again in her old department, the Laboratory of Chemical Crystallography at Oxford, she continues to work on insulin refinement and vitamin B_{12}.

Our sincere thanks go to the Oxford University Press for their interest in publishing this volume and the care with which they have produced it.

March 1980

G. D.
J. P. G.
D. S.

Dorothy: a tribute

ONE always rejoices at expressing feelings of admiration and affection. I know no case in which I have felt this more clearly than at this moment, now that I can contribute to the homage of Dorothy Hodgkin.

Dorothy combines the highest crystallographic achievements with a happy domestic life. In her famous crystal analyses she combines the latest techniques with a unique intuition. The latter enables her, to an amazing extent, to read atomic positions in a nascent electron-density map.

I do not only admire Dorothy, which is self-evident for anyone. I am also, as she knows, grateful to her for much helpful advice and this I should not pass over in silence on this occasion.

I wish Dorothy a happy continuation of the life and work she has fashioned for herself through the years.

J. M. Bijvoet

I History

1. Forty years' friendship with Dorothy

MAX PERUTZ

Example is not the main thing in influencing others,
it is the only thing. (A. Schweitzer)

THE four people who took an interest in my early X-ray work on haemo-globin were J. D. Bernal, W. L. Bragg, D. Keilin, and Dorothy. Bernal would listen intently, make some profound comments, and then suddenly sweep off with the air of having to do something far more important. Bragg would discuss the interpretation of my X-ray patterns, but he knew no protein chemistry and had not taken any X-ray pictures himself for many years. Keilin was a biologist, and X-ray crystallography was a closed book to him. So whenever I obtained exciting new results, or was disheartened by the persistent lack of them, I would take the now extinct branch line from Cambridge via Bedford to Bletchley and hang around at that dismal junction until the ancient rattly train set off on its many stops to Oxford.

Once arrived, I made for Ruskin's Cathedral of Science—the University Museum—walked past the skeletons of extinct species populating its nave to the darkest corner, and descended the stone stairs to Dorothy's crypt-like office, where she laboured on the structure of life in a place that was, but for her vitality, quite dead. Her tables were piled high with structure-factor and Fourier calculations; there were viewing boxes for looking at X-ray pictures. Her X-ray and dark rooms were adjoining. The gothic window was high above as in a monk's cell, and beneath it there was a gallery, reachable only by a ladder, on which stood a table with Dorothy's polarizing microscope. To mount one of her precious crystals of penicillin, Dorothy would climb up there, stick the crystal to a thin glass fibre, stick the fibre to a goniometer head, and descend again, clutching her treasure with one hand while holding on to the ladder with the other. I don't think she ever lost a crystal.

For all its gloomy setting, Dorothy's lab was a jolly place. As Chemistry Tutor at Somerville she always had girls doing crystal structures for their fourth year and two or three research students of either sex working for their

Ph.D.s. They were a cheerful lot, not just because they were young, but because Dorothy's gentle and affectionate guidance led most of them on to interesting results. We looked at their latest organic structures and argued about the meaning of our Pattersons of insulin and haemoglobin until it was time to go home to her house in Bradmore Road.

Some women intellectuals regard their children as distracting impediments to their careers, but Dorothy radiated motherly warmth even while engaged in writing crystallographic papers. Concentration comes to her so easily that she can give all her attention to a child's chatter at one moment and switch

FIG. 1.1. Dorothy when I first knew her.

to lattice transformations the next without any sign of strain. Once she helped her three-year-old son search for a lost toy when he said: 'It must be somewhere; it can't be nowhere.' Not surprisingly he became a topologist. Later in the evening Dorothy's husband, Thomas, would appear if he was not away lecturing to the Workers' Educational Association and he would keep us all in fits of laughter over dinner. Afterwards I would discuss proteins with Dorothy until it finally dawned on me that I was exhausting her and that it was time for bed. On Sundays we would go walking or swimming in the Isis. Many years later the two families, six Hodgkins and five Perutzes, went on holiday together to the Austrian Lakes. Once when we had to take shelter in a storm Thomas kept on inventing such good games and stories that he made us feel sorry when the downpour stopped.

Sometimes Dorothy invited me to Oxford to see her latest results. There was cholesterol iodide. She and Harry Carlisle found the iodine positions from a Patterson and used them to calculate the signs of the $(h0l)$s of the monoclinic crystals. The resulting projection showed the molecule well resolved. To determine its stereochemistry, Harry then recalculated the Fourier along a set of lines drawn through each peak perpendicular to the plane of projection. This Fourier was again done with phase angles based on the iodine positions alone, which imposed a false centre of symmetry on the map. With careful stereochemical reasoning and insight Dorothy picked out the correct one among the two alternative atomic positions indicated for many of the atoms and brought to light the first complete stereo-chemically correct formula of a sterol. Then came penicillin. Dorothy writes that its revolutionary β-lactam structure had actually been the first proposed, but Sir Robert Robinson was strongly against it because β-lactam rings are generally stable, unlike penicillin, and J. W. Cornforth, then a post-doctoral fellow, supported Robinson's view at an early stage, saying: 'If penicillin turns out to have the β-lactam structure I shall give up chemistry and grow mushrooms.' Luckily he did not carry out his threat and won the Nobel Prize for Chemistry eleven years after Dorothy. The difficult step in the initial structure analysis was to pick out the right lumps of density representing one molecule. Dorothy had a shot at it and then asked others, including myself, to find it. Her first guess proved correct, and further refinement brought to light the β-lactam ring. She writes: 'It was a nice day when we could set up the model first precisely in three dimensions and rang up our friends to come and see what penicillin actually was.'

It was the same with vitamin B_{12}. Anyone else looking at her first Fouriers, derived from the cobalt phases alone, and later from the cobalt plus selenium phases of a selenocyanide derivative, would have attributed the close approach of two lumps of electron density to experimental error, but Dorothy combined a firm belief in the significance of even very blurred features of electron-density maps with a profound flair for the chemically reasonable, and she concluded rightly that the lumps represented rings linked directly rather than through a methene bridge as in porphyrins. She re-calculated her phases from the positions of the cobalt plus the atoms of what she later christened the corrin ring, with the result that the next Fourier revealed much of the vitamin B_{12} molecule in outline.

But was it right? On 28 December 1954 Dorothy wrote to me:

We've had one very nice development on the B_{12} front. Lester Smith fed dichlor-benzimidazole to his bugs and got them to synthesize a chlorine substituted vitamin B_{12}. We had encouraged him to make the dibromo compound but he had difficulties with making the dibromobenzimidazole—and anyway we find that we can see what we want without it. On the very first photograph we took of the dichlor

vitamin B_{12}. We had encouraged her
to make the dibromo compound but
they he had difficulty with making
the dibromo benzimiazole — in any
way we find it we can see what
we want — if the dichloro compound On the very first
photograph we took. there were
recognisable small intensity changes
on which I straight away
did a nice little calculation
It runs like this.

032	042	
+ 11	− 16	F_{c} obs. for dry B_{12}　signs as proposed str.
c +10	c + 1	changes : F expected on subst. Cl for Me.
c + 21	c − 15	F_{c} values ∴ expected for dichlor cpd.

observed on first photograph
032 stronger than 042 (both quite weak)
the cl contribution to 032 is almost the
maximum .

Most of the reflections have not
visibly changed to the eye. But
we shall now have to work up
the photographs properly — the
expected differences are, of course,
Cl−Me!
usually very small. But

FIG. 1.2. Facsimile of letter quoted in the text.

compound, there were recognisable small intensity changes on which I straight away did a nice little calculation. It runs like this.

032	042	
+ 11	− 16	F/4 obs. for dry B_{12}; signs on proposed str.
c+ 10	c+ 1	Changes in F expected on substitution of Cl for Me.
c+ 21	c− 15	F/4 values ∴ expected for dichloro cpd.

Observed on first photograph. 032 stronger than 042 (both quite weak). The Cl contribution to 032 is almost the maximum.

Most of the reflections have not visibly changed to the eye. But we shall now have to work up the photographs properly—the expected differences are, of course, usually very small. Cl—Me! But it does seem that this very first observation and calculation confirms that we have quite correctly placed the nucleotide—and this carries with it at least one piece of chemical information, the 3-phosphate binding. Also of course it makes me more than ever confident about everything else.

I have been turning my mind to insulin, thinking B_{12} is drawing to an end and I would like to take up insulin seriously and properly. I am not very happy about one's power of model building—even with all the information we have. I gather Lindley and Rollett at Cal. Tech. have built an α helix model, I imagine with parts of chains in helices—I have asked for details. But I would like to work it out properly and I think it can be done with distributed heavy atom derivatives—and a number of years! We have one or two ideas on the heavy atom front which we hope to try out in the next few months. I find myself wondering if the biosynthetic approach is a possible one—easier perhaps to think of for haemoglobin than for insulin.

Ever since she took her first X-ray pictures of insulin in 1935, Dorothy continued to think about its structure. On 26 April 1949 she wrote to me:

I am very much interested that you have taken up the threefold spiral structure. It certainly is a nice one. Kate [Dornberger-Schiff] and I spent a long time once doing sums on insulin intensities, trying to fit a particular form of it into our picture. Actually when I looked back on those four calculations they seemed not unhopeful— and yet I don't somehow feel in my bones that they are right. And I did some time ago a theoretical Patterson of it—which has quite a lot of good features but doesn't, I think, fit for the gramicidin S.

When Sanger had determined the positions of the disulphide bridges in insulin, I sent her a preprint of his paper. She replied on 20 July 1955:

Thank you ever so much for sending me the insulin paper. It is terribly important— Sanger finds the internal link to be 6 : 11 and this, it seems to me is fatal to simple helical models. Even 7 : 11 I never much liked. Indeed the more one thinks about the geometry of cystine the more one is driven to β type chains (Astbury has said this). I've built a total model for insulin nicely and smoothly and quickly on one of the very oldest schemes anyone has ever had, short folded lengths of β chain (antiparallel pleated sheet type), in the same configuration that I favour for gramicidin S. The scheme is

It was another fourteen years before I had a telephone call from Guy Dodson inviting me to Oxford to celebrate the solution of the insulin structure. I rushed there with a magnum bottle of champagne which someone had given me as a belated present for my Nobel Prize, but when we extracted the cork, all the CO_2 had evaporated and it tasted, as David Phillips remarked in his usual tactful way, just a little peculiar. Then Dorothy showed me that the Patterson of insulin, on which she had spent so much thought, contained near its origin a simple symmetry-duplicated image of the histidines surrounding the central zinc atom which she could easily have interpreted if only she had applied to it her usual faith and courage.

Sometimes I asked Dorothy to visit me in Cambridge. In the summer of 1953 I had just determined the signs of the $(h0l)$ reflections of horse methaemoglobin by isomorphous substitution with paramercuribenzoate and had calculated the first Fourier projection of the molecule. Dorothy came over at once to admire my map, even though she realized that the many overlapping features of the 55-Å thick molecule would render it uninterpretable. Then she said to me: 'If you could get two heavy atom derivatives, you could solve the structure in three dimensions' and drew my attention to Bijvoet's paper on strychnine in which he pointed to the possibility of solving the phase equation by double isomorphous substitution. This remark sent me off on the next stage of the haemoglobin work. Luckily I did not foresee that it would be another six years before Dorothy would ring me up saying: 'I hear you have a fantastic model of haemoglobin. Can I come over and see it?'

I felt embarrassed when I was awarded the Nobel Prize before Dorothy, whose great discoveries had been made with such fantastic skill and chemical insight and had preceded my own. The following summer I said as much to the Swedish crystallographer Gunner Hägg when I ran into him in a tram in

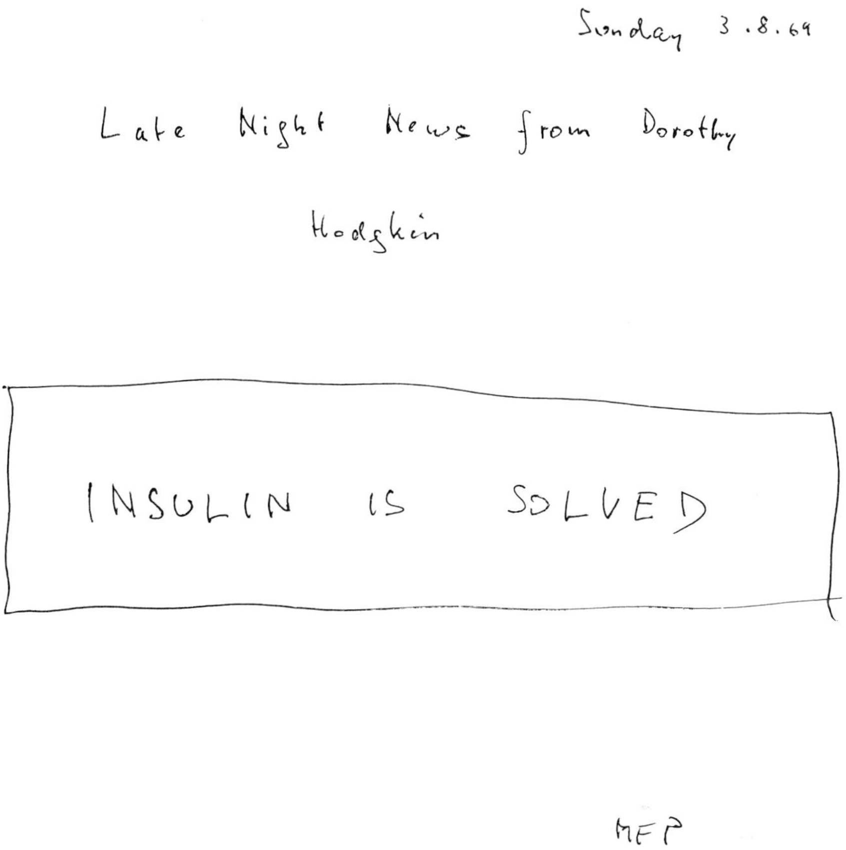

Sunday   3.8.69

Late   Night   News   from   Dorothy

Hodgkin

INSULIN   IS   SOLVED

MFP

FIG. 1.3. Poster at the Laboratory of Molecular Biology, Cambridge, announcing the great news.

Rome. He encouraged me to propose her, even though she had been proposed before. In fact, once there had been a newsleak that she was about to receive the Nobel Prize, but it proved false; Dorothy never mentioned that disappointment to me until long after. Anyway, it was easy to make out a good case for her; Bragg and Kendrew signed it with me, and to my immense pleasure it produced the desired result soon after.

'There are certain letters which I dread to open', Dorothy once told me, 'and when I saw one from Buckingham Palace I left it sealed, fearing that they wanted me to make me *Dame* Dorothy.' I suppose it would have made her feel like a *femme formidable*, which she so happily is not. When she eventually opened the letter she was relieved that instead the Queen offered her the Order of Merit, which is a much greater honour and carries no title. She received it in private audience on the same day as Benjamin Britten. Once when they were both getting honorary degrees, Henry Moore said to her: 'It's really very good of them to give the OM to a simple chap like me.' I suspect that this remark echoed some of Dorothy's own feelings.

2. Dorothy Hodgkin and molecular biophysics in Oxford: a fragment of personal history

DAVID PHILLIPS

THE Laboratory of Molecular Biophysics had its origins at the Royal Institution, London, in the group that Sir Lawrence Bragg assembled there when he became Resident Professor and Director of the Davy–Faraday Research Laboratory in 1954. During the preceding years in Cambridge, Sir Lawrence's main interest had been in the studies of protein crystals by Max Perutz, John Kendrew, and their colleagues, and he had involved himself deeply in the work of the group during the period when very few crystallographers believed that it was likely to be successful (Bragg 1965). When he left Cambridge in January 1954, however, the tide had turned dramatically. A few months earlier Max, with David Green and Vernon Ingram (Green, Ingram, and Perutz 1954), had shown how the method of isomorphous replacement could be used in protein crystallography and the way seemed open to the detailed determination of protein structures—starting with Max's haemoglobin and John Kendrew's myoglobin. Even the computational problems no longer seemed insurmountable with the growing power and availability of digital computers (Bennett and Kendrew 1952).

Bragg would have liked to take Max and John with him to London to continue the work there, but, at this critical time with so much to be done to exploit the breakthrough, neither was willing to leave Cambridge. But they promised to help Bragg assemble a new group at the Royal Institution and to co-operate closely in its work. At this stage Uli Arndt was already at the Royal Institution, where he had been studying proteins by low-angle scattering with Dennis Riley (Arndt and Riley 1955) and developing experimental methods. Early in 1955 he was joined by Helen Scouloudi, who had worked on ribonuclease crystals with Harry Carlisle and now began her study of seal myoglobin. In the autumn, David Green joined the Group from Cambridge and Tony North came from King's College London, where he had worked with J. T. Randall and Pauline (Cowan) Harrison on collagen.

These recruits all had experience of protein work, of one kind or another, but the original team was completed at the end of 1955 by the addition of two more who had not worked directly on proteins. This was Dorothy's doing. Anxious to build up a really strong group at the Royal Institution, Bragg had also asked Dorothy if she would join him there, but she was not prepared to leave Oxford at a time when her family committments were heavy and the work on vitamin B_{12} was promising well. Instead, with her inimitable knowledge of crystallographers around the world, she suggested that Bragg should ask Jack Dunitz and me to join him. At this time Jack was, as Dorothy puts it, in an American phase of his transatlantic oscillations—damped, as it has turned out—and I was working with W. H. Barnes at the National Research Laboratories in Ottawa, Canada. So it came about that I joined the Group at the Royal Institution in January 1956 and began to work on diffractometers with Uli Arndt and on sperm-whale myoglobin with John Kendrew—who came from Cambridge to see us nearly every week.

Jack Dunitz also joined the laboratory, but he did not work directly on proteins and stayed only until 1957 when he left for Zurich—again at Dorothy's suggestion. To complete the catalogue of people who are now in the Laboratory of Molecular Biophysics and were at the Royal Institution, I must mention Colin Blake, who came from industry in 1960; Louise Johnson, who began as a graduate student in 1962; and Schaffi (R. Aschaffenburg), our incomparable grower of crystals, who collaborated from his base in Reading from 1955 onwards.

In the mid 1950s there was still quite a sharp division in crystallography between those who worked on proteins and those who did not. Many crystallographers persisted in the view that protein structures could not be solved and the protein specialists sometimes seemed to neglect what was going on in general crystallography, regarding the determination of protein structures as a special problem requiring the development of new methods of analysis. This applied neither to Sir Lawrence nor to Dorothy, who made no distinction between proteins and the rest, though their individual attitudes differed in an interesting way. Sir Lawrence liked best to apply the methods he had developed during the early 1920s for the analysis of mineral structures and he was fond of quoting his paper with West (Bragg and West 1928) on 'The determination of structures with many parameters'—where 'many' means a dozen or so. Accordingly one of my first assignments, when the earliest semi-automatic diffractometer was ready for operation, was to measure the absolute scale of the seal-myoglobin crystals, using anthracene as the reference crystal to measure the intensity of the incident X-ray beam. Bragg's predilection for considering careful measurements of a few individual reflections is illustrated well by his last research paper, on methods for establishing the positions of heavy atoms in protein crystals (Bragg 1958). Although he had been the first to demonstrate the value of two-dimensional

Fourier syntheses in structure analysis (Bragg 1929), he mistrusted Fourier methods, especially Fourier refinement—which elicited somewhat disparaging remarks about boot straps. Dorothy, on the other hand, is the arch-exponent of Fourier refinement and her papers are full of Fourier maps—many of which seemed meaningless to most other crystallographers when she first described them. (I remember well the scepticism that greeted her early maps of vitamin B_{12} when she presented them at the Stockholm Conference in 1951.) From penicillin (Crowfoot, Bunn, Rogers-Low, and Turner-Jones 1949) through vitamin B_{12} (Hodgkin *et al.* 1957) to insulin (Adams *et al.* 1969) the interpretation of electron-density maps of various kinds has been her principal weapon and delight. Anyone who has seen her at work, surrounded by carefully drawn maps and absorbed in their interpretation, has been privileged to witness an artist in action as, with incomparable insight, she

> Turns them to shapes, and gives to airy nothing
> A local habitation and a name.

Dorothy came often to see us at the Royal Institution, while we contributed to the work on myoglobin and haemoglobin and then branched out into independent studies of β-lactoglobulin (Green *et al.* 1979) and hen-egg-white lysozyme (Blake, Johnson, and Phillips 1965; Blake, Mair, North, Phillips, and Sarma 1965), and she never failed to convey a conviction that proteins would eventually follow vitamin B_{12} in submitting to the most detailed and refined study. But her support went far beyond encouragement and advice. As early as 1958 she was discussing with Sir Hans Krebs the possibility of our helping to build up a protein-structure group in the Biochemistry Department in Oxford. In the event this idea was not realized, but it led on to John Pringle's successful campaign for the establishment of a Molecular Biophysics Laboratory in the Department of Zoology. The happy result was that most of us who had worked with Bragg at the Royal Institution were able to move to Oxford, when he retired in 1966, to set up this new laboratory. Even more happily, Dorothy, and the flourishing group she had built up to intensify her work on insulin, joined us there.

Fourteen years later the traces of Dorothy's influence can be seen everywhere: our biochemical effort under Robin Offord is largely concerned with exploiting the new knowledge of insulin derived from her studies; the study of triose phosphate isomerase came to us from her, since it was naturally to Dorothy that Stephen Waley turned (as have many others) when he first obtained useful crystals of the enzyme; and everywhere, it seems, there are Fourier maps—as larger and larger structures submit to refined analysis just as she knew they would. When I started work on proteins in 1956 not many people expected that we should one day study the apparent thermal motion of the atoms in them (Artymiuk *et al.* 1979) or locate their hydrogen atoms (Sasaki, Sakabe, and Sakabe 1978 and this volume, Chapter 41). But I do not think Dorothy is surprised. When we met the other week she was chuckling over the two additional water molecules she had just seen in the latest map of insulin. Who knows what she may inspire us to see next?

References

ADAMS, M. J., BLUNDELL, T. L., DODSON, E. J., DODSON, G. G., VIJAYAN, M., BAKER, E. N., HARDING, M. M., HODGKIN, D. C., RIMMER, B., and SHEAT, S. (1969). *Nature, Lond.* **224**, 491–5.

ARNDT, U. W. and RILEY, D. P. (1955). *Phil. Trans. R. Soc.* **A247**, 409–39.

ARTYMIUK, P. J., BLAKE, C. C. F., GRACE, D. E. P., OATLEY, S. J., PHILLIPS, D. C., and STERNBERG, M. J. E. (1979). *Nature, Lond.* **280**, 563–8.

BENNETT, J. M. and KENDREW, J. C. (1952). *Acta Crystallogr.* **5**, 109–16.

BLAKE, C. C. F., JOHNSON, L. N., and PHILLIPS, D. C. (1965). *Nature, Lond.* **206**, 761–2.

—— MAIR, G. A., NORTH, A. C. T., PHILLIPS, D. C., and SARMA, V. R. (1965). *Nature, Lond.* **206**, 757–61.

BRAGG, W. L. (1929). *Proc. R. Soc.* **A123**, 537–59.

——(1958). *Acta Crystallogr.* **11**, 70–5.

——(1965). *Rep. Prog. Phys.* **28**, 1–14.

—— and WEST, J. (1928). *Z. Kristallogr. Kristallgeom.* **70**, 475–92.

CROWFOOT, D., BUNN, C. W., ROGERS-LOW, B. W., and TURNER-JONES, A. (1949). In *The chemistry of penicillin* (ed. H. T. Clarke, J. R. Johnson, and R. Robinson), pp. 310–68. Oxford University Press.

GREEN, D. W., INGRAM, V. M., and PERUTZ, M. F. (1954). *Proc. R. Soc.* **A225**, 287–307.

—— ASCHAFFENBURG, R., CAMERMAN, A., COPPOLA, J. C., DUNNILL, P., SIMMONS, R. M., KOMOROWSKI, E. S., SAWYER, L., TURNER, E. M. C., and WOODS, K. F. (1979). *J. molec. Biol.* **131**, 375–97.

HODGKIN, D. C., KAMPER, J., LINDSEY, J., MACKAY, M., PICKWORTH, J., ROBERTSON, J. H., and SHOEMAKER, C. B., WHITE, J. G., PROSEN, R. J., and TRUEBLOOD, K. N. (1957). *Proc. R. Soc.* **A242**, 228–63.

SASAKI, K., SAKABE, N., and SAKABE, K. (1978). *Abst. 6th Int. Biophys. Congr.* **V44–553**, p. 265. IUPAB and Science Council, Japan.

3. Oxford: the early years

DENNIS PARKER RILEY

I WENT up to Christ Church, Oxford, as a freshman in October 1934 to read chemistry. Precisely at that time, Dorothy Crowfoot returned to her old college, Somerville, as a Research Fellow, having spent two to three years in Cambridge working for her Ph.D. under J. D. Bernal's supervision.

So, in a sense, we were exact contemporaries, I starting as an undergraduate and she starting as a young don. Both of us read chemistry at Oxford. Both of us did our Part II at Oxford, Dorothy with H. M. Powell and I, later, with Dorothy herself. I went on to complete my Oxford D.Phil. under her supervision, whereas Dorothy's Ph.D. is a Cambridge one. This was the only time that she left Oxford, apart from a much later absence in Ghana.

Having said that we were both Oxford chemists, which fact gave us much in common, it must be stated that pre-war Oxford was a masculine stronghold and the science faculties even more so. Although women had been admitted into membership of the University in 1920, their numbers were limited by statute to about one-fifth those of men. Women were still barred from taking degrees in Divinity until 1936.

In my first year, the chaperon rule was still in force (it was amended only in 1935). This meant that an undergraduate member of a women's college, if invited by a male undergraduate to luncheon or tea in his rooms in college, had to obtain prior permission from her Dean and be accompanied by a chaperon. (This often had unexpected consequences.) It was only in my first year that an end was put to the practice of segregating women medical students in a separate dissection room in the Department of Human Anatomy. During the same year, a few University women were allowed to take part in the University Opera Society's activities. There was quite a scandal in the press in 1935 when St Hilda's, a women's college, allowed mixed tea parties on Sundays. Women students were not admitted to membership of the Oxford Union. It was in this environment that Dorothy Crowfoot started her career.

I was Dorothy's very first research student. I started work on my Part II with her in September 1937. This, at the time, was quite revolutionary and

FIG. 3.1. Dorothy in her garden with her first research student. Oxford, probably in 1939. It must have been some sort of occasion for we were both dressed with unusual elegance.

FIG. 3.2. View of the X-ray lab taken from the ladder. The author is looking (somewhat bemused!) at the Unicam flat-plate camera with which lactoglobulin was being photographed. The Weissenberg camera is on the other side of the X-ray tube. The set-up is a more professional version with a shock-proof cable which replaced the original one described in the text. The danger notice was retained for no real reason.

The all-purpose table used in the early days is in the foreground. The degree of emaciation of the author (war-time food rationing) would date the photograph probably in 1941.

FIG. 3.3. Harry Carlisle (*left*) and the author (*right*) at their customary places at the large table used for calculating, etc. in the room next to the Porter's Lodge. The photograph was taken from Dorothy's place near the window, probably in 1940.

several eyebrows were lifted. Here was I, a member of a prestigious college, choosing to do my fourth year's research in a new borderline subject with a young female who held no university appointment but only a Fellowship in a women's college. At the House, as Christ Church is always called, there were some misgivings. I was after all an Exhibitioner and by that fact 'on the Foundation' and my tutor, Dr A. S. Russell, was still *in loco parentis*. He was somewhat less than enchanted with the idea but did see Dorothy and discuss it with her. Having done so, he indicated to me the college's acceptance of the proposal.

What led up to all this? As Dorothy gave no University lectures or courses, it could easily have come about that I never would have encountered her. How I first met her makes an interesting story.

The Oxford University Alembic Club is, as its name suggests, the club that unites chemists in the University. It is split into two quite separate parts: Senior (for graduates and dons) and Junior (for undergraduates), each with its own statutes and committees. I was in 1936 a committee member of the Junior Alembic. At this time, women were ineligible for membership of the Senior Alembic and Dorothy had never been invited to address them, although she regularly attended meetings, as did we juniors. It struck me that we, the Junior Alembic, were in no way bound by the stringent limitations of our seniors and that we might invite Dorothy to give us undergraduates a talk on the sterol researches that she had recently completed in Cambridge but which were as yet unpublished. This was duly arranged and was an unprecedented success. Our modest meeting, normally sparsely attended by undergraduates, was honoured by the presence of a sizeable contingent of our seniors, who listened to Dorothy's talk with close attention and evident respect. After this, I was emboldened to approach Dorothy, albeit with diffidence, about the possibility of her taking me for Part II in about a year's time.

I should, for non-Oxford readers, state what 'Part II' signifies. The Honour School of Natural Science (Chemistry) consists of two parts. Part I lasts three years and is devoted to normal undergraduate studies in the subject. At the end of this period, practical and theoretical examinations ('Schools') are held in inorganic, physical and organic chemistry and a plain degree (BA) is awarded, without honours. In order to obtain an honours degree, it is obligatory to proceed to Part II, which consists of a year of research. This, I believe, is unique.

At the end of Part II one has to submit a thesis on the results of the (rather less than a) year's work, which is taken into account at one's *viva voce* (oral) examination in order that one's class (first, second, etc.) for the BA degree already provisionally awarded can be awarded. In other words, it is an essential part of the whole requirement for an honours degree in chemistry. Dorothy did her Part II with H. M. 'Tiny' Powell on the thallium dialkylhalides (using a gas-tube) and I did mine with her on diketopiperazine, or so we thought at the beginning.

A Part II student is normally given a rather simple problem, for only about seven months of actual research time is available. The structure of diketo-piperazine fitted the bill, although at the beginning the lab did not possess a Weissenberg camera and I had to amass data by taking successive oscilla-tion photographs. It should be mentioned that I started absolutely from scratch. I had previously followed no courses whatsoever in crystallography. Now, Dorothy's conception of acquainting a raw research student with the complexities of the subject was to lead him to the deep end and invite him to dive in, without thinking of asking whether he could swim. Thus, she lent me a book on the polarizing microscope and a reprint of Bernal's 1926 paper on the interpretation of rotation photographs. She also gave me a tube containing crystals of dkp and left me to it. In this she was only following traditional Oxford practice. The Oxford tutorial system does not spoon-feed the aspiring student. So, I silently disappeared to give myself a crash course in the library before daring to tackle Dorothy again. This time she showed me how to mount and orientate a crystal on a glass fibre in the Unicam X-ray goniometer and I was off.

I might here interpolate a remark on the lay-out of the lab. The only room that Dorothy and I disposed of at that time was the X-ray lab itself. This was a peculiar room which always gave me a touch of claustrophobia. Owing to the architect's design of the University Museum in which we were housed, an ornate, unpractical Victorian structure, the room allotted to X-ray dif-fraction was a hybrid basement–ground-floor affair. The windows were high up so that from floor level one could not see out. There was a narrow balcony at window level reached by a steep ladder and it was there that our polarizing microscope was placed and where we mounted crystals. Having done so, we had to descend the ladder, more or less precipitously, and fix our precious glass fibre or tube to the X-ray goniometer arcs. When working with wet protein crystals, it was only rarely that I succeeded in orientating the crystal without it having slipped and moved within the tube. This necessitated a further trip, or trips, up and down the 'ship's companionway'. Patience was a necessity. I quote from my D.Phil. thesis:

> When a suitable crystal is seen, it is drawn up with some of its mother-liquor into a capillary-tube by means of capillary attraction. To permit of photography the crystal must be rigidly jammed in the tube at least approximately about the proposed axis of rotation. It is precisely this simple condition which is most difficult to satisfy, and which defines the degree of experimental skill required outside the usual X-ray technique.

I well remember apologizing to Dorothy for lack of progress, saying that I was no good with my hands and her replying, 'No, you're not! You're very good!'

This brings me to the subject of protein crystallography. When Dorothy returned to Oxford from Cambridge in the autumn of 1934, the Professor of Organic Chemistry, Sir Robert Robinson, gave her a sample of insulin crystals. She recrystallized them in order to obtain large enough crystals and, evidence of her training as a chemist, washed them and dried them with alcohol. The importance of water in defining and maintaining the crystalline structure of proteins was then imperfectly understood. In the case of insulin, quite good photographs were obtained from the air-dried crystals and a Patterson projection was calculated and published in 1935. By the time I arrived on the scene, Dorothy was anxious to repeat this work with wet crystals. Now, although we did this with insulin somewhat later, the first protein from which single-crystal X-ray diffraction patterns from both wet and dry crystals were obtained was lactoglobulin. We were stimulated by receiving excellent crystals from several sources. So, my investigations of diketopiperazine were put aside for a time (and later abandoned) and I started what proved to be four years' work on lactoglobulin.

The very first photograph of wet tubular lactoglobulin is dated 1 October 1937 and I well remember that Dorothy and I jigged for joy when we saw spots, albeit weak, extending out nearly to the edges of the film. I see from my Part II thesis that I completed the essential photography of the wet tabular form by the end of that year, 1937—pretty good going for a raw research student. I went on to accumulate data from both the tabular (orthorhombic) and needle (tetragonal) forms, wet and dry, and to calculate Patterson projections and Patterson–Harker sections.

It should be rememberd that the application of these methods to crystal structure determination was then novel and limited. Their use in protein analysis represented, in fact, the largest bulk of such calculations at that time. And calculations they were! We had no computers. Fortunately, Lipson and Beevers had rendered from 1936 on a great service to crystallographers through their strips giving values of cos (sin) $2\pi hx$. The first thing I had to do was to extend the values of h from 20 to 30 and prepare a supplemental box of strips. I was lucky, as when Dorothy performed her first Patterson calculations on dry insulin she had no Beevers–Lipson strips at all.

As we had no other accommodation, the first Patterson maps were calculated sitting at a table in the middle of the X-ray room. Indeed, all our work—indexing of photographs, correction of intensity data, and Fourier summations—was done there in blithe disregard of any health hazards.

Dorothy had no office. Her base was her college, as was mine. It was some time later that a large room was liberated next to the Porter's Lodge and shared by Dorothy, myself, and Tiny Powell's students. Tiny had no room to himself either. He had staked out a claim at the back of our small lecture room which was used only for his own lectures or those of the professor. I attended these, more or less alone, during my first year as a research student.

I might say a little about the equipment we had. In my first year, this was shared between Dorothy and myself and Tiny Powell and his two research students. This often meant queuing. However, as our protein photographs required exposures of at least twelve hours, we settled down to a system whereby Dorothy and I disposed of one X-ray tube and one Unicam rotation–oscillation camera, Tiny's students disposed of another similar combination, and the remaining X-ray tube with its associated Buerger–Weissenberg camera was shared. Not that it was there at the beginning. It was quite an event when this gleaming marvel of American engineering was unpacked and set up. The X-ray generators were Mullard (British Philips) copper target sealed tubes which performed Trojan service but which could not, obviously, give us the intensities really needed for protein crystallography. They were all mounted from start to finish by Frank Welch on good, solid, wooden tables with rather a primitive high tension supply and controls, all assembled by Frank.

Again, safety precautions were slim and mainly consisted of a notice saying 'Danger—60 000 Volts'. It would have been simple to electrocute oneself. Indeed, during the war, an overzealous Home Guard nearly did. I, as usual, was in the lab at around midnight when a local yokel dressed in khaki burst in and menaced me with a rifle and a fixed bayonet which, in his excitement at having actually captured a dangerous spy or saboteur, wavered perilously near a cable at 40 000 volts.

There is also the story of the refrigerator. As we noticed that wet protein preparations tended to grow mouldy when kept at room temperature, we used to cadge refrigerator space in the Biochemistry Department. When I started a systematic study of the crystallization of insulin under various conditions, our need for a small refrigerator of our own became evident and Dorothy undertook to see the professor about it, leaving me to go into the town and choose a suitable model. Having found one marked down in price, I acted fast and the very modest refrigerator (which saw years of use) was delivered within an hour or so. The bill was left on the professor's desk before Dorothy had had a chance to see him. I was summoned and reprimanded about spending the Department's funds without permission and a cloud was cast on the generally sunny atmosphere. Of course, for a classical mineralogist, a refrigerator was hardly a vital item of equipment. This illustrates the peculiar dichotomy of the Department at that time—true mineralogy, including the custody of the Mineral Collection, on the one hand, and structural chemistry and molecular biochemistry, on the other. I should like to stress that Professor Bowman was no fuddy-duddy. It was he who was responsible for bringing X-rays into the Department in the first place and he constantly encouraged the crystal-structure work. He was a keen and talented string-quartet player and our common interest in chamber music soon dissipated the cloud under which my impetuous action had placed me. In fact, he was good enough to give me a private course in his own room on the use of the polarizing microscope.

During this period, 1937–9, we were all very much aware that the war could break out at any moment. When it did, it came as no surprise. The change in Oxford was gradual; the undergraduate strength stabilized at about half the normal numbers. Qualified scientists were by government order told to stay put until further instructions. All new university appointments were suspended. In due course, evacuees from London—individuals, Government Departments, and London University colleges—crowded into Oxford. We provided a haven for Harry Carlisle and Katie Schiff from Bernal's laboratory.

I must say that from then on I continued my scientific work with reluctance and a feeling that it was irrelevant. Contact with Dorothy became intermittent. She was newly married and soon gave birth to her first child. It was also a period of serious illness, for her and for me. I visited her nearly every day she was in hospital and she, me. It was a frustrating period of marking time and finishing off with a view to getting free to do quite other things. I did manage to write up my D.Phil. thesis, but it is not surprising that the considerable amount of research described in it was never published.

It might be of interest to enumerate the Patterson maps that were calculated during this early period as, apart from Dorothy Crowfoot's 1938 dry insulin projections (based on earlier work), they were the first to be derived for protein crystals. Early in 1938, two Patterson projections were calculated from my very imperfect data on wet needle lactoglobulin, and one for the wet tabular variety. They were re-calculated later for inclusion in my D.Phil. thesis. In that thesis, written over the last six months of 1941, the following Patterson patterns were given and interpreted, as far as the coarse resolution permitted:

Lactoglobulin:

Wet tabular	4 projections + 2 recalculations with different temperature factors (2 were dated in July and August 1940)
Partly wet tabular	1 projection
Dry tabular	3 projections
	3 sections through the 3-dimensional map
Wet needle	3 projections

Insulin:

Wet	1 projection

Haemoglobin (horse):

Wet	3 projections

None of the patterns for lactoglobulin, nor the single projection for wet insulin, exhibited peaks at 5 Å, which fact arose from the limited range of spacings of the X-ray reflections. In the case of haemoglobin, however, the reflections extended out to 3 Å, as compared with 5.5 Å at best for lactoglobulin. As a result, 5 Å peaks were clearly seen in two of the projections but were not properly resolved in the third. This was the first instance that detail showing clearly peaks at 5 Å was obtained with protein X-ray data. As a matter of historical record, these patterns were calculated by myself in Oxford during 1939–40 from data supplied by Max Perutz in Cambridge and we both used them in our respective D.Phil. and Ph.D. theses.

A figure in Oxford in the immediate pre-war years was Dorothy Wrinch, who came to Oxford from Cambridge with the prestige of having been a pupil of Bertrand Russell. She turned her talents as a mathematician to the problem of the molecular structure of globular proteins. From 1936 on she developed precisely defined models that she called 'cyclols', which were closed hollow polypeptide cages. Her cyclol hypothesis was worked out *a priori* before the publication of any Patterson diagrams. As the molecular models were beautiful and the reasoning elegant, the cyclol hypothesis attracted exaggerated support from some quarters, and strong opposition from others. At the time, the few experimental workers in the field were mainly concerned with techniques and the laborious accumulation of data which were insufficient to test such a detailed theory. Dorothy Wrinch herself was impatient to have access to our experimental results, inadequate or not, but all she could find at that time was Dorothy Crowfoot's 1938 Patterson maps for air-dried insulin. Enlisting Langmuir on her side, she and he claimed that these substantiated in some detail the cyclol structure for this protein. This bold statement gave rise to a lengthy controversy, in which I became involved.

It is somewhat amusing to mention that, before this, I had been approached by Dorothy Wrinch with a curious proposition over tea and sherry (the traditional Oxford formula). She had become aware of my existence as a budding researcher and, in effect, suggested that I desert one Dorothy for the other, offering to get me a research grant to work with, or rather for, her. When I declined, she asked to have access to all my experimental data on protein crystals anyway, in advance of my own attempts to interpret them. Using some diplomacy, I managed to prevaricate and what might have become a delicate situation was resolved by Dr Wrinch's departure for the USA. Dorothy Crowfoot (the other Dorothy) was highly amused by all this but, as usual, left me quite free to make my own decisions.

It might be mentioned that, apart from the major studies of lactoglobulin and insulin, we had a shot at two other proteins, myoglobin and lysozyme. We had one tiny wet crystal of the first which gave such faint diffraction effects as to be valueless. Lysozyme, being cubic, was photographed as a wet sludge in the hope of obtaining the intensity data from a Debye–Scherrer pattern, again without success. In this case, I think I should have persisted with very much longer exposures, a monochromator and a greater specimen-to-film distance to provide resolution. A similar attempt in 1938 on tomato bushy-stunt virus that I made in the Long Vacation in Bernal's laboratory with Fankuchen was in fact successful.

We were definitely aware, even at a very early stage, that anything approaching a complete structure determination for a protein crystal would only be achieved by making comparative intensity measurements on various crystalline modifications of the same molecule with a view to solving the phase problem, at least partially. I had some hopes of doing this by studying lactoglobulin in different states of hydration. We also took some inconclusive photographs of iodinated insulin.

Looking back, it is evident that the means we disposed of then were grossly inadequate, both on the data gathering side and on the computational. But we were optimistic and inquisitive and had unshakeable faith in the power of X-ray diffraction as a chemical tool. In fact, we could not resist taking a quick look at practically any crystal that came into our hands. We helped Sir Robert Robinson with identifications of sterol derivatives and had a quite active relationship with a number of other organic chemists. I took Laue photographs of diamond dies used for wire drawing. On Dorothy's suggestion, I took some diffraction patterns of camphor, with the intention of studying order–disorder in such crystals. Apart from our main researches, we operated a kind of odd-job workshop in X-ray crystallography, and welcomed all comers. It was a cheerful time in retrospect. I am happy to have been in at the beginning.

4. Reflections

F. WELCH

IT was with some trepidation and yet not a little satisfaction that I accepted an invitation to contribute to this Festschrift for the seventieth birthday of Professor Hodgkin. Inevitably any worthwhile gleanings that I may recall must be mainly an impression of scientific Oxford as seen by a country lad in a small laboratory housed in the University Museum at Oxford, and not a term-by-term report on someone who would later gain a Nobel Prize. The period of which I write began on a September day in 1923 when my presence was required to meet the Professor of Mineralogy with a prospect of employment in the department. The head porter had been alerted to look out for a youth and I was duly ushered in to the learned man. Today I still have the letter sent to my father by Professor Bowman detailing all the subjects which I should have to study—algebra, trigonometry, maths, etc.—all these fearful unknowns to be undertaken at evening classes to ensure a successful future in the laboratory. A promise of ten shillings a week was given with the prospect of an increase later if suitable. An Irishman's rise indeed, as I had been earning twelve and sixpence a week helping on a farm while waiting for such an opening with brighter prospects, but I was much too timid to mention this. Had I acquired the academic skills in these subjects it was very doubtful how they would have been applied in the years ahead as little opportunity arose, as far as I could see, to use them. This did not worry me unduly as my time was occupied with the everyday events in the laboratory, which included stoking of the anthracite boiler in the chemistry laboratory of the department. This form of heating was complemented by several incandescent gas lamps placed at intervals on the walls.

Some indication of the near torpor which prevailed during the vacations of the University year may be formed from the unfamiliar working hours (by today's standards) which I enjoyed, namely 9 a.m. to 1 p.m. and 2.15 p.m. to 4 p.m., combined with the long holidays so envied by my contemporaries in industry. In term time it was, however, the rule to revert to the more normal 'office' hours of 9 a.m. to 5 p.m.

My arrival into the world of crystals, or rather into the world of minerals, began some few years before Professor Hodgkin came to Oxford and X-rays were yet to come, as far as the laboratory was concerned. The Department of Mineralogy, as it was called, under Professor H. L. Bowman and Mr T. V. Barker—later Dr Barker—as University Demonstrator and Mr R. C. Spiller, eventually to become Reader in the department, constituted the staff and the 'Crystallography' was not added to the title until some time after.

In 1924 I had the good fortune to motor up to London with Dr Barker and Mr Spiller to set up a stand in the chemistry section of the Wembley Exhibition. This stand to the best of my knowledge consisted of a series of models and crystals of the sort often found displayed in museums, designed to catch the eye of the visitor. A much more splendid exhibit in the same hall must have enchanted the most non-scientific of the passers-by and today it is recalled in my memory with great satisfaction. This exhibit was displayed by an industrial firm which I understood was Peter Spence Ltd, Alum Works, Manchester, and they had constructed a quantity of potassium alum crystals of octohedra in the form of a huge pyramid. These were so arranged with a base-line some two or three feet long, each layer getting smaller and smaller until the top had a base-line of perhaps half an inch. We were reliably informed that a man spent all day going round the vats in which the crystals were growing to turn each one on to a new face to allow the crystals to grow in a uniform manner. The Chemical Crystallography Laboratory still posseses a small collection of various alums of superb shape which were presented by this same firm.

It was at this time (1924) only six years since the First World War, and University life was not fully geared to the needs and aspirations of the new generation, so that budgets for departmental spending were very low, apparatus expensive or non-existent, and something in the region of £50 a year had to suffice, excluding wages and salaries.

I well remember a particular piece of research going on during this period by a senior researcher investigating the 'Rate of growth of crystals in different directions'. This was carried out with a bow-saw by hand drilling a hole in a suitably seeded crystal, mounting this on a holder and rotating it in its own saturated solution enclosed in a round-bottomed flask for some hours. The rotation was accomplished by gearing with old perambulator wheels driven by a 100-volt d.c. motor, the crystal then being transferred to a microscope for measurement from time to time.

A few years later a more ambitious apparatus was built for the department consisting of two lapidary wheels for polishing mineral specimens. The gearing again for this machine was notable in making use of any second-hand material available. A copy of the weekly magazine *Motor Cycle* was searched and a suitable second-hand motor-cycle gearbox purchased and fitted, noise and all!

The early 1920s saw the beginning of wireless for the amateur and I soon discovered how lucky I was to be working in a mineralogical laboratory containing certain mineral crystals essential for the construction of a simple receiver—the famous 'cat's whisker' crystal set. Not only was there an abundance of small chips of galena available but also some of the rarer minerals such as franklinite, etc., which were solemnly recommended by the weekly magazine *Popular Wireless* as being an alternative means of receiving a louder signal. I had little difficulty in making a synthetic crystal of galena by copying a recipe from the same journal. This I did by taking about equal parts by weight of lead and sulphur and heating them in a crucible buried overnight in the embers of a dying fire, and the resulting crystal on the following day was a finely textured specimen indistinguishable from a natural crystal. Many crystal chips were tried and many cat's whiskers made and coils wound and carefully shellacked in order to 'hold' the faint signal strength beamed from London 2LO. Here I was fortunate since I lived in the country and the necessity for a high aerial created no problem as the highest branch of the tallest tree made an excellent point on which to hang an aerial.

The lectures given in the department were not subject to overcrowding even for a small lecture room, particularly those given by the professor. It was unusual for him to muster more than two or three students and many times this fell to a single scholar. The lectures had to be carefully rehearsed beforehand as many experiments were somewhat temperamental in their behaviour. The production of an image about one foot in diameter between crossed nicols and in red and blue light was from a source of 100 volts d.c.; this was the arc lamp of the early cinema.

It seemed quite natural and dignified to be addressed by my surname when spoken to and it was some twenty years later that the Christian name so common in use today was adopted.

Protective clothing, although universal in the larger University laboratories, had not reached my small domain and it was left to a doting grandmother to kit me out with a white apron and strings, a mode of attire still used by the modern carpenter and joiner. My grandmother had unconsciously pointed to the direction in which I partly developed, as I satisfied many demands in the future for small carpentry jobs even if the apron and strings had a limited life.

It would be unfair of me to conjure up glowing accounts of Professor Hodgkin's student days with stories of long hours of intricate calculations and experiments, but it would be ungallant not to mention how well I remember her as a student, more so because women students were so rare, certainly in crystallography. My lasting impression is of Dorothy Crowfoot, together with a handful of male students, looking down a microscope (of which there were six Swift Student Microscopes) and growing salt, ammonium chloride, etc., advancing later to the refractive indices and birefringence of mounted sections of minerals and simple inorganic substances.

The arrival about 1930 of equipment for the production of X-rays by means of a continuously evacuated tube is a story in itself and one that I leave to others, except to say the impression remains of the gentle flop, flop of the pump and copious amounts of hard wax round the tube to obtain a vacuum. A large notice in red warned of 60 000 volts and a lead shield provided the only safeguard—this was of course before the introduction of any monitoring now considered so essential for the safety of the operator. The apparatus was housed in an upstairs room which had been commandeered by the RAF during the First World War. This same room has experienced a quite remarkable history as in addition to its use by the Forces it was the home of the Radcliffe Science Library in 1914, which in turn had taken over from the original Electrical Department of the University. A plaque outside the door adds further interest and states: 'A Meeting of the British Association held on 30th June 1860 within this door was the scene of the memorable debate between Samuel Wilberforce, Bishop of Oxford, and Thomas H. Huxley.' This room was still making history of a very different kind, when it was used by the fire-watchers as a rest-room while guarding the University Science Area buildings during the Second World War.

Whereas 1924 had seen me motoring to London for the Wembley Exhibition, a visit to the capital in the early 1940s had a more serious purpose. This was the time of American Lend-Lease and X-ray tubes were almost impossible to obtain in the United Kingdom. Some, however, were arriving from the States in spite of the prevailing war conditions. These, having survived U-boats, bombings, and fires at the docks, failed to make the short journey from London to Oxford by railway. When two or three perished in turn by this method of travel it was decided to send me to bring one back alive. Armed with a large box suitably labelled 'Fragile, X-ray tube', and 'This side up' and travelling on a slow train by a branch line to avoid the crowds I returned safely with my capture.

By this time I had completed the first twenty years of my employment and important changes were taking place in the department. Professor Bowman had died and Mr H. M. Powell, now Professor Powell, who had been teaching since the early 1930s, became Reader and head of the laboratory. Many honours were soon to be gained by the small academic staff, numerous students who later graduated, and the Nobel Prize winner whom we honour in these pages.

5. Dorothy and cholesteryl iodide

C. H. CARLISLE

In 1938–9, about the time I joined Birkbeck College, which was then situated between Chancery and Fetter Lanes in Breams Buildings, Dorothy Wrinch with the support of Irving Langmuir had just put forward her cyclol hypothesis for protein structure (Langmuir and Wrinch 1939). It was an imaginative idea and the only protein work available to test it was Dorothy's Patterson projections of insulin. I remember in early 1939 that Dorothy and Dennis Riley visited J. D. Bernal and I. Fankuchen at Birkbeck to discuss, I believe, letters that were to be sent to *Nature* taking Dorothy Wrinch's ideas seriously to task. I was a complete newcomer to crystallography, having just joined Bernal as an M.Sc. research student in the Department of Physics. This was a meeting that stood out in my memory because I was just beginning to realize in a vague way that such things as proteins were crystallizable, but more, I think, because I was impressed by Dorothy and Dennis who produced a Patterson vector projection map of a cyclol skeleton. Its interpretation meant nothing to me, but I was impressed by the young and attractive Dorothy who appeared to be carrying the day in her arguments against Dorothy Wrinch's hypothesis. Little did I realize that Dennis Riley, who also spoke so authoritatively at this meeting, was Dorothy's first research student, and that later that year, after war broke out, I was to be her second. It happened this way.

Just before war broke out, I decided to enlist as a volunteer in one of the services, but Bernal refused to endorse my application form, saying, 'Wait until you hear from me.' In late September he told me that he had made arrangements for me to continue my research work at the University Museum, Oxford, with Dorothy, and that I was to take the Department's X-ray equipment there. And so it was that, late in October, I was to share a large room with Dennis Riley, George Bartindale (who was more interested, I believe, in catching beetles than in carrying out his research), and George Huse. Bartindale and Huse were H. M. (Tiny) Powell's students. Later we were joined by Dr Katie Schiff (now Professor Dornberger), then a refugee from Austria, who was to help Dorothy with the calculations and interpretations of the 3D Patterson maps of insulin. Katie was a mathematical crystallographer of no mean repute who was a great help to me with my crystallographic problems. The whole X-ray group came under the wing of the Mineralogy Department—with Professor H. L. Bowman as the Head— which was then situated in the Museum.

I took my research problem to Oxford. This was the first X-ray study of the synthetic sex-hormone stilboestrol and nine of its derivatives, which Professor Eric Dodds (later Sir Eric) from the Middlesex Hospital Medical School got Bernal interested in, no doubt because it stemmed from Bernal's early X-ray studies on six sterols which put the sterol chemists on to the right path for the structure of the sterol skeleton (Bernal 1932). Naturally Dorothy was interested in the problem as well, and we published a paper in the *Journal of the Chemical Society* on the preliminary investigation of these compounds, entitled 'A determination of the molecular symmetry in the α,β-diethyldibenzyl series (Carlisle and Crowfoot 1941).

These small structures were difficult to solve at that time for many reasons, but principally because they had no heavy atoms attached to them. After all, the heavy-atom technique had just come into use through J. M. Robertson's work on platinum phthalocyanine (Robertson 1940), and so emphasis was given to this method of tackling the X-ray structure determination of organic molecules. It was not surprising, therefore, that Dorothy should suggest that I undertake the X-ray structure determination of cholesteryl iodide (mono-clinic), because it had an iodine atom attached to the steroid skeleton and was capable of being solved, although Dorothy herself had already calculated the Patterson projection maps of cholesteryl chloride and bromide and cholesteryl chloride and hydrochloride which had formed a part of her Ph.D. (Cantab.) thesis. This thesis, and her joint work with Bernal and Fan-kuchen (Bernal, Crowfoot, and Fankuchen 1940) on the crystallographic study of some eighty or more sterol derivatives, provided the basis of some of my background reading for my research with her. Seen in retrospect any of these halogen derivatives of cholesterol could have been tackled by the heavy-atom technique, but it was the small amounts of available material that was the real limitation.

Fortunately I had brought with me from London some of the sterols which Fankuchen had also been investigating and amongst them was a sufficiency of cholesteryl iodide, which Dorothy spotted and which formed the basis of my research. The results of this work, which appeared in the *Proceedings of the Royal Society* (Carlisle and Crowfoot 1945), formed the *first published detailed* three-dimensional X-ray analysis of a complex organic compound, at a time when computers and diffractometers were just being contemplated but certainly not available.

I believe I started the work in early 1941 and completed the collection of the X-ray data in about three months, and it is astonishing to think that the solution was obtained using just over 300 reflections because of the instability of the crystals to X-rays. Of course, the solution was correct but not accurate, and moreover the structure published in our paper was not the absolute configuration because Bijvoet's use of anomalous dispersion to determine the absolute configuration of molecules was to come some seven or eight years later.

I would like to add an anecdote here. I had calculated the electron-density map on (010) of the B form where the two molecules in projection are reasonably well separated except for a slight overlap. At the time I could not make head nor tail of it. After all, I had only used some sixty reflections for this projection and consequently the map was of poor resolution. Dorothy saw it and without difficulty sketched in one molecule; except for minor corrections of the atomic positions in the three-dimensional calculations, her selection of the xz co-ordinates was remarkably close to the correct positions when the work was completed.

I was called to war work, with Bernal, in 1942 at the Ministry of Home Security, which was housed in prefabricated huts in the grounds of the Forest Product Research Station at Princes Risborough about twenty miles away, which could only be reached by train. It was on the journeys to and from Oxford, in cold and ill-lit carriages, that I carried out the one-and-only set of three-dimensional calculations to obtain the whole set of bond lengths and bond angles using a slide rule and logarithmic tables. I've never had the courage since to calculate the R-factor of this structure determination, but Dorothy, unbeknownst to me, had calculated the average bond lengths and bond angles for the structure in preparation for our paper, and it is hard to believe that they were 1.55 Å and 108° 36′ respectively. Fortunately, the statistical deviations were not presented in the paper! Not only that, but if I got tired of calculating—the train was often delayed because it was a single-track line—I would start drafting out parts of my thesis, for what was normally a half-hour journey in daylight could be anything from three to four hours in the blackout.

Nevertheless, I am grateful for Dorothy's help in those hard times, for without it I would not have been able to present my D.Phil. thesis in time. One of my examiners was Professor (now Lord) Alexander Todd, understandably because he had earlier worked on some aspects of the sterols when he was at the Dyson Perrins Laboratory. The other was Tiny Powell. Like every D.Phil. candidate, I was dreading the oral, and just before it was to take place I met Tiny by chance in the Museum, and I suspect he was aware that I was going through mental torture and very tactfully he said to me, 'Don't worry, Harry, it's all there. I don't know what to ask you.'

There is another episode concerning Dorothy which I should like to relate. This was connected with penicillin, for it revealed to me her immense perceptiveness in spotting the correct structure from electron-density maps.

One Sunday morning in early 1944, I walked into the Museum and, very excitedly, Dorothy showed me two electron-density projections. One was the (010) projection of potassium benzyl penicillin, the other was the (010) projection of C. W. Bunn's map of the sodium salt obtained by his 'Fly's Eye' technique. Dorothy asked me to pick out the molecule from the two maps—which I couldn't—and then, with almost childish pleasure, she showed me that both maps presented almost the same view of the projected molecule. These original maps are shown as Figure 12 in *The chemistry of penicillin* (Crowfoot, Bunn, Rogers-Low, and Turner-Jones 1949). This was the time when she integrated her own work, being carried out at the Museum, with Bunn's work. The determination of the structure of penicillin was a greater success story for her then than our work of cholesteryl iodide. I was not involved in this X-ray study, except for some minor calculations, and some of these again I did on the train between Princes Risborough and Oxford.

I relate this story because when later I followed very closely Dorothy's long haul towards her brilliant determination of the structure of vitamin B_{12}, I learned both in conversations with her when she showed me the work at various stages, and from her published papers, that the handling and interpretation of partial electron-density maps were to be treated with extreme caution. In other words, she was a supreme example of the theme that no success story in the borderline fields of X-ray crystallography comes easily.

It has been a pleasure for me to work with Dorothy during the formative years when the X-ray structure determination of the smaller organic molecules presented problems to us, and when each and every one of us was feeling our way. Even when I left her in 1945 to rejoin the late J. D. Bernal as one of the team which set up his Biomolecular Research Laboratory at Nos. 21 and 22 Torrington Square, Dorothy would always visit us whenever she was in London. And when, later, I took up the X-ray study of ribonuclease, she evinced an even keener interest in my work and never failed to encourage me.

I deem it an honour to have been amongst the first of her students and, furthermore, to have been associated with one whose hard work in the crystallographic field and devotion to crystallography was rightfully recognized when she was made a Nobel Laureate in 1964. May she long enjoy her hard-won retirement.

References

BERNAL, J. D. (1932). *Nature, Lond.* **129**, 127.
—— CROWFOOT, D., and FANKUCHEN, I. (1940). *Phil. Trans. R. Soc.* **A239**, 135.
CARLISLE, C. H. and CROWFOOT, D. (1941). *J. chem. Soc.* 6.
——— (1945). *Proc. R. Soc.* **A184**, 64.

CROWFOOT, D., BUNN, C. W., ROGERS-LOW, B. W., and TURNER-JONES, A. (1949). In *The chemistry of penicillin* (ed. Hans T. Clarke), pp. 310–66. Princeton University Press.

LANGMUIR, I. and WRINCH, D. (1939). *Nature, Lond.* **143,** 49.

ROBERTSON, J. M. (1940). *J. chem. Soc.* 36.

6. The early years of X-ray crystallography in the United States

LINUS PAULING

THE science of X-ray crystallography was brought into being in 1912 and 1913 by Max von Laue, W. L. Bragg, and W. H. Bragg. Twenty years later Dorothy Crowfoot Hodgkin, in whose honour I am writing this article, entered the field and began her distinguished career. I had the great good fortune to begin work in this field in 1922, and in the following paragraphs I shall describe the science as it was practised in the California Institute of Technology at that time and the impression that it made on me.

No work was done in X-ray crystallography in the United States until 1917. In that year A. W. Hull reported the determination of the structure of some metals (Hull 1917) by the powder method, which he had invented independently of Debye and Scherrer, and Burdick and Ellis (1917) reported the determination of the structure of chalcopyrite, $CuFeS_2$, by the use of the X-ray spectrometer designed by the Braggs. Hull's work was done in the laboratory of the General Electric Company, and Burdick and Ellis's work was done in Throop College of Technology, which later changed its name to the California Institute of Technology.

I believe that the person principally responsible for getting work in X-ray crystallography started in the United States was Arthur Amos Noyes. Noyes had obtained his Ph.D. in physical chemistry with Ostwald in 1890. After his return to the Massachusetts Institute of Technology he taught courses in organic chemistry, analytical chemistry, and physical chemistry, and wrote textbooks in all three fields. He soon settled down in physical chemistry, however, and in 1902 he founded the Research Laboratory of Physical Chemistry in MIT, and served as its director until 1916, when he began full-time work in Throop College of Technology.

One of his students at MIT was C. Lalor Burdick, an American who had been born in Denver, Colorado in 1892, had obtained the BS degree in Drake University with majors in engineering and chemistry in 1911, and the degree BS from MIT in 1913 and then MS in 1914. Noyes encouraged him to carry on advanced study in Europe, and he sailed in July 1914 from New York on the last German liner to reach Hamburg. He began to study at the Kaiser Wilhelm Institute in Berlin, but then shifted to the University of Basel, Switzerland, where he received the Doctor of Science degree in 1915. Noyes then suggested to him that he go to London, and with some difficulty and the aid of Professor Kammerlingh Omnes of Leiden he reached London and began work in Professor W. H. Bragg's laboratory in University College. At that time, during the war, the only other person working in the X-ray laboratory was E. A. Owen, a convalescent wounded Australian soldier. Burdick and Owen, using the Bragg spectrometer, determined the structure of silicon carbide.

Burdick has reported that the apparatus in Bragg's laboratory was shockingly primitive, with antiquated induction coils, Wehnelt interruptors, gas X-ray tubes with little constancy in spark voltage or output, and with the ionization chamber, apertures, and goldleaf electroscopes leaking so that the values of the intensity of the diffraction maxima were never reproducible.

On his return to MIT in early 1916 he constructed a greatly improved X-ray spectrometer of the Bragg type, with a reliable transformer and rectifier and the newly available Coolidge X-ray tubes, retaining the goldleaf spectrometer with a measuring optical microscope. Before he had a chance to use this instrument, however, he was asked by Noyes to come to Pasadena. During the second half of 1916 he built his second spectrometer, and used it, in collaboration with James Hawes Ellis, to determine the structure of chalcopyrite. Burdick was then inducted into the American army, and until 1918 was involved in war research. In December 1918 he went to Chile for a stretch of ten years in developmental research in metallurgy and Chilean nitrate technology. From 1928 until his retirement he was associated with the DuPont Company. He is now director of the Lalor Foundation.

J. H. Ellis obtained his Ph.D. in physical chemistry at MIT, under direction of Noyes, his thesis being a study of the properties of hydrogen chloride solution. He was invited by Noyes to come to Pasadena, where for several years he held the title of Assistant Professor of Chemical Research. I remember him as a man with wide interests and considerable ingenuity. He enjoyed talking with research students about their work, but he did not seem to have the desire to carry on research himself. After some years he left the Institute and gave up scientific work.

Roscoe Gilkey Dickinson was brought to Pasadena from MIT as a graduate student by Noyes in 1917, and he immediately began work with the X-ray spectrometer that had been built by Burdick. He soon, however, adopted the Nishikawa–Wyckoff technique, making extensive use of Laue photographs, and it was this technique that was taught to me when I arrived in Pasadena in the autumn of 1922, to begin my graduate work.

Dickinson had received the Ph.D. degree in chemistry from the California

Institute of Technology in 1920. He was the first person to receive a Doctor's degree from the Institute. His Ph.D. thesis was on the crystal structures of wulfenite, scheelite, sodium chlorate, and sodium bromate by the Bragg method, but in his next papers he made use of Laue photographs. For several years a majority of the Ph.D. degrees granted by the California Institute of Technology were to men whose research had been in the crystal-structure field.

Ralph W. G. Wyckoff, who has had a distinguished career in X-ray crystallography, electron microscopy, and other fields, received his Ph.D. degree from Cornell University in 1919, when he was twenty-one years old. The Japanece scientist S. Nishikawa, who had made use of Laue photographs and the theory of space groups in a study of the structure of magnetite and spinel in 1915 (Nishikawa 1915), happened to be in Cornell at that time, and Wyckoff had the benefit of help and advice from him. Wyckoff immediately began the job of producing an analytical representation of the results of the theory of space groups, and in 1922 he published the first complete set of atomic coordinates, both general and special cases, permitted by the symmetry elements of the 230 space groups. His first structure determination, that of the rhombohedral crystal cesium dichloroiodide (Wyckoff 1920*a*), constituted an interesting contribution to structural chemistry and valence theory, in that the dichloroiodide ion was found to be linear, with the central iodine atom approximately equidistant from the two adjacent chlorine atoms. Wyckoff made a tremendous contribution to X-ray crystallography during the next few years. In addition to carrying out the difficult and extensive analytical space-group study presented in his book, he had by 1925 published over twenty-five papers in this field, with structure determinations of about fifty crystals. In his 1920 paper on the structure of carbonates of the calcite group (Wyckoff 1920*b*) he gave a detailed discussion of the use of both symmetrical and unsymmetrical Laue photographs and reproduced not only the photographs but also their gnomonic projections (the usefulness of which had already been pointed out by Rinne (1915)).

In the first *Strukturbericht*, covering the years 1913 to 1928, there are more references under Wyckoff's name in the index than under any other except V. M. Goldschmidt's.

As an undergraduate student in chemical engineering in the Oregon Agricultural College and during the year when I was a full-time instructor in quantitative analysis in that college I spent much time reading the chemical literature. I developed an interest in the nature of the chemical bond by reading Irving Langmuir's 1919 papers and Gilbert Newton Lewis's 1916 paper on the shared-electron bond theory. I also was curious about the variation in properties of different chemical substances—properties such as diamagnetism, paramagnetism, ferromagnetism, dielectric constant, hardness, cleavage, and colour.

I arrived in Pasadena to begin my graduate work in late September 1922. My first three months there were a revelation to me. I learned that problems can be attacked in a straightforward way, and that it is often possible to obtain an answer to a question by experimental or theoretical attack on it. In addition, I discovered, of course, that many of the questions that had puzzled me had already been answered.

Noyes had written to me, in Oregon, at the beginning of the summer to suggest that I carry on research with Dickinson in X-ray crystallography, and I had obtained a copy of the book *X-rays and crystal structure* by W. H. and W. L. Bragg and had read it without understanding it very well. I think that I still had, at that time, the ordinary attitude of an undergraduate— that of being satisfied with a rather superficial knowledge of any subject.

The Nishikawa–Wyckoff technique that was then being used by Dickinson involved several consecutive steps. First, a rotating crystal photograph was made, with X-rays from a tube with a molybdenum anticathode reflected from a large developed face or ground face of the crystal. Usually the lines corresponding to the α_1 α_2, β, and γ rays of molybdenum reflected in various orders from the face could be seen on the photographic plate. Their careful measurement gave a reliable value of the interplanar spacing, or a sub-multiple of it, corresponding to that face. No effort was made to mount the crystal in such a way that a principal direction coincided with the axis of rotation, except that this axis lay in the plane of the face under investigation, and the reflections from other crystallographic planes were ignored. For a cubic crystal a single spacing measured in this way was enough to determine the size of the possible unit cells, and for crystals with lower symmetry two or more rotating-crystal photographs were needed.

A thin section of the crystal was then prepared by grinding or sometimes by cleavage, and Laue photographs were taken, using an X-ray tube with tungsten anticathode operating at about 52 000 volts, corresponding to a short-wavelength limit of 0.24A. A print was made of the Laue photograph, and cemented in the centre of a piece of drawing paper about three feet square. With a special ruler that could be rotated about the centre of the photograph, the position of the crystallographic plane corresponding to each of the Laue diffraction spots in gnomonic projection was marked on the paper. The great advantage of the gnomonic projection is that zones are represented by straight lines, and after an initial assignment of coordinates the indices (hkl) could be assigned by checking the zones intersecting at the point representing the plane. The value of θ, the diffraction angle, is given by the distance of the spot from the centre of the photograph, and for each spot the value of $n\lambda/2d$ could be calculated from the Bragg equation.

The first step in the logical sequence of steps consisted in determining the size of the smallest unit that would account for all of the reflections on the Laue photograph. Suppose, for example, that the crystal is cubic, and that the rotating crystal photograph taken from a cube face gave the value 5.00 Å for the spacing. If the value of $n\lambda$ calculated for all of the Laue spots with use of this value of the edge of the unit turned out to be larger than 0.24 Å, the unit was accepted as the correct one. If, however, a value such as 0.15 Å was obtained, it was clear that the edge of the unit structure would have to be doubled, to 10.00 Å. This technique was a powerful one, because faint spots could often be seen on the Laue photographs that would have escaped detection by other techniques. For example, Dickinson (Dickinson 1923, 1926), using Laue photographic data, showed that the cubic unit of structure of tin tetraiodide contained 32 SnI_4 molecules, whereas Mark and Weissenberg (1923) and Ott (1926) assigned to it a unit containing only 4 SnI_4.

Brockway and I decided in 1932 to determine the parameter defining the position of the sulphur atoms in the tetragonal crystal chalcopyrite, which had been studied by Burdick and Ellis, more accurately than had been possible for them. When we examined the Laue photographs we found, to our surprise, that the true unit had twice the volume of the unit assumed by them, being doubled along the c axis (Pauling and Brockway, 1932). The copper and iron atoms are distributed among the available positions at the centres of tetrahedra of sulphur atoms in a way different from that assumed by them. I do not know of any X-ray diffraction study carried out in Pasadena with use of Laue photographs in which it was later shown that too small a unit of structure had been assumed or that any other serious error had been made.

The Laue photographs also have value in determining the possible space groups for the crystal. By orienting the crystal with a principal axis parallel to the X-ray beam, planes and a centre of symmetry can be seen, as well as symmetry axes.

From the size of the unit and the density of the crystal the number of atoms in the unit cell could be calculated. The next step was to check various special zones, to see whether there are systematic absences of reflections of certain kinds. It was assumed that the space group or set of space groups that explained all of the systematic absences was correct, and the possible atomic positions, as listed in Wyckoff's book, were then examined and compared with the intensities of the Laue spots.

At that time, 1922, there was little quantitative knowledge about the scattering factors of atoms as a function of $\sin \theta / \lambda$. It was known, however, that the value of this factor, f, decreased with increase in $\sin \theta / \lambda$, and that various other factors also cause the intensity to decrease with increase in this quantity. It was found often to be possible to determine the atomic positions in the unit cell without making any quantitative assumptions. For example two Laue spots, corresponding to different indices (hkl), were found on a Laue photograph to have the same value of the wavelength λ between 0.24 and 0.48 Å. If the stronger spot was that for the plane with the smaller interplanar distance, then one could conclude that amplitude was larger for it than for the other one. Any structure that gave a smaller calculated amplitude could be ruled out. Often in this way all structures could be eliminated except one. If it involved a variable parameter, the parameter could be determined with considerable accuracy. The task of finding suitable pairs of reflections was simplified by taking asymmetric Laue photographs, so that several reflections from planes of a single form were observed at somewhat different diffraction angles and somewhat different values of the wavelength λ.

For example, just before I had arrived in Pasadena, Dickinson and an undergraduate student, Albert Raymond, had made the first structure determination of a crystal of an organic compound. They had decided to study hexamethylenetetramine, $C_6H_{12}N_4$, because it forms cubic crystals. They found a body-centred unit containing two molecules. The carbon atoms lie along the three principal axes, in positions determined by one parameter, and the nitrogen atoms lie along the diagonals of the cube, in positions determined by a second parameter. It was not difficult to determine the values of these parameters from the intensities of the Laue spots (Dickinson and Raymond 1923). Many years later several refined redeterminations of the structure were carried out, and it was found that the parameter values of Dickinson and Raymond (0.235 for C, 0.120 for N) were essentially correct, the refined values at room temperature being 0.238 and 0.123, with edge of unit 7.021 Å (Dickinson and Raymond 7.02 Å).

Dickinson at this time was working on the structure of tin tetraiodide. He found that he would have to evaluate five parameters to fix the positions in the unit cell of the 32 Sn atoms and the 128 I atoms. He looked at the faint Laue spots that require that the unit be twice as large as that proposed by Mark and Weissenberg. From certain pairs of these faint spots he obtained a set of inequalities in the structure factors that would have to be satisfied by the correct structure. For certain values of the five parameters the small cube is acceptable. He calculated the values of the structure factors for these values of the parameters and then the partial derivatives with respect to each of the five parameters. With use of the observed inequalities he could then evaluate the small deviations of the five parameters from the small-unit values. This was quite a feat in 1922.

Dickinson made a great impression on me through his logical approach to science—not only X-ray crystallography, but also thermodynamics and other subjects in which he was interested. He was also characterized by ingenuity, as was illustrated in his work on tin tetraiodide. To be sure about the number of atoms in the unit cube, he wanted to have a reliable experimental value of the density of the substance. He found that values given in the literature differed considerably from one another, for a reason that soon became clear. Tin tetraiodide is soluble in every common solvent, so that the usual pycnometric method of determining the density could not be used. He resolved this problem by using air as the fluid medium. A bulb was filled with tin tetraiodide powder and with air at atmospheric pressure. The air was then allowed to expand into an evacuated bulb, and the pressure of the air in this increased volume was measured. In this way he could obtain the volume occupied by the known weight of tin tetraiodide, thus giving him the desired value of the density. He later verified this value by using concentrated sulphuric acid as the pycnometric liquid.

During my first couple of months as a graduate student I made fifteen substances, and crystallized them. They were substances in which I was interested for one reason or another. I searched through the volumes of Groth's *Chemische Kristallographie* for cubic crystals, and if I had an interest in the crystal I would try to investigate it. Every one of the cubic crystals that I subjected to X-ray investigation turned out, however, to have such a structure that its determination was not then possible. From aqueous solution I grew crystals of $CaHgBr_4$. The rotating-crystal photograph and the Laue photographs showed the unit to be very large, such that a dozen parameters would have to be determined to locate the atoms. I then built an electric furnace and grew, from the melt, crystals of $K_2Ni(SO_4)_2$. The structure of these crystals also was found to be determined by some dozen parameters. I had an interest in intermetallic compounds, and I found on reading the literature that large octahedral crystals of sodium dicadmide could be obtained by dissolving cadmium in molten sodium allowing the molten alloy to cool slowly, and then dissolving the excess sodium away by treatment with absolute ethanol. I used my furnace for this purpose, and got nice octahedra, about 5 mm on edge. Unfortuately, I soon discovered that the unit of structure is about 30 Å on edge, and that it contains about 1200 atoms. I worked on this structure for about thirty years, without success; it was finally determined by Sten Samson (1962). After I had studied several more cubic crystals, with similar results, Dickinson gave me a sample of the hexagonal mineral molybdenite, which has a pronounced basal cleavage, permitting good rotating crystal photographs to be made and permitting thin sections for Laue photographs to be obtained by cleaving the crystal. The determination of the structure of this crystal involved only a single parameter, and the result was very interesting (Dickinson and Pauling 1923). The coordination polyhedron about the molybdenum atom was found to be a trigonal prism, rather than the octahedron that had been found in cadmium diiodide by R. W. Bozorth and for ammonium hexachlorostannate by Dickinson the preceding year.

Despite the primitive nature of the techniques and methods of analysis, a great amount of information about the structure of crystals, especially of inorganic substances, was gathered during the first fifteen years of X-ray crystallography. By 1929 it had become possible to formulate the most important basic structural principles in this field. During the next decade, as the methods became more powerful and X-ray crystallography was supplemented with electron diffraction of gas molecules, similar progress was made in organic structural chemistry.

The Laue-photograph technique continued to be used in Pasadena during the early 1930s. At about the time that Professor Hodgkin entered the field of X-ray crystallography the procedure was made much more powerful through the formulation by both experimental and theoretical methods of a reliable set of atomic f factors and by striking developments in the methods of analysing the experimental results, such as the use of Patterson diagrams. Dorothy Hodgkin in her work on cobalamin [generic term for B_{12} vitamins] showed how powerful the method can be, even before it was revolutionized by the introduction of the use of giant computers. We owe a great part of our present understanding of the nature of substances, both inorganic and organic, to the X-ray crystallographers, and I am sure that many more discoveries will be made in the future by this ever increasingly powerful technique.

References

BURDICK, C. L. and ELLIS, J. H. (1917). *Proc. natn. Acad. Sci. USA* **3**, 644.
DICKINSON, R. G. (1923). *J. Am. chem. Soc.* **45**, 958.
——(1926). *Z. Kristallogr. Miner.* **64**, 400.
——and PAULING, L. (1923). *J. Am. chem. Soc.* **45**, 1465.
——and RAYMOND, A. L. (1923). *J. Am. chem. Soc.* **45**, 22.
HULL, A. W. (1917). *Phys. Rev.* **10**, 661.
MARK, H. and WEISSENBERG, K. (1923). *Z. Phys.* **16**, 1.
NISHIKAWA, S. (1915). *Tokyo Sugaku-Buturigakkwai Kizi* **8**, 199.
OTT, H. (1926). *Z. Kristallogr. Kristallgeom.* **63**, 222.
PAULING, L. and BROCKWAY, L. O. (1932). *Z. Kristallogr. Kristallgeom.* **82**, 188.
RINNE, F. (1915). *Ber. Verh. Sachs. Akad. Wiss.* **67**, 203.
SAMSON, S. (1962). *Nature, Lond.* **195**, 259.
WYCKOFF, R. W. G. (1920a). *J. Am. chem. Soc.* **42**, 1100.
——(1920b). *Am. J. Sci.* **50**, 317.

7. Personal recollections

TANG YOU-CHI

It was June 1951 when I paid a visit to Dorothy Crowfoot Hodgkin at Oxford. At that time I was on my way to Stockholm to attend the Second International Congress of Crystallography, or more precisely, it was my first stop in a long journey from the United States to my motherland. My first meeting with Dorothy was very enjoyable and impressive. It also recalled to me her presence at a seminar at Cal. Tech. several years before. During the years 1959–77 Dorothy came to China four times. During those long years she helped to facilitate the flow of information and understanding between crystallographers in China and abroad. However, what she would have liked even better would have been to receive us into the big family of Western crystallographers. Such a day has come at long last. In August 1978 the Academia Sinica sent a delegation to Warsaw to attend the Eleventh International Congress and to join the International Union of Crystallography. After the Congress and our visits to Polish institutions we visited several English schools of crystallography at the invitation of Professor Hodgkin on behalf of the Royal Society.

After so many years our meeting in 1951 is still fresh and vivid in my mind. I told her that at Professor Pauling's laboratory my work had begun with studies on the formation of superlattices in alloys, that I had then determined the crystal structure of hexamethylene tetramine with manganous chloride, and that on a postdoctorate George Ellery Hale fellowship I had done X-ray work on some haemoglobin crystals. I thought it necessary to give an excuse for this striking shift from alloys over complex compounds to proteins. To my happy surprise Dorothy gave me an understanding smile and made my explanation for me. This I have always remembered with great appreciation.

I thought it remarkable that Dorothy was still emotionally attached to insulin when occupied with the work on vitamin B_{12}. She was overwhelmed with excitement when she told me about Dr Sanger's work on the amino-acid sequence of insulin. She predicted that Sanger's work would eventually kindle renewed interest in research on insulin.

After I came back to China in 1951, the time seemed to fly by. A few years after my return Sanger determined the primary structure of insulin. On the basis of this discovery several groups of biochemists and organic chemists in China started to synthesize insulin in 1958. By the end of 1965 they succeeded in this total synthesis and obtained a few micrograms of crystalline synthetic bovine insulin. The synthetic work aroused a lot of publicity in China. In the spring of 1966 we were called upon to determine the crystal structure of insulin. Soon after that the Cultural Revolution broke out and we did not start crystallographic work until the beginning of 1967. The structure of pig insulin at a resolution of 2.5 Å was completed in 1971.

Now I should like to look back to Dorothy's four visits to China.

In the 1950s we received few Western visitors, but English crystallographers were an exception. In those years Professor Bernal, Dr Wooster, and Professor Lonsdale came to China relatively often. Dorothy followed in their footsteps and made her first visit in 1959. However, as a latecomer she surpassed her predecessors and established an intimate and lasting friendship with Chinese crystallographers. In this connection thanks are due to the insulin crystal because it has possibly played an important role.

When Dorothy came to China in 1959 she had already accomplished her outstanding work on vitamin B_{12} and we were celebrating the tenth anniversary of the founding of the People's Republic of China. We welcomed her to Peking University to give a lecture on vitamin B_{12} and I was very glad to serve as interpreter for her lecture. At that time the synthetic work on insulin had just begun in China. Dorothy showed enormous interest and made contacts with scientists engaged in the work. Her experience with her first visit to China was, we believe, a pleasant one.

In 1965 Dorothy came to Shanghai, but I was occupied in the countryside and regretfully was not able to meet her there. When she came in 1972 I was still in an awkward situation since our meeting was governed by many restrictions.

By the time Dorothy came to Peking in 1977 we were still overjoyed with the collapse of the 'Gang of Four' in the previous year. All of us became happier and more confident in our work.

Dorothy's fourth visit turned out to be a very fruitful and joyful meeting with Chinese crystallographers. This time she brought with her Guy Dodson and gave us a chance to meet a crystallographer of the younger generation. We discussed the latest work on insulins and what was going on in protein-crystallographic laboratories abroad. When they were about to leave for Shanghai we spent an afternoon together in the Summer Palace. Scores of young crystallographers joined us in the happy and harmonious get-together.

Dorothy will be remembered with great affection and respect by the Chinese crystallographers. Dorothy, I wish you many happy returns!

II Structural chemistry and biochemistry

8. Organic chemistry, X-ray analysis, and Dorothy Hodgkin

J. D. DUNITZ

DURING the early 1960s I used to hear some of my organic chemistry colleagues talking about 'the crisis in Organic Chemistry'. What they were concerned about was the gradual encroachment of X-ray crystallography into an area they regarded as a traditional preserve of organic chemistry, namely the elucidation of the molecular constitution of natural product substances. About that time it was becoming clear that structures of natural products, especially those containing new, unknown features, could be determined a good deal faster and more efficiently by X-ray analysis than by the classical methods involving chemical degradation. Moreover, the structural formulae resulting from X-ray analysis were usually so clear-cut and unequivocal that there was no need to confirm them by chemical synthesis. This was in contrast to the kind of structural conclusions that could be derived by the traditional approach. Here the chain of logical inference, from chemical degradation experiments to structural deductions, was so indirect that any conclusions were necessarily of the nature of constitutional hypotheses, always liable to modification by further experiments or by new interpretations of the existing evidence. Until the arrival of X-ray crystallography, the structural hypotheses derived from degradation could only be regarded as proven once they were backed up by the synthesis of a key, target molecule. The possibility that errors in the interpretation of the synthetic steps might exactly compensate errors in the interpretation of the degradative evidence was regarded as so slight that it could be neglected.

For more than a century, this interplay between degradative and synthetic studies of natural compounds had continued to dominate organic chemistry. Now that the more direct road to structure determination had been opened by X-ray analysis, prospects for the continuing vitality of these two areas of chemistry did not look too encouraging. Was there not a danger that organic chemistry, deprived of the intellectual stimulation and nourishment it had always had from these areas, might degenerate into a branch of applied technology, useful for providing some of the necessities and comforts of modern life, but devoid of the speculative character that had made it so fascinating?

The problems created by the X-ray road to structure analysis were felt not only by organic chemists but also by crystallographers themselves, at least by the more perspicacious among them, as shown by the following citation (Robertson 1962):

> In the past many of the great discoveries of organic chemistry have been made in the course of the long and patient investigations that are required in the elucidation of natural product structures. While solving a structure the chemist does far more than merely find the relative positions of the atoms in space. He makes many discoveries and learns a lot of chemistry ... The X-ray crystallographer can now tell him the positions of the atoms very accurately and often very quickly, but cannot enlighten him about the discoveries that might have been made during a detailed chemical analysis. There is perhaps a real danger that unless serious thought is given to this matter the cause of organic chemistry will not be advanced by this work. It could even be retarded.

Well, as we all know now, organic chemistry has survived the crisis rather well. Relieved of the burden of determining and proving natural product structures, organic chemists could afford to turn their energy and inventiveness to other areas where prodigious advances have resulted. But before we discuss some of the ways in which the crisis was surmounted, it may be of interest to look back a little to see how it gradually came about.

X-ray analysis began to be applied to crystals of organic compounds during the early 1920s. Most of the crystals studied in those days were of compounds of known chemical constitution; most of the conclusions, based on unit-cell dimensions and space-group symmetry arguments, sometimes backed by a few intensity measurements, were wrong. However, this early work was important. Besides paving the way for the advances that were to come, it demonstrated for the first time that the formulae of organic chemistry are not just imaginary constructions having a limited range of validity in explaining the facts of chemical reactivity. These formulae were shown to correspond to objects of rather definite size and shape, even though it was proving to be not such an easy task to determine exactly what these shapes are.

Nevertheless, the potentiality of diffraction methods for solving problems of structural organic chemistry was becoming apparent. In a discussion of the configurational assignment of the two isomeric unsaturated fatty acids, $CH_3(CH_2)_7CH{=\!=}CH(CH_2)_{11}COOH$, Haworth (1923) comments:

> This conclusion is contrary to that drawn from X-ray measurements of the two acids, since the spacing of brassidic acid is longer than that of erucic acid to the extent of about 30 per cent. Consequently, the former should be the trans-, and the latter the cis-modification. The X-ray method furnishes a new instrument for the solution of such problems, and it is abundantly evident that many structural problems may be resolved by this means.

Not all the structures proposed in those early days were wrong. The few exceptions mostly involved atomic arrangements of high symmetry. Hexamethylene tetramine, the condensation product of formaldehyde and ammonia, $C_6H_{12}N_4$, forms cubic crystals where the six carbon atoms must be arranged in a regular octahedron and the four nitrogens in a regular tetrahedron. Thus, only two independent parameters are needed to define the molecular skeleton. Hexamethylenetetramine was the first organic compound whose structure was completely determined by X-ray analysis without the aid of a structural formula (Dickinson and Raymond 1923). A few years later, the X-ray analysis of *sym*-hexachloro- and hexabromocyclohexane, which also crystallize in the cubic system, established the puckered chair form of the cyclohexane ring (Dickinson and Bilicke 1928) with interbond angles close to the tetrahedral value. But most organic molecules are of low symmetry, and the ones that do have high symmetry rarely exhibit their full symmetry in the crystalline state. Sometimes, as in the long-chain hydrocarbons and their derivatives, the internal regularity of the zig-zag carbon chain made it possible to draw reliable structural conclusions (Shearer 1925; Müller 1928), but, in general, the outlook for establishing the crystal structures of organic compounds, even those of known molecular constitution, did not seem encouraging. A progress report by James, West, and Bradley (1927) has a distinctly pessimistic tone:

> The investigation of the detailed structure of organic crystals is particularly difficult. They are in general of relatively low crystallographic symmetry, the molecules are complicated, and a complete determination involves the fixing of a large number of parameters. This can only be done by measuring the intensities of the X-ray spectra, and for organic crystals such measurements are usually not easy to make ... and extremely difficult to interpret. Most of the investigations of the structures of organic molecules have not attempted to go beyond the space group.

The analysis of hexamethylbenzene (Lonsdale 1929) inaugurated a new era, which was to be dominated by the work of the Robertson school on the structures of aromatic benzenoid hydrocarbons. Here the problem was not to provide independent verification of the molecular constitutions of these compounds, but rather to establish the structural details of these molecules as accurately as possible. This venture was only made possible by the introduction of Fourier series calculations of electron-density distributions. These calculations were facilitated by various kinds of computational aids (Robertson 1936*a*; Beevers and Lipson 1936), but for all practical purposes they were limited to two-dimensional projections of the electron density down particular crystal axes. This limitation was imposed by the sheer labour involved in carrying out three-dimensional calculations, but its effect was not too serious for crystals built from flat molecules. There was usually some crystal direction in which the molecules could be viewed to show a recognizable pattern of non-superimposed electron-density peaks. With the Fourier method it became possible to adjust a large number of atomic positions simultaneously, and this was a great advantage over previous methods where the parameters had to be adjusted individually to fit the measured diffraction pattern as well as possible.

There was still the difficulty of guessing the overall arrangement of molecules in order to derive an initial set of signs or phases for the Fourier coefficients—the phase problem, in modern parlance. This still had to be done by a trial-and-error procedure where intuition certainly played a part and where any hints or clues provided by the diffraction pattern or by the physical properties of the crystal (cleavage, refractive indices, etc.) could be essential. The hexamethylbenzene analysis is a good example. The crystals are triclinic, space group $P\bar{1}$. With one centrosymmetric molecule in the unit cell, 18 independent parameters have to be fixed to define the arrangement of the main scattering centres, the carbon atoms, but there are some simplifying features. The crystals cleave along (001), the reflection from this plane is very intense, and the intensities of the (00l) reflections fall off in a very similar way to that from graphite crystals. Hence the molecules must be almost exactly planar, and the molecular planes must lie nearly parallel to (001). In addition, the (hk0) reflections show pseudohexagonal symmetry, suggesting that the carbon atoms lie at the vertices of two concentric hexagons. The projected structure is then defined by only three variables: the common orientation of the two hexagons and their radii, and these could be fixed within narrow limits from the measured intensities of the (hk0) reflections. Lonsdale's work provided one of the first experimental proofs for the planarity of the benzene ring. The structure was later refined by Fourier series methods (Brockway and Robertson 1939), yielding C—C bond lengths of 1.39 Å for the ring bonds and 1.53 Å for the exocyclic ones, values that are probably very close to the correct ones.

The introduction of Fourier series methods led to two important embellishments to the structural formulae of organic chemistry during the next decade or so. First, the electron-density maps provided direct visualization of the molecules, and secondly, they yielded bond distances and angles that were generally much more accurate than those available previously from X-ray diffraction studies (except for a few cases, e.g. diamond, where the C—C distance follows from the unit-cell dimensions). During roughly the same period, structural parameters of even higher accuracy were becoming available from spectroscopic and electron-diffraction studies of molecules in the vapour phase, but these studies were possible only if the molecules in question contained few atoms or if they showed high symmetry. As a result of this work, the various types of bonds in organic molecules began to be distinguishable, not just on the basis of their chemical properties, as previously, but also on the basis of purely geometrical properties—characteristic bond distances and angles, planarity of certain atomic groupings, linearity of others. In addition, various regularities about molecular packing arrangements within crystals were becoming recognized—characteristic non-bonded distances.

But no sooner were these regularities incorporated within the accepted body of knowledge than all sorts of exceptions began to become apparent. For example, although most C—C bond distances were close to the values 1.54 Å (single bond), 1.39 Å (aromatic bond), 1.33 Å (double bond), or 1.20 Å (triple bond) it became clear that there was an almost continuous gradation in observed C—C distances, from about 1.6 Å to about 1.2 Å, depending on special features in the environment of the bond. Bond length could be correlated with bond order, and it was not long until the statement that a bond was shorter than usual came to be virtually synonymous with the statement that it had 'increased double-bonded character'. Some of these newly discovered trends could be rationalized in terms of concepts that were already current in chemical thinking or were soon about to become so—steric effects, conjugation, hyperconjugation, etc., but on the whole they probably had a greater influence on the newly developing branch of theoretical chemistry than on the central body of the subject.

The role of hydrogen bonding as an important factor influencing the structures and properties of crystals (and not only crystals!) was also much studied during this period. Because of the low scattering power of hydrogen atoms for X-rays and the concomitant difficulty of locating these atoms in electron-density distributions, the occurrence of hydrogen bonds in crystals was usually a matter of inference rather than of direct observation. The main diagnostic was the mutual approach of a pair of electronegative atoms (O, N, F), at least one of which could be assumed to carry a hydrogen substituent, to a distance shorter than that typical of a non-hydrogen-bonded situation. The importance of hydrogen bonding in crystals of such substances as carboxylic acids and amides, carbohydrates, etc., pointed to an equally pervasive influence on the properties of these substances in solution. But the seeds planted by this work only came to full fruition in the early 1950s when hydrogen bonding was recognized to be the most significant factor in determining the secondary structure of the important classes of biological macromolecules with a decisive influence on the course of biochemical and physiological events.

Two important methodological advances during the 1930s have to be mentioned. One was the invention of the Patterson synthesis (1934), which made it possible to determine the distribution of interatomic vectors in a crystal directly from the diffraction pattern. Given the distribution of interatomic vectors, the spatial arrangement of the atoms themselves can always be derived in principle (a proof of this is given by Buerger (1950)), but in practice there are formidable difficulties that rapidly become insurmountable as the number of atoms in the unit cell increases beyond a certain limit.

Thus although simple structures (those containing only a few atoms in the unit cell) can be solved by unscrambling the Patterson synthesis, more complex ones can not.

The way around this difficulty was given by the heavy-atom method and its variants, especially the isomorphous replacement method, which were first exploited in the analysis of the phthalocyanines (Robertson 1936*b*; Robertson and Woodward 1940). The heavy-atom method utilizes the presence of a few atoms of high scattering power in the structure. The positions of these atoms are inferred from the distribution of the strongest peaks in the Patterson function. A Fourier synthesis based on the phases of the heavy-atom contributions will be far from correct, but its minor peaks will tend to occur close to the positions of the lighter atoms that were ignored in the first phasing calculations. Some of these peaks are selected to form a provisional, often incomplete, molecular model, which then serves as a basis for further phasing calculations. A few iterative cycles of this kind usually lead to an electron-density distribution in which the entire molecular structure is convincingly reproduced. In the isomorphous replacement method, Fourier coefficients with phases that differ strongly from the heavy-atom phases can be identified by the intensity changes that occur on isomorphous substitution of one heavy atom by another, but otherwise the procedure is very similar.

With the introduction of heavy atom methods, the stage was set for solving crystal structures of very great complexity. If the molecular constitution were known in advance, the recognition of meaningful patterns of peaks in imperfectly phased electron-density maps was made slightly easier—one knew roughly what patterns to look for. But even if the molecular constitution were unknown or uncertain, the additional difficulties that arose were not too serious; one merely had to know enough about general structural rules to distinguish chemically plausible patterns of peaks from unplausible ones. In either case, once one got started on the right track, the iterative phasing process was largely self-correcting.

Why were crystallographers so slow to take advantage of these new possibilities? The answer is not hard to find. In order to locate the atoms of a complex molecule with confidence, its electron-density distribution has to be calculated in three dimensions. The calculation of a three-dimensional Fourier synthesis was such a formidable task that crystallographers were very hesitant to undertake problems that might involve not just a single calculation of this kind but probably several. It took another decade or so before the computational labour could be reduced by the use of punched-card methods, and it took several years more until three-dimensional analyses became commonplace.

During the intervening period (roughly 1940–55), crystallography began its advance into natural product structures, at first slowly and then at an ever increasing rate. The Oxford laboratory soon established itself as one of the main centres of this kind of research. Its reputation certainly did not rest on the number of structures it produced, which was actually quite small. It was rather a matter of the complexity of the structure analyses that were undertaken there. Dorothy Hodgkin was convinced that structures which appeared to others as being of hopeless complexity could be solved by X-ray analysis and she showed relentless determination in pursuing her goal. Three analyses from her laboratory stand as landmarks during this period.

The structural formula of steroids had been a central topic of organic chemistry for many years. Bernal (1932) had shown that the then accepted Wieland–Windaus formula [1] for cholesterol was incompatible with unit-cell dimensions and optical properties of crystals of this and related compounds. Soon afterwards the revised formula [2] was proposed and generally accepted, but many stereochemical details remained to be settled.

Homeless C_2H_4

[1]   [2]

The structure analysis of cholesteryl iodide (Carlisle and Crowfoot 1945) was one of the first analyses based on three-dimensional calculations and it was probably the most complicated X-ray analysis performed up till then. Its chemical importance is that it provided the first complete, detailed stereochemical picture of a steroid. The result must have been particularly gratifying to the Zürich chemists since it confirmed the *cis* relationship of the hydroxyl and methyl groups on ring A, as proposed by Ruzicka, Furter, and Goldberg (1938) a few years earlier.

The penicillin analysis that followed (Crowfoot, Bunn, Rogers-Low, and Turner-Jones 1949) was a far more difficult undertaking, mainly because the structural evidence available from chemical studies was so incomplete and fragmentary. Since much of the chemical work was being done in Oxford, more or less simultaneously with the X-ray work, there was a steady exchange of information between the two groups. Each was continually being influenced by the other—a novel situation that was to recur a decade later in connection with the vitamin B_{12} analysis.

Initially it was hoped that a more or less direct X-ray analysis would be possible by studying the isomorphous potassium and rubidium salts of benzylpenicillin, but this hope was dashed when it turned out that the heavy atoms were placed in such a way that they did not contribute to many of the reflections. Attention was then turned to the non-isomorphous though related sodium salt, using mainly trial-and-error methods and taking advantage of a rapid optical method for testing trial structures, which were mostly based on the thiazolidine oxazolone formula [3]. From the chemical evidence this was considered to be the most likely, although several alternatives could not be ruled out.

The final solution was obtained by comparing what seemed to be the best electron-density projections obtained for the sodium salt with the rough projections derived from the direct but incomplete phases for the isomorphous potassium and rubidium salts. By assuming that the molecule would appear in approximately the same orientation in all three projections, some of their common features could be provisionally identified as corresponding to molecular fragments known to be present in the molecule. Once these were recognized, further phasing calculations, first in projection and then in three dimensions, were possible, leading eventually to a set of atomic positions that corresponded not to the initially assumed structure [3] but to the alternative structure [4] with stereochemistry as shown. This was just as compatible with the chemical evidence but had been disfavoured on account of its β-lactam ring, a previously unknown feature in natural-compound chemistry.

[3]   [4]

The above description of the penicillin analysis may seem too long for a structure which, after all, could be solved by a graduate student using modern methods in a few months. It is nevertheless much too short to give an impression of the intricate way the structure was actually solved, full of false starts and blind alleys. Likewise, with the aid of modern spectroscopic methods, the interpretation of the chemical evidence would not be nearly so difficult as it appeared then. However, by the standards of the 1940s, penicillin was an unusually difficult problem, mainly because of the lability of the molecule, which undergoes extensive structural rearrangements even under mild conditions.

What the penicillin structure was for the 1940s, the vitamin B_{12} structure was for the following decade—the most important achievement of X-ray analysis in the field of natural product chemistry. With allowance for the vastly more complex nature of the vitamin B_{12} molecule and for the much more powerful technical aids that had become available, especially during the latter stages of the B_{12} analysis, there are several points of similarity. When the X-ray analysis was begun, practically nothing was known about the structure of the vitamin except that it was extraordinarily complex. Not just one but several crystalline derivatives were studied. The X-ray work and the chemical investigations were again carried out in close collaboration.

As the chemical work proceeded, several fragments derived by degradation of the molecule could be identified: a nucleotide-like fragment consisting of a benzimidazole-ribose-phosphate, a propanolamine fragment and various amide groupings, besides a large porphyrin-like nucleus containing cobalt. How all these fragments were linked together in the molecule was still unknown.

As the X-ray analysis progressed, these details slowly began to emerge from the three-dimensional density maps; the cobalt-containing nucleus turned out to correspond to an entirely new feature in organic chemistry—the corrin structure, later described (Eschenmoser, 1963) as 'perhaps the finest gift that X-ray analysis has so far bestowed on the organic chemistry of low molecular weight natural products'. The vitamin B_{12} structure (Hodgkin, Kamper, Mackay, Pickworth, Trueblood, and White 1956) was by far the most complex molecule whose structure had been elucidated in atomic detail up to that time, but it was superseded a few years later by the B_{12} coenzyme (Lenhert and Hodgkin 1961), which proved to contain yet another new feature—a cobalt–carbon bond linking the corrin system with the 5′-carbon of an adenosine fragment.

If we exclude DNA and the first crystalline proteins the structure elucidation of vitamin and coenzyme B_{12} was the most spectacular contribution of X-ray analysis to natural-product chemistry during the 1950s and early 1960s, but it was not the only one. The growing availability of electronic computers for crystallographic calculations (not only for three-dimensional Fourier syntheses but also for automatic structure refinement by least-squares analysis) had led to an enormous increase in productivity, so that by the mid-1960s the number of organic crystal structures determined had passed the 2000 mark. While most of these determinations merely confirmed conclusions already reached by organic chemistry or added some needed metrical or stereochemical detail, the list also included many complex natural product molecules whose structures had previously been shrouded in doubt.

One contribution of X-ray analysis to stereochemistry during this period remains to be mentioned—a unique contribution. Ever since the beginnings of stereochemistry there was no way of telling which of a pair of enantiomorphic structures corresponded to a given chiral molecule, and there did not seem to be much hope of ever finding this out. In particular, as far as X-ray diffraction was concerned, Friedel's Law stated that there is nothing to distinguish propagation of X-rays in a direction AB from that in the opposite direction BA. This holds as long as phase differences between different atoms depend only on pathlength differences, i.e. as long as any intrinsic phase shift associated with the scattering process is the same for all atoms. Under these conditions, for two opposite reflections (hkl) and $(\bar{h}\bar{k}\bar{l})$, all phase differences are simply reversed; the two resultant waves then have equal amplitudes and hence equal intensities.

Actually, it had been known for some time that the assumptions behind Friedel's Law were only approximately correct. When the wavelength of the incident X-radiation is close to an absorption edge of one of the atoms in the crystal, the intrinsic phase change associated with scattering by this atom is slightly different from that of the other atoms. There is then a slight 'phase lag' in the wave scattered by this atom. For propagation along two opposite directions, the phase differences due to the relative arrangement of the atoms in space are still reversed, but the same phase lag applies to both. This leads, for non-centrosymmetric crystals, to a breakdown of Friedel's Law, to a slight difference between the intensities of (hkl) and $(\bar{h}\bar{k}\bar{l})$ reflections. Many years earlier, Coster, Knol, and Prins (1930) had demonstrated the break-down of Friedel's Law by producing intensity differences between reflections from the (111) and $(\bar{1}\bar{1}\bar{1})$ faces of a zincblende crystal; from the sign of the difference they were able to derive the sense of polarity of the crystal.

It was Bijvoet who saw that polarity is just one-dimensional chirality and that the same principle could be used for establishing the absolute sense of the reference frame used to describe the three-dimensional structure of a chiral crystal. Once the relative positions are known, the absolute frame of reference can be determined from the sense of the observed intensity difference for the direction of the anomalous phase change can be derived by simple theoretical considerations. In 1951 this principle was applied to crystals of $RbNaC_4H_4O_6 \cdot 4H_2O$ prepared from (+)-tartaric acid, using $ZrK\alpha$ radiation to produce a phase shift in the scattering from the rubidium atoms. The analysis showed that the configuration depicted by the conventional Fischer projection of (+)-tartaric acid happened to be correct. Stereo-chemistry had been transformed from a relative basis to an absolute one.

The Bijvoet method has since been applied in hundreds of cases. At the very least the method seems to be perfectly self-consistent; thus either all absolute configurations determined by X-ray analysis are right or they are all wrong, depending on the rightness or wrongness of the direction ascribed to the anomalous phase shift. In the few cases where absolute configurations can be assigned with reasonable confidence from chiroptical properties the results agree with those determined by the Bijvoet method. A final proof of the correctness of absolute configurations obtained by X-ray analysis was provided by Brongersma and Mul (1973), who determined the polarity sense of zincblende by an independent method involving noble-gas-reflection-mass spectrometry and showed that it was the same as that derived by anomalous scattering experiments.

Twenty years ago, crystal structure analysis still required a high degree of training, skill, and dedication from its practitioners. When it was called in to help with problems of molecular structure, it was usually as a last resort, when all other methods had failed or seemed likely to do so. Things are quite different today for in the meantime diffractometers and high-speed computers have transformed the subject. The diffractometers, now usually on-line and largely automatic, make it possible to measure intensities of several thousand crystal reflections in a small fraction of the time that was formerly involved, and much more accurately as well. With the aid of computers, the calculations have lost their terrors; direct methods of structure analysis have become commonplace, and extensive least-squares refinements that would have taken a lifetime to perform can now be done in a few minutes. As a result, crystal structure analysis, once regarded by most chemists with a mixture of awe and suspicion, has been transformed into a more or less routine method of obtaining information about molecular structure.

Crystal structure analysis has also become much more informative than it was, for it now yields not only atomic positions but also vibrational parameters of sufficient accuracy to enable conclusions to be drawn about dynamic behaviour of molecules. Also, the electron-density deformation maps that are now available reveal convincing pictures of bonding and lone-pair densities that are otherwise not amenable to experimental study.

In any case, the romantic period seems to be over. We need waste no tears about that for the trivialization of X-ray analysis means that we can uncover the facts about crystal and molecular structure without breaking our backs in the process. Indeed, the ease with which crystal structures can be dealt with today leads to a world production running into several thousand structures annually; one result of this is that we are being flooded with structural information faster than anyone can possibly cope with it. Thanks to the labours of the Cambridge Crystallographic Data Centre, the problems of locating and retrieving any desired piece of information from the ocean of publications are well taken care of, at least for organic structures. This is obviously very useful if one wants to know the answer to a specific question, but it still leaves the general problem open: what are we to do with all that information? Surely it can teach us something, but first we must learn how to order the information into meaningful patterns. Disorganized knowledge is no knowledge.

I began this essay by talking about the 'crisis in organic chemistry', which seemed to some people to have arisen in the early 1960s as a result of the intrusion of X-ray analysis into realms that had traditionally been regarded as sacrosanct. As the investment of time and effort in degradative studies began to look less profitable, these studies naturally declined in importance. However, the energy that had formerly been tied up in these studies was simply channelled into other directions, mainly into mechanistic and synthetic (including biosynthetic) studies, which have thereby been advanced prodigiously. Freedom from the task of structure proof meant freedom from the restriction that a synthesis of a specific target molecule had to proceed by steps of known reaction type. Thus, synthetic strategies no longer needed to follow along well-established lines but could be of an outspokenly exploratory nature.

Perhaps the best, certainly the most far-reaching example of this kind of synthesis, is the synthesis of vitamin B_{12}, a co-operative yet competitive enterprise of the Harvard and ETH laboratories under the generalship of R. B. Woodward and A. Eschenmoser respectively. The motivations behind this enterprise have been recently recapitulated by Eschenmoser and Winter (1977):

> When Hodgkin announced the complete structural formula of vitamin B_{12} in 1956, it was clear that this natural product presented an ideal objective for organic synthetic research. It is a compound of great biochemical significance. Its molecular architecture is complex and had not previously been encountered in natural products chemistry. Its structural nucleus had resisted elucidation by means of chemical methods of degradation and had been solved by X-ray crystallographic analysis. The synthetic investigation of vitamin B_{12} would involve a host of new problems in the realm of planning and method and would link 'X-ray island B_{12}' with the mainland of chemical experience. Vitamin B_{12} provided an opportunity to extend the frontiers hitherto established by organic synthesis in the area of low-molecular-weight natural products.

Quite apart from the successful completion of this synthesis (along lines that could hardly be described as being well established), one of its by-products, so to speak, was the recognition of the role of orbital symmetry in governing the stereochemical course of chemical reactions. Woodward (1967) has described the intriguing stereochemical puzzle that arose during his synthetic approach to vitamin B_{12} and how the solution of this puzzle revealed the crucial role of orbital symmetry in directing the stereochemistry of pericyclic reactions in general. Unlike other concepts which take several years to permeate into the general awareness, the impact of the Woodward–Hoffmann rules was immediate and pervasive; their influence on the development of practical and theoretical organic chemistry over the last fifteen years or so can hardly be exaggerated.

These advances and others were only made possible once organic chemists had been liberated from the chore of structure determination; for only then was the necessary leisure time available for less onerous activities. In so far as this liberation process can be personified in any single individual, it is surely Dorothy Hodgkin to whom we should accord this distinction. Of course, Dorothy's work was not done in isolation. It owed much to the example and inspiration set by her contemporaries: Bernal, Pauling, Perutz, Robertson, and others. But her own work had a quite individual direction about it, a direction that soon led her to be acknowledged as the leading crystallographer in the field of natural product research. Dorothy had an unerring instinct for sensing the most significant structural problems in this field, she had the audacity to attack these problems when they seemed well-nigh insoluble, she had the perseverance to struggle onward where others would have given up, and she had the skill and imagination to solve these problems once the pieces of the puzzle began to take shape. It is for these reasons that Dorothy's contribution has been so special.

References

BEEVERS, W. A. and LIPSON, H. (1936). *Proc. phys. Soc., Lond.* **A48**, 772.
BERNAL, J. D. (1932). *Nature, Lond.* **129**, 277.
BROCKWAY, L. O. and ROBERTSON, J. M. (1939). *J. chem. Soc.* 1324.
BRONGERSMA, H. H. and MUL, P. M. (1973). *Chem. Phys. Lett.* **19**, 217.
BUERGER, M. J. (1950). *Acta crystallogr.* **3**, 87.
CARLISLE, C. H. and CROWFOOT, D. (1945). *Proc. R. Soc.* **A184**, 64.
COSTER, D., KNOL, K. S., and PRINS, J. A. (1930). *Z. Phys.* **63**, 345.
CROWFOOT, D., BUNN, C. W., ROGERS-LOW, B. W., and TURNER-JONES, A. (1949). In *The chemistry of penicillin* (ed. H. T. Clarke, J. R. Johnson, and Sir R. Robinson), p. 310. Princeton University Press.
DICKINSON, R. G. and BILICKE, C. (1928). *J. Am. chem. Soc.* **50**, 764.
——and RAYMOND, A. L. (1923). *J. Am. chem. Soc.* **45**, 22.
ESCHENMOSER, A. (1963). *Pure appl. Chem.* **7**, 297.
——and WINTNER, C. E. (1977). *Science, N.Y.* **196**, 1410.
HAWORTH, W. N. (1923). *Rep. Prog. Chem.* **20**, 69.
HODGKIN, D. C., KAMPER, J., MACKAY, M., PICKWORTH, J., TRUEBLOOD, K. N., and WHITE, J. G. (1956). *Nature, Lond.* **178**, 64.
JAMES, R. W., WEST, J., and BRADLEY, A. J. (1927). *Rep. Prog. Chem.* **24**, 289.
LENHERT, P. G. and HODGKIN, D. C. (1961). *Nature, Lond.* **192**, 937.
LONSDALE, K. (1929). *Proc. R. Soc.* **A123**, 494.
MÜLLER, A. (1928). *Proc. R. Soc.* **A120**, 437.
PATTERSON, A. L. (1934). *Phys. Rev.* **46**, 372.
ROBERTSON, J. M. (1936a). *Phil. Mag.*, Series 7, **21**, 176.
——(1936b). *J. chem. Soc.* 1195.
——(1962). In *Fifty years of X-ray Analysis* (ed. P. P. Ewald), p. 170. Oosthoek, Utrecht.
——and Woodward, I. (1940). *J. chem. Soc.* 36.
RUZICKA, L., FURTER, M., and GOLDBERG, M. W. (1938). *Helv. chim. Acta* **21**, 498.
SHEARER, G. (1925). *Proc. R. Soc.* **A108**, 655.
WOODWARD, R. B. (1967). *Special Publication No. 12*, p. 217. The Chemical Society, London.

9. A study of some organic molecules

J. M. ROBERTSON

EXCEPT for a few special cases like hexamethylenetetramine (Dickinson and Raymond 1923) and hexamethylbenzene (Lonsdale 1929), it was a long time before any effective work on organic crystal structures was carried out. Even for the smaller molecules the number of parameters presented formidable difficulties. But by the 1930s trial-and-error methods based on the known chemical structures succeeded in solving the structures of a number of planar or nearly planar aromatic molecules with a high degree of accuracy. It then became possible to place the known chemical structures on an exact metrical basis with accurate bond lengths and bond angles. Many theoretical studies of the chemical bond then became possible and this led to enormous advances in our general understanding of chemistry. However, the organic crystal structures that were solved at that time were generally those of molecules of known chemical constitution.

The late Sir Robert Robinson took a great interest in this work, and I remember when I used to show him some new structure that I had solved, he used to congratulate me, but generally added, 'Of course you are only telling us what we already know!' Many years later it was with great satisfaction that I heard him say, 'In organic chemistry X-ray crystallography is now not only a necessary method, it is sometimes the only method that is necessary.'

In the early days it was, of course, the fundamental phase problem that prevented us solving structures of unknown chemical constitution. The Patterson vector method had been developed, but again, for organic structures with many light atoms, it was too difficult to apply. The big advance began. I think, when I was able to solve the phthalocyanine structures (Robertson 1935, 1936; Robertson and Woodward 1937, 1940) and develop the heavy atom and isomorphous replacement methods. This was a very exciting advance. For the first time it became possible to solve an organic structure without reference to the known chemical constitution and indeed without even assuming the existence of atoms at all. I was so excited about this that in

1939 I suggested that these methods might even be used to solve the structure of the complex proteins and enzymes (Robertson 1939). In an organic structure it is generally possible to insert heavy atoms and find their positions by an application of the Patterson vector method. By successive replacements an isomorphous series can sometimes be achieved. Even although there were no computers at that time, everything seemed hopeful. But at that time, in 1939, we were about to embark on a five-year war, and all scientific work, for me at least, had to switch to something of more immediate application than crystallography.

Afterwards, when things began to settle down, I moved to a new Department in Glasgow. It had very little equipment of any kind, and none in the field of X-ray crystallography. But the University made very generous financial provision for a new Professor, and I was able gradually to make a start again. I was also able, for the first time, to recruit some research students. Two of my earliest, Jack Dunitz and John White, are contributors to this volume. A. McL. Mathieson came to me about the same time, and Sydney Abrahams a little later. However, when you have really good students they soon tend to move on to more senior posts elsewhere and you have to start training others. Jack Dunitz moved to Oxford, London, Pasadena, and finally Zurich. John White to Princeton and then New York. Sandy Mathieson to CSIRO in Melbourne, and Sydney Abrahams to Bell Telephones.

While all this was going on in Glasgow, great progress was being made in the science of X-ray crystallography. Patterson's vector method was universally employed, and this in conjunction with the heavy atom method for phase determination provided a well-established route for elucidating the structures of compounds of unknown chemical constitution. Dorothy was one of the first to make significant use of these methods in her analysis of the cholesterol iodide structure (Carlisle and Crowfoot 1945).† These methods also played an important part in her later brilliant work on penicillin and vitamin B_{12}.

In Glasgow we now turned our attention to some of the outstanding problems in organic chemistry, and especially to those molecules whose structure and constitution had baffled the chemists for many years. In this work we enjoyed the happy co-operation of many organic chemists who willingly prepared suitable derivatives for our work, a task that often involved months of difficult and tedious work. To begin with, the alkaloid field seemed the most attractive, because nicely crystalline hydrochlorides and hydrobromides are often easy to make.

We may note in passing that for an easy application of the heavy-atom method it is desirable that the contribution of the heavy atom to the structure factor should be nearly equal to the average contribution of all the other

---

† See also Crowfoot and Dunitz (1948) and Hodgkin and Sayre (1952).

atoms combined. Much weaker phase determination than this is often possible, although it generally involves a great many successive approximations.
The average structure factor $|F|$ from a group of about 34 randomly placed
carbon atoms is approximately equal to the heavy-atom contribution from
a single bromine atom. An iodine atom would be equally effective for a random group of about 78 carbon atoms, but a chlorine atom would only suffice
for a group of about eight carbon atoms on this basis. However, the heavy-
atom contribution is generally rather more effective than is indicated by this
calculation, when we consider the large number of structure factors of less
than average magnitude. These considerations show how effective the crystalline hydrochlorides and hydrobromides of the alkaloids can be for structure
determination.

Calycanthine (Hamor and Robertson 1962) and echitamine (Hamilton,
Hamor, Robertson, and Sim 1962) were our first targets. Many different
structural formulae had been proposed by various workers over the years,
but they were all unsatisfactory to some degree. With Tom Hamor and others
helping we soon obtained a complete solution in both cases, employing the
bromides, and our structures were different from any that had been proposed.
But they were immediately accepted, and were found to explain all the chemical properties. The unusual complications in these structures [1 and 2] easily
explain the difficulties that the chemists had experienced.

[1] Calycanthine   [2] Echitamine

We then proceeded to obtain solutions for a large number of other hitherto
unknown alkaloid structures, including the toxic curare alkaloids macusine
A (McPhail, Robertson, and Sim 1963) and isocalebassine (Gemmell *et al.*
1969). The latter caused us a great deal of trouble that was ultimately traced
to erroneous indexing of a number of high-order reflections! Other structures
solved included those of hunterburnine (Asher, Robertson, and Sim 1965)
and chimonanthine (Grant, Hamor, Robertson, and Sim 1965).

Most of these structures and those that are described below were solved
by application of the heavy atom and isomorphous replacement methods.
It should, however, be noted that nowadays more direct solutions can often
be obtained using only the measured intensities and mathematical relations
between the structure factors. This development was initiated by the work
of Harker and Kasper in the 1940s and greatly extended by the use of probability methods and other means developed by Sayre, Zachariasen, and

others. The advent of fast computers also greatly facilitated this approach. When derivatives are hard to prepare, heavy-atom methods are now hardly necessary. But where large biological molecules like the proteins and enzymes are concerned, the heavy-atom and isomorphous-replacement method still seems to constitute the only feasible approach.

After the alkaloids, the terpenoids were our favourite field of study in Glasgow. I was biased in this direction by my early work in organic chemistry, when I tried unsuccessfully to solve many of these structures by standard chemical methods. The sesquiterpene caryophyllene and its transformation products were some of those that I struggled with in those early days. It was with great joy that I finally solved these structures many years later by X-ray methods.

The chemistry of caryophyllene is exceedingly complex and had puzzled generations of workers. It was finally solved by the chemical work of Barton and Lindsey (1951) and Sorm, Dolejs, and Pliva (1950), and the X-ray analysis that I carried out with George Todd (Robertson and Todd 1955) on the alcohol caryolan-1-ol, which forms beautiful crystalline and isomorphous chlorides and bromides. One of the remarkable features of caryophyllene is the ease with which it and its derivatives undergo cyclization and molecular rearrangement. This makes their study by ordinary chemical methods very difficult and sometimes almost impossible. Many of these structures could only have been solved by X-ray crystallography. One such molecule is the tricyclic isoclovene, which I first isolated chemically in 1926 (Henderson, McCrone, and Robertson 1929). But it was not until 1960 that I solved this structure [3] by X-ray crystallography with J. S. Clunie (Clunie and Robertson 1961). The fused 5-, 6-, and 7-numbered rings make this an interesting molecule. With George Ferguson, David Hawley and others we solved a number of other tricyclic derivatives in this series, including pseudoclovene A (Hawley, Ferguson, McKillop, and Robertson 1969) [4] and B (Crane *et al.* 1972) [5].

[3] Isoclovene   [4] Pseudoclovene   [5] Pseudoclovene

The other compounds that we worked on in this and related fields are too numerous and complex to be described in any detail here. Perhaps one of the most important, that had far-reaching chemical implications, was the bitter principle of citrus fruits, limonin. I collaborated closely with Derek Barton and George Sim in this work (Arnott, Davie, Robertson, Sim, and Watson

1961). Limonin, $C_{26}H_{30}O_8$, is a tetracyclic triterpenoid. A great deal of detailed chemical work had been carried out by Barton, Arigoni, and others, but the final solution of this important structure had to await our X-ray analysis. The first problem was to find a suitable derivative, and in this search we had the active co-operation of the organic chemists. Many possible compounds were examined and measured, but in the end we chose epiliminol iodoacetate, $C_{28}H_{33}O_9I$. Unfortunately, the asymmetric unit in this crystal structure was found to contain two chemical molecules, thus giving 76 atoms other than hydrogen whose positions we had to determine, 228 parameters in all. But we had the great advantage of the phase-determining power of the iodine atoms to help us, and at a later stage the presence of two crystallographically unrelated but chemically identical molecules was a most important verification of the work. The analysis was a long and intricate process, going through many cycles of three-dimensional Fourier synthesis and structure factor calculation, but in the end an accurate picture of the two molecules emerged. From this, formula [6] could be assigned to limonin without ambiguity.

[6] Limonin

The structures of a number of other bitter principles and heartwood constituents belonging to the triterpenoid family soon followed. They include clerodin (Paul, Sim, Hamor, and Robertson 1962), cedrelone (Grant, Hamilton, Hamor, Robertson, and Sim 1963), and gendunin (Sutherland, Sim, and Robertson 1962). The structures of the latter two are very similar to limonin, and they were also determined as iodoacetates, with phasing based on the iodine atom. The bitter principle clerodin was determined as the bromolactone.

The remaining problems whose solution I wish to mention briefly include a number of important fungal metabolites with unusual and difficult structures. Fumagillin (McCorkindale and Sime 1961) and griseofulvin (Brown and Sim 1963) were known and our work simply elucidated the stereochemistry quantitatively. Byssochlamic acid (Paul, Sim, Hamor, and Robertson 1963), a metabolite from *Byssochlamys fulva*, and glaucanic and glauconic acids (Ferguson, Sim, and Robertson 1962) from *Penicillium purpurogenum* were analysed quantitatively and the results were mostly in agreement with the chemical evidence. A beautifully crystalline bis-*p*-bromophenylhydrazide of byssochlamic acid led to a very detailed and accurate determination of the structure and stereochemistry.

Finally, one of the crystalline pigments isolated from the colouring matter of ergot, which is produced by a fungus grown on rye, is ergoflavin, $C_{30}H_{26}O_{14}$. This compound was first isolated in 1912, and had been the subject of intensive chemical study. Professor Whalley then prepared a di-*p*-iodobenzoate derivative, $C_{48}H_{40}I_2O_{16}$, and this enabled us to carry out a very accurate three-dimensional study which completely defined the structure (McPhail, Sim, Asher, Robertson, and Silverton 1966).

The molecules I have described in this chapter are not all of very direct biological interest. Most of them, however, are of fundamental importance in organic chemistry, and after their structures were determined many of them have been the subject of intensive biogenetic studies aimed at discovering their origin in nature.

References

ARNOTT, S., DAVIE, A. W., ROBERTSON, J. M., SIM, G. A., and WATSON, D. G. (1961). *J. chem. Soc.* 4183.

ASHER, J. D. M., ROBERTSON, J. M., and SIM, G. A. (1965). *J. chem. Soc.* 6355.

BARTON, D. H. R. and LINDSEY, A. S. (1951). *J. chem. Soc.* 2988.

BROWN, W. A. C. and SIM, G. A. (1963). *J. chem. Soc.* 1050.

CARLISLE, C. H. and CROWFOOT, D. (1945). *Proc. R. Soc.* **A184**, 64.

CLUNIE, J. S. and ROBERTSON, J. M. (1961). *J. chem. Soc.* 4382.

CRANE, R. I., ECK, C., PARKER, W., PENROSE, A. B., MCKILLOP, T. F. W., HAWLEY, D. M., and ROBERTSON, J. M. (1972). *J. chem. Soc. Chem Commun.* 385.

CROWFOOT, D. and DUNITZ, J. D. (1948). *Nature, Lond.* **162**, 608.

DICKINSON, R. G. and RAYMOND, A. L. (1923). *J. Am. chem. Soc.* **45**, 22.

FERGUSON, G., SIM, G. A., and ROBERTSON, J. M. (1962). *Proc. chem. Soc.* 385.

GEMMELL, K. W., ROBERTSON, J. M., SIM, G., BERNAUER, K., GUGGISBERG, A., HESSE, M., SCHMID, H., and KARRER, P. (1969). *Helv. chim. Acta* **52**, 689.

GRANT, I. J., HAMOR, T. A., ROBERTSON, J. M., and SIM, G. A. (1965). *J. chem. Soc.* 5678.

——HAMILTON, J. A., HAMOR, T. A., ROBERTSON, J. M., and SIM, G. A. (1963). *J. chem. Soc.* 2506.

HAMILTON, J. A., HAMOR, T. A., ROBERTSON, J. M., and SIM, G. A. (1962). *J. chem. Soc.* 5061.

HAMOR, T. A. and ROBERTSON, J. M. (1962). *J. chem. Soc.* 194.

HAWLEY, D. M., FERGUSON, G., MCKILLOP, T. F. W., and ROBERTSON, J. M. (1969). *J. chem. Soc.* 599.

HENDERSON, G. G., MCCRONE, R. O. O., and ROBERTSON, J. M. (1929). *J. chem. Soc.* 1368.

HODGKIN, D. C. and SAYRE, D. (1952). *J. chem. Soc.* 4561.

LONSDALE, K. (1929). *Proc. R. Soc.* **A123**, 494.

MCCORKINDALE, N. J. and SIME, J. G. (1961). *Proc. chem. Soc.* 331.

MCPHAIL, A. T., ROBERTSON, J. M., and SIM, G. A. (1963). *J. chem. Soc.* 1832.

——SIM, G. A., ASHER, J. D. M., ROBERTSON, J. M., and SILVERTON, J. V. (1966). *J. chem. Soc. B* 18.

PAUL, I. C., SIM, G. A., HAMOR, T. A., and ROBERTSON, J. M. (1962). *J. chem. Soc.* 4133.

———————(1963). *J. chem. Soc.* 5502.

ROBERTSON, J. M. (1935). *J. chem. Soc.* 615.

——(1936). *J. chem. Soc.* 1195.

——(1939). *Nature, Lond.* **143,** 75.

——and TODD, G. (1955). *J. chem. Soc.* 1254.

——and WOODWARD, I. (1937). *J. chem. Soc.* 219.

———(1940). *J. chem. Soc.* 36.

SORM, F., DOLEJS, L., and PLIVA, J. (1950). *Coll. Czech. chem. Commun.* **15,** 186.

SUTHERLAND, S. A., SIM, G. A., and ROBERTSON, J. M. (1962). *Proc. chem. Soc.* 222.

10. Vitamin B$_{12}$: introduction

JENNY P. GLUSKER

PERNICIOUS anaemia, long thought to be an incurable and usually fatal disease, was found by Minot and Murphy in 1926 to be treatable by supplementing the diet with liver. Therefore many research groups attempted to extract the 'anti-pernicious anaemia factor', later called vitamin B$_{12}$, from liver. It was isolated in crystalline form first by Folkers and his co-workers at Merck Laboratories in 1948 (Rickes, Brink, Koniuszy, Wood, and Folkers 1948) and then, later that year, by Smith and Parker at Glaxo Laboratories, and Ellis, Petrow, and Snook (1949) at British Drug Houses. Vitamin B$_{12}$ is required for normal blood formation and the maintenance of neural function and normal growth, probably as a result of its action as a coenzyme in certain metabolic processes.

Small crystals of vitamin B$_{12}$, in the form of deep-red needles and prisms, were grown by Lester Smith and given to Dorothy in late May 1948. Dorothy grew larger crystals and for many years she and members of her laboratory, as well as John White in Princeton, worked on the structure determination of this cobalt-containing molecule. The formula of approximately half the molecule was known from chemical studies, but the rest of the chemical formula was not known. It was eventually revealed by X-ray diffraction methods by this group. This was the largest structure tackled successfully up to that time (93 non-hydrogen atoms), but Dorothy had faith that the heavy atom method of structure determination, with sensible interpretations of electron-density maps, would work.

The result of this project was the determination of a chemical formula for the vitamin that has since been proved to be totally correct (Fig. 10.1(a)). W. L. Bragg, in *Fifty years of X-ray diffraction* (1962), described this study as 'breaking the sound barrier'. Of course, many more complicated structures have been determined since then, but the work on vitamin B$_{12}$ showed that large molecules could profitably be studied by X-ray diffraction methods, and that a careful analysis with good data could reveal a previously unknown structure. This analysis also led to many discussions on the possibility of phasing larger structures with heavy atoms. This matter had first been remarked on by J. M. Robertson (1939), and the work on vitamin B$_{12}$

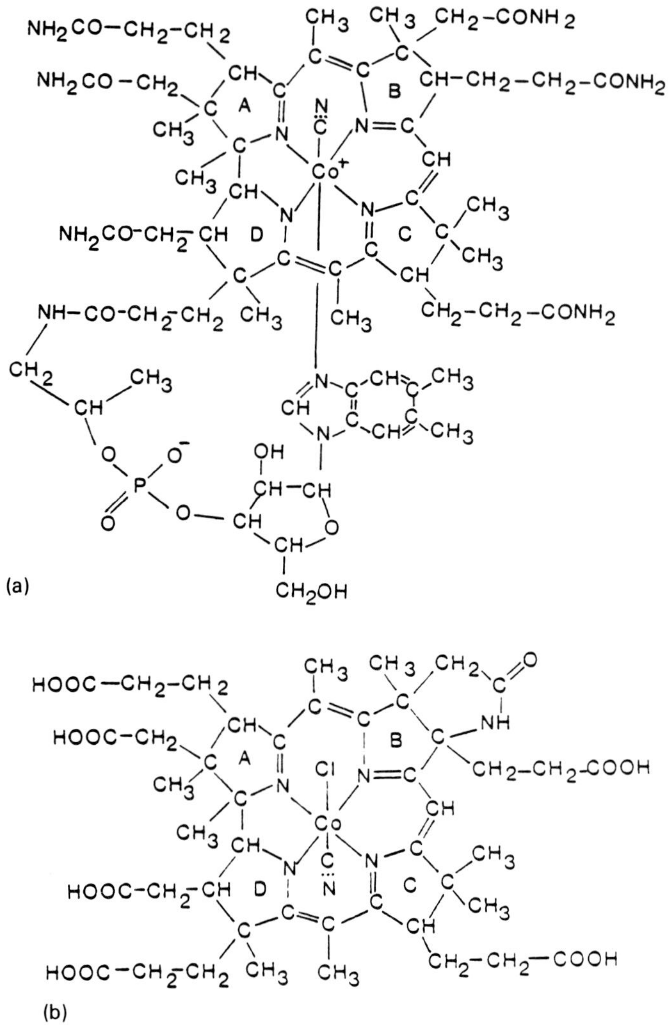

FIG. 10.1. Chemical formulae of (a) vitamin B$_{12}$ and (b) the hexacarboxylic acid derived from it.

reinforced current ideas, emerging at that time, that very heavy atoms could be used to phase X-ray data for macromolecules such as proteins and lead to the determination of their structures.

It **was** known by 1951, from chemical studies, that the vitamin had the approximate formula $C_{61-4}H_{83-92}N_{14}O_{13-20}PCo$ and contained a nucleotide-like group 5,6-dimethyl-1-(α-D-ribofuranosyl) benzimidazole-2' or 3'-phosphate. In addition, a cyanide group, one or two propanolamine groups, and a number of amide groups were identified. The structure of the rest of the molecule was assumed, from the very high pleochroism of vitamin B$_{12}$ crystals, to contain a highly absorbing planar group. The nature of this was found by X-ray analyses (Hodgkin, Porter, and Spiller 1950; Brink *et al.* 1954; Hodgkin *et al.* 1956; Hodgkin *et al.* 1957; Hodgkin *et al.* 1959; Hodgkin *et al.* 1962; Brink-Shoemaker, Cruickshank, Hodgkin, Kamper, and Pilling 1964; Hodgkin, Kamper, Trueblood, and White 1960) starting in 1948 of 'air-dried' and 'wet' vitamin B$_{12}$ (Fig. 10.1(a)), a selenocyanide derivative of vitamin B$_{12}$ (CN replaced by SeCN in Fig. 10.1(a)), and a hexacarboxylic acid fragment of the vitamin prepared by Cannon, Johnson, and Todd in 1953 (Fig. 10.1(b)). These three compounds all crystallized in the ortho-rhombic space group $P2_12_12_1$.

The studies of vitamin B$_{12}$ and its selenocyanide presented experimental difficulties but provided some of the general features of the ring system. Starting in 1949, Dr June Broomhead (now Lindsey), working in Dorothy's laboratory in Oxford, collected three-dimensional intensity data for a crystal of the vitamin, coated to inhibit drying, but described as 'air-dried'. Similar data were collected by Dr John G. White at Princeton. However, since crystals of vitamin B$_{12}$ deteriorate in time owing to loss of solvent of crystallization, Dr Clara Brink (now Shoemaker) collected data at Oxford in 1951 for the crystals surrounded by their mother liquor. The unit-cell dimensions of the 'wet' crystals were larger since the crystals were more highly hydrated. Dr John H. Robertson, also in Dorothy's laboratory, worked on the seleno-cyanide starting in 1952, hoping that the uncertainties of phase determination from the cobalt atom position alone would be overcome by the introduction of a second and heavier atom into the molecule. The cobalt and selenium atoms were clearly located from Patterson maps. Around the cobalt–cobalt vectors in this map it was possible to recognize a pattern of peaks that could be ascribed to the cyanide and nucleotide-like part of the molecule with, surprisingly, a Co–Se bond. The selenium atom was found to be at the position postulated for the cyanide group in the vitamin itself, and this was the first confirmation that the choices made for atomic positions were reliable. Dr John Callomon studied oriented crystals with polarized infrared radiation and confirmed the direction of the cyanide group in crystals of vitamin B$_{12}$. Later, in 1955, M. Jennifer Harrison (now Kamper) studied a vitamin B$_{12}$-like factor synthesized by micro-organisms in a culture medium containing 5,6-dichlorobenzimidazole instead of 5,6-dimethylbenzimidazole (Kamper and Hodgkin 1955). The positions of the chlorine atoms were those expected for such a substitution in isomorphous crystals.

The innermost part of the molecule appeared to be porphyrin-like in dimensions and nearly planar. Detailed sorting of peaks, especially from the electron-density distribution for the selenocyanide derivative, suggested that this group around the cobalt atom might contain four five-membered rings with bridge atoms, as in a porphyrin, but with one bridge atom missing. However, because of the position of the cobalt atom in the unit cells of crystals of vitamin B$_{12}$ and its selenocyanide, the cobalt atom did not contribute appreciably to the phasing of many reflections. Therefore maps phased on the cobalt atom could not be interpreted with any certainty.

In October 1953 a red cobalt-containing product of alkaline hydrolysis was obtained in crystalline form by Dr J. R. Cannon working in Professor Alexander Todd's group. Crystals of this hexacarboxylic acid were obtained from an acidified aqueous solution to which Dr Cannon had, in frustration before leaving for a vacation, added a large variety of solvents (including water, ether, and acetone). On his return rock-like crystals had deposited in the bottom of the flask. This preparation of chunky crystals has not yet been repeated. By the time A. W. Johnson and co-workers had decided the degradation product was inhomogeneous, X-ray diffraction data were being collected on a crystal cut out of the chunks. This crystal of degradation product diffracted to give much higher resolution data than those obtained for the vitamin. The structure of this hexacarboxylic acid, which contained the porphyrin-like portion of vitamin B$_{12}$, was determined by Jenny Pickworth (now Glusker), John H. Robertson, Ken Trueblood (who calculated the numerous necessary structure-factor and electron-density maps on the National Bureau of Standards Western Automatic Computer, SWAC, at UCLA), and Dorothy. The cobalt and 'planar' group lay in general positions in the unit cell (again space group $P2_12_12_1$) and therefore the cobalt atom position could be used to phase most reflections. The entire structure (Fig. 10.1(b)) was determined from electron-density maps, phased on successively increasing portions of the structure. The first map was phased on the cobalt atom position alone, and the second on the cobalt and chlorine atoms and much of the 'planar' group. Then, in further calculations, the positions of side chains, and eventually even solvent molecules, were revealed.

The resulting structure could then be used to interpret with much more certainty the maps for vitamin B$_{12}$ and its selenocyanide. The formula found for vitamin B$_{12}$ corresponded to $C_{63}H_{88}N_{14}O_{14}PCo$. There are several molecules of water of crystallization per molecule of B$_{12}$ in the crystalline state (Brink *et al.* 1954) (18 molecules of water per asymmetric unit in 'air-dried' crystals, 25 in 'wet' crystals). Several features not previously observed in naturally occurring chemical structures were found in the analysis. The most important of these was the existence of a ring system which was later dubbed a 'corrin-ring' system. This looks like a porphyrin, but there is a direct link between rings A and D (Fig. 10.1).

The four papers that now follow include accounts of the laboratory in Oxford in the early days, the work that was simultaneously being carried out in Princeton, the transatlantic collaboration on the refinement of the structure of the hexacarboxylic acid, and recent studies on the mode of action of the vitamin.

References

BRAGG, W. L. (1962). In *Fifty years X-ray diffraction* (ed. P. P. Ewald). N.V.A. Oosthoek's Uitgeversmaatschappij, Utrecht.

BRINK, C., HODGKIN, D. C., LINDSEY, J., PICKWORTH, J., ROBERTSON, J. H., and WHITE, J. G. (1954). *Nature, Lond.* **174,** 1169.

BRINK-SHOEMAKER, C., CRUICKSHANK, D. W. J., HODGKIN, D. C., KAMPER, M. J., and PILLING, D. (1964). *Proc. R. Soc.* **A278,** 1.

ELLIS, B., PETROW, V., and SNOOK, G. F. (1949). *J. Pharm. Pharmac.* **1,** 60.

HODGKIN, D. C., PORTER, M. W., and SPILLER, R. C. (1950). *Proc. R. Soc.* **B136,** 609.

—— KAMPER, M. J., TRUEBLOOD, K. N., and WHITE, J. G. (1960). *Z. Kristallogr.* **113,** 30.

—— LINDSEY, J., SPARKS, R. A., TRUEBLOOD, K. N., and WHITE, J. G. (1962). *Proc. R. Soc.* **A266,** 494.

—— KAMPER, J., MACKAY, M., PICKWORTH, J., TRUEBLOOD, K. N., WHITE, J. G. (1956). *Nature, Lond.* **178,** 64.

—— PICKWORTH, J., ROBERTSON, J. H., PROSEN, R. J., SPARKS, R. A., and TRUEBLOOD, K. N. (1959). *Proc. R. Soc.* **A251,** 306.

—— KAMPER, J., LINDSEY, J., MACKAY, M., PICKWORTH, J., ROBERTSON, J. H., SHOE-MAKER, C. B., WHITE, J. G., PROSEN, R. J., and TRUEBLOOD, K. N. (1957). *Proc. R. Soc.* **A242,** 228.

KAMPER, M. J. and HODGKIN, D. C. (1955). *Nature, Lond.* **176,** 551.

MINOT, G. R. and MURPHY, W. P. (1926). *J. Am. Med. Ass.* **87,** 470.

RICKES, E. L., BRINK, N. G., KONIUSZY, F. R., WOOD, T. R., and FOLKERS, K. (1948). *Science, N.Y.* **107,** 396.

ROBERTSON, J. M. (1939). *Nature, Lond.* **141,** 523.

SMITH, E. L. and PARKER, L. F. J. (1948). *Biochem. J.* **43,** viii.

Author supported by grants from National Cancer Institute.

11. Memories of Dorothy Hodgkin and of the B₁₂ structure in 1951–4

JOHN H. ROBERTSON

THE University Museum in Oxford was a large, imposing building, remarkable in any case for its architecture. Once inside, visitors found themselves in a large roofed enclosure filled with a maze of zoological exhibits. Around the perimeter were endless cases of insects, beetles, moths, and butterflies. The pillars of this cloister walk-way were each of a different polished stone and each decorated with different carved floral designs. The well-lit central area was particularly impressive. The eye was captured by stark skeletons of large mammals of all kinds and sizes, solemnly pedestalled and labelled. Between these ghoulish things, an array of narrow pillars with radiating arms of ornamental wrought iron, whose function was to support the roof structure above, seemed only to mimic the rib-cages of the displays below and add to the eerie atmosphere of the place. At the corner, a flight of wide, well-worn stone steps led to the upper floor balcony where visitors could study more glass-fronted cases and could look down, or rather across at the skeletal monsters filling the main enclosure. What an improbable setting for any sort of crystallography, much more a world-famous crystallographer and a crystal structure breakthrough that was going to be honoured, in due course, with a Nobel Prize.

It was here, near the top of the stone staircase, that a side door opened into the room that was the crystallographers' working place: a long barn-like room with gothic windows and a high arched roof. Plain, rough-wood tables, a few stools and chairs constituted almost the only furniture, but spread out over those table surfaces and bench tops at the side were all the familiar signs of intense crystallographic activity. Wire models, especially of the rod-and-cork-mat variety, alternated with X-ray photographs, viewing stands (for the visual measurement of intensities), carefully stacked heaps of Hollerith cards, papers covered with meticulously copied figures, and, most important of all, large multiple sheets of tracing paper bearing grids of numerals and the hand-drawn contours of Patterson or electron-density functions, which were the prime objects of study.

This was in fact one-half of the large room which had originally been the

University Museum Library, and which had been made famous (or rather, infamous) by the great debate that took place here on 28 June 1860, between Bishop Samuel Wilberforce and Professor T. H. Huxley. An artistically lettered plaque, just outside the door, commemorates that occasion. Inside, however, all thought of the past history of that room, or indeed of any of its very peculiar features, was absent. Crystal structure and the architecture of molecules was the wholly dominating consideration. This was the working room. About six or eight people were here, working on the various research projects of Tiny Powell and Dorothy Hodgkin. X-ray photographs were obtained downstairs. A basement room below was dedicated to that activity, with a cellar-like chamber for the darkroom and, up an extraordinary ladder, a bench for the optical microscope and its accessories. As for computing (on a Hollerith machine), that was carried out in a different building, by courtesy of the Mathematical Institute. Photographs, computations, and all other materials then gravitated to that upstairs room for interpretation.

There was a family atmosphere in this room upstairs. Each member of the community took his or her turn, weekly, to provide the little cakes that went with the afternoon's cup of tea. When anyone had a birthday, or a new baby, or anything comparable to celebrate, it was, by unwritten rule, that person's duty to provide a large iced cake, free, for that occasion. Each person had his or her own desk, of course, but everyone knew, at least in outline, what everyone else was doing. All the problems, and everything that was going on, were interesting. Mutual assistance was frequent; animated, even heated discussions were normal. The motivation was the interest of the subject. Everyone worked hard. It came naturally to do so.

And Dorothy?

As her office and study, Dorothy used what was merely the other half of this former Library. A high wooden partition separated her portion from the rest. The small door in this partition was open as often as shut, and through this doorway Dorothy would always appear, quietly, when the cups of tea were being poured in the middle of the afternoon. Dorothy's room was even more barn-like because it used to double as a teaching area where, occasionally, a class of undergraduates would be given crystallographic exercises to work on, seated at the bare wooden benches. Beyond these bare benches there was no other furniture but Dorothy's desk. That, however, was anything but bare. Hardly any part of its surface could be seen, but heaps of scientific papers, Royal Society letters, and reprints of articles mingled with lecture-notes, diagrams, tables of figures, and her children's drawings. It looked totally disorganized. Yet Dorothy seemed always instantly able to find anything she needed.

The greater part of her time Dorothy would spend with her research students, drifting inconspicuously from one to another, sometimes settling at one table for hours at a time, sometimes working downstairs with the

X-rays or the polarizing microscope. Dorothy was extremely skilled in the use of the polarizing microscope. She knew so many tricks of that trade and was never at a loss to interpret the observations. She was, in any case, a skilled micromanipulator, as good as any in the laboratory, and better than most—and this despite the horrid distortion of her finger joints and wrists which her long-standing arthritis had inflicted on her. Some of the B_{12} materials, especially those of the SeCN derivative, were available to us only in the form of a few tiny solid clumps; single crystals were present but all hopelessly welded to one another. These clumps were themselves smaller than a millimetre, yet this was all there was. But they were not hopeless. Dorothy chopped out a usable single crystal from them. Her wizardry with those fingers of hers was astonishing—to mention another example: persuading solutions of B_{12} to be saturated and to crystallize inside a thin-walled capillary, depositing a nicely shaped single crystal where it was wanted inside the tube and beautifully free from adhering crystallites.

Dorothy's outstanding ability was, however, most evident upstairs where there took place that continuous struggle to interpret the correct meaning of X-ray photographs, Patterson functions, and electron-density distributions. Her combination of wide experience and extensive knowledge was an inexhaustible resource, enabling her often to see her way through problems that appeared insoluble to the rest of us. When shown an oscillation photograph, still wet, of some brand new crystal, she was capable, after an apparently casual glance, of gaily pronouncing what the space group must be, to the consternation (even humiliation) of the young man who had just developed the picture; yet, two or three days later, after the appropriate zero and first-layer Weissenberg photographs had been made, dried, indexed, and analysed. Dorothy's remark would turn out to have been correct.

This ability to jump to the correct conclusion when the evidence was incomplete, or before the evidence had been properly sifted, was uncanny. Sometimes it seemed almost illegitimate. 'Women's intuition', it was often said; but really it was the product of her phenomenal knowledge of relevant chemistry and physics, her long experience, her marvellous memory for detail and her tirelessly active mind. Here is a factual example: readers of this chapter who are familiar with the B_{12} problem will remember that there is a cyanide group attached to the cobalt atom, projecting outwards at right-angles to the plane of the corrin nucleus. Now when the entire structure has been fully determined, this CN group and its position seem obvious enough. But in those early days, nothing was completely obvious. Very little indeed was known of the B_{12} molecule beyond its chemical composition and the size of the crystal unit cell. There were approximately 110 non-hydrogen atoms in each of the four asymmetric units in the unit cell, and we knew that the Patterson function must contain some $(400)^2$ vector peaks altogether: an average density of approximately 20 Patterson peaks superposed in every

FIG. 11.1. Section at $z=\frac{1}{2}$ in the three-dimensional Patterson distribution calculated for air-dried B_{12} crystals. Four strong symmetry-related peaks indicate cobalt-to-cobalt vectors. The two lines drawn in show the orientation of the molecule suggested by the pattern surrounding these peaks. (After Hodgkin *et al.* 1957.)

cubic Ångstrom. Cobalt vector peaks would be some four times higher than the rest but, even so, could be washed out by quite minor fluctuations of the mass of other vectors behind. However, when the first Patterson summation for B_{12} was drawn out, back in 1950, that cyanide group was immediately obvious to Dorothy. She saw it in the $z=\frac{1}{2}$ Harker section. This section is reproduced here (Fig. 11.1). Of this, she later wrote, 'It is noticeable that in one direction, *d*, the peak is extended as if two atoms are directly attached in a straight line to the cobalt atom at this point. When, a few weeks later [i.e. a few weeks after this section had been drawn], the discovery of a CN group in the vitamin was reported, it was tempting to associate this extended peak with CN' (Hodgkin *et al.* 1957). It is easy to agree now, when looking at this Harker section long after the event, that there is something suggestive of a CN group close to the Co–Co Harker vector; but the confidence with which Dorothy drew her conclusions about the cyanide group from this 'evidence' was at that time almost an embarrassment. Nevertheless, she proved in due course to be absolutely right, as usual.

This intuitive, apparently unerring recognition of what were correct clues, as opposed to false ones, was continually operative when the electron-density maps of vitamin B$_{12}$ were being studied. The earliest maps, especially of the air-dried crystal, were full of all sorts of distortions. They had been phased with the cobalt atom, representing only a small fraction of the total scattering power of the unit cell, but it was confidently believed that these maps would lead us to the structure of the whole molecule.

Our problem was to set up a chemically credible structure consistent with the distribution of electron density in these three-dimensional Fourier maps. The technique used for doing this was to have a full-sized three-dimensional model of the unit-cell contents alongside the large heap of contoured sections of the Fourier summation. (At the start there was one such heap of tracing sheets, viz. the summation for the air-dried crystal; later there were two, one for dry B$_{12}$, the other, rather better, for the wet crystal, and these two were used in conjunction with one another.) Atomic positions were plotted in space by fixing small markers on vertical rods. Each rod, pointed at its end, was pressed into the cork base at the desired x, y position and the marker on it moved to the desired z level. Whenever any atom in the structure was placed, or had its position altered, this had to be done both in the model and on the Fourier sheets. The compatibility of that position with the electron-density calculations could be checked on the sheets, while its compatibility, stereochemically, with its neighbours in space, both as to distances and angles, could be checked in the model. The process called for careful choice of the best compromise at every point, with continual readjustments as more and more atoms were added. But these electron-density peaks in the Fourier summations were nearly always misshapen, sometimes very much deformed, always the wrong height, and sometimes just not there. Furthermore, various sorts of 'false' peak were present as well. All this was due to the gross inaccuracy of the phases in the Fourier summation. Some of these false peaks could be recognized, by common consent, at certain crystallographically special positions in the unit cell. However, others were not in special positions; nor were all special positions occupied by these errant peaks. So, how to distinguish rogue peaks from proper ones? Nothing in those fields of figures and contour lines could tell us. The ubiquity of error in the calculated function was only too evident. On the face of it, almost any one of the observed peaks could be the result merely of an accumulation of positive errors, whilst any observed flat region, or even a valley, could be a genuine peak overlaid by random negative error. How to tell truth from falsehood? Such depressing reflections did not seem ever to affect Dorothy, not even when, one winter, a sepulchral cough and prolonged attack of laryngitis added a pitiful burden to her physical condition. She had a happy knack of pressing ahead, slowly, unruffled, with buoyant cheerfulness, often with sheer merriment, through these three-dimensional thickets of uncertainty. Her manner could appear extraordinarily casual. Working on these Fourier sections and the models, she could be heard to be humming away (hymn tunes, oddly enough) while pouring over the problems, quite unconcerned by the world about her, but deep in reverie. And so, in regard to this little matter of the unreliability of the electron-density maps, Dorothy had no qualms, was conscious of no great dilemma. On the contrary, she showed a mischievous enjoyment of the absurdity that it was sometimes necessary to strike out some offending large peak as 'false', while placing atoms on vague hillocks or on the steeply sloping foothills of other peaks nearby. We all used to laugh a lot together about this. It became an accepted joke, and it was really quite funny, especially when the condemned peak had been a nice round one.† But amongst the rest of us there was often that awkward awareness that the exercise was not exactly a game in reality. We were searching for what was genuinely true in the B$_{12}$ structure, and the validity or otherwise of our findings would soon enough be publicly evident. Of course, the steady improvement of quality in the succession of Fourier summations, phased by increasing numbers of atoms, together with the improving value of the reliability index, R, gives this evidence, as every crystallographer knows. Fortunately, it proved to be the case, as time went on, that Dorothy's hunches were nearly always right.

So, while the skeletons still stood, silent and static in the cavernous hall below, receiving scant attention from the crystallographers, real progress was being made in that room upstairs, to the point where, eventually, all the component parts of the B$_{12}$ molecule could be described reliably and in detail. The composition of the team that Dorothy had working with her altered from time to time as the years passed; a number of different avenues were explored, not all with equal success; a lot of water flowed under the bridge. But Dorothy's quiet, friendly confidence continued unfaltering, an inspiration to all who were associated with her. Her satisfaction with the outcome, when the full structure of the vitamin, inclusive of its stereochemistry, could be reported to the scientific world, must have been immense. Certainly, her unaffected enjoyment of every phase of that endeavour was infectious and truly memorable.

Anyone who has heard Dorothy Hodgkin give even one lecture—on

---

† Readers of the published papers on the B$_{12}$ analysis will have seen this aspect of the work referred to in several places: for example, the sentence 'The detailed interpretation of this [the electron-density distribution in the first three-dimensional summations] was obscured, as expected, by many spurious maxima, and by the fact that the peaks were much less precise in shape ...' (Hodgkin *et al.* 1957). Actually, of course, Dorothy *did* have her qualms and *was* conscious of the dilemma, as her letter of October 1954 to Ken Trueblood reveals; see page 96.

penicillin, or vitamin B_{12}, gramicidin-S, insulin, or anything else—will have been made aware of the prodigious memory that she has for even the finest detail of the vastly complex subjects that she handles. This wonderful memory, which seems to be so effortless for her, together with her encyclopaedic knowledge of chemistry as a whole, and her breadth and energy of vision, these attributes give her the key to her repeated triumphs. Her successes are no accident. These attributes, together with her delightful temperament, gentle humanity, kindliness, and good humour, endear her to us, commanding both our highest respect and our deepest affection.

Reference

HODGKIN, D. C., KAMPER, J., LINDSEY, J., MACKAY, M., PICKWORTH, J., ROBERTSON, J. H., SHOEMAKER, C. B., WHITE, J. G., PROSEN, R. J., and TRUEBLOOD, K. N. (1957). *Proc. R. Soc.* **A242**, 228.

12. The Princeton work on the structure of vitamin B₁₂

J. G. WHITE

Princeton, 1948–50

DURING the time of my Ph.D. research work in Glasgow (1944–7) crystallography was a very optimistic field. It was clear that the power of X-ray structural methods had not yet been exploited to anything approaching the limit, although practically nobody would have believed then the successes actually to be achieved in the next thirty years. In September 1947 I moved to Princeton University, and although involved for a time in other areas of research, I was looking for a crystallographic problem which would serve as a real challenge, one much more difficult than any crystal structure so far completed. I found such a problem in the structure of vitamin B_{12}.

Dr T. J. Webb had taught physical chemistry at Princeton during the 1930s before moving to the research laboratory of Merck and Co. in Rahway. He was interested in the post-war development of the Princeton Chemistry Department, and also in the application of X-ray crystallographic methods in the solution of organic structural problems. The solution of the crystal structure of penicillin had, of course, aroused great interest among the Merck chemists, many of whom had participated in the war-time collaborative effort to establish the chemical formula of the antibiotic. During my first year or so in Princeton Jeff Webb and I had many discussions on these matters, usually at dinner in the Nassau Club. It was therefore quite natural that after the isolation of crystalline B_{12} at the Merck Laboratories (Rickes, Brink, Koniusky, Wood, and Folkers 1948) I should suggest an attempt on the crystal structure at Princeton.

A visit to Merck was therefore arranged. Of particular interest was my meeting with Dr Karl Folkers, who was in charge of the ongoing chemical work on B_{12}, and, more briefly, Dr Randolph Major, then Research Director. The discussions were rather general, but one point kept recurring. This was basically the protection of Merck proprietary rights on work done in their laboratories, should the crystal and molecular structure of the vitamin be investigated in a university, with full freedom of publication. The agreement, which was arrived at some months later, was that I should work on the crystal structure without receiving any confidential information on the Merck organic research, but that in exchange for the crystals I should keep Merck informed on the progress of my investigation.

I received the actual crystals of vitamin B$_{12}$ (in their mother liquor) in the autumn of 1949. These were beautifully formed but rather small for data collection, and the crystals actually used for the X-ray analysis were regrown by slow evaporation from water solution. Crystallization was very easy, probably because of the high purity of the sample. The data collection by Weissenberg photographs, visual estimation of intensities, and correlation and reduction to a single set of *F*s, occupied nine months. This period seems long, but a great deal of repetition was involved. I had taken the decision to work on the air-dried crystals, since these gave resolution clearly at the atomic level ($d_{min} \sim 1$ Å), and the smaller number of water molecules in the dry, as compared to the wet, crystals might make the cobalt atom more effective in phase determination. On drying, the crystals reached a metastable state which might last for days or weeks. When any deterioration became evident in the quality of the X-ray photographs, a new crystal was mounted and the reciprocal lattice level repeated. Later on, when my data were compared with the corresponding set taken in Oxford, it was found that I had an appreciably larger number of reflections, and it was assumed that my crystals were slightly wetter on average, due to more frequent changing to fresh crystals. Much later, on carefully checking representative cell dimensions for final publication, it became apparent that my crystals were actually *drier*, due to the higher average room temperature in America. The more complete series that I had obtained was probably because of the use of rather larger crystals.

One day during this period I had a visit from Cecily Darwin who, having completed her degree in Chemical Crystallography in Oxford, had begun post-doctoral work with Dr A. L. Patterson in Philadelphia. On learning that I was working on B$_{12}$, Cecily told me that work on the vitamin had also begun in Dorothy Hodgkin's laboratory in Oxford, and communication began shortly. Dorothy felt that the problem was certainly large enough for two groups, but that we should keep in touch on progress, and this was done.

Before the completion of data collection, I had begun to look ahead to the mechanics of the actual structural work. From the dimensions of the unit cell it was clear that the structure must be attacked three-dimensionally from the beginning, a rarity in those days, and high-power computing was therefore desirable. Ray Pepinsky and his co-workers at Pennsylvania State University had constructed an extremely fast and efficient electronic analogue computer for Fourier series computation, X-RAC. I wrote to Ray and asked if I might come to Penn State for a few months and use X-RAC, to which he enthusiastically agreed. At the same time I applied for a sabbatical from

Princeton for the first semester of 1950–1. During the late summer of 1950 June Broomhead, who had been collecting the B$_{12}$ X-ray data in Oxford, visited me in Princeton during a personal trip to America, and we compared notes briefly. June had also visited Penn State, and shortly afterwards Ray Pepinsky wrote to Dorothy and suggested that June come and use X-RAC while I was there. This was one factor leading to collaboration. The other was the very encouraging nature of the first structural results obtained.

State College and Princeton, 1950–1

I arrived in State College in September 1950, and immediately started on the calculation of a three-dimensional Patterson map. This was in progress when I received an extremely optimistic letter from Dorothy. The Patterson map had been computed from the Oxford data and, as I was also to find almost immediately, the cobalt-other atom vectors seemed to be dominant in the central part of the structure to a much greater degree than we had anticipated. Dorothy suggested that since it now appeared that we should each solve the structure independently, time might be saved by close collaboration, with joint or simultaneous publication of the results. Of course I agreed, and suggested that June come and work in collaboration with me as soon as the current calculations and interpretations in Oxford were complete. In the meantime, rather than spending time at this point in detailed Patterson interpretation, I calculated a three-dimensional electron-density map based on the phase angles calculated from the co-ordinates of the cobalt atom alone. The most striking feature was a porphyrin-like group surrounding the cobalt atom.

June arrived at the beginning of 1951 with all the Oxford results to date, including the corresponding electron-density map, and we compared the co-ordinates of postulated atoms which we had derived separately. Almost all of these were in excellent agreement and therefore we simply averaged the co-ordinates of some 35 atoms in both our lists. These included the 'porphyrin' nucleus, some first substituents, the cyanide group, and, after some discussion, the phosphorus atom in what actually turned out to be the correct position. There was quite a lot of difference in detail in the electron-density maps, since it appeared that in the criteria for inclusion of reflections, I had been too cautious and June too bold. (Later these maps were recalculated both in Princeton and in Oxford using intermediate selection criteria.) Much later, when the corrin nucleus was fairly well established, I went back to these first co-ordinates and found that two of my five-membered rings and the other two of June's were very close to the final co-ordinates. Apparently we had inserted the additional bridging group by rather different movements. In principle, it seems that we should have interpreted the combined maps rather than combining the separate interpretations, but, in fact, it is very doubtful whether the correct nucleus would have been deduced at this stage.

During the next few months we carried out many calculations on X-RAC, but the principal one was a new electron-density map for which the new postulated atoms were added into the phase calculations. Since we calculated the structure factors by hand we gained a feeling of the sensitivity of the phases to inclusion or omission of groups of atoms. Also, the very strong bias in the non-centrosymmetric case towards atoms included in the phase calculations had become apparent. In the Penn State laboratory at this time were, among many others, Bill Cochran with whom June had worked at Cambridge, and who was helpful in discussion, and John Robertson who was to take over June's position in Oxford at the end of the coming summer, when she left to marry and live in Canada.

Cambridge and Oxford, 1951

I spent the summer of 1951 in Britain, mainly in Scotland, but included visits to both Cambridge and Oxford. Bill Cochran had invited me to spend a few days in Cambridge and I visited the Cavendish laboratory. Soon after I arrived I received a request to meet Professor A. R. Todd (now Lord Todd) who had taken over the organic chemical work on vitamin B_{12} begun at Glaxo Laboratories. Personally he was very pleasant (I had met him before when I was a research student in Glasgow, of which he was, of course, a very distinguished graduate), but he said very bluntly that he was not going to give me any information which I might relay back to the Merck chemists. This puzzled me, since I had not asked for any, nor had I initiated the meeting. However, the situation was to be clarified shortly.

A conference on protein structure, which I attended, was being held in Cambridge. Of the proceedings, I remember principally Pauling's α-helix paper which was read by Dr E. W. Hughes of Cal. Tech. However, Dorothy had come to Cambridge for the meeting, and somehow it was arranged that Todd, Dorothy, and I would get together at a party in the garden of Peterhouse College. We had a brief discussion and I gathered that the crystallographic collaboration had caused complications for the British and American organic chemists, who were definitely in competition. Dorothy was very diplomatic and finally Todd indicated that if his group succeeded in isolating and crystallizing B_{12} with the nucleotide removed, she would receive it for crystallographic work.

I spent a week or two in Oxford going over some of the work there and playing with models a bit, but nothing very serious was done. As I remember, in the last discussion I had with Dorothy, she now felt that B_{12} would take a long time, and that it would be advisable for her to utilize the data on wet B_{12} which Clara Brink (later to marry David Shoemaker) had collected and work on both crystals together. With my more limited resources in Princeton it was obviously more sensible for me to stick to the air-dried crystal data, and use this set as a guinea pig for exploring various crystallographic approaches, and I so indicated.

Princeton, Paris, and Oxford, 1951–4

During the major part of the next three years I continued to work on the air-dried crystal structure in Princeton. Most of the operations carried out, mainly by hand computation, have been described (White 1962), and only the most relevant will be mentioned here. At that time, in all the 'complex' non-centrosymmetric structures which had been solved, centrosymmetric projections had been used extensively, although three-dimensional data had been used in refinement. In B$_{12}$ all three crystallographic axes were too long for detailed information to be obtained from projections, and therefore the structure determination itself had to be carried out in the non-centrosymmetric situation. The electron-density maps, as noted in the early work, were phase rather than amplitude dominated, to a much greater degree than in past experience. The main questions to be answered were whether real positional refinement would take place once a much larger fraction of the structure was included in the phase calculations, and if the inclusion of false atoms could be demonstrated by internal evidence, and if so how corrected. These matters, of course, are well understood today, but at that time we were in a state of considerable ignorance of the requisite crystallographic methodology as well as of the chemical formula of the molecule.

A straightforward attempt was made to fill up space by adding in to the phasing calculations blocks of new postulated atoms suggested by successive electron-density maps. Owing to computing limitations small positional corrections were not made even when these became evident. When the process was nearly complete some major changes did appear, the most important chemically being in the region of two of the five-membered rings which had been connected by a bridging atom as would be the case in a porphyrin. The electron density became extremely distorted in this region, clearly necessitating a change in the chemical bonding (see White (1962), Fig. 13(a)). Parallel to this series of calculations I carried out a large number of approaches which were experimental in nature. It was clear that in using as light an atom as cobalt for initial phasing, the electron-density maps must have a high noise level, and that some errors in selecting groups, particularly where the chemistry was unknown, were unavoidable. I demonstrated that withdrawal from the phasing calculations of all the atoms in a region where some problem had arisen would give an 'objective' map of the region. This became standard procedure.

The Third Congress of the International Union of Crystallography was scheduled for Paris in the summer of 1954. I planned to spend the summer in Britain and Europe and therefore submitted a brief abstract on the air-dried B_{12} work just before the deadline. I mentioned in this that one part of the original 'porphyrin' was changing on refinement (White 1954). Very shortly after I had a letter from Dorothy saying that she had been invited to give the evening lecture at the Paris meeting and that she intended to discuss the B_{12} work. If I were going she would concentrate on the Oxford research, but if I were not she would like to include any information I could send. In the same letter there was a note from John Robertson, giving me a brief summary of the recent work in Oxford. They had attempted to postulate a complete structure for the selenocyanate derivative of B_{12}, on which they had been working for some time, but the value of R was so high that many details must be incorrect. They had experimented with modification in the nucleus, particularly omitting one bridging CH group, which seemed sug-gested by comparison of the electron-density maps from all three crystals (wet, dry, and CNSe). In addition they had received from Cambridge crystals of a 60–70-atom fragment of B_{12}, one of which gave excellent photographs.

During the next few months I concentrated on completing the operations then in hand, and sailed for Britain in July 1954 with my wife and ten-day-old daughter. I carried in my luggage a large number of electron-density maps, and after trying to explain what they were, finally accepted the customs classification of 'art work of no commercial value'. After settling my family in Scotland I made a very quick trip to Oxford to compare notes before the Paris meeting. The immediate point of interest was that our deviations from the porphyrin structure were, in fact, in the same place, which was immediately resolved positively. After Paris, towards the end of the summer, we paid a more leisurely visit of a few days to Oxford. By this time the first electron-density map, computed by Ken Trueblood from data for a crystallized hexacarboxylic acid derived from vitamin B_{12}, had been received from Los Angeles. This map gave strong support to the direct linkage between two five-membered rings. I had in the meantime confirmed this by the subtraction technique with my own data, but the evidence from the smaller structure, with better phasing, was much more powerful. It was during this visit that we drafted the first note to *Nature* on the structure of vitamin B_{12} (Brink *et al.* 1954).

Princeton, 1954–5

On returning to Princeton, I used all my previous calculations to obtain the most accurate co-ordinates possible for the parts of the molecule which now seemed most certain. I then began a new cycle of calculations aimed at finding the remaining side-chain atoms and water molecules. It was soon apparent that the structure was refining. New peaks appeared or strengthened in chemically reasonable positions, and co-ordinate shifts could also be interpreted in a rational way. In the final electron-density map of this cycle all but three of the atoms in the molecule itself, the head of one amide group, were either included in the calculation or were quite obvious. These three atoms were represented by weak positive density, but another refinement would have been necessary to establish them. In addition, one or two false additional atoms had been included, but had begun to appear anomalous in the context of the emerging chemistry of the rest of the molecule, and I was now quite familiar with the methods of handling this situation.

During this period (September 1954–June 1955) I had no information on the refinement of the hexacarboxylic acid fragment then in progress in Oxford. In June I received a letter from Dorothy giving a complete chemical formula for the fragment as postulated from their refinement. I immediately returned my B_{12} co-ordinates, and very shortly a copy of the final electron-density map. Between the fragment and B_{12} crystal structures, there seemed no reasonable doubt as to the chemical formula of the vitamin. In September, Jenny Pickworth, who had carried out most of the work on the hexacarboxylic acid fragment, and her fiancé Don Glusker visited me on their way to Cal. Tech. for a year. We discussed progress and future plans. After further correspondence with Dorothy it was decided that further refinement of the air-dried B_{12} structure would be carried out using my data set, with Ken Trueblood providing computations from UCLA.

Princeton, 1955–61 (!)

The refinement, including location of the remaining water molecules, many of them disordered, took much longer than originally envisaged. After the publication of the chemical structure (Hodgkin *et al.* 1955), and the submission of the first full paper on the X-ray analysis (Hodgkin *et al.* 1957), the pressure seemed somewhat lifted. At the end of 1955 I left Princeton University for RCA Laboratories, where I began applying crystallography to problems in solid-state physics, and I was able to work on B_{12} only at weekends. The others involved also had other commitments. The Royal Society actually published the final papers with which I was involved (White 1962; Hodgkin, Lindsey, Sparks, Trueblood, and White 1962) on my fortieth birthday, which was rather ironical in view of the elapsed time since the first photographs were taken.

Retrospect

While the complete success of the X-ray structural work on vitamin B_{12} was very gratifying, it was slightly disappointing that the original attempt to solve a crystal structure of about 110 independent, non-hydrogen atoms and of largely unknown chemistry, was modified by the use of information from a much smaller crystal structure (i.e. the hexacarboxylic acid). Nevertheless, the work on the air-dried B$_{12}$ crystals proceeded far enough for one to be reasonably confident that the structure would have been completely solved alone.

However, there are other important considerations. The fragment provided a greater degree of certainty about the structure than could have been obtained from any of the crystals of the larger molecule. At that time X-ray crystallography was regarded by chemists much more as black magic than is the case today. Any error in chemical formulation, even if only a detail in the structure as a whole, would have aroused doubts about the reliability of crystallographic methods. Such mistakes have indeed been made since, in detailed interpretation of some protein structures, but only after the power of X-ray methods had been established conclusively. A great deal of the credit for this state of affairs is due to Dorothy Hodgkin. Not only was she very frequently right in the early stage of the 'landmark' structures with which she was associated, but possibly more important, she showed restraint in publication until the necessary confidence about detail had been achieved.

References

BRINK, C., HODGKIN, D. C., LINDSEY, J., PICKWORTH, J., ROBERTSON, J. H., and WHITE, J. G. (1954). *Nature, Lond.* **174**, 1169.
HODGKIN, D. C., LINDSEY, J., SPARKS, R. A., TRUEBLOOD, K. N., and WHITE, J. G. (1962). *Proc. R. Soc.* **A266**, 494.
—— PICKWORTH, J., ROBERTSON, J. H., TRUEBLOOD, K. N., PROSEN, R. J., and WHITE, J. G. (1955). *Nature, Lond.* **176**, 325.
—— KAMPER, J., LINDSEY, J., MACKAY, M., PICKWORTH, J., ROBERTSON, J. H., SHOEMAKER, C. B., WHITE, J. G., PROSEN, R. J., and TRUEBLOOD, K. N. (1957). *Proc. R. Soc.* **A242**, 228.
RICKES, E. L., BRINK, N. G., KONIUSZY, F. R., WOOD, T. R., and FOLKERS, K. (1948). *Science, N.Y.* **107**, 396.
WHITE, J. G. (1954). *3rd Congr. Int. Union Crystallography*, Paris.
—— (1962). *Proc. R. Soc.* **A266**, 440.

13. Structure analysis by post and cable

K. N. TRUEBLOOD

ONE of my problems is that I almost never throw anything away. On this occasion, at least, that proved a blessing—for in thinking about Dorothy, I remembered those yellowing files. There I found every letter Dorothy had sent in those exciting days when the B_{12} structure was unfolding, many in that distinctive hand we all know—and copies of most of my letters to her too. These exchanges tell the story so well, conveying the flavour of the times and of Dorothy at work, that I can do no better than to quote a few excerpts from some of them.

First, though, how did it happen that we at UCLA were in a position to help on the B_{12} work? The US National Bureau of Standards had established a small computing laboratory on the UCLA campus in the late 1940s and Harry Huskey put together there the NBS Western Automatic Computer, SWAC, which had a cathode-ray tube memory (Huskey, Thorensen, Ambrosio, and Yowell 1953). When we first used it, in 1951, SWAC had only a high-speed memory, for 256 36-bit numbers; Stan Mayer, my first graduate student, none the less managed to program it for three-dimensional Fourier summations (Mayer and Trueblood 1953). Soon thereafter a magnetic drum that held 8192 such words was added, and Dick Prosen and Bob Sparks wrote an elegant program for three-dimensional structure factor and Fourier calculations in any space group (Sparks, Prosen, Kruse, and Trueblood 1956). I had urged them to do a quicker and simpler job, just for the common space groups and for crystals we were then studying, but they ignored me.

The Prosen–Sparks structure factor-Fourier program could handle up to 1000 atoms with individual isotropic temperature factors, and an unlimited number of reflections. It was nearing completion in the early summer of 1953, just as I was leaving with my parents for a holiday in Europe. An early stop was Oxford, where I visited Jack Dunitz, a friend from Cal Tech days who was back with Dorothy again. We came into that large room in the Museum, so well described by John Robertson elsewhere in this volume, and there talked with John and Dorothy about a number of the large structures in progress, including calciferol and B_{12}. I mentioned that we had a computer, and a program almost ready, that might be useful in the kinds of calculations needed.

But now I will quote from the file, the first entry being in my hand, on a piece of stationery of the Royal Hotel, Ashby de la Zouch, our first stop after Oxford. This note marks the moment when SWAC and B_{12} started to come together.

Oxford, 8 July 1953

Talked with Dorothy Hodgkin, John Robertson and Jack Dunitz. Offered help on 3-dim Fourier & SF calcs for calciferol (40 atoms, 2500 refl) and Vit B12 (112 atoms?) if SWAC seemed feasible.

KNT to DCH, 23 November 1953

Dear Dorothy:

It doesn't seem almost five months since I visited your laboratory all too briefly, leaving you with vague promises of the miracles SWAC could perform ...

Briefly the current situation is this. Budget cuts in the National Bureau of Standards, which operates the computing service on our campus, have necessitated the removal of virtually all IBM punched card equipment. We still have a tabulator, sorter, and reproducer, but none of the calculating units. However, SWAC, our high-speed digital electronic computer, is improved and is actually available to us more than when there was a larger staff working there. Furthermore, since we must of necessity use it for all machine computing now, we have spent some time working out programming and coding. As yet, we have not tested our new, more general programs, but we expect approximately the following:

(1) Three-dimensional Fouriers in times of the order of one-half hour to five hours. The latter time would be for about 1500 independent reflections and calculations in 60ths along all three edges of a complete unit cell; the program is written for P1, and uses all eight of the most general quadruple products of trigonometric terms (including phase) ...
(2) We also have a program for structure factor calculations, including the incorporation of temperature factors (at the moment only isotropic ones, but that will not be hard to modify) ...

The above descriptions sound promising. The proof of them all will come when we manage to apply them to data and turn out some structures. I'm sure that for a while we will be removing errors and modifying in various ways, and then we will get back to collecting data ...

To get back to the point of this letter, my original offer to see just what we might do for you still stands ... One highly desirable feature of this computer is that we are allowed to use it free of charge; the charge to air-plane plants, the Navy, and others who use it is something of the order of $150 an hour. I shudder to think where we might be if they ever start to charge us; I've been spoiled badly by never having less than a fairly good complement of IBM equipment to work with.

DCH to KNT, 18 December 1953

Dear Kenneth,

Your letter was extremely welcome and I am sending herewith details for the next stage of the calciferol investigation. We need structure factors cal-culated for some 3000 reflections and 37 out of the 41 atoms. I enclose the co-ordinates we propose and the limits of h, k, and l ...

After the structure factor calculation is complete and has been checked against the intensities, we shall need the Fourier series calculated on the observed F values, broken up according to your calculation of A and B. But I will not send you the full list of F values immediately as it takes a bit of time to get it copied out.

I am sending calciferol first because the data are all ready for this next stage of calculation and also there is Jack [Dunitz] near at hand in Pasadena to tell you immediately about any points I have left obscure. But if this goes well and you can stand it I will pretty certainly follow on with B_{12}.

Your offer of free calculations of this order of magnitude is a very marvel-lous one and I do thank you very greatly for making it—and look forward to results.

We are all hoping to see you again next summer.

DCH to KNT, 2 January 1954

Dear Kenneth,

I enclose Jack's F values for the calciferol compound. As you'll see, they do not cover as much of the sphere of reflection as I suggested covering in my previous letter ...

We have an extremely interesting compound coming on now—data should be ready in a couple of months. So I'll be interested in many ways about what you make of this compound.

With best wishes for the New Year.

KNT to DCH, 20 January 1954

Dear Dorothy,

I've had your letter and the data for some three weeks now but I can see that it will be at least that long, if not longer, before I even approach the problem in detail. Everything seems perfectly clear and it is just a matter of planning and carrying out the work. It looks as though I'll be doing most of it myself, and inasmuch as my teaching load is moderately heavy this coming semester it will take some time. Our structure factor routine is still not completely checked out; on the other hand we've successfully done several Fouriers now and are getting that pretty well under control, although of course we continually see ways it might have been improved. The running time is perhaps 40% longer than I guessed, but this may be able to be cut, and in any event isn't serious, for it is still pretty efficient.

In any event, I can promise that you'll hear from me with more positive results sometime during the next several months. I'm looking forward to doing it and am only sorry that I can't get at it sooner ...

KNT to DCH, 16 May 1954

Dear Dorothy,

I'll start this communication to you now, though I just looked at my watch and discovered it was 2:30 am, because these moments come rarely. We did the structure factors for calciferol about a week ago or so and did the Fourier Thursday night (and corrected it Friday—my naive idea of how to get 120ths turned out to be only partially correct, but on a little reflection we saw what to do and it required only another three hours of computing time, roughly) ...

But the chief point is that every one of the 37 atoms which we put in came at virtually the right position and with the proper density—I'll enclose a list—and the four missing ones are represented by peaks of about 0.3–0.5 of the right height. I seemed to have found a great number of them at first— just by going through and tabulating all above 2 electrons per A^3—but then on plotting them I found that most were grouped around the iodine which of course has all sorts of waves around it ...

DCH to KNT, 24 May 1954

Dear Ken,

I am very much delighted with your news about calciferol. It is a great achievement on your part to get this through. And one of the things I like most is that, of the last four atoms which appeared, one is *not* in the expected place (or at least not where Jack and I expected it). It is in a perfectly reasonable place and one which actually had a higher electron density in Jack's very first projection than that of the little peak we preferred. Also it is in a position which resolves sterochemical difficulties I found in fitting the projection as we thought it was to our first three dimensional series. And it fits with the configuration of the calciferol side chain preferred by David Sayre from his projections of lumisterol. (Actually there is a little peak at this position in the first three dimensional series too—but somehow at that stage we were just convinced of the high reliability of the projection and only seriously pursued lines suggested by this) ...

Can you face more calculations? We have an awful lot of possibilities for you (or anyone else who could take them on). We have a fragment of B_{12} now going strong—through early stages—and want a three dimensional series for this calculated—phases based on Co positions. It's a fragment containing about 70 atoms—all the unknown parts of the molecule. We're doing one type of series ourselves now—expect it to take three weeks (!)—but we hanker after a second variety, preferably coming off at the same time as the first.

Don't let me take up too much of your time though. Anyway, perhaps the machine isn't free. I expect it will take a couple of months or so before I would know where to go next on calciferol.

And very many thanks and congratulations.

KNT to DCH, 28 May 1954

Dear Dorothy,

This will just be a brief answer to your letter of Monday (which made very good time). I was certainly glad to hear that one of the carbons was not quite where anticipated—somehow that makes one feel better about it all.

We'll be doing lots of computing anyway, and would be glad to work on B_{12}, though as before I can't promise just when we'll have any answers. But actually there is a good chance it would be appreciably faster for we know much better how to proceed now—just a matter of finding the time. In any event let me know what you want done and send us the necessary data—and we'll someday send you the calculations based on them. I know so little about B_{12} that the idea of the *un*known part having 70 atoms in it is a little staggering—but it would be all the better for demonstrating the method. I imagine that this will entail quite a large fragment of calculation (120ths in how many directions?). When you do send the data, send a few references too, or a little chemical information, if you could, so that we'll be able to understand a little.

I'll be here until mid-August, when I'll go east for a month. If you get the calciferol data to us before then, we'll try to run through another cycle before I leave—and also to do what you want on B_{12} possibly before then too.

DCH to KNT, 16 July 1954

Dear Ken:

I must thank you first for sending all the massive calciferol data. I have not had time to plot out the series completely but the print I enclose shows the new series and the old one compared on a representative section ...

The B_{12} situation has in the meantime developed in a most promising way. We have found atomic positions for the atoms immediately surrounding the cobalt in the red fragment which are essentially similar to those we derived originally from B_{12}. The agreement is so good we think we might drop our idea of doing a vector convergence series—sharpened F^2 values multiplied by Co A–B values—and proceed direct to the next stage of refinement. We therefore send you all the F values for the red fragment + coordinates of some 26 atoms. What we should like are structure factors calculated for all observed reflections on these 26 atoms, and then a three dimensional series calculated on the observed Fs phased according to the structure factor calculation.

We are extremely interested in the results of this calculation. But owing to the intervention of the Paris conference next week and the shutting of our lab. during August, we would not be able to begin it ourselves till early September. So that if you were able to do it before then it would be pure gain for us. But if you can't we shall probably do it ourselves in September. So let us know how you are placed.

With very many thanks and best wishes.

KNT to DCH, 15 August 1954

Dear Dorothy,

This will be a hasty letter as I'm leaving in a few hours to fly to New York, and thence to Maine, for a month's vacation. But news is again so good and so promising that I want to send it off to you now.

Briefly, we've managed to complete one Fourier for B_{12} with apparent confirmation of everything you gave us and all sorts of other reasonable looking semi- (and smaller) peaks showing up too. I'll enclose a list which I just made, and since I've had no time to proofread I won't guarantee it but the Fourier itself will go at about the time this does so you'll be able to check it. All 26 of your atoms have heights of about 1000 or more; at first we started listing only peaks above 400 on our arb. scale (which is very roughly 200 times the e/A^3) but near the end of the list you can see we stopped much lower. We have plotted them all roughly and lots of bonds, rings, etc seem apparent. I'll not send you the graph because we've only one copy and I've no time to copy it out now and I'm sure you'll want to do a more careful job yourself. But even some of the feeblest of these peaks (and I'm by no means sure we found all for this has been very rushed) seem to be in very reasonable positions ...

We of course calculated all the structure factors but I don't see much point sending them since there is so much missing. If you want them I'll mail them to you later—let me know. This has been a lot of fun to do but took of course most of the last week and a half—even though you may not need it right away it seemed like such a promising structure study and such a worthwhile one to hurry on that I ran it off anyway. Summer school just finished here August 1, at just the time this arrived, so it was timed very well.

I'll be back here around Sept. 18. I'm not sure how fast we'll be able to do things for you in the Fall since I'll be busier then and not able to devote full time to research as I can in summer. In any event I'll await word from you and a list of positions as accurate as you care to make them for the next stage. If you wish we can just go ahead with the list we have here but of course you can probably get more precise positions—how significant they'll be is of course another question.

DCH to KNT, 28 August 1954

Dear Ken,

It is marvellous that you have been able to do this calculation—and more. I have plotted out your listed atomic positions—some of the new ones correspond well with bits of structure we suspected earlier. But we will know better just how well we stand when we have the whole lines plotted out, which is now being done. Indeed by Sept. 18th, when your holiday finishes, we should be in a good position to know what we ought to do next—we are to have a serious discussion with our chemical colleagues next week, which may also help. So I will write again in two or three weeks' time.

DCH to KNT, 2 October 1954

Dear Ken:

This B_{12} fragment series you calculated seems astonishingly good. We have gone all through it carefully, compared details with our B_{12} series and, as a result, have written out a whole structure for B_{12}. There are parts of this structure that are definitely more uncertain than others, and we favour a further round in the structure factor—Fourier series calculation before we attempt to put all the atoms in. For this purpose we are sending you a list of co-ordinates of atoms, to be added to the list of 26 atoms you already have. Our present list takes the total up to 64. But we think that the first 54—26 you have already and 28 new ones—are more reliable than the rest—

better peaks in the last fragment series and also in the B_{12} series. So we rather favour your calculating all structure factors on these 54 and repeating the 3-dimensional Fourier series and seeing how this turns out—whether all the remaining 20 odd atoms are at that stage clear. Then one more round might effectively establish the whole chemical structure.

Now first would you like to—and can you— continue these B_{12} fragment calculations with us? It seems to us that a very remarkable situation here is developing—the whole structure of a large molecule, chemically unknown, coming out by X-ray analysis. And the existence of you and SWAC quite changes our ideas of what we can do in the way of finding and making sure of this structure. If you can't carry on, we'll be extremely sorry—and move on ourselves in a pedestrian way, but we do hope you can continue as a full collaborator with us ...

The new set of atomic positions we send you do make chemically reasonable structures—mostly side chains ending in carboxyl groups, so that it would be quite possible to guess the identity of the various atoms. But we incline to the idea you should give them all an even weight—say 7—and hope they show their chemical nature in the new series.

About this structure as a whole. On the results of our earlier B_{12} work and the first fragment series we calculated here, we wrote a letter for *Nature*, a copy of which I am sending separately. Your fragment series, as you will realise, confirms the outline of the ring system proposed and checks one point which we were a little hesitant over in this letter—the existence of an atom at X in the figure. (We deliberately left it out in the list of co-ordinates we sent you and the atom has appeared in your series as a peak of 334). We thought it best not to modify our letter in the light of your calculation but to let it go in as a first round. So much more is now coming out that in a few months you and ourselves should be able to give an account of something approaching the whole structure. While sending you the fragment calculations we are returning to the data on B_{12} itself and hoping to do a round on it, perhaps on the Manchester machine.

KNT to DCH, 11 October 1954

Dear Dorothy,

Your letter arrived in the middle of last week and I have waited this long to answer it merely so I could collect the answers. First though let me say that we'll be very glad to go on doing the calculation on the B_{12} fragment (and, if you can't get them done elsewhere, on B_{12} itself—or even if you can if you'd like us to—we're still not having to pay more than an occasional month's salary to the key-punch operator, and to contribute cards occasionally, both of which we can easily manage out of department funds).

I must confess that we didn't wait for your letter to come to go through another cycle. Dick Prosen had calculated an xy-projection while I was away (as I mentioned in my letter that he might) but it of course showed little. But since we had the structure factor parameters all ready and the routine all done, we just continued to get all the rest of the hkl's in mid-September, and then between the 25th and 30th we calculated another Fourier. The input of course didn't agree with what we would have done had we followed your letter, nor was it very wisely chosen—we just arbitrarily took all 62 of the peaks (of the 66 we had, the last four of which were too feeble) and went ahead, counting the last 36 as N's. The idea was of course to see what we could do with no chemical information at all.

The results seemed excellent at first, and not too bad even on reflection. However, one aspect doesn't please us—or rather two aspects. One is trivial, viz. that the symmetry isn't quite right (as you'll note in the copy you'll get)—it is however nearly right. The night we did it SWAC was a little erratic, and it may be that it made a slight error, or just as likely we did something wrong in reproducing and sorting beforehand. However, at this stage the error from this cause is negligible. The other point however which bothers us is that *every* atom which we put in came out considerably enhanced, nearly to a reasonable value, and without a great deal of pattern to the peak heights in terms of the relative probabilities of reality ...

I must say that I am most impressed with the B_{12} fragment as a molecule—it surely is fascinating to watch it develop. Some of those positions near Co must be warped—but not unrecognizably so. The symmetry of the two side chains (four) gives one more faith too. Incidentally we've not yet figured out what your atoms 54 and 53 belong to, though we think we've tried all possible positions. Perhaps the next Fourier will tell us—but we had them in the last one, and though they came out very well, little else developed to help. We suspect another angular methyl, on C_7, which fits beautifully by symmetry also with the situation at C_{12}. This peak was only at about 79 on the first Fourier, is up to 180 or so in the second. Tomorrow actually we may put in a few more than your first 54, though not any of which we are in serious doubt—but perhaps your 54 would be safest. You might let us know what you think best—that is, whether we should alter your plan in light of this intermediate Fourier—since perhaps we won't get started tomorrow, and the following week and a half will be pretty busy. In any case however, we'll put in fewer atoms than in the second Fourier, and adopt your parameters for all we do use. I'm toying with the idea of putting in one atom at a false position, known to be wrong and not serious in the structure—e.g., at say about $1\frac{1}{2}$ or 2 Å from the CN group—just to see what happens to a false atom in a structure with this many atoms ...

This is certainly an exciting business and we appreciate the chance of being in on the most thrilling phase of the investigation. It certainly represents still another feather in your cap, and should cause quite a stir.

DCH to KNT, 19 October 1954

Dear Ken:

It is very interesting that you have done this new series on 62 atoms and will certainly help in assessing the reliability of our methods ... Our past experience on B_{12} itself has shown us that in these asymmetric series, and particularly those phased on only part of the structure, it is extremely easy to manufacture good looking atoms. While John White and ourselves derived a number of atomic positions that were the same for B_{12}, we also deduced many that were different. On our separate calculated series in many cases—not all—nice peaks appear where we each put them though our two placings are often quite incompatible. John White's or our porphyrin stage of the inner ring system looked quite respectable seen alone—only less respectable than our new arrangement. We don't think you need try the experiment of putting in an atom in an arbitrary wrong position—we've done it in the past and we're pretty sure you did it in the last series in placing your atoms 50, which is in the middle of a five membered ring, and 60— too near a nicely defined carboxyl group.

When working on B_{12} alone we got very cold feet about using series phased on half the atoms and tried in the end to find a possible whole atomic arrangement before our latest three dimensional series. We were scared stiff we were just inventing a molecule.† Our change of strategy in the fragment investigation is partly due to the existence of yourself which makes a more experimental attitude possible, partly to the fact that our survey of the first cobalt phased fragment Fourier series convinced us that we did know and had derived *correctly*, the structure of the planar group. Our idea now is that a series phased on a *correctly* placed group of atoms should have peaks in it corresponding to the other atoms (whether all or most, we are not sure) +spurious peaks, accidentally arising, through the many terms inserted with incorrect phases. Our hope is that, from a comparison of the different series and from considerations of chemical geometry, we may gradually eliminate the spurious peaks. At present we are hopeful that this process is happening quite rapidly.

Western Union to KNT 20 October 1954

PLEASE REMOVE WRONG ATOM.
DOROTHY.
            515A ...

---

† Because of the phenomena described above.

KNT to DCH, 27 October 1954

Dear Dorothy,

Your letter arrived yesterday and I had hoped to be able to answer it with the results of the 54-atom structure factor calculation and Fourier, which we did during the last six days, finishing at a late hour last night. But, and nothing could be more crushing, we have just discovered, as you will see when you look at the last list of parameters of the atoms which we used for atoms 27–54, that a mistake was made in x of atom 41, by a factor of ten— 0.717 was used instead of 0.072. Consequently all that labor was totally wasted. I proofread the list myself and God only knows how I overlooked it— naturally at the moment I'm too discouraged to care how. Sometime soon we'll start again, but since I'm leaving for Pittsburgh next week it will probably be at least three weeks before we send it.

We have been particularly careful to establish controls of various sorts (by card counts and comparing of various sorts) to ensure the absence of trivial errors ...

Twenty minutes ago the Fourier and Structure Factor lists were on a desk upstairs about to be put in the mail to reach you at once. I doubt now that you want them. We'll send them however if you do. In the meantime we'll start over as soon as our morale recovers from this shock—and proofread even more often, though blindness is hard to cure.

Your tales of your past experiments putting in false atoms were quite sufficient to answer all our questions—had we known of course that you had done all that we would not have proposed trying it.

Please pardon my depression—the blow, coming on top of only about 14 hours sleep in the last three nights, is just too much. It is conceivable that we may try to get this done much sooner than I indicated above—but don't count on it—it depends on who else is working graveyard shifts, what Dick's wife thinks about his never being home at night, how tired we are tomorrow, etc.

Western Union to KNT, 1 November 1954

CHEER UP. SEND EVERYTHING AIRMAIL.
DOROTHY.

KNT to DCH, 1 November 1954

Dear Dorothy:

We rebounded rapidly and actually have finished the calculation of the structure factors and the sorting for the Fourier. Dick will run the Fourier tomorrow night (while I am in flight to Pittsburgh) and see that both sets of

F's and the corresponding Fouriers (with a beautiful wrong-atom peak in the first, and I trust, nothing anomalous in the second—numbers 4 and 5, incidentally) are airmailed to you on Wednesday. This cycle gets easier each time—we know just what to watch for now—but I shall not crow until it is actually done; after all, last time I thought we were error-free too. We reconverted every parameter from binary to decimal this time before starting and proofread them again—that is a check on what is actually in the machine. All card counts are correct too, so we'll hope.

Thanks for the cheering message, I probably should not have written when so digusted with myself, but then at the time I really didn't know when we would be able to go on.

Perhaps within a week or so I'll finish the bond-distance bond-angle routine I've been at occasionally—designed to calculate all the distances and angles of interest in any structure (though not yet for triclinic ones, though that wouldn't be hard) ...

KNT to DCH, 2 November 1954

Dear Dorothy:

Tragedy! The drum has been scored in some unexplicable manner, and it will probably be a few weeks before SWAC is working again. We are sending off the fourth set of structure factors and the corresponding incorrect Fourier, and the fifth structure factor set. I hope it will be of use to you until we can get the fifth Fourier done.

DCH to KNT, 1 November 1954

Dear Ken,

My heart bled for you when I read your letter this morning. We felt pretty sorry for ourselves too, but nothing like so sorry as for you.

We think it best we have a look at these series and see whether in spite of the misplacing of atom 41 or other atoms the regions we feel most uncertain about have cleared up. If they have, it may be reasonable to go straight on to the next Fourier, omitting the strictly 54 atom stage.

I wouldn't mind atom 41 just being removed from the calculation—but in our experience any correction synthesis is as much trouble, if not more, as a complete repeat. Atom 41 has very odd implications, particularly so in relation to the formation of this fragment compound from B_{12} ...

I do hope you and Dick have a good rest, anyway. There's still hope for the future.

All our sympathy.

KNT to DCH, 11 November 1954

Dear Dorothy,

I've returned from a week in the eastern half of the country to find SWAC still being repaired; it will probably be another two weeks, plus or minus a few days, before it is ready to go again. We are of course all ready to use it at the earliest opportunity. Apparently the bearings on the drum, which rotates at 3600 rpm (hence an access time of 17 milliseconds, one revolution), went bad.

Your letter of November 1 was much appreciated; of course we recovered from our low state comparatively rapidly and I'm only sad now that the current delay is holding us up. We've told the engineers they are helping the progress of science by every minute of progress they make in returning it to working order.

KNT to DCH, 13 December 1954

Dear Dorothy,

I had hoped to have at the very least a Christmas present for you but things are still as bad as they were just six weeks ago when the failure occurred. They have been trying to resurface the drum but it doesn't turn out well—usually warps when they think they are finished. The last word was that it would be in last Tuesday and that we could use it this week again (after appropriate testing). But it still has not been returned from the shop which is working on it. B_{12} of course will be the first thing we do and we'll do it at the first opportunity, but I fear that two months have just been thrown away. When I think of the fact that we were only 12 hours away when it failed—and that we could easily have kept it running over the week-end then to finish it—all I can do is shake my head sadly. Of course perhaps the drum would have scarred that much sooner if we had kept it running.

In spite of this disappointment, I do hope that you have a very pleasant Christmas and New Year and that we will be able to surprise you very soon with Fourier No. 5.

DCH to KNT, 28 December 1954

Dear Ken,

I meant to write to you a special Christmas letter to report progress, but just before Christmas we had a new excitement. Lester Smith sent us crystals of a chlorine substituted vitamin B_{12} made by feeding dichlorbenziminazole to the appropriate bug. We took a few quick photographs, confirmed that the material is isomorphous with B_{12}, that on the weaker reflections small changes of intensity are visible to the naked eye and that on a few test reflections the changes are just as expected for the effect of chlorine atoms replacing methyl groups at the positions we placed them: B_{12} itself. So this confirms our skeleton picture of B_{12} itself. After Christmas, we'll have to work up at least one projection of this compound properly—and then go fully back to B_{12} itself.

Now about the red fragment and your work. We've gone through both of your series you sent us, plotted a good part of one, and we've also slowly calculated one example of this F^2 type of series for comparison. We've convinced ourselves that we know several more atoms than we suggested before but we have still a bit of final comparing to do before we send you another list. Perhaps your next series will be calculated by then—we should much like that, of course, to cross check our conclusions.

One thing we liked about your last calculations—the 'wrong' atoms do all tend to be lower than the 'right' atoms. Particularly this is true of the atom that went astray in the 54 atom series. I think this is good indication that the series is refining. Another point that is formally interesting—that I had not expected—is that 'right' atoms omitted from the calculation don't seem to improve much through these different series—1 atom, 26 atom, 54 atom— one becomes convinced of their reality by their persistence more than by anything else.

KNT to DCH, 29 January 1955

Dear Dorothy,

I mailed the fifth Fourier to you airmail yesterday—you should have received it by the time you get this. We had to try four times with SWAC to get it out and you'll discover, as you probably already have, that the symmetry is not quite right. The machine is still erratic unfortunately; last Monday night it wouldn't do anything right, Tuesday night we got through all of the $l = 0$ reflections in the first two dimensions of the summation but not further; Wednesday we started again and got through $l = 10$ before it failed and on Thursday it seemed to work well until very near the end, and then failed but we were able to circumvent the trouble, we thought. In any event, the Fourier seems generally pretty good but it is certainly by no means all it should be. SWAC is just still a little in need of exercise after its long layoff and is a little better every day, as you can see from the above account. We'll try this again as soon as it seems to be reliable...

DCH to KNT, 2 February 1955

Dear Ken:

The 54 atom distribution arrived today to our great delight. We have, in the meantime, from a study of your other sums, made small modifications in many of the first set of co-ordinates and added a number of atoms more. Jenny [Pickworth] is checking quickly through the series to see how far our conclusions are confirmed, or can be extended immediately. Then we will send you back co-ordinates for another round to be concluded as soon as you can...

I expect you have written a formula for the fragment from the various atomic positions you and we have postulated. We enclose one of ours. We gather that Merck has a paper in the *J.A.C.S.*—out in the States but not yet arrived here—which describes the isolation of products

which might well be derived from a bit of our structure. So I think we might aim soon to write jointly a little note about what has happened in refining this fragment structure without waiting for the rest of B_{12}. There's going to be a lot of organic chemistry of odd kinds eventually—certainly as far as we can see various things have happened in the conversion of B_{12} to the fragment.

There are some nice bits of pure crystallography involved in the relation of your different series—I enclose one little item—in case you have not drawn the figures out.

DCH to KNT, 2 February 1955

Dear Ken,

Here are the various items I promised. The coordinates are only for 66 atoms—I had somehow hoped for the lot, but there is a region of space we cannot make up our minds about yet. We have not quite properly studied your latest series in detail—it may help. I'll send word if it does in case you have not yet put through these lists...

I do think the events shown in Jenny's little drawings attached give one great confidence in the reality of the structure we find. This carboxyl group is a rather important one to us—it is the one which in B_{12} ties back with the phosphate group...

Jenny has a theory that the last two atoms written in—our 68, 69 your, I forget what, are part of acetone of crystallisation—I don't know. It's difficult to make them a side chain, otherwise there's a great hole.

KNT to DCH, 10 February 1955

Dear Dorothy,

I trust that the false alarms are past. SWAC has functioned pretty well for the past two nights—after being extremely erratic ever since the Thursday just two weeks ago when I ran off the first trial of the fifth Fourier—we tried every night last week without success and were getting pretty discouraged, but gradually they have ironed out most of the sources of trouble. Both the drum and the high speed memory circuits are just a little sensitive now, or were; but we've had as much as six hours of computing with only trivial troubles, and no appreciable delay, each of the last two nights. It is now 8:15 AM and I'll not take time to look over the latest version in detail. I have verified that in a number of different ways the symmetry seems good...

JP to KNT, 7 February 1955

$$h \ k \ l$$

I have by accident sent you an F value for the 12 1 0 reflection for the B_{12} fragment which is 10 times too great. The value should be 17.0 not 170.0 as you have it at present. If you have already started the next calculation don't worry about it. I intend to measure some of the innermost reflections from oscillation photographs as there seem to be considerable discrepancies between some of the calculated and observed structure factors which may be due to experimental error as well as the fact that the structure is not yet completely known....

I hope SWAC is up [to] standard in the next calculation. Good luck.

KNT to DCH, 12 February 1955

Dear Dorothy,

The correct (we hope) version of the fifth Fourier (labelled 'Trial III'—trial II consisted of a great number of abortive attempts finally abandoned in favor of a completely fresh start when the machine finally started behaving reasonably) must be in your hands by now. Meanwhile I have looked it over reasonably carefully and found no serious discrepancies from Trial I. The errors in that version were certainly not great but they do make some differences in the lower peak heights, though seldom in positions. Because the positions are so nearly the same in the two, and because you have already looked over the first one, we'll assume that you would come to the same list of parameters from the correct version and proceed accordingly. ...

Jenny's letter arrived yesterday. We have not yet started the next round, because I wanted to double check this Fourier before doing it, and we will correct the F for 12, 1, 0 before we do. Actually I'm not sure how fast things will go in the next few weeks because the semester is just starting, and also Dick is getting pretty nearly finished with his thesis work—now at a crucial stage with one structure—and so we may not have much time. However, it shouldn't be more than three weeks at the most if we run into no snags. ...

Now that SWAC is again healthy—we had three good nights in a row last week—we can hope to be sending you a set of SF's and a Fourier in a few weeks.

DCH to KNT, 13 February 1955

I think I've sorted out another end of side chain. So if you have not yet put the last lot through you might add these in. I hope I've started numbering at the right point—anyway here are the four. ...

... as I said earlier, we are having various consultations with chemists this week—I'll let you know the outcome. If you don't put this one through immediately I may add yet another one or two after these talks—a bit more considerations of past sums. ...

KNT to DCH, 16 February 1955

Your letter of February 13 arrived today, in excellent time. Meanwhile you must have received the corrected copy of the fifth Fourier and the letter which followed it. ...

It seems to me that this latest group of four is on pretty shaky grounds; it certainly looks much better in the wrong copy than in the correct one. Furthermore, though I've not calculated the angles, just on inspection it appears that the angle subtended by the two branching atoms at the end of the chain is much too large. What I'm leading up to is that I'd prefer to omit them from this next SF calculation. We started to do it last night but then found that we would have some recoding to do because there are a few channels on the drum that we cannot now use. Consequently it will be at the very earliest next Tuesday and perhaps later—actually that day is a legal holiday here (Washington's birthday, new style—actually February 11 but Gregory changed all that) before we get started. ...

If you do want them included, let us know at once. Otherwise we'll just do a 66-atom set. I trust you won't object to my omitting them—but actually we won't start before you've had a chance to cable (on Monday presumably) and may not get started for some time anyway. I'm just wary of artificial atoms after our earlier experience. It might be entirely safe to include your #70 and #73, at 7.7, 18.5, 48.5 and 6.0, 26.8, 47.0 respectively, making a 68-atom set, for those two look quite respectable. ...

If you do cable you can just say 'two' or 'four' and I'll interpret this to mean 'include 70 and 73' and 'include all four mentioned', respectively.

DCH to KNT via Western Union, 21 February 1955

DR. K. N. TRUEBLOOD

TWO.

DOROTHY

OXFORD

DCH to KNT, 17 February 1955

Your new 54 atom series arrived safely. I have only had time to look over it rather superficially. My newly selected atoms—the last I sent you—are disappointingly small, but I still think they may be right. They are particularly favored by the projections which Jenny calculated here on the signs given by your 54 atom stage structure factors. Never mind if you don't do the calculation for a week or two. A little delay may give us time for this chemical consultation and a few more serious thoughts about your last calculations.

In the meantime I send you the proposed coordinates for 49 atoms in air dried B_{12} crystals and the F values June Lindsey collected. The latter are on $\frac{1}{4}$ absolute scale. They look a little rough but the first series calculated on them on Co phases really gave surprisingly good evidence—seen in retrospect. . . . We have wet B_{12} in hand ourselves but may send you a final sum— if we get there! I do hope this isn't too much for you. It is marvellous for us.

DCH to KNT, 24 February 1955

Dear Ken,

Jenny and I have been brooding over the situation since I cabled you and our present opinion is rather different. We therefore propose two alternative courses of action according to where you now are.

1) If you have already calculated the series on the atoms proposed in my letters + cable modification, well and good. We will proceed accordingly, when you send us the structure factors and next sums, to maneuver the figures and hope to find the full list of atoms. This may be the best plan any way <u>but</u>

2) If you have <u>not</u> yet put through the structure factor calculation then we think it best to add in all the remaining atoms that we believe are there. We send you a list with revised numbering to add to Jenny's original lists. Our reason for this different view is a, perhaps mistaken, confidence that these positions really do represent the structure. And we think that if they don't, then our best hope is to have a structure factor calculation on the full number of atoms which we can use to adjust scale factors, and then to remove such wrong atoms as there are, with difference syntheses on the projects. If you've already done the other thing, course 1), <u>don't bother to follow this plan.</u> . . .

Also we were much strengthened in our view of the structure by Todd's discussion with us last Saturday. He showed how all the little bits of oxidation products they had found from B_{12} dropped straight out of our structure. He thinks we should all set about preparing papers on the fragment structure...

DCH to KNT, 25 March 1955

Dear Ken,

By all means go ahead with the plan for discussing the B_{12} fragment at the A.C.A. meeting. The authors this end are Jenny Pickworth, John H. Robertson and myself. Jenny is giving a little account of the present position at the X-ray analysis group meeting next week—annual meeting—but by late June the position ought to be more favourable...

Jenny has made some drawings for slides of which we'll send you prints, probably next week. We could send you over whatever you need, practically, for June, by Bill Lipscomb who will be making the direct journey Oxford—Pasadena for the A.C.A. meeting...

DCH to KNT, 19 April 1955

The air dried B_{12} series has come and I have begun to look at it, in the way one does, casually, before having it plotted out. It looks all right to me so far. Some bits of this molecule we now are so sure about—I look for certain particular atoms, not put in, and they've come up all right. This gives one a lot of confidence. But there are a lot of regions still rather obscure to us which we hope now to sort out...

... Jenny's paper at Bristol XRAG meeting went very nicely. Bill Cochran, opening the discussion, said we might have borrowed a title from Booth and called it "Crystal Structures determined by Magic"...

References

HUSKEY, H. D., THORENSEN, R., AMBROSIO, B. F., and YOWELL, E. C. (1953). *Proc. Instn Radio Engrs, N.Y.* **41**, 1294.

MAYER, S. W. and TRUEBLOOD, K. N. (1953). *Acta Crystallogr.* **6**, 427.

SPARKS, R. A., PROSEN, R. J., KRUSE, F. H., and TRUEBLOOD, K. N. (1956). *Acta Crystallogr.* **9**, 350.

14. Vitamin B$_{12}$: then and now

JENNY P. GLUSKER

THE chemical formula of vitamin B$_{12}$, found from the X-ray crystallographic studies of Dorothy Hodgkin and her co-workers (Hodgkin 1965), and described in the previous papers, was first reported in 1954, approximately twenty-five years ago. What has been learned about the vitamin since that date? How has knowledge of the three-dimensional structure aided in our understanding of the mode of action of the vitamin? These questions, together with some reminiscences, form the subject of this review.

The years 1952 to 1955 were very exciting ones for the X-ray crystallographers involved in the study of vitamin B$_{12}$. The first structures obtained, described in the previous articles by John H. Robertson, John G. White, and Kenneth N. Trueblood, were vitamin B$_{12}$ ('wet' and 'air-dried') and a hexacarboxylic acid derived from it. These structure determinations were far from routine. The methods used were still being developed. The X-ray and computing equipment were tedious and time-consuming. Great patience and a firm belief that the structure must ultimately be found were essential. Betty Patterson wrote:

> I'll never forget my introduction to Dorothy Crowfoot Hodgkin. After the Stockholm meeting in the summer of 1951, Lindo and I went to see Dorothy at Oxford. She had told us to meet her in one of the rooms of the Oxford Museum and we walked in at the end of what seemed to be a very large room with parallel rows of long tables. At a table at the far side sat Dorothy and John White with their heads bowed studying a contour map in complete silence. We quietly walked up to them and it seemed as if it were minutes before she looked up and greeted us. What typical concentration!

Our realization, from a study of the peaks around the cobalt atom in a map of the hexacarboxylic acid phased on the cobalt atom alone, that the two five-membered (pyrrole) rings were directly joined, was a memorable event for John Robertson, Dorothy, and me. The maps from the selenocyanide of B$_{12}$ had given some inkling that this might be the case (although I had been kept in the dark about this so that I could interpret the hexacarboxylic acid data in an unbiased way). But this time there was no question about the location of atoms, no possibility of error.

The calculation of the cobalt-phased Fourier map of the hexacarboxylic

acid involved six weeks of hard work day and night, for a graduate student
(me) using a Hollerith adding machine with punched cards. This machine
was housed in the basement of the Mathematical Institute in Oxford. Since
the machine could not subtract, the punched cards (which effectively contained entries from Beevers–Lipson strips, i.e. cosine or sine values at intervals of 1/60 of the cell) had to be turned upside down to obtain negative
values. To keep a check on this the tops of the cards were painted with differently coloured inks to indicate how they should be turned at various stages
of the calculation. All three-dimensional maps were prepared by copying the
numbers on to a grid and then contouring them by hand. If Ken Trueblood's
offer to do calculations on swac had not been made at this point I don't
know how we would have been able to solve such a large structure in three
dimensions in a reasonable period of time.

A day-by-day account of the analysis is contained in the letters I wrote
from June 1954 to June 1955 to my then fiancé, Don Glusker, who was doing
postdoctoral work at Cal.Tech. On 9 July 1954 I wrote about results from
the analysis of the cobalt-phased map, 'We are thinking of sending initial
results for calculation of a second Fourier on swac at UCLA by a man called
Ken Trueblood' and by 9 September could write that we were 'very excited
about the way this new Fourier calculation from UCLA is turning out. All
the atoms I chose parameters for have come up beautifully in the right positions and now I must look for a structure in the remaining peaks.' By 23
September a possible chemical formula was being written out and a model
was built which David Lawton said 'looks like Battersea Funfair'. It was
Dorothy's belief that it was essential that everything should be calculated
in three dimensions, not two, in spite of the magnitude of the task. Most
crystallographers were, at that time, studying three two-dimensional projections of their structure because the computations involved were much
simpler. The structure of vitamin B$_{12}$ was, however, so large that two-
dimensional results would probably have been impossible to interpret. The
only exception to this was that, to everyone's surprise, the co-ordinates of
the cobalt atom in the hexacarboxylic acid could be determined from the
two-dimensional Patterson projections (which saved a lot of time).

An indication of the general spirit of the laboratory is contained in a portion of a letter dated 11 October 1954. 'A conversation in the lab ran like
this on Saturday. Hugh Cardwell: Well, Dorothy, how long do you think
it will take to finish B$_{12}$? Dorothy: If Ken does this next calculation we should
have the structure tied up by next year. Anyway Jenny'll see it finished. She
had interests over in California.' But by 3 November 1954 the excitement
had waned: 'All this week work has gone wrong. Ken Trueblood wrote to
say he had done the calculation as we asked and then made a mistake
(changed the position of one atom by 12 Å). He was heartbroken as nights
of toil were then useless. He said he had had 14 hours sleep in 3 nights and

was going off to Pittsburgh in despair so he wouldn't do the calculation again for 3 weeks at least. Then the Hollerith machines here have been breaking down continuously.' (We were not to know at that time that the correct calculation would not be done for three months because the SWAC computer drum had to be repaired.)

But despair did not last long, either. Dorothy realized that this calculation, with an atom in the wrong place, could be used as a check. After all, that was one of our worries—that we had used too much imagination in interpreting maps. As Ken Trueblood notes in his article (p. 96), Dorothy wrote to him on 19 October 1954 and said, 'Our past experience on B_{12} itself has shown us that in these asymmetric series, and particularly those phased on only part of the structure, it is extremely easy to manufacture good-looking atoms.' We were not sure what happened in wrongly phased maps of large non-centrosymmetric structures such as this one, and the analysis of the 'erroneous' map, done after some prodding by Dorothy, showed that a peak appeared in the 'wrong place' but that a small peak also appeared in the 'right place'.

Refinements to the 1954 structure are still being made. There were a few atoms that were hard to place in the original maps of the hexacarboxylic acid. We assumed that this was simply because of high thermal motion. But later Dorothy had a neater explanation which we are now investigating. Dorothy wrote on 26 July 1971: 'A rather nice and surprising thing just happened (well you may not think it nice, but I do). Helen Evans, one of my present research group, worked recently on the structure of a compound called neovitamin B_{12}, which has proved to be the epimer of B_{12} at C_{13}. The crystal structure is really very similar to that of B_{12}, the group just takes the place of certain water molecules in the original structure and the additional water molecules take its place. It made me think suddenly about the old hexacarboxylic acid structure and the troubles we had in placing the side chain at C_{13}.' On looking at the old original map phased on the cobalt atom alone at C_{13} she continued 'we can run the chain in both directions, and so we now think the original crystal is likely to have contained some epimer and that this is the reason for our old troubles'. Neovitamin B_{12} (Bonnett *et al.* 1971; Stoeckli-Evans, Edmond, and Hodgkin 1972) (cyano-13-epicobalamin) is formed when vitamin B_{12} is subjected to high acidity. It has 14 per cent of the biological activity of vitamin B_{12}. In my notebooks a diagram, dated 3 February 1955, accounts tentatively for all the side chains of the hexacarboxylic acid except for that on what is now called C_{13}. By symmetry we presumed we were looking for a propionic acid side-chain.

As recently as 22 May 1979, Dorothy wrote that the hexacarboxylic acid 'still exists on its original mounting ... I had it on show at the Bryce lecture at Somerville ... It *would* be interesting to see how it diffracts after all these years, and you may find it practically impossible to recrystallize the old preparation—grossly inhomogeneous as Alan Johnson thinks it was. I find it difficult now to believe that there is not a little of the epimer in our original crystal—I don't think there can be a lot as Bob Woodward would like to believe.' Since we were not able to repeat the original crystallization, the crystal of the hexacarboxylic acid, the one from which data had been collected in 1953, was hand-carried to the United States after being examined on its mount by airport security officials in London. Bud Carrell found that it still diffracted, although not very well, but enough for the general features of the structure to be obvious from a cobalt-phased Fourier map. We will look to see what happens in the area of C_{13}.

The absolute configuration of the hexacarboxylic acid fragment was settled in 1959 by the work of Aafje Vos. The anomalous dispersion effect had been seen on all early higher level Weissenberg photographs and it should have been obvious to us that this would yield structural (and even phasing) information. But Bijvoet had only recently (1951) reported the use of anomalous dispersion to find the absolute configuration of tartaric acid as its rubidium salt (Bijvoet, Peerdeman, and van Bommel 1951). Therefore, we initially missed the importance of these intensity differences. Dorothy, in writing about this recently (1975), remarked, 'It is very curious how blind we often are.' The assignment of configuration for the vitamin itself was known because the configuration of D-ribose is known.

On the determination of the absolute configuration, Aafje Vos sent the following note:

> Having heard Dorothy lecturing on the vitamin B$_{12}$ work in Groningen, I very much wanted to see how she and her group had succeeded in interpreting the complicated electron density maps. During a two months stay in Oxford in the spring of 1956, Dorothy asked me to have a look at the photographs of the hexacarboxylic acid to find out why they did not satisfy the orthorhombic symmetry assumed for the crystals. She gave me three possibilities to consider. Either there were slight deviations from orthorhombic symmetry in the structure, or symmetrically related spots were affected in a different way by absorption, or there was an effect due to the anomalous scattering of the cobalt atom in the noncentrosymmetric crystal. It was not too hard to detect that the latter possibility was likely to be true. I had a marvelous time finding out which reflections were *hkl*'s and which *h̄k̄l̄*'s. This involved, for instance, running down to the basement to see in which way the Weissenberg cameras had coupled the crystal rotation with the film translation. After visual estimation of the intensities of a series of Friedel pairs for which the photographs clearly showed the intensities to be unequal, and some calculations, the absolute configuration of the hexacarboxylic acid was found. Later John Robertson used the estimated intensities to see how the anomalous effect could have been used for phasing the structure factors in earlier stages of the structure determination. In 1963 I heard Dorothy lecturing about that aspect of the work during the IUCr Congress in Rome.

Later, measurements of intensity differences of *hkl* and *h̄k̄l̄* reflections were used to determine phase angles for a B$_{12}$ derivative (Venkatesan *et al.* 1971).

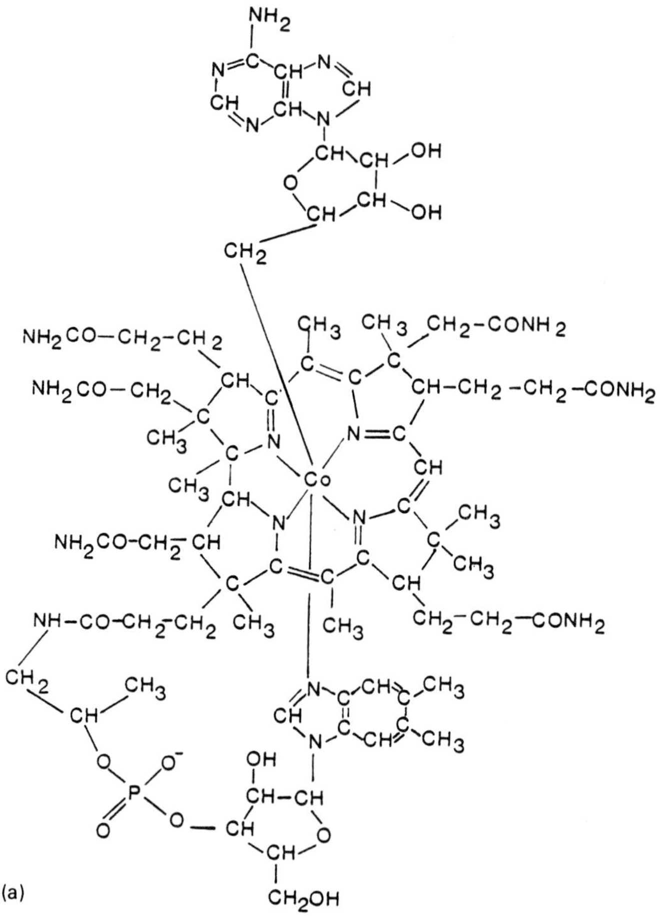

FIG. 14.1. Chemical formula of the B$_{12}$-coenzyme (a) and the numbering system used in this article (b—facing page).

The first electron-density map computed with these phases revealed the structure almost completely.

The shape of the ring system was unexpected in 1953 and has been refined by later studies. The corrin ring, shown in the formulae in Fig. 14.1, is structurally very like that of a porphyrin but is more highly reduced. An approximately planar system of double bonds extends from N_{21} through N_{22} and N_{23} to N_{24} (Fig. 14.1). The corrin ring lacks a bridge carbon atom between rings A and D, which are directly connected by two saturated carbon atoms, and contains an additional seven 'extra' methyl groups. Bond lengths and

(b)

angles were obtained for wet and dry vitamin B_{12} and for the hexacarboxylic acid, but much better values had been obtained from studies of cobyric acid, also called Factor V l*a* (Venkatesan *et al.* 1971) (the aquocyanide of the vitamin, isolated from sewage sludge and lacking the benzimidazole, ribose, phosphate, and propanolamine), and later (1961–8) from the B_{12} coenzyme (Lenhert and Hodgkin 1961; Lenhert 1968) (the vitamin with an adenosyl group replacing the cyanide group), and a monocarboxylic acid in 1967 (Nockolds, Ramaseshan, Waters, Waters, and Hodgkin 1967) (the vitamin with one amide group, that on C_{13}, hydrolysed to an acid group). Neutron-diffraction studies aided the analysis of the monocarboxylic acid (Moore, Willis, and Hodgkin 1967) by showing hydrogen atom positions and a neutron analysis of the B_{12} coenzyme is now in progress at Grenoble. The corrin ring system is nearly flat, which is surprising because the direct junction of rings A and D means that there are two tetrahedral carbon atoms in the ring. However, instead of assuming a helical form, the ring system adjusts to an approximately, but not exactly, planar form with only atoms near the A–D ring junction well out of the plane.

There may be more variation in corrin ring planarity when the cobalt atom is replaced or when the corrin ring is modified. Several corrins containing metals such as nickel (Dunitz and Meyer 1971), rhodium (Dresow, Koppenhagen, and Sheldrick 1979), and palladium (Bartlett and Dunitz, personal communication), and a metal-free synthetic corrin (Edmond and Hodgkin 1975) in which ring A lies well out of the plane of the other three rings, have been investigated crystallographically. In the rhodium corrin (Dresow *et al.* 1979) the larger size of the rhodium cation is accommodated by widening of the methine bridge angles rather than by ring buckling. When the angular methyl group, C_{20}, is removed to give a corrole with ten double bonds and a porphyrin-like conjugation, the ring system is still not planar, probably because of close contacts between ring nitrogen atoms (Harrison, Hodder, and Hodgkin 1971; Anderson, Bartczak, and Hodgkin 1974; Hitchcock and McLaughlin 1976). When alkali is added to vitamin B_{12}, a colour change takes place and compounds called 'stable yellow corrinoids' are formed. X-ray studies (Gossauer, Grüning, Ernst, Beck, and Sheldrick 1977) showed that the conjugation of the corrin ring system is broken in these compounds by stereo-specific addition of a hydroxyl group at C_5 (but not the symmetrical C_{15}). At low temperatures Co(II) corrinoids react with molecular oxygen to give a monomeric Co(III) corrin and superoxide (O_2^-). EPR studies (Jörin, Schweiger, and Günthard 1979) suggest that the oxygen is bound to cobalt with a Co–O–O bond angle of $111°$ (similar to the accepted model for oxygenation of haemoglobin). Nath has described another form of vitamin B_{12} which he calls vitamin B_{12}' (Katada, Tyagi, Nath, Petersen, and Gupta 1979). This is prepared by heating, and is also found as an impurity in the commercially available vitamin. He has suggested, from proton magnetic-resonance spectra and emission Mössbauer spectra, that the difference may be due to a different conformation of the corrin ring. X-ray crystallographic studies, if the compound can be crystallized, may help solve this enigma.

In the initial phases of the work the existence of a cobalt—carbon bond was also unexpected and later led to fundamental experiments on Co–C breakage and formation. Dorothy had guessed that the carbon atom of the cyanide group was co-ordinated to the cobalt atom in vitamin B_{12}, where the Co—C—N bond is linear. But the X-ray analysis of B_{12} coenzyme (Fig. 14.1), with the carbon atom of the ribosyl part of an adenosyl group co-ordinated to the cobalt atom, showed that a cobalt—carbon bond was really formed in B_{12} systems. This coenzyme was discovered in 1958 by Barker

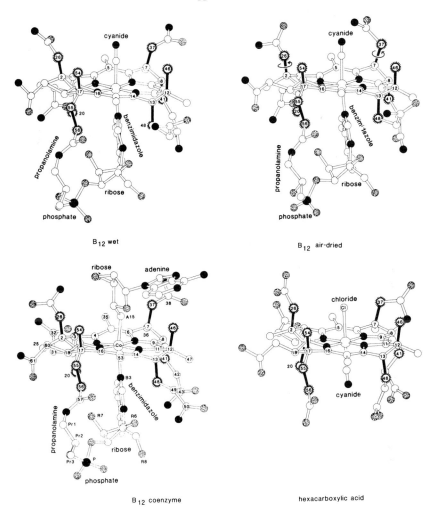

FIG. 14.2. Views of the structure of vitamin B$_{12}$ (both 'wet' and 'air-dried') compared with those of the coenzyme and the hexacarboxylic acid, with hydrogen atoms omitted for clarity. Nitrogen atoms are black, oxygen atoms stippled, and carbon atoms white. All molecules are viewed in approximately the same direction and show the restraints on free rotation about the axial bonds to the cobalt atom. These restraints are imposed by methylene or methyl groups 26, 37, 46, and 54 above the corrin ring and 20, 41, 48, 55, and 56 below the corrin ring. These atoms, and the axial bonds from the corrin ring to them, are indicated. Note the similarities and few differences (indicated by arrows around rotated bonds) between the 'wet' and 'air-dried' vitamin.

(Barker, Weissbach, and Smyth 1958) when he tried to find out why a certain anaerobic bacterium could ferment glutamate. He found that the catalyst was a mutase enzyme and that a cocatalyst was a derivative of vitamin B_{12}. The structure of this orange–yellow derivative, B_{12} coenzyme, was later determined by Galen Lenhert, working with Dorothy (Lenhert and Hodgkin 1961; Lenhert 1968). It was the first naturally occurring organometallic complex to be discovered. The X-ray analysis showed that the Co—CH_2—R group is distorted so that its bond angle is 125°, much greater than either trigonal or tetrahedral carbon values, possibly for steric reasons. Also, the Co—N bond to benzimidazole is longer in the coenzyme than in the vitamin.

A comparison of the structures of the B_{12} coenzyme (using co-ordinates provided by Galen Lenhert from a refinement of diffractometer data (Lenhert, private communication)), vitamin B_{12} ('wet' and 'air-dried'), and the hexacarboxylic acid is shown in Figs. 14.2, 14.3, and 14.4. In these figures are shown the general shapes of the molecules drawn using the VIEW computer program written by Bud Carrell in 1977. The features to note are the corrin ring system (C_1 to C_{19}) which is very roughly planar and the axial ligands on cobalt. The short side-chains (methyl and acetamide side-chains) extend above the plane of the corrin ring while the long side-chains (propionamide side-chains) extend below the plane of the ring. The axial ligands do not have freedom of rotation about the bond to the cobalt ion. They are constrained, as shown in Figs. 14.2 and 14.3, by axial substituents on the corrin ring (methyl groups 46 and 54 and acetamide methylene groups 26 and 37 above the plane of the corrin ring; methyl group 20 (the methyl group at the A-D ring junction) and methylene groups 41, 48, 55, and 56 of propionamide groups below the plane of the corrin ring). These hydrophobic axial groups, marked in Fig. 14.2, stand sentinel and block rotation of either axial group. In addition these groups form a hydrophobic pocket and so serve a protective function, particularly for the organometallic Co—C bond in the B_{12} coenzyme. In Fig. 14.2 the differences in chain conformations in the 'wet' and 'dry' vitamin are shown, although these changes are few. In Fig. 14.4 more general views of the compounds are shown to indicate the natures of the axial ligands and the extent of puckering of the corrin ring system.

During various biochemical reactions involving the B_{12} coenzyme the Co—C bond is broken and formed again. In methylcobalamin the adenosyl group in the coenzyme is replaced by a methyl group. Some biochemical reactions utilize the B_{12} coenzyme and others utilize methylcobalamin. In both cases a Co—C bond is broken and theoretically this bond cleavage could be either homolytic (to give a radical) or heterolytic (to give either a carbanion or a carbonium ion). It seems generally believed that the Co—C bond is broken homolytically to give reduced vitamin (B_{12r} (Co(II))) and the C_5 methylene radical (Babior and Gould 1969; Finlay, Valinsky, Mildvan, and

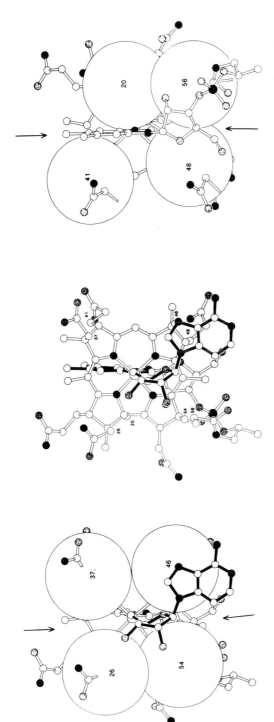

Fig. 14.3 Views of B$_{12}$ coenzyme from above the plane of the four nitrogen atoms of the corrin ring system. Circles radii 3 Å are drawn around the sentinel atoms 26, 37, 46, 54 above the plane and 20, 41, 48, and 56 below the plane. These show in the diagram on the left that there are only two possible orientations for the ribose of the adenosyl group (as observed and 180° from this). In the diagram on the right, it can be seen that there are similar restraints on the orientation of the 5,6-benzimidazole group below the plane of the ring. The entire molecule is shown in the middle with the axial groups indicated by black bonds.

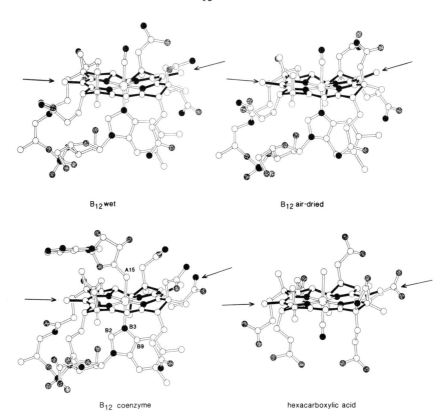

B₁₂ wet   B₁₂ air-dried

B₁₂ coenzyme   hexacarboxylic acid

FIG. 14.4. Views of vitamin B_{12} ('wet' and 'air-dried'), the coenzyme and the hexacarboxylic acid, all in approximately the same orientation, showing the corrin ring system
(with black bonds), the folding of these corrin ring systems about the Co—C_{10} line
(indicated by arrows), and the asymmetry of the co-ordination of the 5,6-dimethylbenzi-
midazole to the cobalt atom (the angle Co—B3—B2 is less than the Co—B3—B9 angle).

Abeles 1973). It has been suggested (Abeles and Dolphin 1976) that the $C_{5'}$
methylene radical abstracts hydrogen from a substrate (XH, below) to give a
product with a methyl group (three equivalent hydrogen atoms) at $C_{5'}$ and
a substrate radical. This latter substrate radical may then either rearrange
directly or combine with vitamin B_{12r} and undergo rearrangements with loss
of hydrogen. In the latter case, when the Co—C bond to the rearranged product is cleaved again, the product radical (P, below) removes a hydrogen
atom from the $C_{5'}$ methyl group leaving two free radicals that combine to
regenerate coenzyme (Abeles and Dolphin 1976).

An alternative hypothesis, that has not been as widely accepted, is that
the mechanism of action of B_{12} coenzyme proceeds via cleavage to give a

$$\underset{\overset{|}{\text{Co(III)}}}{\overset{\overset{R}{|}}{\text{CH}_2}} + \text{XH} \rightleftharpoons \underset{\overset{|}{\text{·Co(II)}}}{\overset{\overset{R}{|}}{\text{CH}_2}} + \text{XH} \rightleftharpoons \underset{\overset{|}{\text{·Co(II)}}}{\overset{\overset{R}{|}}{\text{CH}_3}} + \text{X·} \rightleftharpoons \underset{\overset{|}{\text{·Co(II)}}}{\overset{\overset{R}{|}}{\text{CH}_3}} + \text{P·} \rightleftharpoons \underset{\overset{|}{\text{Co(III)}}}{\overset{\overset{R}{|}}{\text{CH}_2}} + \text{PH}$$

$$\underset{\text{CH}_3 \;\; \text{Co(III)}}{\overset{\text{R} \;\; \text{X}}{| \quad + \quad |}} \rightleftharpoons \underset{\text{CH}_3 \;\; \text{Co(III)}}{\overset{\text{R} \;\; \text{P}}{| \quad + \quad |}}$$

$$\underset{\text{CH}_3 \;\; \text{·Co(II)}}{\overset{\text{R} \;\; \text{·P}}{| \quad + }} \rightleftharpoons \underset{\text{CH}_2 \;\; \text{·Co(II)}}{\overset{\text{R} \;\; \text{PH}}{| \quad + }} \rightleftharpoons \underset{\overset{|}{\text{Co(III)}}}{\overset{\overset{R}{|}}{\text{CH}_2}} + \text{PH}$$

1,19-*seco*-Co(I) corrin (Corey, Cooper, and Green 1977). It has also been suggested that when the coenzyme binds to an enzyme the benzimidazole becomes protonated and swings away from its co-ordination position on the cobalt atom (Pratt 1972). Such a modification could change the polarization of the Co—C bond and its tendency to be cleaved. It has also been suggested (Abeles and Dolphin 1976) that activation of the cobalt–carbon bond may be caused by an interaction of the amide groups on the edge of the corrin group of the coenzyme with groups on the protein. Several isomers of B$_{12}$ coenzyme, with the amide groups of the propionamide side chains individually modified by chemical means, were tested for coenzyme activity with the enzyme dioldehydrase (Toraya, Krodel, Mildvan, and Abeles 1979). The results show that an increase in the steric bulk of the propionamide side chain on C$_3$ (*b*, numbered 30–32), and a decrease in hydrogen bonding capability of the propionamide side chain on C$_{13}$ (*e*, numbered 48–52), eliminate coenzyme activity. It is suggested that the interaction of these propionamide side chains with the enzyme dioldehydrase facilitate the homolytic cleavage of the Co—C bond, and any obstruction to this interaction renders the coenzyme inactive. The side chain on C$_{13}$ has again appeared in a discussion and perhaps its conformational variability, observed in the crystalline state, is important for the activity of the coenzyme. In the structure of B$_{12}$ coenzyme (Lenhert 1968) the nitrogen atom (N$_{52}$) of this propionamide group lies 3.0 Å from the ether oxygen atom (R6) of the ribose group, and the methylene group (C$_{48}$) is 3.7 Å from the benzimidazole group (B2) attached to this ribose.

It was pointed out by Lenhert (1968) that the corrin ring in B$_{12}$ coenzyme and in vitamin B$_{12}$ is bent about the Co—C$_{10}$ line (19° in wet B$_{12}$ and 15° in the coenzyme, marked by lines in Fig. 14.4). This bending is attributed to the bulkiness of the benzimidazole which binds asymmetrically in order to lessen the repulsion between C$_6$ and the hydrogen atom on B4 (angle Co—B3—B2 is less than Co—B3—B9). These distortions contrast with the unbent conformation of the nickel corrin (Dunitz and Meyer 1971) and cobyric acid (Venkatesan *et al.* 1971). Thus the benzimidazole group co-ordinated to the cobalt atom causes a folding of the corrin ring about the non-methylated methine carbon atom, C_{10}. In B_{12} coenzyme the bulky adenosyl group reduces this effect. It therefore seems likely that flexibility in the corrin ring is important during coenzyme activity and may be a controlling factor in the cleavage of the Co—C bond.

A crystallographic model for Co—C bond cleavage has been found in studies of some cobaloximes with an *S*-α-methylbenzylamine group in one axial position and an *R*- or *S*-α-cyanoethyl group on the other. On exposure to X-rays, racemization of the cyanoethyl group occurs with changes in cell dimensions and disordering of groups. At very low temperatures this racemization does not occur, but there is high thermal motion for the carbon atom attached to the cobalt atom. It is presumed that the cobalt—carbon bond is cleaved and that the planar cyanoethyl radical may realign before recombination (Ohashi and Sasada 1977; Ohashi, Sasada, and Ohgo 1978).

Several biochemical reactions are catalysed by derivatives of vitamin B_{12}. B_{12} coenzyme (adenosylcobalamin) and methylcobalamin function as coenzymes in enzymatic reactions involving transfer of hydrogen and methyl groups respectively (Stadtman 1971). There are two B_{12}-dependent reactions found in mammals. One is the methylation of homocysteine to give methionine, mediated by enzyme-bound methylcobalamin, and the other is the isomerization of L-methylmalonyl-coenzyme A to succinyl-coenzyme A, mediated by B_{12} coenzyme. In the first case a hydrogen atom is interchanged with a methyl group on an adjacent carbon atom, and in the second case a -CO-S-coenzyme A group is interchanged with a hydrogen atom on an adjacent carbon atom. Failure of the rearrangement of methylmalonyl coenzyme A to succinyl coenzyme A results in the excretion of large amounts of methylmalonic acid in patients with pernicious anaemia. This problem is relieved when the vitamin is given. The vitamin is obtained for pharmaceutical use as a product of fermentation by the micro-organism *Streptomyces griseus*. However, it is not synthesized by animals or plants although it accumulates in the liver. It catalyses many reactions in micro-organisms and variations in B_{12} are found, for example, in micro-organisms that have no scheme for the synthesis of 5,6-dimethylbenzimidazole (for example, adenine may replace this to give pseudovitamin B_{12}). Micro-organisms can, in some cases, use vitamin B_{12} to provide themselves with building blocks for DNA synthesis. The B_{12} coenzyme, in this case, aids in the reduction of ribonucleotide triphosphates to deoxy-analogues.

During its cocatalytic activity it is believed that the cobalt atom in B_{12} coenzyme can take up more than one valence state (Co(III), in B_{12}, Co(II) in B_{12r}, and Co(I) in B_{12s}). To date only one corrin with cobalt in a reduced state, Co(II), has been studied crystallographically (Werthemann 1968).

Dorothy wrote on 14 August 1968 that 'B_{12} continues to produce fancy structures. Our latest is reduced cobyric heptamethyl ester iodide, which proves to be a dimer, acetic residues in, rings staggered to avoid collision!' This was a Co(II) derivative with a Co—I—Co bridge, prepared by Professor Eschenmoser and Lucius Werthemann and in which the sixth co-ordination position of the cobalt atom is empty.

Robert Woodward, who, sadly, died earlier this year, was interested in the structure of vitamin B_{12} from the moment the structure was announced and wrote to Dorothy immediately in order to check out some minor details that he was not happy about. However, finally he was persuaded to accept the X-ray evidence, particularly when he (1968, 1973) and, independently but co-operatively, Eschenmoser (1974) succeeded in synthesizing the vitamin. Woodward joined two rings (A and D) directly and two rings (B and C) with a methine bridge. Then he joined these two pairs of rings together. In case there had been any doubt, the structure of the vitamin was thus proven! As he remarked in a publication on the synthesis, vitamin B_{12} is more interesting to synthesize than haeme, because the latter contains only trigonal carbon atoms while vitamin B_{12} contains nine asymmetric carbon atoms. Eschenmoser also prepared the vitamin via a cadmium complex of a seco-corrin lacking an A–D bridge (for which the X-ray structure clarified the stereochemistry (Bartlett and Dunitz, private communication)).

How is this vitamin synthesized in the bacterial cell? The sequence of side-chains in vitamin B_{12} has the same unsymmetrical sequence (short, long, short, long, short, long, long, short) as in porphyrins such as chlorophyll and haeme. This immediately made us wonder if porphyrins and corrins were made as branches of the same biosynthetic pathway. Sir Robert Robinson guessed that the route must be the same as for porphyrins and, when first presented with this problem, wrote (15 June 1955):

> My theory for the turnround of a pyrrole unit in the porphyrins cannot be employed for an *independent* synthesis of B_{12}. It seems to require that this vitamin is a trans-formation product of a porphyrin first formed. That is the methine carbon first goes in and is later removed, for which several mechanisms can be considered. Probably an origin from a porphyrin will be biologically more acceptable than an independent synthesis.

He was, of course, proved right since corrins and porphyrins initially share a common biosynthetic pathway. There has been a lot of work done on what A. I. Scott calls the 'porphyrin–corrin connection' (Scott 1978). It seems well established that the corrin nucleus is built up from δ-aminolevulinic acid which gives uroporphyrinogen III, as for chlorophyll and haeme. But at that point C-methylation of uroporphyrinogen III leads to divergence of further pathways of haeme and corrin. The methyl group at the bridge, C_1, does not, it appears, correspond to the missing methine bridge carbon atom, but is simply added later (Brown, Katz, and Shemin 1972). In fact methionine, from S-adenosyl methionine, provides all the methyl groups (at C$_1$, C$_2$, C$_5$, C$_7$, C$_{12}$, C$_{15}$, and C$_{17}$). It is still not clear how one porphobilinogen unit (ring D) is switched when uroporphyrinogen III is formed so that the sequence of side chains is not symmetrical. It is also not clear when the cobalt atom is introduced, but it is believed that this occurs at the end of this synthesis.

The mechanism of absorption of the vitamin into the body is complex. Castle (Castle, Townsend, and Heath 1930) suggested that two factors were needed to prevent pernicious anaemia. One was taken from food (extrinsic factor) and turned out to be vitamin B$_{12}$, the 'anti-pernicious anaemia factor'. The other is found in the stomach and intestines (intrinsic factor) and is a glycoprotein (Grasbeck, Simons, and Sinkkonen 1966) that assures assimilation of vitamin B$_{12}$ into the body. The vitamin is then transported to the peripheral tissues by transcobalamin, another protein. Deficiencies in these proteins lead to pernicious anaemia, and other anaemias (Hitzig, Fräter-Schröder, and Seger 1979). It has been suggested, from circular dichroism studies, than when vitamin B$_{12}$ is bound to intrinsic factor, the 5,6-dimethylbenzimidazole is displaced by a histidine residue of the protein (Lien, Ellenbogen, Law, and Wood 1973). The details of the recognition of vitamin B$_{12}$ by intrinsic factor are still to be investigated.

One of the most interesting points about vitamin B$_{12}$ is that it is nearly, but not quite, like many other significant biological molecules. It is like a porphyrin but has two pyrrole rings joined directly and, unlike haeme, in which all carbon and nitrogen atoms are trigonally bonded, contains nine asymmetric carbon atoms. It contains trivalent cobalt rather than divalent magnesium or iron which are found in porphyrins, and the valence state of the cobalt atom may change during biological activity. Unlike nucleotides of nucleic acids the vitamin generally contains no purines or pyrimidines (except in some micro-organisms) but an unusual base, 5,6-dimethylbenzimidazole. The sugar is ribose but the glycosidic link is α- in vitamin B$_{12}$, rather than β- as in the nucleic acids. Thus nature has insured that vitamin B$_{12}$ can be distinguished from these other molecules. This may prevent it from being broken down by enzyme systems designed for related molecules.

Everyone, especially Dorothy, knew when they worked on the structure that the result would be exciting and interesting. But we did not expect so much new chemistry and biochemistry to result.

References

ABELES, R. H. and DOLPHIN, D. (1976). *Acc. Chem. Res.* **9**, 114.

ANDERSON, B. F., BARTCZAK, T. J., and HODGKIN, D. C. (1974). *J. chem. Soc., Perkin Trans.* **2**, 977.

BABIOR, B. and GOULD, D. C. (1969). *Biochem. biophys. Res. Commun.* **34**, 441.

BARKER, H. A., WEISSBACH, H., and SMYTH, R. D. (1958). *Proc. natn. Acad. Sci. U.S.A.* **44**, 1093.

BIJVOET, J. M., PEERDEMAN, A. F., and VAN BOMMEL, A. J. (1951). *Nature, Lond.* **168**, 271.

BONNETT, R., GODFREY, J. M., MATH, V. B., EDMOND, E., EVANS, H. and HODDER, O. J. R. (1971). *Nature, Lond.* **229**, 473.

BROWN, C. E., KATZ, J. J. and SHEMIN, D. (1972). *Proc. natn. Acad. Sci. U.S.A.* **69**, 2585.

CASTLE, W. B., TOWNSEND, W. C. and HEATH, C. W. (1930). *Am. J. med. Sci.* **180**, 305.

COREY, E. J., COOPER, N. J., and GREEN, M. L. H. (1977). *Proc. natn. Acad. Sci. U.S.A.* **74**, 811.

DRESOW, B., KOPPENHAGEN, V. B., and SHELDRICK, W. S. (1979). *Proc. 3rd Eur. Symp. Vitamin B_{12} and Intrinsic Factor*, p. 46.

DUNITZ, J. D. and MEYER, Jr, E. F. (1971). *Helv. chim. Acta* **54**, 77.

EDMOND, E. and HODGKIN, D. C. (1975). *Helv. chim. Acta* **58**, 641.

ESCHENMOSER, A. (1974). *Naturwissenschaften* **61**, 513.

FINLAY, T. H., VALINSKY, J., MILDVAN, A. S., and ABELES, R. H. (1973). *J. biol. Chem.* **248**, 1285.

GOSSAUER, A., GRÜNING, B., ERNST, L., BECK, W., and SHELDRICK, W. S. (1977). *Angew. Chem., Int. Ed.* **16**, 481.

GRASBECK, R., SIMONS, K., and SINKKONEN, I. (1966). *Biochim. biophys. Acta* **127**, 47.

HARRISON, H. R., HODDER, O. J. R., and HODGKIN, D. C. (1971). *J. chem. Soc. B.* 640.

HITCHCOCK, P. B. and McLAUGHLIN, G. M. (1976). *J. chem. Soc., Dalton Trans.* 1927.

HITZIG, W. H., FRÄTER-SCHRÖDER, M., and SEGER, R. (1979). *Proc. 3rd Eur. Symp. Vitamin B_{12} and Intrinsic Factor*, p. 53.

HODGKIN, D. C. (1965). *Science, N.Y.* **150**, 979.

——(1975). In *Anomalous scattering* (ed. S. Ramaseshan and S. C. Abrahams), p. iii. Munksgaard, Copenhagen.

JÖRIN, E., SCHWEIGER, A., and GÜNTHARD, Hs. H. (1979). *Proc. 3rd Eur. Symp. Vitamin B_{12} and Intrinsic Factor*, p. 45.

KATADA, M., TYAGI, S., NATH, A., PETERSEN, R. L., and GUPTA, R. K. (1979). *Biochim. biophys. Acta* **584**, 149.

LENHERT, P. G. (1968). *Proc. R. Soc.* **A303**, 45.

——and HODGKIN, D. C. (1961). *Nature, Lond.* **192**, 937.

LIEN, E. L., ELLENBOGEN, L., LAW, P. Y., and WOOD, J. M. (1973). *Biochem. biophys. Res. Commun.* **55**, 730.

MOORE, F. H., WILLIS, B. T. M., and HODGKIN, D. C. (1967). *Nature, Lond.* **214**, 130.

NOCKOLDS, C. K., RAMASESHAN, S., WATERS, T. N. M., WATERS, J. M., and HODGKIN, D. C. (1967). *Nature, Lond.* **214**, 129.

OHASHI, Y. and SASADA, Y. (1977). *Nature, Lond.* **267**, 143.

————and OHGO, Y. (1978). *Chem. Lett.* 457.

PRATT, J. M. (1972). *Inorganic chemistry of vitamin B_{12}*. Academic Press, London.

SCOTT, A. I. (1978). *Acc. Chem. Res.* **11**, 29.

STADTMAN, T. C. (1971). *Science, N.Y.* **171**, 859.

STOECKLI-EVANS, H., EDMOND, E., and HODGKIN, D. C. (1972). *J. chem Soc., Perkin Trans.* **2**, 605.

TORAYA, T., KRODEL, E., MILDVAN, Q. S., and ABELES, R. H. (1979). *Biochemistry* **18**, 417.

VENKATESAN, K., DALE, D., HODGKIN, D. C., NOCKOLDS, C. K., MOORE, F. H., and O'CONNOR, B. H. (1971). *Proc. R. Soc.* **A323**, 455.

VOS, A. (1959). *Proc. R. Soc.* **251**, 346.

WERTHEMANN, L. (1968). Untersuchungen an Kobalt (II)- und Kobalt (III)-Komplexen des Cobyrinsaure-heptamethylester. Diss. No. 4097, ETH.
WOODWARD, R. B. (1968). *Pure appl. Chem.* **17**, 519.
——(1973). *Pure appl. Chem.* **33**, 145.

Author supported by grants from National Cancer Institute (CA-10925, CA-06927, and CA-22780).

15. The crystal and molecular structure of the oxo-bridged binuclear iron (III) complex containing the methoxide derivative of the tetrabenzo-[b, f, j, n] [1, 5, 9, 13]-tetraazacyclohexadecine

B. KAMENAR AND B. KAITNER

Introduction

IN the past decade considerable attention has been given to the complexes containing metal–oxygen–metal bridging species. The iron complexes of the type LFe—O—FeL are of particular interest due to their importance as model compounds in studying and understanding some biological systems. The Fe—O—Fe bridging core appears to be present in some haem proteins and haemerythrins (Murray 1974). This type of complex is also interesting because of its antiferromagnetic properties which result from a strong spin–spin exchange through the Fe—O—Fe bridge (Lewis, Mabbs, and Richards 1967). To date complexes with Schiff bases, porphyrins, amines, imines, HEDTA, and quinolinates or chlorine atoms as ligands L, have been structurally characterized.

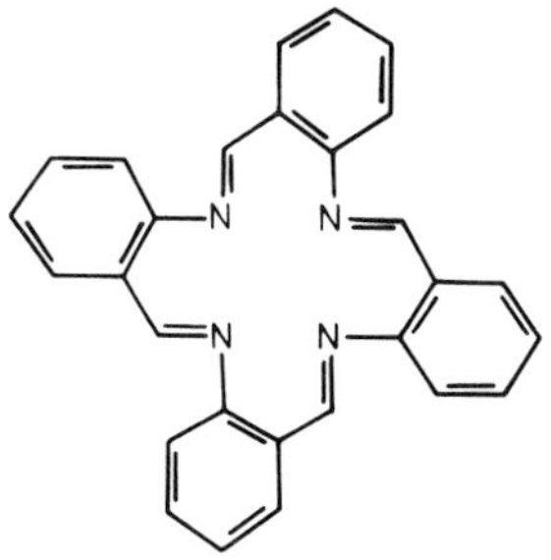

[1]

Busch and coworkers have shown that the 16-membered tetradentate macrocyclic ligand tetrabenzo-[b,f,j,n,] [1,5,9,13]-tetraazacyclohexadecine (TAAB)[1], is a product of a self-condensation reaction of *o*-aminobenzalde-hyde in the presence of different metal ions. In its metal complexes the TAAB ligand is a subject of a two-electron reduction, producing the anion [2] which can be considered as a structural analogue of the porphine ligand of haem proteins (Katović, Taylor, and Busch 1971; Takvoryan *et al.* 1974; Skurato-wicz, Madden, and Busch 1977).

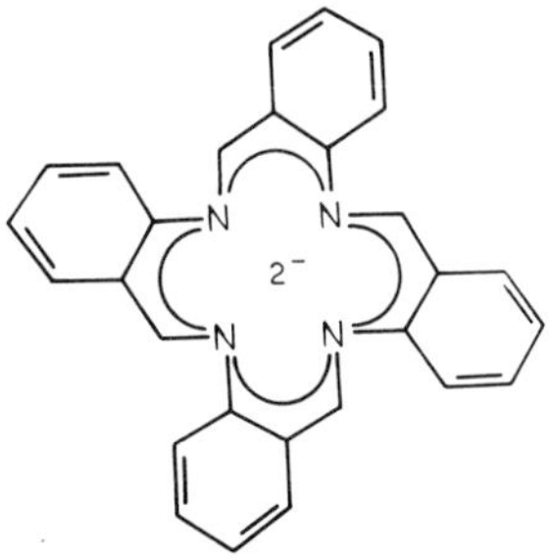

[2]

The synthesis and characterization of TAAB complexes with Cu^{2+}, Ni^{2+}, Co^{2+}, Zn^{2+}, Pd^{2+}, Fe^{2+}, Co^{3+}, and Fe^{3+} ions have been reported (Katović, Vergez, and Busch 1977). A stable iron(III)-oxo-bridged dimeric $Fe_2(TAAB)_2O^{4+}$ cation is the oxidation product of the Fe(II)–TAAB com-plex. The reaction of the dimeric $Fe_2(TAAB)_2O^{4+}$ with the methoxide ion results in the neutral dimeric species [3] with the characteristic Fe—O—Fe moiety[4]. The Fe—O—Fe linkage is very strong and its cleavage is achieved only with hydrofluoric acid.

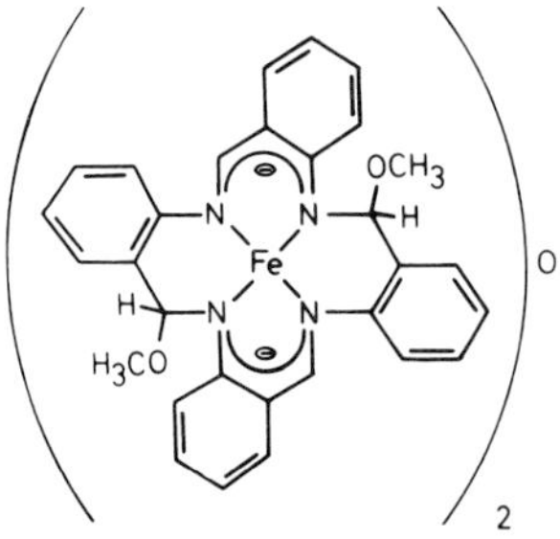

[3] 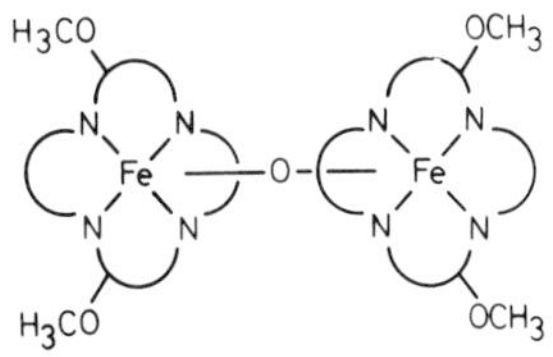   [4]

The structural analysis of Ni(TAAB)I$_2$·H$_2$O and Ni(TAAB)(BF$_4$)$_2$ was originally carried out over ten years ago, but owing to disorder in the molecular packing of the crystal structure the accuracy of the data is questionable (Hawkinson and Fleischer 1969). Recently the crystal structure of the $Ni(TAAB)(CH_2COCH_3)_2$ complex has been determined (Kamenar, Kaitner, Katović, and Busch 1979) and we now report the structure of $[Fe(TAAB)(OCH_3)_2]_2O$.

Crystallographic data

$[Fe(TAAB)(OCH_3)_2]_2O$ forms monoclinic, holohedral crystals which belong to the space group $P2_1/c$. The unit-cell dimensions (in Å) are: $a = 20.764$, $b = 11.959$, $c = 20.433$ and $\beta = 94.69°$. There are four formula units in the unit cell, the measured and calculated densities are 1.41 and 1.403 g cm^{-3}, respectively.

The crystal and molecular structure

The crystal structure consists of a packing of independent [Fe(TAAB) $(OCH_3)_2]_2O$ molecules held together by van der Waals forces. All intermolecular distances are normal, the closest molecule-to-molecule contacts vary from 3.32 to 3.50 A. The molecular structure is shown in projection along the c-axis in Fig. 15.1, and the interatomic distances and angles in Figs. 15.2(a) and (b).

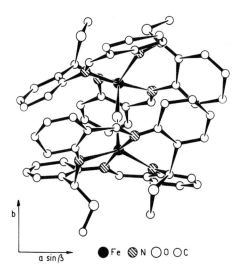

b

a sin β

● Fe ◈ N ○ O ○ C

FIG 15.1. Molecular structure of the $[Fe(TAAB)(OCH_3)_2]_2O$ complex projected along the crystallographic c-axis.

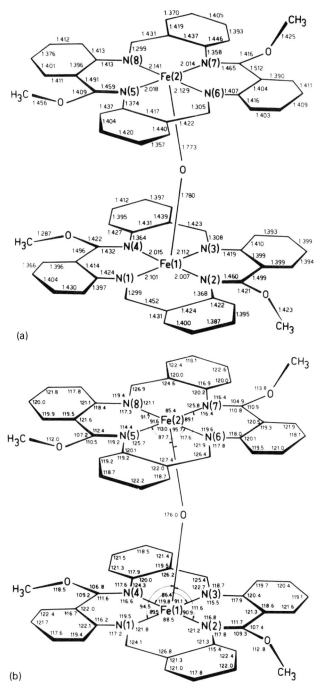

(a)

(b)

FIG 15.2. Skeletal diagrams of the [Fe(TAAB)(OCH$_3$)$_2$]$_2$O complex with (a) bond lengths, and (b) bond angles. The estimated standard deviations of the distances and angles including the iron atom are approximately 0.005 Å and 0.2°, while those between light atoms are approximately 0.01 Å and 0.6°, respectively. The interatomic parameters are given at the final $R=0.058$.

The geometry of the macrocyclic ligand

Two tetradentate $(TAAB)(OCH_3)_2$ ligands co-ordinated to two iron atoms are linked through the oxo-oxygen atom into a dimeric molecule. Both ligands are saddle-shaped with methoxy groups in *cis* positions to each other with respect to the planes of the ligands. The 16-membered ring structure of the TAAB ligand is similar to that of porphyrin so that it is reasonable to compare complexes with both ligands. The TAAB ligand is considered more flexible than porphyrin. The inflexibility of porphyrin is ascribed to the aromatic character of its ring; consequently it has been almost universally accepted that the size of the site that accommodates the metal ions in the porphyrin ring is more or less constant and cannot significantly expand if a larger metal ion enters its core (Busch *et al.* (1971). Thus the porphyrin ring accepts the effectively smaller low-spin nickel(II) and low-spin iron(III)

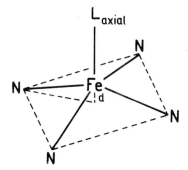

FIG 15.3. Square pyramidal arrangement in the 5-coordinated complexes with TAAB and porphinato ligands.

ions without any, or at least with insignificant, displacement of the metal ion from the plane of the four nitrogen donor atoms, while the effectively larger high-spin iron(III) ion is usually displaced out of this plane. The displacement d (Fig. 15.3) varies from 0.4 to 0.5 Å (see Tables 15.1 and 15.2). This is true for the 5-coordinated iron complexes, whereas the 6-coordinated iron(III) porphyrin complexes with mixed-spin electronic ground state or intermediate-spin state (or even high-spin state) exhibit quite insignificant out-of-plane displacement of the iron atoms. So, for example, Einstein and Willis (1978), examining the equatorial Fe–N distances with a mean value of 2.034 Å in the 6-coordinated iron complex $(Fe(OEP)ClO_4 2EtOH$, claim that this complex cannot be considered as a low-spin complex, but rather as one with a mixed- or simply high-spin electronic ground-state. In this complex the displacement of the iron ion amounts to only 0.008 Å, about the same value as the estimated standard deviations of the Fe—N bond lengths. The same applies to the structure of $[Fe(TPP)(OH_2)_2]ClO_4 \cdot 2THF$ with the average

TABLE 15.1

Comparison of some structural data for nickel(II) and iron(III) complexes with TAAB and porphinato ligands

Complex	CN of M	M—N (Å)	M—L$_{axial}$ (Å)†	d (Å)	Comments	Reference
Ni(TAAB)(BF$_4$)$_2$	4	1.90		0	low-spin, disordered structure	Hawkinson and Fleischer (1969)
Ni(TAAB)(CH$_2$COCH$_3$)$_2$	4	1.902		0	low-spin	Kamenar *et al.* (1979)
Ni Etio I	4	1.96		0		Fleischer (1963)
NiDeut	4	1.96		0		Hamor, Caughey, and Hoard (1965)
Ni(TAAB)I$_2$·H$_2$O	6	2.09	Ni—OH$_2$ 2.20	0	high-spin, disordered structure	Hawkinson and Fleischer (1969)
			Ni—I 2.903	0		
[Fe(TAAB)(OCH$_3$)$_2$]$_2$O	5	2.067	Fe—O 1.777	0.50	high-spin	This work
[Fe(TPP)]$_2$O	5	2.087	Fe—O 1.763	0.50	high-spin	Hoffman *et al.* (1972)
[Fe(TPP)]$_2$N	5	1.991	Fe—N 1.661	0.32	low-spin(?)	Scheidt *et al.* (1976)
MeOFeMeso	5	2.073	Fe—O 1.842	0.455	high-spin or intermedi-ate-spin state	Hoard *et al.* (1965)
Chlorohemin	5	2.062	Fe—Cl 2.218	0.475	high-spin or intermedi-ate-spin state	Koenig (1965)
Fe(TPP)Cl	5	2.049	Fe—Cl 2.192	0.383	high-spin, orien-tational disorder	Hoard, Cohen, and Glick (1967)

Complex					d	Spin state	Reference
Fe(TPP)Br	5	2.069	Fe—Br	2.348	0.49	high-spin	Skelton and White (1977)
Fe(TPP)I	5	2.066	Fe—I	2.554	0.46	high-spin	Hatano and Scheidt (1979)
Fe(TPP)ClO$_4$	5	1.997	Fe—O	2.025	0.27	mixed-spin state	Kastener et al. (1978)
Fe(PPIXDME) (SC$_6$H$_4$-p-NO$_2$)	5	2.064	Fe—S	2.324	0.434	high-spin	Tang et al. (1976)
[Fe(TPP) (Im)$_2$]Cl·CH$_3$OH	6	1.989	Fe—N	1.974	0.009	low-spin	Collins, Countryman, and Hoard (1972)
[Fe(1-ME-Im)$_2$(PPIX)]·CH$_3$OH·H$_2$O	6	1.990	Fe—N	1.977	0.011	low-spin	Little, Dymock, and Ibers (1975)
Fe(OEP) (ClO$_4$)·2EtOH	6	2.034	Fe—O	2.135	0.008	high-spin or intermediate-spin state, partially disordered structure	Einstein and Willis (1978)
[Fe(TPP) (OH$_2$)$_2$]ClO$_4$·2THF	6	2.041	Fe—O	2.090	0	high-spin or mixed-spin state	Kastner et al. (1978)
Fe(TPP) (C(CN)$_3$)	6	1.995	Fe—N	2.317	0	intermediate-spin state	Summerville et al. (1978)
[Fe(TMSO)$_2$(TPP)]ClO$_4$	6	2.045	Fe—O	2.078	0	high-spin	Mashiko et al. (1978)

TAAB = tetrabenzo [b,f,j,n] [1,5,9,13] tetraazacyclohexadecine; Etio = etioporphyrin; Deut = 2,4-diacetyldeuteroporphyrin-IX dimethyl ester; TPP = tetraphenylporphyrin; Meso = mesoporphyrin-IX dimethyl ester; PPIX = protopophyrin IX; DME = dimethyl ester; IM = imidazole, 1-Me-Im = methylimidazole; OEP = octaethylporphyrin; TMSO = tetramethylene sulphoxide.

d = displacement of the metal ion out of the plane defined by the four nitrogen atoms from TAAB or porphinato ligand.

† Only the average values of the M—L$_{axial}$ bond lengths are given in 6-coordinated complexes.

TABLE 15.2

Some structural data for iron(III) complexes containing Fe—O—Fe bridging system

Complex	Fe—O_b (Å)	Fe—L (Å)	CN of Fe	d (Å)	Fe—O—Fe angle (°)	Fe...Fe separation (Å)	Reference
(pyH)$_2$[FeCl$_3$]$_2$O·py	1.755	2.213 (Cl)	4		155.6	3.43	Drew, McKee, and Nelson 1978
[Fe(TAAB)(OCH$_3$)$_2$]$_2$O	1.777	2.067 (N)	5	0.50	176.0	3.55	This work
[F3(TPP)]$_2$O	1.763	2.087(N)	5	0.50	174.5	3.53	Hoffman et al. 1972
[Fe)sal-N-p-chlorophenyl)$_2$]$_2$O	1.76	2.16 (N) 1.89 (O)	5	0.53	175	3.53	Davies and Gatehouse 1973b
[Fe(salen)]$_2$O·2py	1.80	2.087 (N) 1.918 (O)	5	0.56	139.1	3.36	Gerloch, McKenzie, and Towl 1969
[Fe(N-n-propyl-sal)$_2$]$_2$O	1.77	2.14 (N) 1.93(O)	5	0.53	164	3.51	Davies and Gatehouse 1972
[Fe(salen)]$_2$O	1.78	2.12 (N) 1.92 (O)	5	0.58	144.6	3.39	Davies and Gatehouse 1973a
[Fe(2-Mequin)$_2$]$_2$O·CHCl$_3$	1.780	2.190 (N) 1.918 (O)	5	0.64	151.6	3.45	Mabbs, McLachlan, McFadden, and McPhail 1973
[Fe(salen)]$_2$O·CH$_2$Cl$_2$	1.794	2.105 (N) 1.923 (O)	5	0.56	142.2	3.40	Coggon, McPhail, Mabbs, and McLachlan 1971
[Fe(C$_{22}$H$_{22}$N$_4$)]$_2$O·CH$_3$CN	1.792	2.054 (N)	5	0.698	142.8	3.40	Weiss and Goedken 1979
[Fe(OOC)$_2$pyCl(H$_2$O)$_2$]$_2$O·4H$_2$O	1.773	2.094 (O) 2.056 (OH$_2$) 2.105 (N)	6	0.17 (Fe 1) 0 (Fe 2)	180	3.55	Ou, Wolimann, Hendrickson, Potenza, and Schugar
[Fe(TEPA)]$_2$OI$_4$	1.77	2.33 (N) 2.12 (N)	6	0.33	172	3.53	Coda, Kamenar, Prout, Carruthers, and Rollett 1975
(enH)$_2$[Fe(HEDTA)]$_2$O·6H$_2$O	1.79	2.25 (N) 2.03 (O)	6	0.36	165.0	3.56	Lippard, Schugar, and Walling 1967
[FeB(H$_2$)]$_2$O(ClO$_4$)$_4$	1.80	2.20 (N) 2.15 (OH$_2$)	7		178	3.6	Fleischer and Hawkinson 1967

(OOC)$_2$pyCl=4-chloro-2,6-pyridinedicarboxylate; TAAB=tetrabenzo[b,f,j,n][1,5,9,13]tetraazacyclohexadecine; TPP=tetraphenylporphyrin; sal=salicylaldehyde; 2-Mequin=8-hydroxy-2-methylquinolinate; salen=N,N'-ethylenebis (salicylideneiminate); TEPA=tetraethylenepentaamine; HEDTA=N-hydroxyethyl-ethylenediaminetriacetate; B=2,13-dimethyl-3,6,9,12,18-pentaazabicyclo[12,3,1]-octadeca-1(18),2,12,14,16-pentaene; C$_{22}$H$_{22}$N$_4$=[7,16-dihydro-6,8,15,17]-tetramethyldibenzo[b,i][1,4,8,11]tetraazacyclotetradecinate.
d=displacement of the iron atom out of the plane defined by the equatorial donor atoms.

Fe–N bond length of 2.041 Å, again substantially longer than the values found for these bonds in the low-spin iron porphyrins (Kastner, Scheidt, Mashiko, and Reed 1978). The authors suggest that such an increase in the Fe—N bond length, with the corresponding expansion of the porphinato core, is in accordance with the significant population of the $3d_{x^2-y^2}$ orbital and consequently with either a high-spin $(S = 5/2)$ or a quantum mechanically mixed-spin $(S = 3/2, 5/2)$ ground-state. A similar 6-coordinated Fe(TPP) (C(CN)$_3$) complex with the average Fe–N distance of 1.995 Å is considered as an intermediate $(S = 3/2)$ spin-state species. The major difference between the two complexes is ascribed to the apparent strength of the axial ligand interaction with the central metal atom (Summerville, Cohen, Hatano, and Scheidt 1978). In both structures the iron atoms are by symmetry requirements in the plane of the porphinato ligand. Kastner *et al.* (1978) suggest that in the 5-coordinated complex Fe(TPP)ClO$_4$, with an average Fe—N bond length of 1.997 Å and with the displacement of the iron atom $d = 0.27$ Å, the iron atom has the $3d_{x^2-y^2}$ orbital either unoccupied or partially occupied, which would be consistent with either an intermediate- or mixed-spin state. A similar situation is encountered in the presumably low-spin 5-coordinated iron(III) complex $[Fe(TPP)]_2N$ in which the average iron-to-porphinato-nitrogen distance is only 1.991 Å and the displacement of the iron atom is 0.32 Å (Scheidt, Summerville, and Cohen 1976). Indeed, if we consider 2.065 Å and 1.990 Å as typical values for the equatorial Fe—N bonds in high-spin and in low-spin iron(III) porphyrins respectively, then the values found for such bonds in the above 6-coordinated complexes are also in some senses intermediate. Finally, a very recent structural investigation of the 6-coordinated high-spin iron(III) complex $[Fe(TMSO)_2(TPP)]ClO_4$, with the mean Fe—N bond length of 2.045 Å, and again with the iron atom by symmetry requirement in the plane of the porphinato ligand, has shown that the porphinato core can accommodate a high-spin iron atom without its displacement (Mashiko, Kastner, Spartalian, Scheidt, and Reed 1978). Consequently, it seems that the earlier conclusions concerning the possibility for the expansion of the porphyrin core should be re-examined.

Although the TAAB ligand is more flexible than the porphinato ligand and folds when the metal ion is too large, in the case of the 5-coordinated $[Fe(TAAB)(OCH_3)_2]_2O$ complex the iron ions are displaced from the plane of the four nitrogen atoms by about 0.50 Å towards the bridging oxo-oxygen atom. As shown in Table 15.2 this displacement is common in all complexes containing the Fe—O—Fe system. If we compare this structure with that of the high-spin Ni(TAAB)I$_2 \cdot$H$_2$O containing a 6-coordinated nickel ion in which the average value of the equatorial Ni—N bond lengths amounts to 2.09 Å, it seems reasonable to suppose that the metal ion displacement is mainly connected with its coordination, type of the axial ligands, and steric interactions between non-bonded atoms.

The coordination of the iron atom

Both iron atoms are almost identically surrounded by four nitrogen atoms from the TAAB ligands and by one bridging oxo-oxygen atom, thus forming an approximately square pyramid about each iron atom. The donor nitrogen atoms are not exactly planar but exhibit distortion towards a tetrahedral arrangement. The N(1) and N(3) atoms are $+0.39$ while N(2) and N(4) are -0.39 Å out of the least squares best plane defined by all four nitrogen atoms. Similarly, N(5) and N(7) are $+0.36$ (mean value) and N(6) and N(8) are -0.36 Å out of their least-squares best plane. The N(1)N(2)N(3)N(4) and N(5)N(6)N(7)N(8) planes are not strictly parallel to each other, they deviate from parallelism by an angle of $2.86°$.

The mean value of the Fe—N bond lengths is 2.067 Å, although it might be considered that around both iron atoms two iron-to-nitrogen bond lengths are slightly shorter (mean value 2.014 Å) than the other two (mean value 2.121 Å). The average Fe—N distance is comparable with the values found for such bonds in other high-spin iron complexes, particularly with those in chlorohemin (Koenig 1965) and in methoxyiron(III) mesoporphyrin-IX dimethyl ester (Hoard, Hamor, Hamor, and Caughey 1965) (see Table 15.1). The N—Fe—N and O—Fe—N bond angles are again in pairs. Two N—Fe—N angles (mean value $87.0°$) deviate, while the other two (mean value $90.3°$) are close to a right angle. Similarly, at both iron atoms, two O—Fe—N angles have an average value of $93.24°$, while the other two, with an average value of $115.51°$, are significantly different. These values are substantially different when compared with similar iron(III) porphyrin complexes, e.g. $[Fe(TPP)]_2O$. All four N—Fe—N as well as O—Fe—N angles in $[Fe(TPP)]_2O$ are more or less equivalent and their mean values amount to 86.8 and $103.7°$, respectively (Hoffman *et al.* 1972).

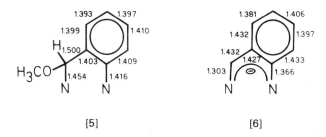

[5]   [6]

The $(TAAB)(OCH_3)_2$ ligands in the present structure contain two pairs of dissimilar chelate rings analogous to the $(TAAB)(CH_2COCH_3)_2$ ligand in the structure of $Ni(TAAB)(CH_2COCH_3)_2$. The like rings appear *trans* to each other [3]. The chelate ring [5], to which the —OCH_3 group is attached, contains C—N single bonds, while in the ring [6] the π-electrons are more delocalized. This has already been observed in the structure of the Ni(TAAB) derivative (see above). Some stereochemical data between non-substituted TAAB ligands in Ni(TAAB)$I_2 \cdot H_2O$ and Ni(TAAB)$(BF_4)_2$ and substituted TAAB ligands in Ni(TAAB)$(CH_2COCH_3)_2$ and $[Fe(TAAB)(OCH_3)_2]_2O$ are compared in Table 15.3.

Fe—O—Fe bridge

The Fe(1)-oxo- and Fe(2)-oxo-oxygen atom bond lengths of 1.773 and 1.780 Å are equivalent and correspond to individual $(\sigma + \pi)$ bond orders of 1.5. These values for Fe—O bonds are in very good agreement with those found in the other oxo-bridged binuclear iron(III) complexes (see Table 15.2). The average Fe—O (bridging) bond length of the examples given in Table 15.2 amounts to 1.78 Å. This value is 0.14 Å shorter than that usually taken as the Fe—O bond length (~ 1.92 Å) and indicates some degree of π-bonding associated with the Fe—O—Fe bridge.

The angle at the bridging oxygen atom is of particular interest and although it has been the subject of many discussions the factors affecting the geometry of the Fe—O—Fe bridge are still not well understood. In [Fe(TAAB)$OCH_3)_2]_2O$ this angle is 176.0° while in other complexes of the same type it ranges from 139.1° in [Fe(salen)]$_2$O·2py to 180° in [Fe(OOC)$_2$pyCl$(H_2O)_2$]O·4H_2O. It is assumed that the Fe—O—Fe bridge without any degree of π-bonding character would have a tetrahedral or nearly tetrahedral angle at the oxygen atom, similar to the water molecule or other bi-covalent σ-bonded compounds. Consequently, the increasing angle could be explained by the increasing degree of π-bonding with the bridging oxygen atom. At the same time the relative Fe—O (bridging) bond length should also be a measure of the relative degree of multiple bonding. Since in all known structures of the LFe—O—FeL type the Fe—O bond lengths are more or less constant, it seems that the steric interactions between ligands attached to the iron atoms as well as the crystal packing effects are predominant factors in determining the magnitude of the Fe—O—Fe angle.

All complexes containing the Fe—O—Fe bridging system have a magnetic moment at room temperature substantially lower than the spin-only value of $5.09\mu_B$ for high-spin iron(III)$(S = 5/2)$. The room-temperature μ_{eff}, values for the complexes given in Table 15.2 range from 1.82 to $1.94\mu_B$. The μ_{eff} for $[Fe(TAAB)(OCH_3)_2]_2O$ is an even lower $1.65\mu_B$ and is strongly temperature dependent. The temperature dependence of the magnetic susceptibility is a general characteristic of the LFe—O—FeL complexes. The calculated values of the exchange integral J for the oxo-bridged binuclear iron(III) complexes vary from -95 to $-105\,\mathrm{cm}^{-1}$ (for $[Fe(TAAB)(OCH_3)_2]_2O$, $J = -117.6\,\mathrm{cm}^{-1}$ (Katović *et al.* 1977)).

TABLE 15.3

Average bond lengths and angles in the nickel(II) and iron(III) complexes with TAAB ligand

Atoms	$Ni(TAAB)I_2 \cdot H_2O$†	$Ni(TAAB)(BF_4)_2$‡	$Ni(TAAB)(CH_2COCH_3)_2$†	$[Fe(TAAB)(OCH_3)_2]_2O$†
Distances (Å)				
M—N	2.09	1.90	1.917 / 1.887	2.013 / 2.120
N—$C_{benzene}$	1.42	1.41	1.43 / 1.35	1.42 / 1.37
C—$C_{benzene}$	1.65‡	1.48	1.50 / 1.42	1.50 / 1.43
C—N	1.16‡	1.32	1.49 / 1.29	1.45 / 1.30
$C_{benzene}$—$C_{benzene}$	1.40	1.38	1.40 / 1.41	1.40 / 1.41
Angles (°)				
N—M—N	90	90	93.2 / 90.8	90.3 / 87.0
M—N—C	130‡	123	119.6 / 127.5	116.1 / 121.9
N—C—$C_{benzene}$	115‡	119	109.9 / 124.3	111.6 / 125.7
C—$C_{benzene}$—$C_{benzene}$	135‡	130	120.6 / 122.1	121.4 / 126.2
$C_{benzene}$—$C_{benzene}$—N	113	113	117.1 / 122.0	117.6 / 120.4
$C_{benzene}$—N—M	122	121	115.9 / 124.3	118.0 / 124.2
C—N—$C_{benzene}$	107‡	116	115.0 / 116.4	117.7 / 118.3
$C_{benzene}$—$C_{benzene}$—$C_{benzene}$	120	120	120.0 / 119.9	120.0 / 120.0

† The upper values refer to the substituted chelate ring (see text).
‡ These values are unreliable due to the disorder in the crystal structure (Hawkinson and Fleischer 1969).

There is also another unusual characteristic of the Fe—O—Fe bridge in $[Fe(TAAB)(OCH_3)_2]_2O$. As Katović *et al.* (1977) pointed out, the Fe—O—Fe linkage is unexpectedly difficult to cleave. The cleavage was successfully achieved only in an acidic fluoride medium.

Acknowledgements

We thank Professor V. Katović for supplying us with crystals and for reading the manuscript as well as Professor D. Grdenić for his interest in this research. This work has been supported by Scientific Research Council of the S.R. Croatia, Zagreb.

References

BUSCH, D. H., FARMERY, K., GOEDKEN, V., KATOVIĆ, V., MELNYK, A. C., SPERATI, C. R., and TOKEL, N. (1971). *Adv. Chem. Ser.* **100**, 44–78.
CODA, A., KAMENAR, B., PROUT, K., CARRUTHERS, J. R., and ROLLETT, J. S. (1975). *Acta crystallogr.* **B31**, 1438–42.
COGGON, P., McPHAIL, A. T., MABBS, F. E., and McLACHLAN, V. N. (1971). *J. chem. Soc. A* 1014–19.
COLLINS, D. M., COUNTRYMAN, R., and HOARD, J. L. (1972). *J. Am. chem. Soc.* **94**, 2066–72.
DAVIES, J. E., and GATEHOUSE, B. M. (1972). *Cryst. Struct. Commun.* **1**, 115–20.
—— —— (1973a). *Acta crystallogr.* **B29**, 1934–42.
—— ——(1973b). *Acta crystallogr.* **B29**, 2651–8.
DREW, M. G. B., McKEE, V., and NELSON, S. M. (1978). *J. chem. Soc., Dalton Trans.* 80–4.
EINSTEIN, F. W. B. and WILLIS, A. C. (1978). *Inorg. Chem.* **17**, 3040–5.
FLEISCHER, E. B. (1963). *J. Am. chem. Soc.* **85**, 146–8.
—— and HAWKINSON, S. (1967). *J. Am. chem. Soc.* **89**,720–1.
GERLOCH, M., McKENZIE, E. D., and TOWL, A. D. C. (1969). *J. chem. Soc. A* 2850–2858.
HAMOR, T. A., CAUGHEY, W. S., and HOARD, J. L. (1965). *J. Am. chem. Soc.* **87**, 2305–2312.
HATANO, K. and SCHEIDT, W. R. (1979). *Inorg. Chem.* **18**, 877–9.
HAWKINSON, S. W. and FLEISCHER, E. B. (1969). *Inorg. Chem.* **8**, 2402–10.
HOARD, J. L., COHEN, G. H., and GLICK, M. D. (1967). *J. Am. chem. Soc.* **89**, 1992–6.
—— HAMOR, M. J., HAMOR, T. A., and CAUGHEY, W. S. (1965). *J. Am. Chem. Soc.* **87**, 2312–19.
HOFFMAN, A. B., COLLINS, D. M., DAY, V. W., FLEISCHER, E. B., SRIVASTAVA, T. S., and HOARD, J. L. (1972). *J. Am. chem. Soc.* **94**, 3620–6.
KAMENAR, B., KAITNER, B., KATOVIĆ, V., and BUSCH, D. H. (1979). *Inorg. Chem.* **18**, 815–18.
KASTNER, M. E., SCHEIDT, W. R., MASHIKO, T., and REED, C. A. (1978). *J. Am. chem. Soc.* **100**, 666–7.
KATOVIĆ, V., TAYLOR, L. T., and BUSCH, D. H. (1971). *Inorg. Chem.* **10**, 458–62.
—— VERGEZ, S. C., and BUSCH, D. H. (1977). *Inorg. Chem.* **16**, 1716–20.
KOENIG, D. F. (1965). *Acta crystallogr.* **18**, 663–73.
LEWIS, J., MABBS, F. E., and RICHARDS, A. (1967). *J. chem. Soc. A* 1014–18.
LIPPARD, S. J., SCHUGAR, H., and WALLING, C. (1967). *Inorg. Chem.* **6**, 1825–31.

LITTLE, R. G., DYMOCK, K. R., and IBERS, J. A. (1975). *J. Am. chem. Soc.* **97,** 4532–4539.

MABBS, F. E., McLACHLAN, V. N., McFADDEN, D., and McPHAIL, A. T. (1973). *J. Chem. Soc., Dalton Trans.* 2016–21.

MASHIKO, T., KASTNER, M. E., SPARTALIAN, K., SCHEIDT, W. R., and REED, C. A. (1978). *J. Am. chem. Soc.* **100,** 6354–62.

MURRAY, K. S. (1974). *Co-ord. Chem. Rev.* **12,** 1–35.

OU, CHIA CHIH, WOLIMANN, R. G., HENDRICKSON, D. N., POTENZA, J. A., and SCHUGAR, H. J. (1978). *J. Am. chem. Soc.* **100,** 4717–24.

SCHEIDT, W. R., SUMMERVILLE, D. A., and COHEN, I. A. (1976). *J. Am. chem. Soc.* **98,** 6623–8.

SKELTON, B. W. and WHITE, A. H. (1977). *Aust. J. Chem.* **30,** 2655–60.

SKURATOWICZ J. S., MADDEN, I. L., and BUSCH, D. H. (1977). *Inorg. Chem.* **16,** 1721–5.

SUMMERVILLE, D. A., COHEN, I. A., HATANO, K., and SCHEIDT, W. R. (1978). *Inorg. Chem.* **17,** 2906–10.

TAKVORYAN, N., FARMERY, K., KATOVIĆ, V., LOVECCHIO, F. V., GORE, E. S., ANDERSON, L. B., and BUSCH, D. H. (1974). *J. Am. chem. Soc.* **96,** 731–42.

TANG, S. C., KOCH, S., PAPAEFTHYMIOU, G. C., FONER, S., FRANKEL, R. B., IBERS, J. A., and HOLM, R. H. (1976). *J. Am. chem. Soc.* **98,** 2414–34.

WEISS, M. C. and GOEDKEN, V. L. (1979). *Inorg. Chem.* **18,** 819–26.

16. X-ray crystallographic studies on the conformational features and the packing modes of amides

K. VENKATESAN AND S. RAMAKUMAR

A STUDY of the conformation of the peptide unit and the energetics of its deformation is of considerable importance in understanding the structural features of polypeptides and proteins. The amide unit, in addition to being the building block of biological macromolecules such as proteins, is also a constituent of a variety of biologically active small molecules, e.g. penicillins and cephalosporins. In fact a rationalization has been given for the high reactivity of the β-lactam in penicillins in terms of a destabilization of the amide bond due to a decreased delocalization from the adjacent carbonyl group (Woodward 1949). Fusion of the four unit at the five ring in these compounds results in a pyramidal geometry of the β-lactam nitrogen (Crowfoot, Bunn, Rogers-Low, and Turner-Jones 1949; Sweet and Dahl 1970). Some of the observations made by us on the crystal and molecular structures containing the amide group (*cis* as well as *trans*) investigated in our laboratory and those reported in the literature by other investigators are discussed in this chapter. The peptide unit is associated in the crystalline state and is involved in interactions such as hydrogen bonding with other peptides which is similar to the conditions prevailing in biological macromolecules. We discuss briefly (i) geometry of the amide group under various conditions, (ii) influence of the side chains on the conformation of the amide group, (iii) possible effects of intermolecular interactions on the conformation of the amide group, and (iv) some features of the packing modes of amides in crystals.

Geometry of the amide group

Accurate X-ray crystallographic investigations of simple peptides and other molecules containing the amide group have revealed that the amide group cannot be considered to be planar. The notation and conventions followed in describing the conformation of the peptide unit are those proposed by

TABLE 16.1

Out-of-plane deformation ($\Delta\omega$, θ_N), hydrogen bonding parameters (θ_{NA}, $\Delta\theta_{NA}$), and amide bond lengths

Compound	$C'{-}N$ (Å)	$\Delta\omega$ (°)	θ_N (°)	θ_{NA} (°)	$|\theta_N - \theta_{NA}| = |\Delta\theta_{NA}|$ (°)	Reference
(1) t-Boc-Sar-Gly-benzester	1.340	12.4	-2.4	1.6	4.0	Itoh et al. (1976)
(2) Biurea	1.361	-12.6	26.6	17.1	9.6	Brown and Russell (1976)
(3) Cyc-(L-Thr-L-His)-dihydrate	1.321	-4.8	6.6	4.8	1.8	Cotrait et al. (1976)
(4) Cyc-(L-Thr-L-His)-dihydrate	1.332	-5.7	8.3	1.0	7.3	Cotrait et al. (1976)
(5) Benzyloxy carbonyl-Gly-Pro-Leu	1.334	-4.1	0.8	-0.6	1.4	Yamane, Ashida, Shimonishi, Kakudo, and Sasada (1976)
(6) Benzyloxy carbonyl-Gly-Pro-Leu	1.333	-3.1	2.5	2.6	0.1	Yamane et al. (1976)
(7) Uridine-5-oxyacetic acid methyl ester monohydrate	1.382	-1.9	20.7	37.0	16.3	Morikawa, Torii, Iitaka, and Tsuboi (1975)
(8) 9-Ethyl adenineparabanic acid	1.353	1.7	-0.1	5.2	5.3	Huey-Sheng Shieh and Voet (1976)
(9) 9-Ethyl adenineparabanic acid	1.376	3.2	6.3	6.3	0	Huey-Sheng Shieh and Voet (1976)
(10) K salt of N-(Purin-6-yl carbonyl) glycine monohydrate	1.341	9.4	-12.9	-4.8	8.1	Parthasarathy, Ohrt, and Chheda (1976)
(11) Uridine	1.387	-5.1	1.2	8.6	7.4	Green, Rosenstein, Shiono, and Abraham (1975)
(12) 2-Acetamido-2,3-dideoxy-D-threo-hex-2-enono-1,4-lactone	1.369	-1.9	4.8	-12.6	17.4	Ziva Ruzic-Toros and Biserka Kojic-Prodic (1976)

	Compound						Reference
(13)	7-Methyl xanthine HCl·H$_2$O	1.395	−4.0	3.9	4.2	0.3	Kistenmacher and Sorrell (1975)
(14)	7-Methyl xanthine HCl·H$_2$O	1.373	−0.5	12.1	26.1	14.0	Kistenmacher and Sorrell (1975)
(15)	Primidone	1.330	1.2	−6.4	2.4	8.8	Yeates and Palmer (1975)
(16)	Primidone	1.328	0.5	0.1	−10.2	10.3	Yeates and Palmer (1975)
(17)	Urea and 5,5-diethyl barbituric acid complex	1.371	−5.4	1.7	−0.2	1.9	Gartland and Craven (1974)
(18)	Urea and 5,5-diethyl barbituric acid complex	1.376	9.3	−0.6	3.6	4.2	Gartland and Craven (1974)
(19)	Imidazole and 5,5-diethyl barbituric acid complex	1.362	0.6	3.8	8.8	12.6	I-Nan Hsu and Craven (1974a)
(20)	N-methyl-2-pyridone and barbital	1.369	−0.4	−4.0	−14.3	10.3	I-Nan Hsu and Craven (1974b)
(21)	Bz-DL-Leu-Gly-Et-ester	1.331	−9.2	−2.9	−11.5	8.6	Timmins (1975)
(22)	Bz-DL-Leu-Gly-Et-ester	1.329	−10.5	7.1	3.6	3.5	Timmins (1975)
(23)	Bz-DL-Leu-Gly-Et-ester	1.328	1.0	−3.9	−6.5	2.6	Timmins (1975)
(24)	Bz-DL-Leu-Gly-Et-ester	1.340	9.7	2.5	5.9	3.4	Timmins (1975)
(25)	Thiazolidine-2,4-dione	1.372	0.2	−6.2	1.7	7.9	Form, Raper, and Downie (1975)
(26)	N-acetyl-α-D glucosamine	1.346	−10.1	20.0	29.5	9.5	Mo and Jensen (1975)
(27)	9-Ethyl guanine hemihydrochloride	1.396	−1.2	1.9	4.5	2.6	Mandel and Marsh (1975)
(28)	L-Ala-L-Ala-L-Ala	1.327	−4.8	−4.7	−12.4	7.7	Fawcett et al. (1975)
(29)	L-Ala-L-Ala-L-Ala	1.324	0.8	−3.2	−12.4	9.2	Fawcett et al. (1975)
(30)	L-Ala-L-Ala-L-Ala	1.316	3.4	2.1	−5.0	7.1	Fawcett et al. (1975)
(31)	L-Ala-L-Ala-L-Ala	1.330	7.0	6.2	−13.0	19.2	Fawcett et al. (1975)
(32)	5-N(L-Leu) amino uridine	1.371	−2.0	−0.2	−3.9	3.7	Narayanan and Berman (1977)

TABLE 16.1 contd

Compound	C'—N (Å)	Δω (°)	θ_N (°)	θ_{NA} (°)	$\lvert\theta_N-\theta_{NA}\rvert=\lvert\Delta\theta_{NA}\rvert$ (°)	Reference
(33) 5-N(L-Leu) amino uridine	1.343	-6.8	2.2	-8.6	10.8	Narayanan and Berman (1977)
(34) Nialamide-NF	1.323	-5.3	6.7	11.8	5.1	de Lerma, Garcio Blanco, and Fayos (1977)
(35) Nialamide-NF	1.338	2.2	9.2	0.3	8.9	de Lerma et al. (1977)
(36) 6-Endo hydroxy-3-endo amino-methyl bicyclo[2.2.1]heptane-2-endo-carboxylic acid lactam	1.336	-1.2	4.7	16.3	11.6	Olson, Templeton, and Templeton (1977)
(37) N-acetyl-L-tryptophan	1.348	-7.8	11.4	11.1	0.3	Yamane, Andon, and Ashida (1977)
(38) Acetamide–barbital complex	1.372	-2.5	6.8	-0.2	7.0	I-Nan Hsu and Craven (1974c)
(39) Acetamide–barbital complex	1.368	-5.8	5.5	1.6	3.9	I-Nan Hsu and Craven (1974c)
(40) Aza-isotwistanone	1.333	1.2	10	-5.0	15	Blaha et al. (1978)
(41) Enantholactam HCl	1.298	5.6	-2	-1.8	0.2	Winkler and Dunitz (1975a)
(42) Caprylolactam	1.334	-31.6	23	22.9	0.1	Winkler and Dunitz (1975b)
(43) L-Ala-Gly	1.334	-6.1	3	2.9	0.1	Koch and Germain (1970)
(44) N-Ac-Gly	1.326	2.2	1.9	2.5	0.6	Mackay (1975)
(45) Gly-Gly-phosphate monohydrate	1.332	-4.5	-10	-9.0	1	Freeman et al. (1972)

Compound						Reference
(46) N-Ac-L-Phe-L-Tyr	1.335	3.5	0	−2.6	2.6	Stenkamp and Jensen (1973)
(47) L-Ala-L-Ala	1.344	−4.3	−4	−1.2	2.8	Fletterick, Tsai, and Hughes (1971)
(48) N-Ac-L-Glu	1.322	8.2	−6	−8.9	2.9	Narashimha Murthy et al. (1974)
(49) Aza-twistanone	1.341	14.5	−14	−10.7	3.3	Ramakumar et al. (1977)
(50) Caprolactam	1.340	−4.1	7	3.5	3.5	Winkler and Dunitz (1975c)
(51) α-Gly-Gly	1.326	3.6	8.1	4.5	3.6	Freeman et al. (1970)
(52) Ac-L-Pro-L-methyl acetamide	1.337	2.7	−9	−4.4	4.6	Matsuzaki and Iitaka (1971)
(53) N-Ac-L-Phe-L-Tyr	1.336	−17.7	11	5.9	5.1	Stenkamp and Jensen (1973)
(54) Caprylolactam HCl (mol. B)	1.302	−9.6	7	12.6	5.6	Winkler and Dunitz (1975d)
(55) Caprylolactam HCl (mol. A)	1.299	−9.1	−2	−10.1	8.1	Winkler and Dunitz (1975d)
(56) Gly-L-Leu	1.331	−11.4	24	32	8	Pattabhi et al. (1974)
(57) t-Boc-Gly-L-Pro-benz	1.344	−7.9	20	8.2	11.8	Marsh, Narasimha Murthy, and Venkatesan (1977)
(58) Gly-Gly-HCl·H_2O	1.331	−3.3	−3.4	−16.6	13.2	Koetzle et al. (1972)
(59) Gly-L-Phe	1.334	9.8	−1	13.5	14.5	Marsh et al. (1976)
(60) N-methyl dipropyl acetamide	1.320	−0.1	15	−0.7	15.7	Grand and Addad (1973)

Winkler and Dunitz (1971) and Ramachandran, Lakshminarayanan, and Kolaskar (1973).

From the data recorded in Table 16.1 there is no obvious correlation between C′—N bond length versus the twist angle $\Delta\omega$, for $\Delta\omega$ varying from about 0 to 15°. However, an increase in the C′—N bond length has been reported in the structures of tricyclic and tetracyclic spirolactams which contain highly non-planar peptide units (van der Helm, Ealick, and Washecheck 1975). These authors found a correlation of the C′—N length with $\Delta\omega$ angle which varies from 12.9 to 32.0°. An increase in $\Delta\omega$ causes an increase of the C′—N length and to a lesser degree a corresponding decrease in the C′—O length (Table 16.2). Such a variation is to be attributed to the decrease of the double bond character of the C′—N bond following decrease of the overlap between the Π orbitals in the C′ and N atoms. The overlap integral

TABLE 16.2

Correlation of C′—N and C′—O lengths ($\mathring{A}$) with the torsion angle about the peptide bond(°) in tricyclic and tetracyclic spirolactams (van der Helm *et al.* 1975)

$\Delta\omega(°)$	$d_{C'-N}$	$d_{C'-O}$
12.9	1.351	1.220
17.7	1.339	1.239
31.0	1.361	1.216
31.5	1.362	1.223
32.0	1.372	1.220

is related to the energy of stabilization and owing to the overlap of the Π orbitals being a function of $\cos\theta$ (Roberts 1962), where θ is the angle between the Π orbitals on C′ and N atoms, we do not observe any significant change in C′—N bond for values of $\Delta\omega$ less than about 15°.

Among the 14 diketopiperazine (DKP) molecules investigated using X-ray crystallography, viz. DKP (Degeilh and Marsh 1959), C(gly-L-tyr) (Chi-Fan Lin and Webb 1973), C(gly-L-try) (Morris, Geddes, and Sheldrick 1974), C(L-pro-gly) (Von Dreele, personal communication), C(3,4-dehydroproline anhydride) (Karle, Ottenhyem, and Witkop 1974), C(D-ala-L-ala) (Sletten 1970), C(L-ala-L-ala) (Sletten 1970), C(L-ser-L-tyr) (Chi-Fan Lin and Webb 1973), C(L-pro-L-leu) (Karle 1972), C(L-pro-L-pro) (Benedetti, Goodman, Marsh, Rapoport, and Musich 1976), C(L-thr-L-his) $2H_2O$ (Cotrait, Ptak, Bussetta, and Heitz 1976), C(L-pro-D-phe) (Ramani, Venkatesan, Marsh, and Hu Kung 1976), C(L-his-L-asp)$3H_2O$ (Ramani, Venkatesan, and Marsh 1978) and C(L-val-D-val) (Benedetti, Goodman, Easter, and Marsh, unpublished results), $\Delta\omega$ range from 0.4 to 8.1° with no correlation between

C′—N length and $\Delta\omega$ being observed. There is another factor which may complicate any attempts at finding the correlation between $\Delta\omega$ and C′—N. It has been shown quantitatively (Ramani and Venkatesan 1973) that inter-molecular interactions such as hydrogen bonding can significantly affect in-tramolecular bond lengths, within the carbonyl group. In fact in the structure of C(L-his-L-asp)3H$_2$O (Ramani *et al.* 1978) it has been observed that the carbonyl bond of one peptide group with the oxygen involved in a stronger intermolecular hydrogen bond is longer [1.245(4) Å] than the other carbonyl bond [1.226(4) Å] of the second peptide group accepting a weaker hydrogen bond. Further the two C′—N bonds exhibit the expected differences, namely the C′—N bond being shorter when the C′—O is longer [C′—N = 1.320(4) Å, C′—O = 1.245(4) Å; C′—N = 1.336(4) Å, C′—O = 1.226(4) Å].

Ramachandran and Sasisekharan (1968) have obtained the dimensions of the *cis* peptide unit from those of the *trans* peptide unit (Corey and Pauling 1953) by retaining the bond lengths of the *trans* peptide unit and by allowing for an increase of the angle N—C′—C$_1^\alpha$ and C′—N—C$_2^\alpha$ to take into account the non-bonded repulsion between the two C$^\alpha$ atoms. An increase of 4° in the N—C′—C$_1^\alpha$ angle and 3° in C′—N—C$_2^\alpha$ angle is, according to these authors, a good compromise between non-bonded repulsion and bond-angle strain. The most recent values for the *cis* peptide arrived at by Benedetti (1977) on the basis of crystallographic measurements on cyclic dipeptides are in good agreement with those of Ramachandran and Sasisekharan (1968) except the C$_1^\alpha$—C′ bond length. Table 16.3 shows the bond lengths and bond angles of the *cis* amide group when it forms part of four-, five-, six-, and seven-membered rings. The angles at C′ and N within the ring show the expected pattern in their variation as we go from one ring size to another. In the C′—N and C′—O bonds there are small perturbations which could be partly due to the effect of intermolecular hydrogen bonding involving the oxygen of the carbonyl group. It is interesting to note that the bond length distributions of the *cis* amide group in phenoxymethyl Δ^2-desacetoxyl cepha-losporin (Sweet and Dahl 1970) agree with the average values recorded by Benedetti (1977). But in cephaloridine (Sweet and Dahl 1970) the C′—N bond is significantly increased [1.382(8) Å] and in this molecule the nitrogen atom is observed to be highly pyramidal ($\Delta\omega = 9.9°$; $\theta_N = -33.1°$), the deviation of this atom from the plane through the three surrounding atoms being as high as 0.24 Å, which is far more than in any of the other molecules containing the amide group.

There has been considerable theoretical and experimental work carried out to learn more about the energetics of deformation of the amide unit. The poten-tial energy surface describing the deformation of the amide group based on structural and spectroscopic evidence has been derived by Dunitz and Winkler (1975*a–d*). CNDO/2 calculations on *N*-methyl acetamide (Rama-chandran *et al.* 1973), INDO calculations (Kolaskar, Lakshminarayanan,

TABLE 16.3

The bond lengths (Å) and angles (°) in the cis peptide unit when it forms a part of rings of different sizes. E.S.D. in $\sigma(l)$ ranges from 0.002 to 0.008 Å; $\sigma(\theta)$ from 0.2 to 0.6°

Molecule	Number of atoms in the ring	$C'-N$	$C'-O$	$N-C_2^\alpha$	$C'-C_1^\alpha$	$C'-N-C_2^\alpha$	$N-C'-C_1^\alpha$	$N-C'-O$	$C_1^\alpha-C'-O$	Reference
Cephaloridine hydrochloride Monohydrate	4	1.382	1.214	1.463	1.499	94.0	92.4	131.1	136.3	Sweet and Dahl (1970)
Cephalosporin	4	1.339	1.223	1.453	1.536	95.6	92.2	133.2	134.5	Sweet and Dahl (1970)
Pyroglutamic acid	5	1.354	1.258	1.472	1.532	113.6	109.4	124.6	126.0	Pattabhi and Venkatesan (1974)
4-azatricyclo[4,4,0,0^{3.8}]-decan-5-one	6	1.341	1.237	1.469	1.512	115.3	110.4	124.5	125.0	Ramakumar et al. (1977)
DKP	6	1.325	1.24	1.449	1.499	126.0	118.9	122.6	118.5	Degeilh and Marsh (1959)
Caprolactam	7	1.327	1.242	1.470	1.501	125.5	118.5	120.9	120.6	Winkler and Dunitz (1975)
Average dimensions cis peptide	—	1.32	1.24	1.47	1.53	126.0	118.0	123.0	119.0	Ramachandran and Sasisekharan (1968)
Average dimensions cis peptide	—	1.329	1.233	1.457	1.508	126.5	118.4	123.0	118.6	Benedetti (1977)

Sarathy, and Sasisekharan, 1975), semi-empirical molecular orbital and *ab initio* study of *N*-methylacetamide and *N*-acetyl-L-alanine *N*-methylamide (Renugopalakrishnan and Rein 1976), and *ab initio* study of aliphatic amides (Christensen, Kortezeborn, Bak, and Led 1970; Perricaudet and Pullman 1973; Armbruster and Pullman 1974) are some of the series of theoretical operations carried out. According to the CNDO/2 calculations (Ramachandran *et al.* 1973) the minimum energy conformation for the amide is

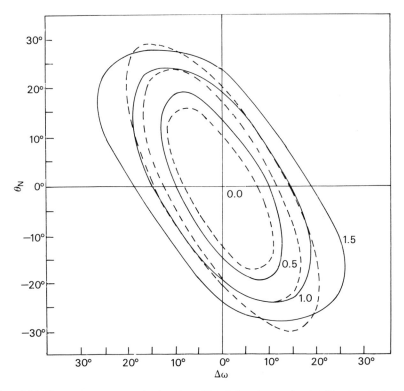

FIG. 16.1. Isoenergy contours in the plane $(\Delta\omega, \theta_N)$ at intervals of $0.5\,\text{kcal mol}^{-1}$. The energy is $0.5\,\text{kcal mol}^{-1}$ for the planar conformation. -----Potential based on structural and spectroscopic data. —— Potential from INDO-method (*N*-methyl acetamide).

the non-planar form with $\Delta\omega$ between 10 and $15°$ and θ_N between -22 and $-25°$. According to the INDO calculations (Kolaskar *et al.* 1975) the minimum energy conformation corresponds to the planar form with $\Delta\omega = \theta_N = 0°$. However, distortions of the order of $10°$ in $\Delta\omega$ require an energy expenditure of the order of $0.5\,\text{kcal mol}^{-1}$ only (Fig. 16.1). Based on low frequency Raman spectra of crystallized *N*-methylacetamide, Fillaux, Baron De Loze, and Sagon (personal communication) have proposed a potential function for amide deformation which involves two minima ($\Delta\omega = \pm 10°$) separated by a small barrier of $0.6\,\text{kcal}\,\text{mol}^{-1}$ indicating a non-planar molecular skeleton. Hence the non-planar peptide unit may be expected to occur frequently in small molecules as well as in proteins.

We thought it would be instructive to compare the potentials, derived from different approaches, by Winkler and Dunitz (1971) and by Kolaskar *et al.* (1975). It may be seen (Fig. 16.2) that the agreement between the results is remarkably good, especially in view of the fact that both the calculations are subject to assumptions and approximations. The Winkler–Dunitz potential corresponding to the amide in the solid state is found to be harder than that due to Kolaskar *et al.* (1975). This is to be expected, since according

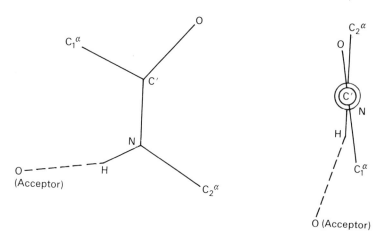

FIG. 16.2. Hydrogen bonding environment in an amide crystal. θ_{NA} is the angle of rotation (positive when clockwise looking from C′ towards N) from the plane C′, N, C_2^{α} to C′, N, acceptor atom. θ_{N} is the angle of rotation (positive when clockwise looking from C′ towards N) from the plane C′, N, C_2^{α} to C′, N, H.

to Winkler and Dunitz (1975) the carbonyl oxygen is effectively protonated in the solid phase to an extent of 15–20 per cent. The protonation implies a weakening of the C′—O bond and a corresponding strengthening of the C′—N bond, making it more difficult to introduce any distortions about the amide bond.

Even though the planar configuration has the minimum energy, the peptide unit in the solid state is found to be non-planar with the bonds attached to the nitrogen atom being highly susceptible to out-of-plane distortions. We shall now investigate the possible reasons for such non-planar distortions; in particular to find out whether the non-planar deviations are purely statistical in nature or intermolecular interactions such as hydrogen bonding and the intramolecular interactions between the side group and the peptide unit could influence the peptide geometry.

Influence of the side-chains on the conformation of the amide group

To take up the question of backbone influence first, we notice that the observed $\Delta\omega$ values are small (about 4°) in the various derivatives of the dipeptide gly—gly: gly—gly 3.6° (Freeman, Paul, and Sabine 1970), gly—gly PO_4 4.5° (Freeman, Hearn, and Bugg 1972), gly—gly HCl $-3.2°$ (Koetzle, Hamilton, and Parthasarathy 1972). It is interesting to note that in dipeptides with one or both bulky side-chains the observed $\Delta\omega$ values are larger than in gly—gly: gly-L-leu $-11.4°$ (Pattabhi, Venkatesan, and Hall 1974); L-ala-gly $-6.1°$ (Koch and Germain 1970); gly-L-ala HCl $-10.7°$ (Naganathan and Venkatesan 1972); gly-DL-phe 9.8° (Marsh, Ramakumar, and Venkatesan 1976); *N*-acetyl glutamine 8.2° (Narasimha Murthy, Venkatesan, and Winkler 1974); and *N*-acetyl L-phe-L-tyr 17.7° (Stenkamp and Jensen 1973). There are large variations in polyamino acids. For example, in the structures of tosyl (α-aminoisobutyric acid) $_5$OMe (Shamala, Nagaraj, and Balaram 1978) and (L-ala)$_3$ (Fawcett, Camerman, and Camerman 1975) there are large variations in $\Delta\omega$ within each molecule. In the former, $\Delta\omega$ varies from -0.9 to $-8.3°$ whereas in the latter the variation is from -0.8 to $-7.0°$. Any correlation between $\Delta\omega$ and (φ, ψ) is not easily discernible.

The question whether a bulky group at the C-terminal has a larger influence on the amide deformation or the one at the N-terminal deserves consideration. From model building and assuming folded conformation for the side chain in linear peptides, it seems that when the bulky group is at the C-terminal, the side-chain atoms are nearer to the N atom, which is more susceptible to distortion, whereas when the substitution is at the N-terminal, the side group is closer to the C′ atom, which is much less susceptible to distortion. However, the quantitative aspects of the side-chain-backbone interaction need to be further investigated. The influence of bulky side group on the deformation of the amide group is more easily recognized in cyclic dipeptides with different side-groups. From the nuclear magnetic resonance studies of cyclic dipeptides containing aromatic side-chains (Kopple and Marr 1967) it was first concluded that the aromatic ring folds over the DKP ring. The folded conformation has been confirmed in the solid state by X-ray diffraction studies of C(gly-L-tyr) (Lin and Webb 1973), C(L-ser-L-tyr) (Lin and Webb 1973), C(L-pro-D-phe) (Ramani *et al.* 1976), C(L-his-L-asp)3H$_2$O (Ramani *et al.* 1978), and in the N-methylated derivatives of C(L-phe-L-phe) and C(L-phe-D-phe) (Benedetti, Marsh, and Goodman 1976). The conformational angles which define the amide deformation in the cyclic dipeptides with bulky aromatic side-groups are:

$$\omega_1 = -4.0°; \omega_2 = -7.0° \quad \text{C(gly-L-tyr)}$$
$$\omega_1 = 5.0°; \omega_2 = 6.0° \quad \text{C(L-ser-L-tyr)}$$
$$\omega_1 = -5.6°; \omega_2 = -4.7° \quad \text{C(L-asp-L-his)}$$

In C(NMe-L-phe)$_2$ one of the phenyl groups is folded over the DKP ring, whereas in C(NMe-L-phe-D-phe) both the rings fold over the DKP ring. It is significant that in the first case $\omega_1 = -7°$ and $\omega_2 = -12°$, whereas in the second case $\omega_1 = 14°$ and $\omega_2 = -14°$. In C(gly-L-trp) in which the aromatic side-chain does not fold over the DKP ring, the ω_1 and ω_2 are 2.7° and $-2.7°$ respectively. Thus from the results recorded above, the influence of the bulky side-chain on the peptide bond distortion is established convincingly.

From the limited number of examples discussed above it appears that $\Delta\omega$ distortion of the order of 5–10° must be introduced when model building of polypeptides involving bulky side-chains is performed in situations where the side-chain folds over the backbone.

Possible effects of intermolecular interactions on the conformation of the amide group

According to CNDO/2 calculations the θ_N is such that a relation of the form $\theta_N = (-1.5$ to $-2.0)\Delta\omega$ is valid whereas the INDO calculations suggest that for a given $\Delta\omega$ the peptide conformation with $\theta_N = -\Delta\omega$ leads to the minimum energy. Our aim is to investigate whether the experimentally observed deviations from a configuration dictated by the relation $\Delta\omega = -\theta_N$ could be due to intermolecular hydrogen bonding. That a hydrogen bond can affect molecular geometry is recognized and in support of this the following observations may be quoted:

(i) The O—H bond length in the O—H---O bond is found to increase from 0.97 Å to 1.07 Å as the O...O distance decreases from 2.85 Å to 2.50 Å (Hamilton and Ibers 1968).

(ii) The three H—N—H bond angles at the NH$_3^+$ group in the structure of glycyl-DL- phenylalanine (Marsh *et al.* 1976) are in a direction so as to improve the linearity of the N–H---O hydrogen bonds.

(iii) It has been found (Ramani and Venkatesan 1973) that the C—O lengths in amino acids and peptides are correlated with the strength of the hydrogen bond involving the oxygen atoms.

(iv) The hydrogens of the NH$_3^+$ group in the structure of glycyl-L-leucine (Pattabhi *et al.* 1974) are observed to be in an eclipsed conformation so that linear N—H----O hydrogen bonds is achieved in the crystals.

It is well known that a hydrogen bond in which the donor—H----acceptor angle is close to 180° is energetically more favourable compared to a bent bond. We define an angle θ_{NA} with reference to the acceptor atom, as the angle of rotation (positive when clockwise looking from C' towards N) from the plane C', N, C$_2^\gamma$ to C', N, acceptor atom. The definition of the angle θ_{NA} is illustrated in Fig. 16.2 depicting the hydrogen bonding environment in an amide crystal. It can be shown that with the constraint that the N—H length and the angle C'—N—H are invariant the most linear hydrogen bonding is realized when the angle θ_N is equal to θ_{NA}. If the proton bonded to the amide nitrogen moves in the direction of the acceptor to improve linearity of the hydrogen bond, then θ would be distributed about θ_{NA}, i.e. $\Delta\theta_{NA} = \theta_{NA} - \theta_N$ will be distributed around $0°$. A survey of sixty accurately solved crystal structures was made to examine this possibility. Both the *cis* and *trans* amides were included in the analysis and the acceptor atom is either oxygen or chlorine. Results based on neutron diffraction studies would have been most valuable. However, so far neutron measurements have been carried out for only three peptides and these are used in the analysis. The accuracy in

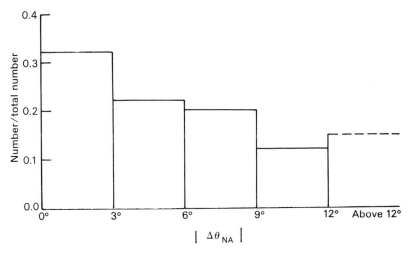

FIG. 16.3 Histogram depicting the distribution of $|\Delta\theta_{NA}|$.

θ_N and in the quantities derived from it is expected to be about $3°$ for X-ray diffraction and about $0.6°$ for neutron diffraction results. The accuracy associated with θ_{NA} is expected to be better than $0.6°$. The results are recorded in Table 16.1. A histogram depicting the distribution of $\Delta\theta_{NA}$ is shown in Fig. 16.3. It is observed that $\Delta\theta_{NA}$ is distributed around $0°$, the distribution being approximately Gaussian. The fact that the distribution is not very sharp could be due to that the energy required to make the hydrogen bond non-linear is only of the order of 0.3 kcal mol^{-1}. The effect of the hydrogen bond on the electronic structure of the peptide group is not considered. The extent to which the experimentally determined hydrogen position is favourable for the formation of a hydrogen bond is judged by a single parameter, θ_{NA}. Any improvement in the hydrogen bond caused by bond angle distortion at the nitrogen could not be introduced as the analysis is based on X-ray measurements. In spite of the above limitations it seems fairly clear that the hydrogen atom position is strongly correlated with the position of the acceptor atom. The correlation coefficient as defined by Topping (1961) between θ_N and θ_{NA} is 0.73. The conclusion arrived at depends critically on the accuracy with which a hydrogen atom of the peptide group is located. Thus single crystal neutron diffraction studies of a large number of peptides preferably at low temperatures would be fruitful. Another result of a subtle nature that accrues from the crystallographic studies of amides may be pointed out. From Table 16.1 it is observed that $\Delta\omega$ can take both positive and negative values. In this connection the N-methylated amides (NMAs) would provide far more convincing evidence, and our analysis of some of these compounds reported in the literature again show that $\Delta\omega$ can be both positive and negative. Looking at the results obtained from the X-ray analysis of different amides in its entirety one may conclude that inversion at the amide nitrogen could take place in the solution state. Indeed Fillaux and De Loze (1976) have concluded from the infrared studies of N-methylacetamide that the NMA is pyramidal and inverts through a tunnel effect. Crystal structure data (Table 16.1) show there are more cases with negative $\Delta\omega$ than positive $\Delta\omega$.

Some features of the packing model of amides in crystals

We wish to make a brief mention of our observations on the packing of amides in a few crystals solved in our laboratory. In the crystal structure of glycyl-DL-phenylalanine (Marsh *et al.* 1976) the molecules are packed such that two molecules of the same configuration are paired and the configuration of such pairs alternates along rows and columns (Fig. 16.4).

DD	LL	DD
LL	DD	LL
DD	LL	DD
LL	DD	LL

In this crystal the hydrogen bonding between the molecules of the same configuration is much stronger than those between the molecules of opposite configuration. Three N—H----O bonds are formed between molecules of the same configuration with N----O (2.833), H----O (2.05); N----O (2.798), H----O (1.94) and N----O (2.986 Å), H----O (2.22 Å). Two hydrogen bonds between molecules of opposite configuration are of length N----O (3.135), H----O (2.30), and N----O (2.827 Å), H----O (2.28 Å) (weak bifurcated hydrogen bond). This intermolecular feature in gly-DL-phe does not seem to occur in all the peptide crystals containing both D- and L-isomers. However, it seems worth examining whether this observation has implications with the organization of biological molecules which consists exclusively of one type of (L-type) amino acids.

Some characteristics of packing of molecules with *cis* amide group are

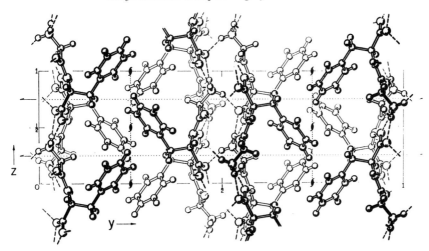

FIG. 16.4. Packing of the gly-DL-phe molecules looking down *c*-axis.

worth recording. In the crystals of 4-azatricyclo[4,4,0,0^{3,8}]decan-5-one (Ramakumar, Venkatesan, and Weber 1977) which belongs to the space group $P2_1/n$, the molecules organize themselves to form centrosymmetrically related hydrogen-bonded dimers—a feature observed in some molecules containing a *cis* group crystallizing in a centrosymmetric space group, e.g. saccharine (Okaya 1969) and caprolactam (Winkler and Dunitz 1975). In the crystal structure of 4-azatricyclo[4,3,1,0^{3,7}]decan-5-one which is an isomer of the one discussed above and which crystallizes in the space group $P2_1/c$, the twofold screw related molecules are hydrogen bonded (Blaha *et al.* 1978). A similar situation exists in the crystals of C(L-pro-D-phe) which crystallize in the space group $P2_1$ (Ramani *et al.* 1976), glutarimide ($P2_1/c$) (Peterson 1971), and C(L-his-L-asp)3H$_2$0 ($P2_12_12_1$) (Ramani *et al.* 1978). The repeat distances along the direction of the chain in the above-mentioned crystals are close to each other, varying from 6.143 to 7.416 Å.

Acknowledgements

We wish to record our thanks to Dr M. R. Narasimhamurthy and Miss R. Usha for useful discussions and help in the preparation of this article. We thank the Council of Scientific and Industrial Research of India for financial support.

References

ARMBRUSTER, A. M. and PULLMAN, A. (1974). *FEBS Lett.* **49**, 18.
BENEDETTI, E. (1977). In *Peptides. Proc. 5th Am. Peptide Symposium* (ed. M. Goodman and J. Meienhofer) p. 257. John Wiley, New York.

——MARSH, R. E., and GOODMAN, M. (1976). *J. Am. chem. Soc.* **98**, 6676.

——GOODMAN, M., MARSH, R. E., RAPOPORT, H., and MUSICH, J. A. (1976). *Cryst. Struct. Comm.* **4**, 641.

BLÁHA, K., MALOŇ, P., TICHÝ, M., FRIČ, I., USHA, R., RAMAKUMAR, S. and VENKATE-SAN, K. (1978). *Czech. Chem. Comm.* **43**, 3241.

BROWN, D. S. and RUSSELL, P. R. (1976). *Acta crystallogr* **B32**, 1056.

CHI-FAN LIN and WEBB, L. E. (1973). *J. Am. chem. Soc.* **95**, 6803.

CHRISTENSEN, D. H., KORTZEBORN, R. N., BAK, B., and LED, J. J. (1970). *J. chem. Phys.* **53**, 3912.

COREY, R. B. and PAULING, L. (1953). *Proc. R. Soc.* **B141**, 10.

COTRAIT, M., PTAK, M., BUSETTA, B., and HEITZ, A. (1976). *J. Am. chem. Soc.* **98**, 1073.

CROWFOOT, D., BUNN, C. W., ROGERS-LOW, B. W., and TURNER-JONES, A. (1949). In *The chemistry of penicillin* (ed. H. T. Clarke, J. R. Johnson, and R. Robinson) p. 440. Princeton University Press.

DEGEILH, R. and MARSH, R. E. (1959). *Acta crystallogr.* **12**, 1007.

DUNITZ, J. D. and WINKLER, F. K. (1975). *Acta crystallogr.* **B31**, 251.

FAWCETT, J. K., CAMERMAN, N., and CAMERMAN, A. (1975). *Acta crystallogr.* **B31**, 658.

FILLAUX, F. and DE LOZE, C. (1976). *J. chim. Phys.* **73**, 1010.

FLETTERICK, R. J., TSAI, C. C., and HUGHES, E. R. (1971). *J. phys. Chem., Ithaca* **75**, 918.

FORM, G. R., RAPER, E. S., and DOWNIE, T. C. (1975). *Acta crystallogr.* **B31**, 2181.

FREEMAN, G. R., HEARN, R. A., and BUGG, C. E. (1972). *Acta crystallogr.* **B28**, 2906.

FREEMAN, H. C., PAUL, G. L., and SABINE, T. M. (1970). *Acta crystallogr.* **B26**, 925.

GARTLAND, G. L. and CRAVEN, B. M. (1974). *Acta crystallogr.* **B30**, 980.

GRAND, A. and ADDAD, C. C. (1973). *Acta crystallogr.* **B29**, 1149.

GREEN, E. A., ROSENSTEIN, R. D., SHIONO, R., and ABRAHAM, D. J. (1975). *Acta crystallogr.* **B31**, 102.

HAMILTON, W. C. and IBERS, J. A. (1968). In *Hydrogen bonding in solids*, p. 16. W. A. Benjamin, New York.

VAN DER HELM, D., EALICK, S. E., and WASHECHECK, D. (1975). *Acta crystallogr.* **B31**, 104

HUEY-SHENG SHIEH and VOET, D. (1976). *Acta crystallogr.* **B32**, 2361.

I-NAN HSU and CRAVEN, B. M. (1974a). *Acta crystallogr.* **B30**, 988.

——(1974b). *Acta crystallogr.* **B30**, 998.

——(1974c). *Acta crystallogr.* **B30**, 974.

ITOH, H., YAMANE, T., ASHIDA, T., SUGIHARA, T., IMANISHI, Y., and HIGASHIMURA, T. (1976). *Acta crystallogr.* **B32**, 3355.

KARLE, I. L. (1972). *J. Am. chem. Soc.* **94**, 81.

——OTTENHEYM, H. C. J., and WITKOP, B. (1974). *J. Am. chem. Soc.* **96**, 539.

KISTENMACHER, T. J. and SORRELL, T. (1975). *Acta crystallogr.* **B31**, 489.

KOCH, H. J. and GERMAIN, G. (1970). *Acta crystallogr.* **B26**, 410.

KOETZLE, T. F., HAMILTON, W. C., and PARTHASARATHY, R. (1972). *Acta crystallogr.* **B28**, 2083.

KOLASKAR, A. S., LAKSHMINARAYANAN, A. V., SARATHY, K. P., and SASISEKHARAN, V. (1975). *Biopolymers* **14**, 1081.

KOPPLE, K. D. and MARR, D. H. (1967). *J. Am. chem. Soc.* **89**, 6193.

DE LERMA, J. L., GARCIA BLANCO, S., and FAYOS, J. (1977). *Acta crystallogr.* **B33**, 2311.

LIN, C. F. and WEBB, L. E. (1973). *J. Am. chem. Soc.* **95**, 6803.

MACKAY, M. F. (1975). *Cryst. Struct. Comm.* **4**, 225.

MANDEL, G. S. and MARSH, R. E. (1975). *Acta crystallogr.* **B31**, 2862.

MARSH, R. E., NARASIMHA MURTHY, M. R., and VENKATESAN, K. (1977). *J. Am. chem. Soc.* **99**, 1251.

—— RAMAKUMAR, S., and VENKATESAN, K. (1976). *Acta crystallogr.* **B32,** 66.

MATSUZAKI, T. and IITAKA, Y. (1971). *Acta crystallogr.* **B27,** 507.

MO, F. and JENSEN, L. H. (1975). *Acta crystallogr.* **B31,** 2867.

MORIKAWA, K., TORII, K., IITAKA, Y., and TSUBOI, M. (1975). *Acta crystallogr.* **B31,** 1004.

MORRIS, A. J., GEDDES, A. J., and SHELDRICK, B. (1974). *Cryst. Struct. Comm.* **3,** 345.

NAGANATHAN, P. S. and VENKATESAN, K. (1972). *Acta crystallogr.* **B28,** 552.

NARASIMHA MURTHY, M. R., VENKATESAN, K., and WINKLER, F. (1974). *Cryst. Struct. Comm.* **3,** 743.

NARAYANAN, P. and BERMAN, H. M. (1977). *Acta crystallogr.* **B33,** 2047.

OKAYA, Y. (1969). *Acta crystallogr.* **B25,** 2257.

OLSON, A. J., TEMPLETON, D. H., and TEMPLETON, L. K. (1977). *Acta crystallogr.* **B33,** 2266.

PARTHASARATHY, R., OHRT, J. M., and CHHEDA, G. B. (1976). *Acta crystallogr.* **B32,** 2648.

PATTABHI, V. and VENKATESAN, K. (1974). *J. chem. Soc. Perkin II* 1085.

—— and HALL, S. R. (1974). *J. chem. Soc. Perkin II* 1722.

PERRICAUDET, M. and PULLMAN, A. (1973). *Int. J. Peptide Protein Res.* **5,** 99.

PETERSEN, C. S. (1971). *Acta chem. scand.* **25,** 379.

RAMACHANDRAN, G. N. and SASISEKHARAN, V. (1968). *Adv. Protein Chem.* **23,** 283.

——LAKSHMINARAYANAN, A. V., and KOLASKAR, A. S. (1973). *Biochim. biophys. Acta* **303,** 8.

RAMAKUMAR, S., VENKATESAN, K., and WEBER, H. P. (1977). *Helv. chim. Acta* **60,** 1691.

RAMANI, R. and VENKATESAN, K. (1973). *Indian J. Biochem. Biophys.* **10,** 297.

———— and MARSH, R. E. (1978). *J. Am. chem. Soc.* **100,** 949.

———— and HU KUNG, W. J. (1976). *Acta crystallogr.* **B32,** 1051.

RENUGOPALAKRISHNAN, V. and REIN, R. (1976). *Biochim. biophys. Acta* **434,** 164.

ROBERTS, J. D. (1962). In *Notes on molecular orbital calculations,* p. 83. W. A. Benjamin, New York.

SHAMALA, N., NAGARAJ, R., and BALARAM, P. (1978). *J. chem. Soc. Chem. Comm.* 996.

SLETTEN, E. (1970). *J. Am. chem. Soc.* **92,** 172.

STENKAMP, R. E. and JENSEN, L. H. (1973). *Acta crystallogr.* **B29,** 2872.

SWEET, R. and DAHL, L. F. (1970). *J. Am. chem. Soc.* **92,** 5489.

TIMMINS, P. A. (1975). *Acta crystallogr.* **B31,** 2240.

TOPPING, J. (1961). In *Errors of observation and their treatment,* 3rd edn. Chapman & Hall, London.

WINKLER, F. K. and DUNITZ, J. D. (1971). *J. molec. Biol.* **59,** 169.

———— (1975*a*). *Acta crystallogr.* **B31,** 273.

———— (1975*b*). *Acta crystallogr.* **B31,** 276.

———— (1975*c*). *Acta crystallogr.* **B31,** 268.

———— (1975*d*). *Acta crystallogr.* **B31,** 278.

WOODWARD, R. B. (1949). In *The chemistry of pencillin* (ed. H. T. Clarke, J. R. Johnson, and R. Robinson) p. 440. Princeton University Press.

YAMANE, T., ANDON, T., and ASHIDA, T. (1977). *Acta crystallogr.* **B33,** 1650.

—— ASHIDA, T., SHIMONISHI, K., KAKUDO, M., and SASADA, Y. (1976). *Acta crystallogr.* **B32,** 2071.

YEATES, D. G. R. and PALMER, R. A. (1975). *Acta crystallogr.* **B31,** 1077.

ZIVA RUJIC-TOROS and BISERKA KOJIC-PRODIC (1976). *Acta crystallogr.* **B32,** 2333.

17. Nucleoside-5'-diphosphates: crystal and molecular structures of adenosine-5'-diphosphoric acid, monopotassium adenosine-5'-diphosphate, and monosodium cytidine-5'-diphosphoethanolamine

M. A. VISWAMITRA, S. K. KATTI, AND M. V. HOSUR

Introduction

NUCLEOSIDE-5'-DIPHOSPHATES have an essential role in many metabolic processes and are the starting substances for the biosynthesis of a variety of activated compounds in the synthesis of biological molecules. Enzymatic reactions involving these diphosphates generally require various metal ions as cofactors for catalytic activation. The detailed geometries of the molecules in their free and metal-bound states will therefore be of interest. We report here the results of our recent X-ray crystallographic studies on adenosine-5'-diphosphoric acid (ADP), monopotassium adenosine-5'-diphosphate (ADP-K), and monosodium cytidine-5'-diphosphoethanolamine (CDP ethanolamine). The ADP–ATP system plays a vital role in the energy transfer processes of the cell and the phosphorylation of ADP to ATP by the enzyme pyruvate kinase specifically requires the presence of monovalent potassium ions along with magnesium ions. CDP derivatives are key intermediates in the metabolism of phospholipids and CDP ethanolamine is involved in the biosynthesis of cephalin. The present study is in continuation of our earlier X-ray studies of the nucleotide coenzymes monorubidium (ADP, Viswamitra, Hosur, Shakked, and Kennard 1976), tris ADP (Shakked, Viswamitra, and Kennard 1979), CDP and CDP choline (Viswamitra, Seshadri, Post and Kennard 1975), and dipotassium UDP (Viswamitra, Post, and Kennard 1979).

ADP free acid

Single crystals of ADP free acid, $C_{10}N_5O_{10}H_{14}P_2 \cdot 3H_2O$, were obtained during our attempts to crystallize ADP as its monopotassium salt. One of the crystallization tubes containing acetone layered over 1 ml of an equimolar solution of ADP free acid and potassium chloride at pH 3.5 was found to have needle-shaped crystals showing good optical extinction. These crystals were thought to be ADP-K. However, their mass spectrometric analysis showed that potassium ions were not present in the crystals. The crystals were of ADP free acid as confirmed later by the detailed X-ray study. The crystals displayed considerable mosaicity and the one that showed the least mosaic spread of X-ray spots was used for three-dimensional intensity data collection on a CAD-4 diffractometer with Cu Kα radiation to a 2θ limit of 120°. The crystal was stable to X-rays and the fluctuations in the intensities of monitor reflections were less than 5 per cent. The crystal data are: $a = 6.714(3)$, $b = 10.980(4)$, $c = 26.320(8)$ Å, $V = 1941.30$ Å^3, $Z = 4$, $D_{meas} = 1.64 \, \text{g cm}^{-3}$, $D_{calc} = 1.64 \, \text{g cm}^{-3}$ and space group $P22_12_1$.

Structure solution

A partial structure for ADP free acid was obtained through the application of MULTAN (Main, Woolfson, and Germain 1971) and then developed into the complete structure using Fourier methods. This process was slightly complicated as the ribose group was found statistically disordered between C2′-*endo* C1′–*exo* and C3′-*endo* conformations. The structure was refined to a final *R*-factor of 7.8 per cent using the block diagonal approximation. The average estimated standard deviations in the C—C (O, N) and P—O bond lengths are about 0.02 and 0.01 Å respectively; the e.s.d.s in bond angles are about 1.5° in C—C—C, C—O—C, C—C—N, C—N—C and about 0.9° for O—P—O and P—O—P angles.

Molecular geometry

The atomic numbering scheme is given in Fig. 17.1. The molecule is composed of the three basic units—heterocyclic adenine base, five-membered ribose ring, and the pyrophosphate group. The conformation of the ADP molecule is described below in terms of the geometries of these components and their relative orientations.

Pyrophosphate

The P—O bonds to unesterified oxygens in the two phosphate groups are nearly equal within experimental errors suggesting that the monoanionic charge on each phosphate is distributed between its two unesterified oxygens (P1—O11 = 1.47, P1—O12 = 1.49, P2—O21 = 1.47, P2—O22 = 1.50 Å). The average length of 1.48 Å found for these bonds agrees well with that observed

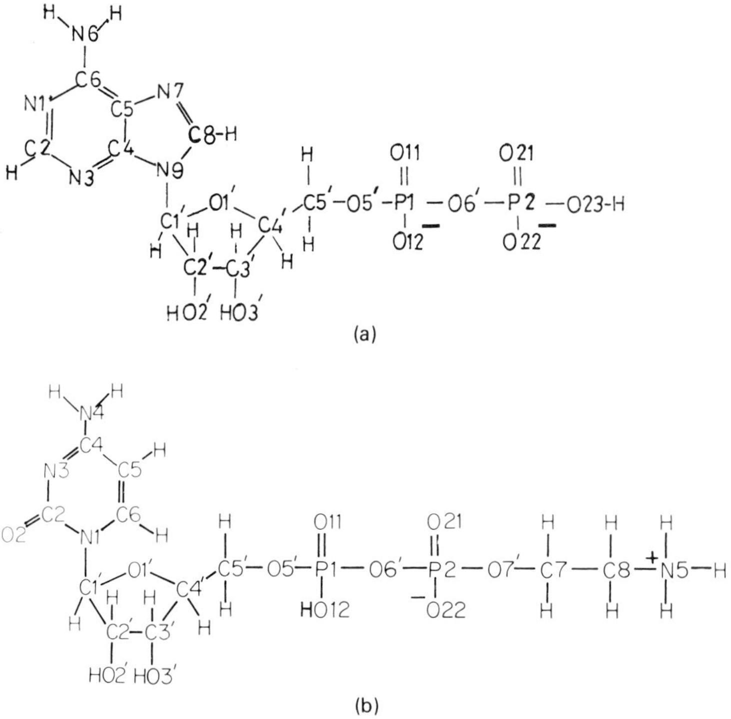

(a)

(b)

FIG. 17.1. Atomic numbering scheme in (a) ADP and (b) CDP ethanolamine.

in many other nucleotide structures. The P—O23 bond of 1.58Å is longer than other terminal P—O bonds indicating that O23 is a hydroxyl oxygen. The expected hydrogen on this oxygen atom was located from a difference Fourier map.

The similarity of the bridging bonds in the pyrophosphate group

P1—O6′ and P2—O6′ in the high-energy pyrophosphate linkage are equal (1.60 Å). This is in significant contrast to their inequality found in other nucleotide coenzymes (ADP-tris 1.585 and 1.626 Å; ADP-Rb (Shakked *et al.* 1979) 1.588 and 1.629 Å; CDP (Viswamitra *et al.* 1975) 1.583 and 1.613, CDP choline (Viswamitra *et al.* 1975) 1.600 and 1.640 Å; UDP-K$_2$ (Viswamitra *et al.* 1979) 1.615 and 1.566; NAD (Saenger, Reddy, Muhlegger, and Weimann 1977) 1.66 and 1.56Å). These P—O6′ bonds are much shorter than 1.71 Å calculated for a pure P—O single bond on the basis of Schomaker and Stevenson's (1941) rule, suggesting a partial double-bond character for these bonds probably because of the participation of the *d*-orbitals of phosphorus in bonding. The bridging angle P1—O6′—P2 is 131.6° and is close to the values found in many other nucleoside-5′-diphosphate structures.

O12—P1—O11 and O21—P2—O22 are the largest (117.2°) and O5′—P1—O6′ is the smallest (99°). Other O—P—O angles range from 103.9° to 112.4°. These O—P—O angles show a definite correlation to P—O bond lengths. The angles involving the shortest P—O bonds are the largest and that involving the longest bonds is the smallest. Such a correlation has been seen in many organic monophosphate esters (Watson and Kennard 1973).

The pyrophosphate conformation with respect to the P1—P2 vector is in between the nearly eclipsed conformation found in ADP-tris (Shakked *et al.* 1979) and the staggered geometry found in ADP-Rb (Viswamitra *et al.* 1976).

Ribose

The ribose ring is found disordered between two different conformations which have C3′-*endo* and C2′-*endo* C1′-*exo* pucker geometries. The energy barrier between different modes of sugar pucker in nucleotides is small (Wood 1973; Levitt and Waschel 1978) and there can be easy interconversions between different isomers in solution. This flexibility appears to be present in the solid state also as seen in the crystal structures of ADP free acid and ATP-Na$_2$ (Kennard *et al.* 1971) which have two different sugar conformations in the asymmetric unit.

Adenine

The purine ring is planar within experimental error, the deviations of the exocyclic amino nitrogen and glycosidic C1′ atoms being 0.03 and 0.05 Å, which are in the range normally found in nucleotide structures. Adenine is found protonated at N1 as in the monoclinic structure of AMP (Kraut and Jensen 1963). The bond lengths and angles in the base are normal.

The folded shape of ADP

The adenine base is *anti* with respect to ribose, the torsional angle about the glycosidic bond N9—C1′ being 25°. The dihedral angle between the sugar and base planes is 85.3°. The orientations about the backbone C4′—C5′ and C5′—O5′ bonds are *gauche–gauche* and *trans* respectively. The relevant dihedral angles are O5′—C5′—C4′—C3′ = 52°, O5′—C5′—C4′—O1′ = −65.9°, and P1—O5′—C5′—C4′ = 152.4°. The conformation of the α phosphate is *gauche trans*, the dihedral angles C5′—O5′—P1—O6′ and P2—O6′—P1—O5′ being −69.3 and 163.8° respectively. The molecule adopts a folded shape. Fig. 17.2 gives a view of the molecule perpendicular to the base.

Molecular interactions

Adenine forms two pairs of hydrogen bonds

The nature of hydrogen bonds found from adenine is shown in Fig. 17.3. The protonated N1 atom makes a hydrogen bond to the unesterified

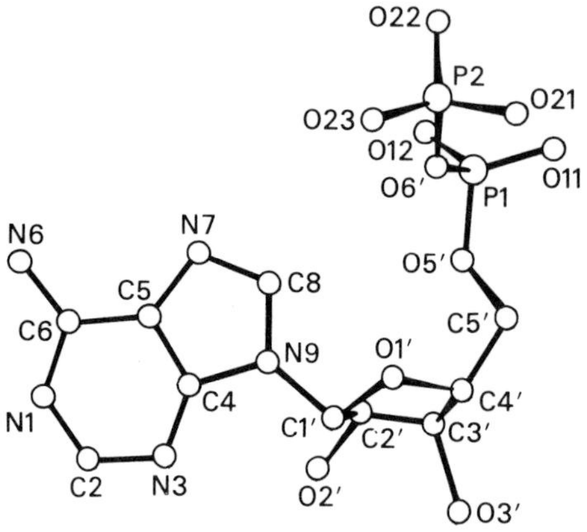

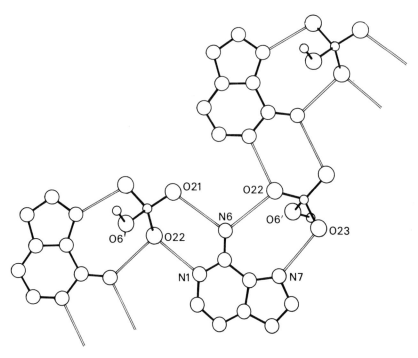

FIG. 17.2. ADP free acid. View perpendicular to base.

FIG. 17.3. ADP free acid showing pairs of hydrogen bonds from adenine base to phosphate groups.

phosphate oxygen O22 of a neighbouring molecule (N1 ... O22=2.69 Å, H1 ... O22=1.96 Å, and N1—H1 ... O22=146.5°). The other hydrogen bonds are: N6 ... O21=2.85 Å (H61 ... O21=1.91 Å, N6—H61 ... O21=176.1°); N6 ... O22=2.77 Å (H62 ... O22=1.75 Å, N6—H62 ... O22=142.1°), and N7 ... O23=2.72 Å (N7 ... H23=1.78 Å, N7 ... H23—O23=152°). Thus there are two pairs of hydrogen bonds between the adenine base and the β phosphate oxygens of neighbouring ADP molecules. The α phosphate group, however, is bound only by water molecules.

Monopotassium adenosine-5′-diphosphate

Needle-shaped crystals of ADP-K, $C_{10}H_{14}N_5O_{10}P_2K\cdot 2H_2O$, were grown by liquid diffusion of isopropyl alcohol into aqueous solutions of the sample (Boehringer Mannheim chemicals). X-ray intensity data to a 2θ limit of 100° were collected using a suitable crystal sealed inside a Lindemann capillary along with a drop of mother liquor, using crystal monochromatized Cu Kα radiation on a CAD-4 diffractometer. The space group is $P2_12_12$ with $a=$ 28.470, $b=11.449$, $c=6.325$ Å, $Z=4$, $D_{meas}=1.81$ g cm^{-3}, $D_{cal}=1.82$ g cm^{-3}.

Structure solution

The cell dimensions of ADP-K are found to be very close to those of ADP-Rb (Viswamitra *et al.* 1976) in the same space group. Hence the structure was solved using the co-ordinates of ADP-Rb crystal structure with K$^+$ replacing Rb$^+$. Block-diagonal least-squares refinement with individual isotropic temperature factors has led to a current *R*-factor of 16.6 per cent. Further refinement is in progress.

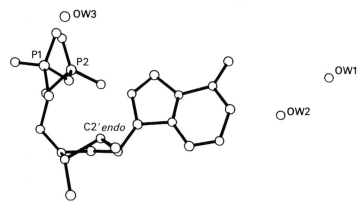

F IG. 17.4. ADP-K. View showing C2′—*endo* sugar pucker.

Molecular conformation

The·orientation of the adenine base about the glycosidic C1′—N9 linkage is *anti* (C8—N9—C1′—O1′=42.4°). The ribose ring shows C2′—*endo* puckering (Fig. 17.4). The conformation about C4′—C5′ bond is *gauche–gauche* (C3′—C4′—C5′—O5′=66.5°, O1′—C4′—C5′—O5′= −67.4°) and that about C5′—O5′ bond is *trans* (C4′—C5′—O5′—P1=144.6°). The pyrophosphate has a staggered geometry down the P–P vector (Fig. 17.5). The overall shape of the ADP molecule is folded and is compatible with that found in ADP free acid, ADP-tris and ADP-Rb salt forms. It may be mentioned that ADP is found to take an extended form when bound to either lactate dehydrogenase (Chandrasekhar, McPherson, Adams, and Rossmann 1973) or phosphoglycerate kinase (Bryant, Watson, and Wendell 1974; Blake and Evans 1974).

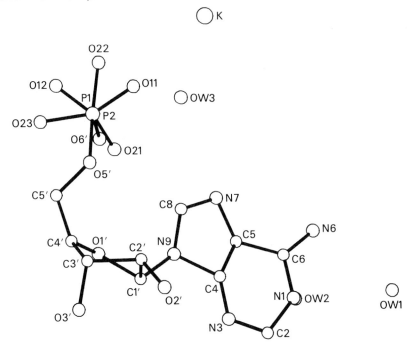

FIG. 17.5. ADP-K. View down P1—P2 vector showing the staggered pyrophosphate geometry.

Molecular interactions

Adenine forms pairs of hydrogen bonds as in ADP free acid

There are two pairs of hydrogen bonds linking the adenine base to neighbouring β phosphate groups as found in the ADP free acid structure. N6---O22=2.77, N6---O21=2.85, N1---O21=2.69, N7---O23=2.72 Å. The

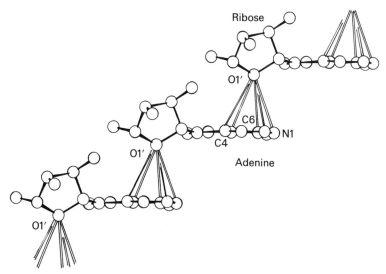

FIG. 17.6. ADP-K. View of ribose O1′ stacking on adenine base.

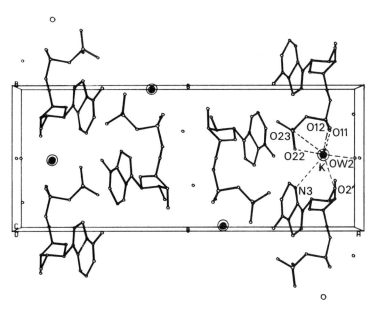

FIG. 17.7. ADP-K. Extended crystal structure showing potassium co-ordination.

formation of a pair of hydrogen bonds rather than a single bond is suggested as a basis for the recognition of base pairs by proteins (Seeman, Rosenberg, and Rich 1976). The two pairs of hydrogen bonds seen in the ADP structures could reflect possible directional interactions by which proteins can recognize the adenine and phosphate portions of the ADP molecule. The environment of ribose O1′ atom is similar to that found in ADP free acid with O1′ stacking on the adenine base (O1′—N1 = 3.04, O1′—C6 = 2.91, O1′—C5 = 3.30, O1′—C4 = 3.14, O1′—C2 = 3.29 Å) (Fig. 17.6). Such sugarbase interactions have been earlier documented and could be important in the binding of proteins to nucleic acids through aromatic amino-acid residues (Bugg, Thomas, and Sundaralingam 1971).

The potassium ion is in contact with three ADP molecules and is co-ordinated to seven nearest neighbours with distances ranging from 2.8 to 3.3 Å (Fig. 17.7).

Monosodium cytidine-5′-diphosphoethanolamine

Single crystals of the monosodium salt of CDP-ethanolamine, $C_{11}H_{19}N_4O_{11}P_2Na\cdot7H_2O$ (sample from Boehringer Mannheim), were obtained from water–acetone solutions by the liquid diffusion technique. The crystal data are $a = 6.946$, $b = 12.503$, $c = 28.264$ Å, $Z = 4$, $D_{meas} = 1.61$ g cm^{-3}, $D_{cal} = 1.61$ g cm^{-3}, space group $P2_12_12_1$. A crystal, $1.9 \times 0.15 \times 0.03$ mm^3, sealed inside a glass capillary with some mother liquor was used for Cu Kα data collection on a CAD-4 diffractometer.

Structure solution

The structure was solved by a combination of MULTAN and Fourier methods. The final R factor after least squares refinement is 10.4 per cent for 1454 observed reflections. The atom numbering scheme is given in Fig. 17.1.

Molecular conformation

Nucleoside geometry

The orientation of the cytosine base about N1—C1′ bond is *anti* (C6—N1—C1′—O1′ = 62.8°). The sugar pucker is C1′—*exo* C2′—*endo*, which differs from the C2′—*endo* and C3′—*endo* geometries observed in CDP free acid and CDP choline respectively (Viswamitra *et al.* 1975).

The molecule has typical *gauche–gauche* and *trans* conformations about the C4′—C5′ and C5′—O5′ bonds (O1′—C4′—C5′—O5′ = −66.3°, C3′—C4′—C5′—O5′ = 55.6°, C4′—C5′—O5′—P1 = 175.7°).

Pyrophosphate

The pyrophosphate has the characteristic staggered conformation about the P1—P2 vector (Fig. 17.8). The bond lengths from the bridging oxygen O6′

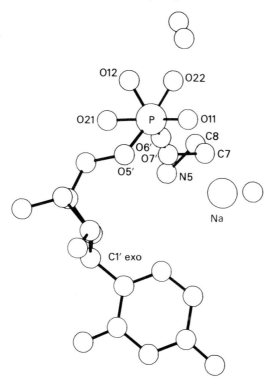

FIG. 17.8. CDP-ethanolamine. View of the molecule down P1—P2 vector, showing staggered geometry.

are P1—O6′ = 1.59 and P2—O6′ = 1.62 Å. Most of the torsional angles are similar to those of CDP choline. However, the orientations of the phosphorylcholine and the phosphoryl ethanolamine moieties are significantly different in the two structures. The three relevant dihedral angles C7—O7′—P2—O6′, P2—O7′—C7—C8, and O7′—C7—C8—N5 are −102.7, −101.7, and −54.3° in CDP ethanolamine and 71.3, −166.4, and 68.8° in CDP choline. These conformational angles lead to a comparatively extended shape for CDP ethanolamine with the hydrophilic N5⁺ ammonium group pointing away from the hydrophobic cytosine base which is shown in Fig. 17.9, a view perpendicular to the sugar plane.

Molecular interactions

The intramolecular nonbonded N5—O7′ distance is 2.74 Å as a result of the *gauche* conformation about the C7—C8 bond (O7′—C7—C8—N5 = −54.3°). All the three protons on N5 are engaged in hydrogen bonds to water molecules. The sodium ion has five ligands (Fig. 17.10) provided by the α

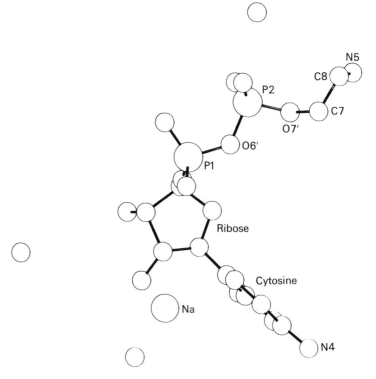

FIG. 17.9. CDP-ethanolamine. View perpendicular to ribose ring.

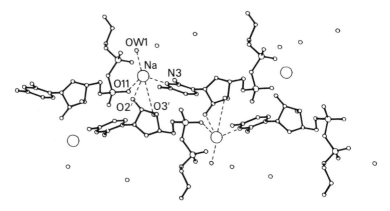

FIG. 17.10 CDP-ethanolamine. Extended crystal structure showing Na⁺ binding.

phosphate oxygen O11, the imidine nitrogen N3, the ribose hydroxyls O2′ and O3′ and a water molecule (Na—O11 = 2.22, Na—N3 = 2.45, Na—O2′ = 2.39, Na—O3′ = 2.38, and Na—OW1 = 2.3 Å). The crystal structure is so built that the CMP—5′ portions of the CDP ethanolamine molecules are tightly bound by metal ions while the phosphorylethanolamine part is only loosely held by water molecules. This mode of interaction is interestingly also found in CDP choline and could be significant to the group transfer functions of these coenzymes.

Acknowledgements

We thank the DAE and DST for financial support.

References

BLAKE, C. C. F. and EVANS, P. R. (1974). *J. molec. Biol.* **84**, 585.

BRYANT, T. N., WATSON, H. C., and WENDELL, P. L. (1974). *Nature, Lond.* **247**, 14.

BUGG, C. E., THOMAS, J. M., and SUNDARALINGAM, M. (1971). *Biopolymers* **10**, 175.

CHANDRASEKHAR, K., MCPHERSON, A., ADAMS, M. J., and ROSSMANN, M. G. (1973). *J. molec. Biol.* **76**, 503.

KENNARD, O., ISAACS, N. W., MOTHERWELL, W. D. S., COPPOLA, J. C., WAMPLER, D. L., LARSON, A. C., and WATSON, D. G. (1971). *Proc. R. Soc.* **A325**, 401.

KRAUT, J. and JENSEN, L. H. (1963). *Acta crystallogr.* **17**, 672.

LEVITT, M. L. and WARSHEL, A. (1978). *J. Am. chem. Soc.* **100**, 2607.

MAIN, P., WOOLFSON, M. M., and GERMAIN, G. (1971). 'MULTAN', a multisolution tangent formula refinement programme. University of York.

SAENGER, W., REDDY, B. S., MUHLEGGER, K., and WEIMANN, G. (1977). *Nature, Lond.* **267**, 225.

SCHOMAKER, V. and STEVENSON, D. P. (1941). *J. Am. chem. Soc.* **63**, 37.

SEEMAN, N. C., ROSENBERG, J. M., and RICH, A. (1976). *Proc. natn. Acad. Sci. U.S.A.* **73**, 804.

SHAKKED, Z., VISWAMITRA, M. A., and KENNARD, O. (1980). *Biochemistry* (in press).

VISWAMITRA, M. A., POST, M. L., and KENNARD, O. (1979). *Acta crystallogr.* **B35**, 1089.

—— HOSUR, M. V., SHAKKED, Z., and KENNARD, O. (1976). *Nature, Lond.* **262**, 234.

—— SESHADRI, T. P., POST, M. L., and KENNARD, O. (1975). *Nature, Lond.* **258**, 497.

WATSŎN, D. G. and KENNARD, O. (1973). *Acta crystallogr.* **B29**, 2358.

WOOD, D. J. (1973). *FEBS Lett.* **34**, 323.

18. Crystallography responds to pharmacological questions

ELI SHEFTER

THE experiments of P. Ehrlich and J. N. Langley at the turn of this century demonstrated that a high degree of selectivity and specificity are associated with the action of drugs. It was observed that slight chemical modifications of a drug could have a dramatic effect on its activity. This led to the concept of receptors; the idea being that a drug interacts in a very characteristic manner with certain tissue elements, termed receptors, to induce or inhibit a series of biochemical events which result in a pharmacological response. The response observed is, therefore, intimately associated with the molecular structure of the drug.

One of the major goals of pharmacological research is the delineation of the structural requirements for a particular action. With information on the relationship between structure and activity, the synthetic chemist is able to tailor molecules having optimum pharmacological behaviour. The types of structural data required for exploring these relationships is three-dimensional in nature, i.e. stereochemical, as well as electronic. Since crystallographic procedures are able to provide such data, it is not surprising to find crystallographers making significant contributions to the understanding of drug activity.

The measurements made by Bernal on sterols in the early thirties (Bernal 1932a, b) represent the first significant contribution of X-ray diffraction to the understanding of biological molecules. At the time of his crystallographic experiments the skeletal formula for steroids was in doubt. From the unit-cell dimensions and the crystal optics of various steroids, he approximated the overall dimensions of the steroid nucleus. His findings indicated that the Windaus–Wieland sterol formula [1] was incorrect, but his measurements were consistent with the structure proposed by Rosenheim and King [2]. Though this work did not in itself have a profound influence on the understanding of the biological function of steroid molecules, it did contribute, so that a combination of chemists, biochemists, and pharmacologists were able to continue their efforts in discovering the function of these molecules.

[1]

[2]

The development of the methodology to solve crystal structures parallels the efforts of crystallographers to join forces with other scientists studying drug activity. The early work in this regard was to elucidate the three-dimensional structure of pharmacologically active molecules whose structures were unknown. Pharmacologists screened materials and crystallographers along with chemists provided the basic ground-work for further studies, the molecular structure. A classical example of this is provided by the work on the antibiotic penicillin. Crystallographic studies commenced almost as soon as the antibiotic was isolated. A major difficulty faced by those working on the elucidation of the structure was to prove which of two alternative structures was correct, a thiazolidine-oxazolone or a beta-lactam ring system. The X-ray efforts showed the beta-lactam system to be correct (Crowfoot, Bunn, Rogers-Low, and Turner-Jones 1949).

There are many other instances where a substance is screened for pharmacological activity before its structure has been determined. The literature is replete with examples where X-ray diffraction is the principal technique used to determine the molecular structure. Probably the earliest of these is strychnine, where the crystal structure analysis was initiated before the chemical structure was established (Bokhoven, Schoone, and Bijvoet 1951; Robertson and Beevers 1951).

The search for drugs from the most exotic sources continues. With modern X-ray techniques capable of establishing the structure and stereochemistry of a molecule in a relatively short time, crystallography will be increasingly called on to establish their molecular frameworks.

When drugs act as metabolic antagonists, their three-dimensional similarity with the agonist can be exceedingly important. The competition for the same receptor site or active site of an enzyme might require a certain degree of structural superimposability. In the case of the antibacterial activity of sulphonamides, Wood (1940) proposed that the normal utilization of para-amino benzioc acid by bacteria is prevented by sulphonamides through competitive antagonism. Crystallographic studies on para-aminobenzoic acid [3] (Lai and Marsh 1967) and sulphanilamide [4] (O'Connell and Maslen 1967) provide structural verification of the high degree of similarity between these molecules.

One of the principal approaches used to explore structure–activity relationships has been by synthesizing and testing numerous analogues of a known pharmacologically active molecule. In recent years crystallographers have joined forces with the synthetic chemist and pharmacologist in determining the three-dimensional atomic pattern of these molecules. Through such combined efforts, the relationships developed have greater meaning in terms of understanding the active site of the receptor surface.

Many pharmacologically active molecules have a certain degree of conformational flexibility. The conformation of the receptor-bound ligand is that which determines activity. This conformation may be quite different from that adopted by the molecule in the solid state or solution. This could especially be the case where low rotational barriers exist about bonds in a particular drug molecule. The conformation of the molecule at the receptor may even be in a thermodynamically unstable arrangement compared to that in the solid or solution. One approach that has been used to overcome this problem is through the examination of structurally 'rigid' analogues, in which the possibilities of conformational flexibility are eliminated or greatly reduced. It should be noted that a significant alteration in the molecular architecture of a drug to fix it in a particular conformation might enhance its antagonistic properties and/or alter the mode by which the molecule binds to the receptor to produce its activity.

In certain instances these types of studies have led to very useful results. An example is furnished by the crystallographic studies of the thyroid hormones and analogues. The two major thyroid hormones, thyroxine (T4) and 3,5,3′-tri iodo-L-thyronine (T3), have a high degree of conformational flexibility. There are five bonds in these molecules whose rotational flexibility influences the overall topology of the molecule (see Fig. 18.1). The problem of finding the conformation or conformations responsible for the observed bioactivity is, therefore, a multidimensional problem.

From an examination of 17 thyro-active molecules, certain generalizations have evolved (Cody 1976, 1978). Depending upon which of the two iodines is lost from the outer ring in the conversion of T4 to T3, the remaining iodine bond could either point away from the inner ring (distal) or towards it (proximal). The thyronine structures determined to date have not shown any proclivity for either conformation. Energy calculations (Kollman, Murray, Nuss, Jorgensen, and Rothenberg 1973) and nuclear magnetic resonance studies (Emmett and Pepper 1975) suggest that the barrier for rotation about the O_4—C_1' bond (ϕ') is small.

The relationship of the amino-acid side-chain and the outer phenyl ring are described as being cisoid or transoid. The cisoid form is characterized by the outer phenyl ring and the alanine moiety being on the same side of the inner phenyl ring. The value of the torsion angle ϕ describes this conformation (Cody 1976).

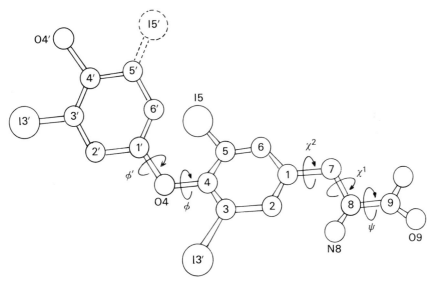

FIG. 18.1. The torsional parameters which describe the conformation of the thyroid hormones.

The values found for ϕ (C5—C4—O4—C1′) and ϕ' (C4—O4—C1′—C6′) describe the orientation of the two rings. The structures determined suggest that the preferred conformation is the skewed model; with $\phi = 90°$ and $\phi' = 0°$.

Cody (1978) also observed that in all the iodine-substituted analogues examined, the iodine atoms are involved in short intermolecular contacts. This is felt to be an indication that the iodines making up these hormones could participate in a charge transfer interaction with the receptor.

Combining structural data on the thyroid hormone analogues with information on their binding affinities with cellular binding tissue, Cody (1976, 1978) was able to construct a model for the active site. The postulated model is shown in Fig. 18.2. Blake and Oatley (1977) were also able to use structural information gathered on the thyroid hormones to show how they fit into

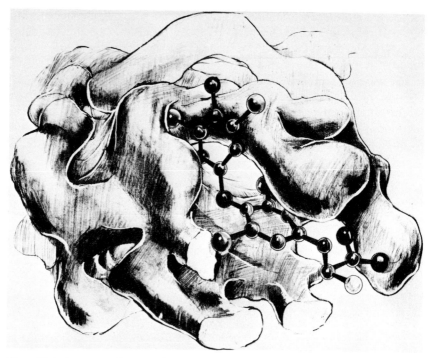

Fɪɢ. 18.2. A proposed model for the active site of a thyroid receptor (Cody 1978).

thyroxine-binding prealbumin (a transport protein) whose structure they determined by X-ray methods.

The enkephalins are endogenous analgesics with morphine-like activity. These molecules bind to the same receptor site in the nervous tissue as do the opiates. Common to the opiates and enkephalins [5] is the 'tyramine' moiety. This group of atoms is known to be important for the analgesic activity of opiates. Horn and Rodgers (1976, 1977) used crystallographic data obtained for opiate analgesics and antagonists which have somewhat constrained conformations to describe a possible binding mechanism for the enkephalins. They showed that the spatial arrangements of the 'tyramine' residue in these molecules are very similar. The nitrogen was found to be on average 7.0 Å from the phenolic oxygen and 4.4 Å from the centre of the

$$
\text{HO} \underset{\tau_1}{\overset{\tau_2}{\bigcirc}} \quad \overset{H \diagdown N \diagup H}{\underset{\overset{\|}{O}}{\overset{C}{\underset{C}{|}}}} \text{C} \longrightarrow \text{Gly-Gly-Phe-Met-Leu}
$$

[5]

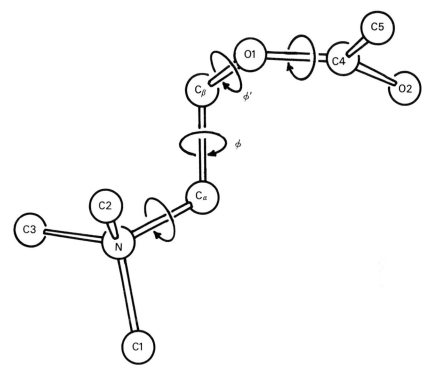

FIG. 18.3. Principal torsional parameters for acetylcholine.

phenyl ring. The torsion angles τ_1 and τ_2 [see 5] are 173° and −89° for the opiate. Though these parameters may not represent the preferred conformation of the 'tyramine' fragment of the enkephalins it may be the orientation that it adopts at the analgesic receptor site.

With the hope of elucidating the requirements of the cholinergic receptors, a variety of acetylcholine analogues were synthesized and tested. As with the thyroid hormones, there exists a great deal of conformational flexibility in the parent molecule, acetylcholine. The principal rotational parameters in this case are the torsion angles about the C_α—C_β bond and the O_1—C_β bond (see Fig. 18.3).

Crystal structure analyses were carried out on many analogues exhibiting muscarinic activity as well as those devoid of this activity (Shefter 1971; Shefter and Smissman 1971; Pauling 1975). The rotation about C_β—O_1 is found to vary considerably for cholinergic ligands, including those which are most potent (circled numbers in Fig. 18.4(a)). The rotational barrier about this bond is relatively low, and the gamut of values observed in these crystal structures no doubt reflects a composite of steric factors and crystal packing forces. In regard to the other torsion angle, ϕ, there are two basic regions where the ligands exhibit muscarinic activity (see Fig. 18.4(b)). To try to ascertain *a particular* conformational requirement for the muscarinic receptor from these data could be misleading. One must consider the possibility that the receptor can accommodate a number of conformations in its active site. This makes it difficult to draw any definitive conclusions from the X-ray data about the receptor surface (Shefter and Triggle 1970).

It has been demonstrated that cholinergic receptors have a polyfunctional active site (Moran and Triggle 1970, 1971; Belleau 1970). The possibility that a number of 'keys' could open the 'door' to cholinergic activity makes it difficult to ascribe a particular conformational requirement for activity. It should be noted that the general concept of multiple modes of ligand interaction

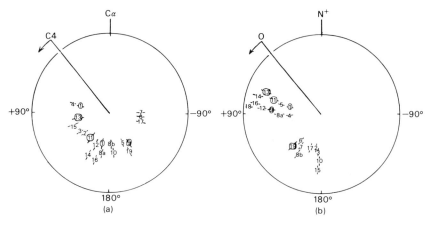

FIG. 18.4. The ϕ' (a) and ϕ (b) torsion angles for a number of acetylcholine analogues (Shefter and Smissman 1971).

at a single catalytic site receives strong support from studies on a number of enzyme systems. The active sites of lysozyme (Beddell, Moult, and Phillips 1970) and ribonuclease (Richards, Wyckoff, and Allewell 1970) are able to accommodate equally well molecules with markedly different conformations. The crystallographers examining the structures of drugs for structure–activity relationships should be aware of this possibility.

Aside from the conformations of drug ligands, differences in their electronic structure and the polarizability of substituent atoms can markedly influence their activity. For example, both acetylthiolcholine and acetylselenolcholine are isosteric, both prefer a *trans* conformation for ϕ, but they are not isoelectronic (Shefter and Mautner 1969). This shows up in their measured cholinergic activities, with the thio-analogue being active as a depolarizing agent in the electric-eel electroplaque preparation while the selenocompound is relatively inactive (Mautner, Bartels, and Webb 1966).

There are a number of drugs which exert their effect by direct interaction with the nucleic acids. This has encouraged numerous types of studies to delineate the nature of these interactions. X-ray fibre-diffraction studies of nucleic acid–proflavin complexes by Lerman (1961, 1963) led to the intercalation model. The model received support and further refinement by Fuller and Waring (1964), who studied ethidium–DNA complexes. In this model the drug molecule binds by inserting itself between adjacent base pairs of the DNA helix. The model requires that the plane of the bound drug molecules be parallel to that of the base pairs. Intercalation is believed to be responsible for the observed activity of many drugs as well as frameshift mutagens.

The crystallographic studies of Sobell and co-workers (Jain and Sobell 1972; Sobell and Jain 1972) on the binding of the antibiotic actinomycin D to DNA provided a precise model for the basis of this drug's activity. This antibiotic consists essentially of an aromatic phenoxazone chromophore linked to two peptide rings each containing five amino-acids. The action of actinomycin D can be explained by its specific binding to DNA via the guanosine–cytosine base pairs which in turn interferes with the readout of the information stored in the DNA. The X-ray study of an actinomycin–deoxyguanosine complex (1:2) demonstrated the sterochemical nature of the interaction. The X-ray results showed the manner in which the phenoxazone ring intercalates the DNA helix (see Fig. 18.5). It also pointed out that specific hydrogen bonds exist between the peptide rings and the stacked guanines. Through model building it was demonstrated that actinomycin could bind analogously to DNA, with the phenoxazone chromophore inserting itself between consecutive G:C and C:G base pairs. Since actinomycin D itself has an approximate twofold axis of symmetry, the sequence specific insertion will not distort the symmetry of DNA.

Sobell (1973, 1974) was able to postulate possible modifications to the actinomycin structure which might enhance its therapeutic selectivity towards tumour cells and extend its range of chemotherapeutic usefulness to include treatment of viral infections. It was proposed, for example, that the introduction of a long alkyl chain attached to the D-valine residue of each peptide would probably not interfere with the binding to DNA and might provide a 'tail' by means of which the molecule could be endowed with tailor-made cell permeability characteristics.

Other crystallographic studies on co-crystallized drug–nucleic acid fragments are also providing valuable data about the nature of these interactions. The subject has been reviewed by Tsai (1978). The prospects of designing specific agents of chemotherapeutic value from the knowledge gained from such investigations is excellent.

Protein crystallographers have made major strides in gaining insight into the mechanisms of enzymes and hormones at the molecular level. Through

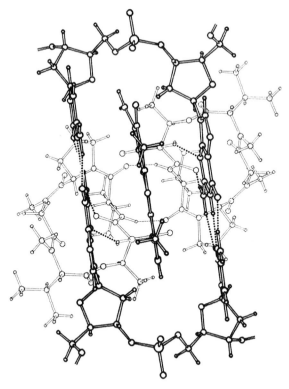

Fig. 18.5. Actinomycin-D–GpC complex (Sobell and Jain 1972).

their studies we now have a much better understanding of many drugs, most notably enzyme inhibitors. Some of their efforts and results are discussed in other chapters and therefore will not be mentioned here.

In recent years significant gains have been made in the identification, isolation, and characterization of a variety of neurotransmitter receptors (Wollemann 1975). It probably will not be long before sufficient quantities of these receptors can be obtained in a crystalline state. This will open a whole new avenue for X-ray crystallographers, which will enable them to explore the nature of drug activity. There is little doubt in my mind that clues to the many unanswered questions of pharmacology will continue to be provided by crystallographers.

References

BEDDELL, C. R., MOULT, J., and PHILLIPS, D. C. (1970). In *Molecular properties of drug receptors* (ed. R. Porter and M. O'Connor) p. 85. Churchill, London.
BELLEAU, B. (1970). In *Fundamental concepts in drug receptors interactions* (ed. J. Danielli, J. F. Morgan, and D. J. Triggle) p. 121. Academic Press, New York.

BERNAL, J. D. (1932*a*). *Nature, Lond.* **129**, 277, 721.
——(1932*b*). *Chemy Ind. Rev.* **10**, 466.
BLAKE, C. C. F. and OATLEY, S. J. (1977). *Nature, Lond.* **268**, 115.
BOKHOVEN, C., SCHOONE, J. C., and BIJVOET, J. M. (1951). *Acta Crystallogr.* **4**, 275.
CODY, V. (1976). In *Thyroid research* (ed.. J. Robbins and L. E. Braverman) p. 290. Excerpta Medica, Amsterdam.
——(1978). *Recent Prog. Horm. Res.* **34**, 427.
CROWFOOT, D., BUNN, C. W., ROGERS-LOW, B. W., and TURNER-JONES, A. (1949). *The chemistry of penicillin*, p. 310. Princeton University Press.
EMMETT, J. C. and PEPPER, E. S. (1975). *Nature, Lond.* **257**, 334.
FULLER, W. and WARING, M. J. (1964). *Ber. Bunsenges Physik Chem.* **68**, 805.
HORN, A. S. and RODGERS, J. R. (1976). *Nature, Lond.* **260**, 795.
————(1977). *J. Pharmac.* **29**, 257.
JAIN, S. C. and SOBELL, H. M. (1972). *J. molec. Biol.* **68**, 1.
KOLLMAN, P. A., MURRAY, W. A., NUSS, M. E., JORGENSEN, E. C., and ROTHENBERG, S. (1973). *J. Am. chem. Soc.* **95**, 8518.
LAI, T. F. and MARSH, R. E. (1967). *Acta Crystallogr.* **22**, 885.
LERMAN, L. S. (1961). *J. Molec. Biol.* **3**, 18.
——(1963). *Proc. natn. Acad. Sci. U.S.A.* **49**, 94.
MAUTNER, H. G., BARTELS, E., and WEBB, G. D. (1966). *Biochem. Pharmacol.* **15**, 187.
MORAN, J. F. and TRIGGLE, D. J. (1970). In *Fundamental concepts in drug receptor interactions* (ed. J. Danielli, J. F. Moran, and D. J. Triggle) p. 133. Academic Press, New York.
————(1971). In *Cholinergic ligand interactions* (ed. D. J. Triggle, J. F. Moran, and E. A. Barnard) p. 119. Academic Press, New York.
O'CONNELL, A. M. and MASLEN, E. N. (1967). *Acta Crystallogr.* **22**, 134.
PAULING, P. (1975). In *Cholinergic mechanisms* (ed. P. G. Waser) p. 241. Raven Press, New York.
RICHARDS, F. M., WYCKOFF, H. W., and ALLEWELL, N. (1970). In *The neurosciences: second study program* (ed. F. D. Schmitt). Rockefeller University Press, New York.
ROBERTSON, J. and BEEVERS, C. A. (1951). *Acta Crystallogr.* **4**, 270.
SHEFTER, E. (1971). In *Cholinergic ligand interactions* (ed. D. J. Triggle, J. F. Moran, and E. A. Barnard) p. 83. Academic Press, New York.
——and MAUTNER, H. G. (1969). *Proc. natn. Acad. Sci. U.S.A.* **63**, 1253.
——and SMISSMAN, E. E. (1971). *J. Pharm. Sci.* **60**, 1364.
——and TRIGGLE, D. J. (1970). *Nature, Lond.* **227**, 1354.
SOBELL, H. M. (1973). *Prog. Nucleic Acid Res. molec. Biol.* **13**, 153.
——(1974). *Cancer Chemother. Rep.* **58**, 101.
——and JAIN, S. C. (1972). *J. molec. Biol.* **68**, 21.
TSAI, C. (1978). *Ann. Rep. Med. Chem.* **13**, 316.
WOLLEMANN, M. (ed.) (1975). *Properties of purified cholinergic and adrenergic receptors.* Elsevier, New York.
WOOD, D. (1940). *Br. J. exp. Path.* **21**, 74.

19. Crystallographic studies of histamine and some H₂-receptor antagonists in relation to biological activity

KEITH PROUT AND C. ROBIN GANELLIN

THE pharmacological actions of histamine are considered to be mediated by at least two distinct classes of receptor, designated H_1 and H_2 (Ash and Schild 1966; Black, Duncan, Durant, Ganellin, and Parsons 1972), distinguished by the action of specific histamine antagonists that selectively block histamine responses. The receptor differentiation is also reflected in the selective agonist action of certain histamine analogues and this poses an interesting question for the medicinal chemist whether histamine acts in chemically different ways at its two receptor sites (Durant, Ganellin, and Parsons, 1975). Structure–activity studies of conformation and tautomerism for histamine and congeners indicate that this may be the case; evidence comes from theoretical calculations (various molecular orbital procedures), n.m.r. spectroscopy, ionization constants, and crystallography. This paper describes the contribution made by X-ray crystallography.

Histamine can exist as the neutral molecule (free base), or the monocation, or the dication (Fig. 19.1). The neutral molecule and monocation have two tautomeric forms, designated N^τ-H (proton on the ring-nitrogen distal to the side chain) and N^π-H (proton on the adjacent ring-nitrogen).† All forms may exist in various conformations, defined by two torsion angles θ_1 and θ_2 (Fig. 19.2).

Molecular orbital calculations predict that, as with other disubstituted ethanes, the energetically most favoured conformations are the *trans* (anti-periplanar, fully extended side-chain, $\theta_2 = 180°$) and *gauche* (syn-clinal, folded side-chain, $\theta_2 = 60°$ or $300°$) in which the hydrogen atoms in the side chain have a staggered arrangement. The various calculation procedures disagree, however, in their predictions of the relative stability of the conformers. Extended Huckel Theory (EHT) (Kier 1968; Ganellin, Pepper, Port, and

† Nomenclature according to Black and Ganellin (1974).

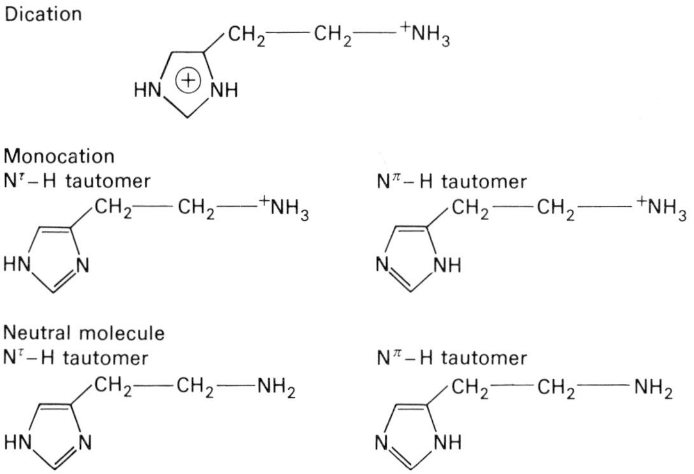

Dication

Monocation
N^τ–H tautomer N^π–H tautomer

Neutral molecule
N^τ–H tautomer N^π–H tautomer

FIG. 19.1. Histamine species.

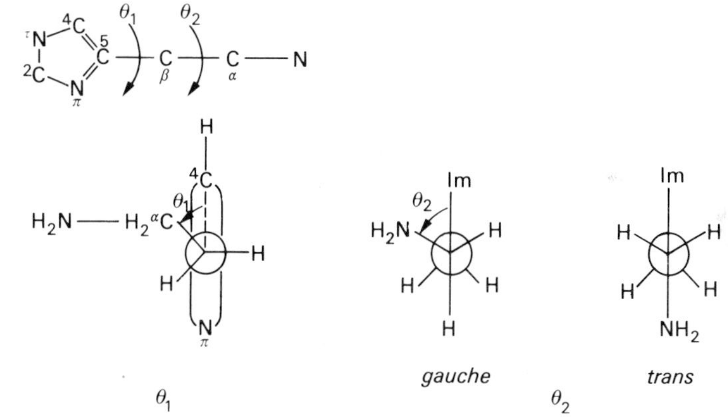

gauche   *trans*

θ_1 θ_2

θ_1 is the angle formed between the planes of the imidazole
ring and the side chain; for the system ^4C–^5C–$^\beta C$–$^\alpha C$
it is the torsion angle about the bond ^5C–$^\beta C$ and is defined
as the angle through which bond $^\beta C$–$^\alpha C$ is rotated from the
$^4C^5C^\beta C$ plane. Thus $\theta_1 = 0°$ where ^4CH and $^\alpha CH_2$ are eclipsed;
θ_1 is positive when, on looking from $^\beta C$ to ^5C the
rotation is anticlockwise.

θ_2 is the torsion angle about the bond $^\beta C$–$^\alpha C$ and is defined
analogously.

FIG. 19.2. Histamine numbering, conformers, and torsion angles.

Richards 1973) and empirical treatments (Kumbar 1975) predict stability for both *trans* and *gauche* conformations in histamine monocation and dication, with the *trans* conformer being slightly more stable by approximately 1 kcal mol^{-1}. These conclusions agree with experimental results derived from nuclear magnetic resonance (n.m.r.) studies in aqueous solution (Ganellin, Pepper, Port, and Richards 1973; Ham, Casey and Ison 1973). Surprisingly, the more sophisticated PCILO (perturbative configuration interaction using localized orbitals) (Pullman and Port, 1974), INDO (Green, Kang, and Margolis 1971), and CNDO/2 (Ganellin, Pepper, Port, and Richards 1973) methods give strikingly different results and predict conformation to be dependent upon the type of cation, viz. the dication is entirely *trans* but the monocation is overwhelmingly *gauche*.

A possible explanation for the discrepancy between these predictions and the n.m.r. results is that the calculations relate to isolated molecules whereas the n.m.r. data were obtained from aqueous solution. All known structures for histamine dication show a *trans* conformation (Decou 1964; Bonnet, Jeannin, and Laaouini 1975; Veidis, Palenik, Schaffrin, and Trotter 1969; Yamane, Ashida, and Kakudo 1973; Bonnet and Jeannin 1972), but at the time of this controversy there had been no crystallographic study of histamine monocation; Pullman (Pullman and Port 1974) used histidine as an analogy and argued that the PCILO predictions were supported by the crystal structures of zwitterionic histidine base (with —NH_3^+ hence resembling the histamine monocation) which showed a *gauche* relationship between the imidazole and —NH_3^+ groups. According to Pullman the discrepancy between the PCILO prediction and the n.m.r. result was due to the absence or the presence of water. However, histidine is not a very satisfactory analogy since the histidine carboxylate anion may influence the crystal structure. Furthermore, histidine monohydrochloride (analogous to histamine dication) also has a *gauche* relationship between the imidazolium and ammonium residues (Bennett, Davidson, Harding, and Morelle 1970). It became clear that a study of histamine monocation would be helpful.

It was surprising that the crystal structure of salts of the histamine monocation had not been determined since the monocation is the main species under physiological conditions (pK$_a$ studies indicate 96 per cent of molecules are monoprotonated at pH 7.4 (Durant *et al.* 1975)) and is considered most likely to be the physiologically active form of histamine. Initial attempts to obtain a crystalline form of the monocation were thwarted, as the monohydrochloride disproportionated on recrystallization and gave a mixture of solid mono- and di-hydrochlorides. Eventually the monohydrobromide was obtained with the required stoichiometry, suitable crystals were grown and the structure solved (Prout, Critchley, and Ganellin 1974). In the crystal the cation takes the *trans* conformation, $\theta_1 = 90°$, $\theta_2 = 180°$ and hydrogen-bonded dimers are formed (Fig. 19.3). The cations in these dimers have the

N^τ-H tautomeric form with two hydrogen bonding sites, one H-donor, the other H-acceptor, 4.5 Å apart, forming parallel hydrogen bonds. The crystal structures do not, of course, identify the relative preference for the various conformations, but in this case the crystallographic result together with the n.m.r. studies in aqueous solution indicate that predictions of the conformation of the histamine monocation in a solution or crystal environment based on PCILO molecular orbital calculations on isolated molecules are in error. Similar discrepancies exist for histamine base; the INDO (Green *et al.* 1971) and PCILO (Coubeils, Courriere, and Pullman 1971) procedures predict overwhelming preference for the *gauche* conformation, yet crystalline histamine base has been found to have a *trans* conformation (Bonnet and Ibers

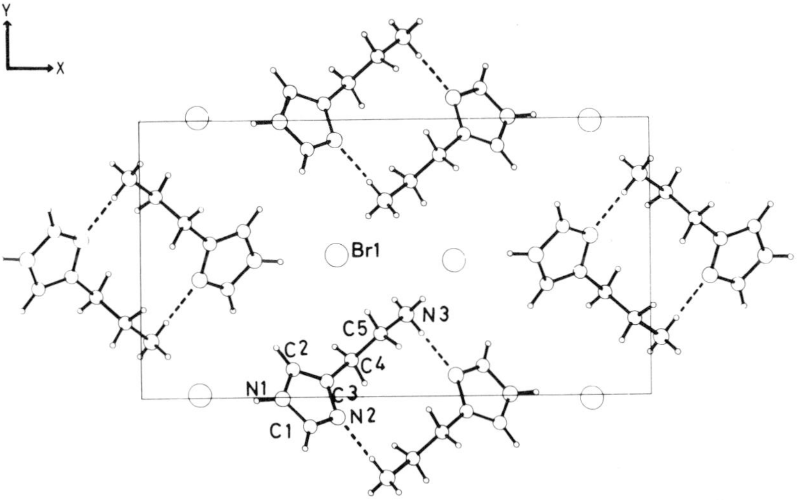

FIG. 19.3. Histamine hydrobromide: the crystal structure seen in projection down *c*. (Reproduced with permission of the editor of *Acta Crystallographica*.)

1973). It appears that these molecular orbital procedures fail to be predictive with histamine, but that the less sophisticated EHT method makes reasonable predictions, probably through a fortuitous cancellation of errors.

The predicted orientation θ_1 of the imidazole ring also varies according to the molecular orbital calculation procedure, but the only physical measures are crystallographic; various orientations of the imidazole ring have been found, depending on the salt used for crystallization. The values of θ_1 occur at approximately 30° increments, viz. near 0° (e.g. 4°, 7°, 9°) (Yamane *et al.* 1973; Bonnet and Jeannin 1972), 30° (e.g. 26.7°, 30°) (Bonnet *et al.* 1975; Decou 1964), 60° (e.g. 66.3°) (Bonnet and Ibers 1973), and 90° (e.g. 89.7°) (Prout *et al.* 1974). The molecular orbital calculations predict substantial energy barriers to interconversion between *trans* and *gauche* side-chain conformations but less restriction in rotation of the imidazole ring, especially for *trans* conformers. These predictions are in keeping with the crystallographic findings of a single value for θ_2 but a multiplicity of values for θ_1.

Structure–activity studies of a series of methylhistamines using EHT calculations (Ganellin, Port, and Richards 1973) have indicated three regions of conformational space which are possibly essential for H_1 receptor activity (Ganellin 1973; Farnell, Richards, and Ganellin 1975), one *trans*, $\theta_1 = 50 \pm 20°\,\theta_2 = 180 \pm 30°$, and two *gauche*, $\theta_1 = 50 \pm 20°$, $\theta_2 = 65 \pm 25°$, or $\theta_1 = 90 \pm 10°$, $\theta_2 = 300 \pm 30°$. In this analysis studies of 4-methylhistamine (Fig.

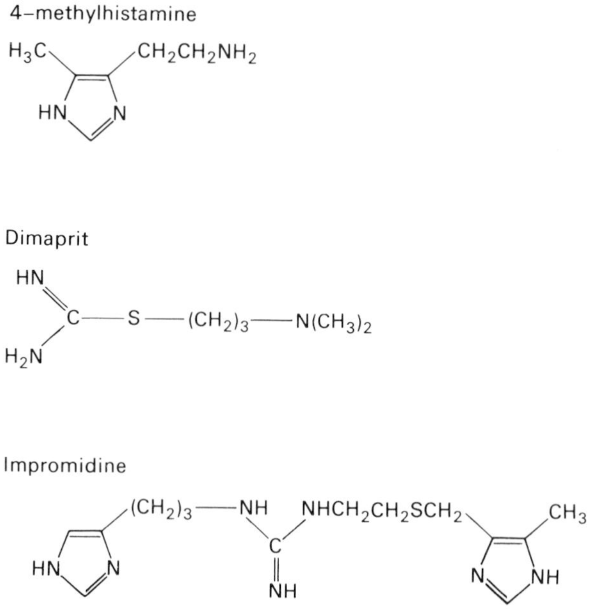

4–methylhistamine

H_3C＼　　　＼$CH_2CH_2NH_2$

HN　　N

Dimaprit

HN＼

C—S—$(CH_2)_3$—$N(CH_3)_2$

H_2N／

Impromidine

$(CH_2)_3$—NH　$NHCH_2CH_2SCH_2$＼　　CH_3

HN　N　　　　C

　　　　　　　‖　　　　　N　NH

　　　　　　　NH

FIG. 19.4. Selective H_2-receptor agonists.

19.4) were of particular importance; this compound shows a dramatic receptor selectivity, having 40 per cent of the potency of histamine as an H_2-receptor stimulant, but only 0.2 per cent at H_1 receptors (Black *et al.* 1972). The EHT calculations suggest (Ganellin, Port, and Richards 1973) that the *trans* monocation ($\theta_2 = 180°$) has a broad energy well for stability in the range $\theta_1 = 80–280°$ and that it cannot easily achieve a ring orientation $\theta_1 = 0 \pm 60°$. As a possible test of these predictions, the crystal structure (Fig. 19.5) of 4-methylhistamine monohydrobromide was examined and found to be isostructural (Critchley 1979) with histamine monohydrobromide, having $\theta_1 = 91.5°$, $\theta_2 = 178.1°$, thus fitting into the range predicted for stability.

Structure–activity studies have also implicated histamine tautomerism in its H_2-receptor actions (Durant *et al.* 1975). Studies on aqueous solution suggest that about 80 per cent of histamine monocation in water is in the N^r-H tautomer (Ganellin 1973; Reynolds and Tzeng 1977) and, according to ^{13}C n.m.r. results, histamine base appears to resemble the monocation (Wasylishen and Tomlinson 1977). Theoretical studies indicate that predicted tautomer stability is very sensitive to the ring geometry used as input data (Kang and Chou 1975). Crystal structures therefore provide vital information for these studies. Interestingly, both histamine and 4-methylhistamine mono-hydrobromides (monocation) crystallize as the N^r-H tautomer (Prout *et al.* 1974; Critchley 1979), but histamine base crystallizes as the N^π-H tautomer

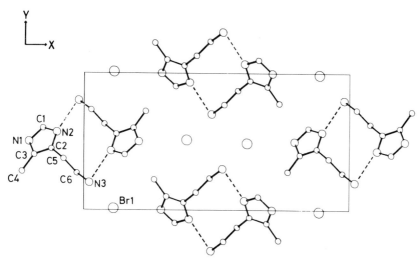

FIG. 19.5. 4-Methylhistamine hydrobromide: the crystal structure seen in projection down *c*.

(Bonnet and Ibers 1973). Crystal structure analyses of a wide range of imidazole derivatives show that the imidazole ring is not symmetric. Specifically the endocyclic angle (105.3° average) at the unprotonated nitrogen atom is less than the endocyclic angle (107.3° average) at the protonated nitrogen atom. Hence a change in tautomeric form is accompanied by a rearrangement of the atoms of the imidazole skeleton. Only when this ring asymmetry is taken into account are predictions from *ab initio* molecular orbital calculations (Richards, Wallis, and Ganellin 1979) of the tautomeric form of histamine found to be in accord with the crystal structure findings. The analysis (Richards *et al.* 1979) suggested that water has a much greater influence on tautomer stability for histamine monocation than for the base. It appears that knowledge of the crystal geometry of only one tautomer may be adequate to permit calculation of imidazole tautomer preference for the isolated molecules (Richards *et al.* 1979).

Histamine agonists continue to be studied. Two recently described selective H_2-receptor agonists, dimaprit (Durant, Ganellin, and Parsons 1977) and impromidine (Durant *et al.* 1978) Fig. 19.4), are of special interest. Both involve—$(CH_2)_3$—side-chains, and provide interesting questions of conformation, e.g. the higher and lower homologues of dimaprit are not active (Durant, Ganellin, and Parsons 1977) and it should be possible to compare the conformational properties of these molecules with those of dimaprit and histamine to indicate a likely active conformation for stimulation of H_2 receptors.

Antagonists of histamine at H_1 receptors have been known and used clinically for many years. Their structures, structure–activity relations, and crystallography have been extensively reviewed (Witiak 1970). The structures of these antagonists bear little resemblance to histamine itself, but each has one or more aromatic rings and an aliphatic amino-group which will be protonated at physiological pH and may interact with an anionic site on the receptor. To explain why stricter structure–activity relationships exist for histamine analogues than for the H_1-receptor antagonists it has been suggested that these do not occupy the total receptor area but only the anionic site, the rest of the molecule being necessary for interactions with regions adjacent to the site (Ariens and Simonis 1966). However, the observation that in the H_1-receptor antagonist triprolidine the separation distance between the aliphatic nitrogen atom and the nitrogen of the aromatic (pyridine) ring approached 4.8 Å encouraged Kier (1968) to associate H_1-receptor activity with the extended *trans* conformation of histamine, and H_2-receptor activity with the *gauche* form.

This proposal of Kier directed our interest to the conformations of H_2-receptor antagonists. The first H_2-receptor antagonist, burimamide, was developed by Black and co-workers (1972). A slight but definite antagonism had been observed in N^z-guanylhistamine (Fig. 19.6) which is a partial agonist. This molecule, possessing an imidazole ring and a cationic side chain, was considered to have sufficient chemical resemblance to histamine to mimic it biologically and thus have agonist properties. Therefore attempts were made to replace the guanidinium group by a similar but non-basic polar hydrogen bonding group which would not be charged (Brimblecombe *et al.* 1978). This led to the development of burimamide (Fig. 19.6) in which the thione S replaces imino NH of the guanidine and the side-chain is extended. In the crystal the structure of burimamide (Kamenar, Prout, and Ganellin 1973) is best described in terms of three planes, the imidazole ring plane, the tetramethylene side-chain plane perpendicular to the ring, and the thiourea plane. The molecule is in the N^r-H tautomeric form and the thiourea group has the staggered E,Z, configuration. The crystal structure (Fig. 19.7)

(a) N^{α}–guanylhistamine

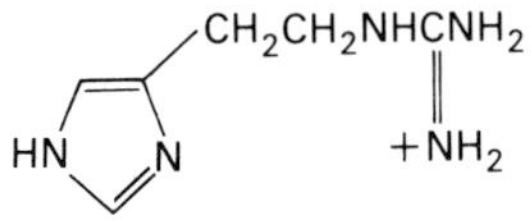
CH2CH2NHCNH2
+NH2
HN
N

(b) Burimamide

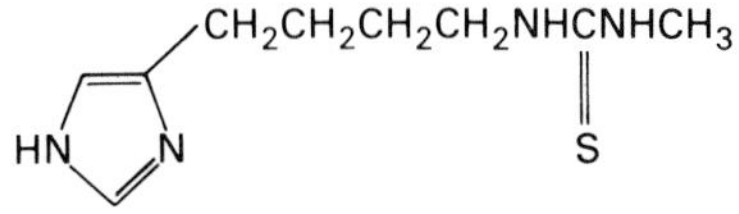
CH2CH2CH2CH2NHCNHCH3
S
HN
N

(c) Thiaburimamide

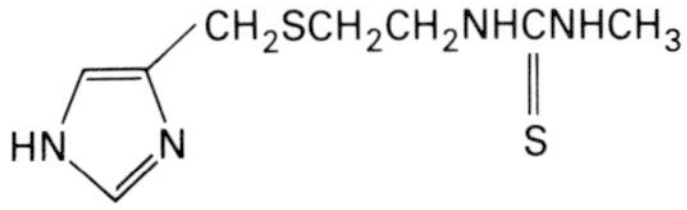
CH2SCH2CH2NHCNHCH3
S
HN
N

(d) Metiamide

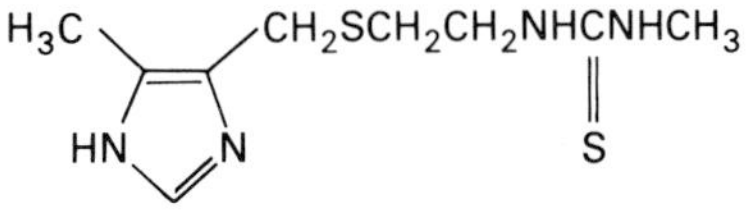
H3C
CH2SCH2CH2NHCNHCH3
S
HN
N

(e) Cimetidine

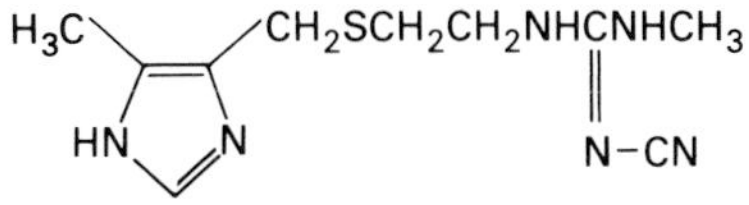
H3C
CH2SCH2CH2NHCNHCH3
N–CN
HN
N

(f) N–[2{(4–bromoimidazol–5–yl) methylthio} ethyl]–N′–methylthiourea

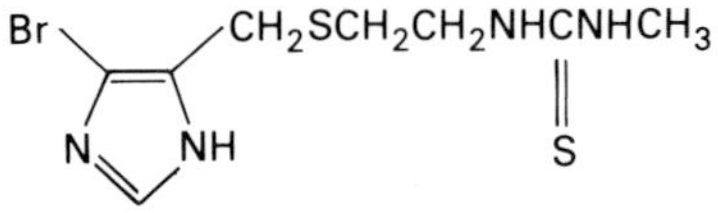
Br
CH2SCH2CH2NHCNHCH3
S
N
NH

(g) N–{3–(imidazol–4–yl)propyl}–N′–methylthiourea

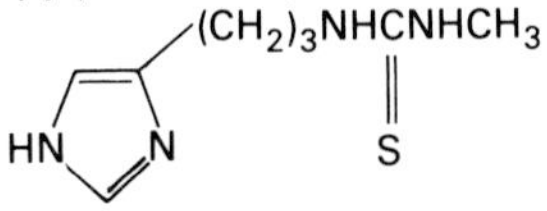
(CH2)3NHCNHCH3
S
HN
N

(h) N–{5–(imidazol–5–yl)pentyl}–N′–methylthiourea

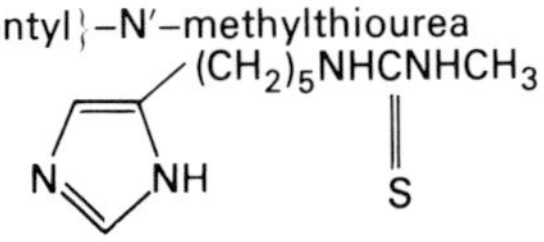
(CH2)5NHCNHCH3
S
N
NH

(i) 2–[{2–|(5–methylimidazol–4–yl)methylthio|ethyl} methylamino]–2′–methylamino–1–nitroethene

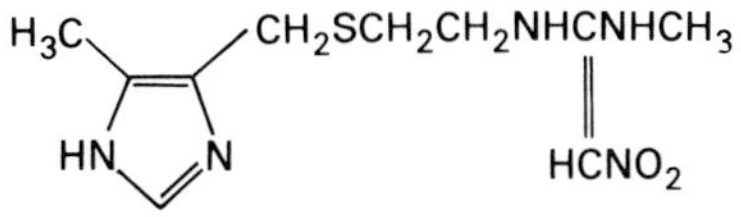
H3C
CH2SCH2CH2NHCNHCH3
HCNO2
HN
N

FIG. 19.6. Some selective H_2-receptor antagonists and related compounds.

shows the imidazole rings linked to form ribbons by N----H—N H-bonds
as in imidazole itself. The ribbons are linked to form sheets by pairs of
S---H—N bonds about centres of symmetry such that the side-chains meet
side to side rather than end to end. Further S---H—N bonds link the sheets,
thus there is a complete absence of contact, inter- or intramolecular, between
the imidazole rings and the thiourea groups. The intramolecular separation
of the two residues is 6 Å. Thus the crystal structure, while not disproving
Kier's (1968) suggestion that implies that an H_2-receptor antagonist should
have an inter-nitrogen distance in the region of 3.6 Å, does not provide any
evidence in support of it.

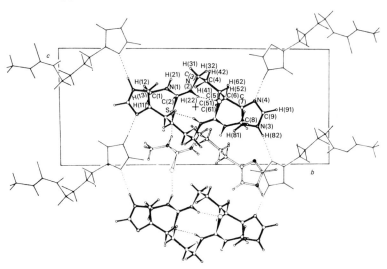

FIG. 19.7. Burimamide: the crystal structure projected down *a*, showing the N—
H---N hydrogen-bonded ribbons and the N—H----S hydrogen-bonded pairs.

Burimamide, although active in man, is not sufficiently active to be clinically useful. To increase the activity Black, Durant, Emmett, and Ganellin
(1974) sought to enhance the stability of the N^τ-H tautomer by decreasing
the electron releasing power of the side chain and at the same time introducing an electron releasing group on to the neighbouring carbon atom. The
result was metiamide (Fig. 19.6).

Considerations of toxicity made the replacement of the thiourea residue
desirable. Cyanoguanidine(=N—CN)was introduced as a possible isosteric
group and resulted in the clinically successful drug cimetidine (Durant,
Emmett, Ganellin, Miles, Parsons, Prain, and White 1977) (Fig. 19.6). The
crystal structures of thiaburimamide (Prout, Critchley, Ganellin, and Mitchell 1977) (Fig. 19.6), metiamide (Prout *et al.* 1977), and cimetidine (Critchley 1979; Hädicke, Frickel, and Franke 1978) have been reported. In the crys-

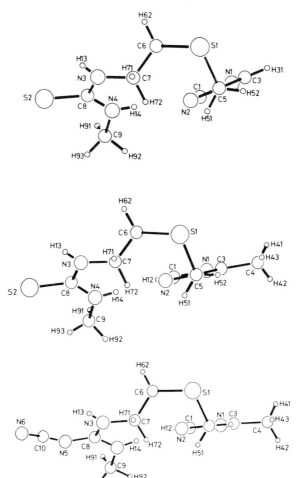

FIG. 19.8. (a) Thiaburimamide, (b) metiamide, (c) cimetidine, each viewed down the bond C(5)—C(2) looking from C(5)—C(2).

tal these molecules like burimamide exist as the N$^\tau$-H tautomer and the end-group has the E,Z configuration, but in contrast with burimamide's semi-extended conformation all three new compounds are folded to form ten-membered H-bonded rings (Fig. 19.8) as predicted from the i.r. spectra (Prout *et al.* 1977). This raised the question of the significance of the intramolecular H-bond for biological activity, but it was recognized that other factors were more important. There is an apparent inverse correlation between biological activity and the lengths of these intramolecular H-bonds, 2.97, 2.89, and 2.87 Å. As Fig. 19.8 shows, the degree of tilt of the imidazole ring with respect to the rest of the molecule is virtually the only conformation difference; the angle between the plane of the imidazole ring and the C(2)–C(5) and S(1) plane adopts values of 86.5° in thiaburimamide, 78.2° in metiamide, and 69.1° in cimetidine, which suggest that the methylene hydrogen atoms at C(5) are more strongly repelled by the nitrogen lone pair than by the hydrogen H(31) at C(3) and much more strongly than by the methyl group at C(3). The length of the intramolecular H-bond is determined by the degree of tilt. If burimamide were to take up the intramolecular hydrogen-bonded conformation not only might the H-bond be fairly long, but models suggest that in this con-

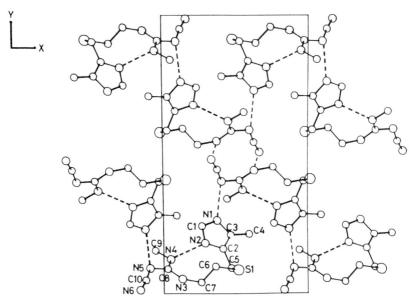

FIG. 19.9. The crystal structure of cimetidine in projection down *c*. (Monoclinic P2₁/*n*; *a*=10.696, *b*=18.813, *c*=6.816Å, β=111.30°, cf. metiamide.)

formation a tetramethylene chain would be hindered by repulsions between the pair of hydrogens H(51) and H(72) (Fig. 19.7). The repulsion is reduced by the replacement of the second methylene group of the chain by a thioether linkage because C—S bonds are much longer than C—C although the reduction of bond angle in the thioether to 102° tends to act against this steric relief. A major consequence of the supposed isosteric replacement of the side-chain —CH₂—by a thioether linkage is to encourage the side-chain to fold by the introduction of a *gauche* conformation at the S—C(5) bond (Fig. 19.8).

The formation of the intramolecular H-bond in thiaburimamide, metiamide and cimetidine prevents the formation of the H-bonded imidazole ribbons characteristic of burimamide, but the linkage of the molecules into centrosymmetrically related S---H—N H-bonded pairs persists in thiaburima-

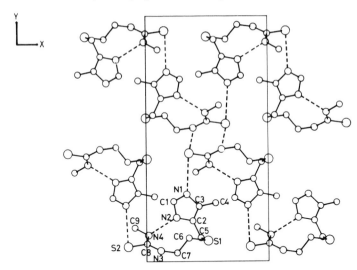

FIG. 19.10. The crystal structure of metiamide in projection down c. (Monoclinic $P2_1/n$; $a=10.229$, $b=19.045$, $c=6.701$ Å, $\beta=107.76°$, cf. cimetidine.)

mide and metiamide. So closely does the $=$N—CN group mimic $=$S that not only do similar dimer pairs form, but the structure of the cimetidine (Fig. 19.9) crystal is almost exactly the same as that of metiamide (Fig. 19.10). Indeed these two structures differ from that of thiaburimamide (Fig. 19.11) only in that the imidazole N: to HN (terminal) H-bond links the dimer pairs into chains in thiaburimamide but sheets in metiamide and cimetidine to

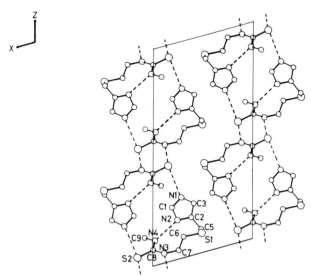

FIG. 19.11. The crystal structure of thiaburimamade in projection down b.

Crystallographic studies of histamine

accommodate the 4-methyl group. The extraordinary similarity of the environments and steric properties of $=$S and —N—CN in the crystal begs for an extrapolation to the receptor site where it might be assumed that the two functional groups will play equipollent roles.

In the early days of crystallography isomorphous replacement of methyl by bromo- was used as a possible source of an isomorphous derivative. The replacement of the 4-methyl group in metiamide by bromine might, for steric reasons, be expected to yield an isomorphous compound, but electronically the 4-bromo derivative tends to favour the N^π-H tautomer. However, the

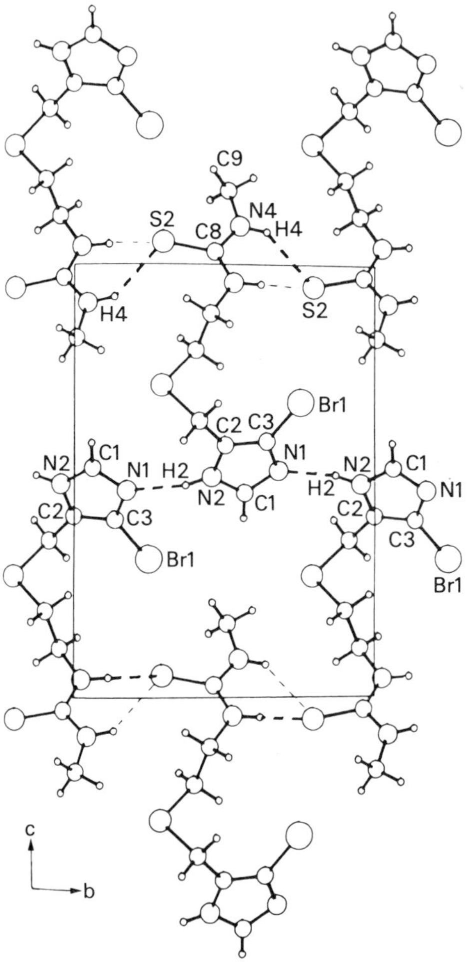

FIG. 19.12. Hydrogen bonding between imidazole rings in N-[2{(4-bromoimidazol-5-yl)methylthio}ethyl]-N′-methylthiourea (the bromo analogue of metiamide) in projection down *a*.

4-bromo derivative (Fig. 19.6) is an antagonist (Durant, Emmett, and Ganellin 1973). A crystal structure determination (Glover 1979) reveals that in the solid state it is indeed the N^π-H tautomer (Fig. 19.12) confirmed by both the hydrogen atom location and the ring shape. The bond lengths and angles of the side-chain are very similar to those in the cimetidine/metiamide structures, but the conformation of the side-chain differs in that, although the conformation about $S(1) - C(5)$ is still *gauche*, C(6) lies on the same side of S(1)—C(5) as C(3) and Br(1). The C(2)—C(5)—S(1)—C(6) torsional angle is equal in magnitude and opposite in sign to that found in metiamide. This change results in a semi-extended conformation, but a different one to that

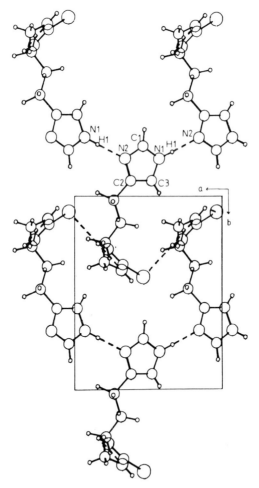

FIG. 19.13. Hydrogen bonding between imidazole rings in N-{3-(imidazol-4-yl)propyl}-N′-methylthiiourea in projection down *c*.

F IG. 19.14. The crystal structure and hydrogen-bonding network in N-{5-(imidazol-5-yl)pentyl}-N′-methylthiourea in projection down *c* showing the planar heavy atom skeleton.

of burimamide where the carbon skeleton of the tetramethylene chain is planar. The terminal thiourea group has the Z,Z configuration. In the crystal the imidazole groups link to form H-bonded ribbons as in burimamide negating any possible argument that the N^{π}-H tautomer is a requirement of the crystal packing. Both N—H groups of the thiourea residue are involved in S---H—N H-bonds but to two sulphur atoms of different molecules (Fig. 19.12). The H-bonded dimer pairs characteristic of the earlier structures are not found.

The effective antagonists tend to have a four-atom chain between the imidazole ring and the end group, but the trimethylene and pentamethylene homologues (Durant, Emmett, Ganellin, and White 1973) (Fig. 19.6) of buri-mamide are available as crystalline solids and the former at least is weakly active as an antagonist (Brimblecombe *et al.* 1978). In the crystal there are substantial differences between the conformation found in burimamide and those of the higher and lower homologues (Glover 1979). In the molecules described thus far the torsion angle about the bond C(2)—C(5) has been in the region of 90°, a conformation thought to minimize repulsions between the imidazole ring and the hydrogen atoms of the first side-chain methylene group. However, in the tri- and pentamethylene homologues these torsion angles (C(3)—C(2)—C(5)—C(6) and C(3)—C(2)—C(5)—C(11) (Figs. 19.13 and 19.14)) are about 0° and there are no short contacts to the hydrogen atoms at C(5). Furthermore, before these structures were determined there were strong indications that a single alkane-like substituent in the imidazole ring

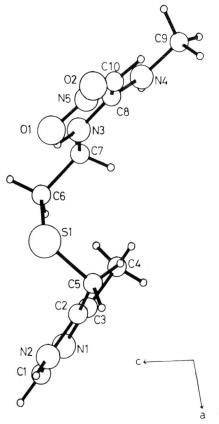

FIG. 19.15. Molecule of 2-|{2[(4-methylimidazol-5-yl)methylthio]ethyl}amino|2'-methylamino-1-nitroethene, projected down *b*.

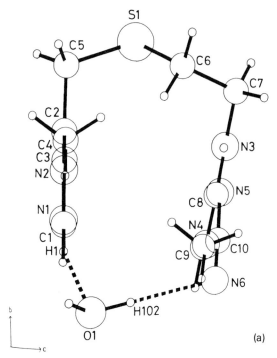

FIG. 19.16. Cimetidine hydrate in projection down the *c* and *a* axes showing one molecule of cimetidine and its attendant water molecule viewed approximately perpendicular
and parallel to the plane of the imidazole ring.

was always associated with the N$^\tau$-H form, but these determinations show
the trimethylene homologue in the N$^\tau$-H form but the pentamethylene homologue in the N$^\pi$-H form. Both homologues have an E,Z configuration at the
thiourea group. The pentamethylene homologue has a fully extended planar
skeleton whereas the trimethylene compound can be described by two planes
related by a *gauche* conformation at the C(7)—N(3) bond. The crystal
structure of the trimethylene homologue shows a now familiar H-bond
pattern with imidazole ribbons and centrosymmetrically related S - - - H—N
systems, using this time the terminal N(4)—H group and further linkage by
a second type of S - - - H—N(3) bond. In the pentamethylene compound the
S - - - H—N(4) H-bonds link the molecules into centrosymmetric dimers; the
the dimers are linked into chains by N(3)—H - - - N bonds between the
non-terminal thiourea N—H and N of the imidazole, and the chains
into sheets by S - - - H—N(2) H-bonds using the imidazole N(2)—H group.

    The observation that the nitroethene analogue of metiamide (Fig. 19.6)
was active as an antagonist (Durant, Emmett, Ganellin, and Prain 1976) introduced the possibility that the N—H groups were not both required for

(b)

activity since in this new compound one N—H would most probably be blocked by intramolecular H-bonding to the nitro group. The structure analysis (Glover 1979) (Fig. 19.15) showed the end group to be planar with an intramolecular H-bond from O(1) of the nitrogroup to N(3)—H, the non-terminal amidine nitrogen. The C(8)—C(10) bond of the nitroethene residue though formally double is 1.428 Å long, i.e. little shorter than an sp^2—sp^2 single bond, suggesting that the molecule might have a greater configurational flexibility than is suggested by the structural formula. The molecule is in the N^{π}—H tautomeric form and has a conformation up to C(7) similar to the bromo analogue of metiamide, but the bromo compound is *trans* at C(6)—C(7) and the new nitroethene gauche. The resulting molecule can be described in terms of two planar groups, the imidazole and the nitroethene in a *trans* conformation linked by a flexible chain. As in cimetidine, one type of H-bond is no longer available for intermolecular bonding but the crystal structure is more reminiscent of the pentamethylene homologue of burimamide than that of cimetidine.

An interesting *cis* variant of the conformation of the nitroethene antagonist occurs in cimetidine hydrate. The cimetidine (Critchley 1979; Hädicke *et al.*

1978) structure is of crystals obtained from cyanomethane, but from aqueous solution a monohydrate is obtained. In contrast to the form from cyanomethane there is no intramolecular H-bond in cimetidine hydrate (Glover 1979), but H-bonding to a water molecule links the imidazole and cyanoguanidine ends of the molecule (Fig. 19.16(a) and (b)). The molecule has the N^τ-H tautomeric form and the cyanoguanidine residue the Z,E configuration. The conformations about all the bonds in the side-chain are *gauche*.

X-ray diffraction studies of crystalline drug molecules rarely lead to firm hypotheses concerning the relationship between molecular structure and biological activity. However, clues and pointers often emerge. In the group of histamine H_2-receptor antagonists model building indicates that any given molecule is able with little gain in energy to take up the conformations found for its analogues. The antagonists occur for the most part in the N^τ-H tautomer form. The N^π-H tautomer is found in only three out of nine crystals studied.

The nature of the interaction of histamine with the H_2-receptor site is uncertain, but this interaction is almost certainly H-bonding/coulombic in character and at physiological pH the likelihood is that the interacting form of histamine is the monocation. The separation distances between a given imidazole nitrogen and the ammonium atom for various conformations are given in Table 19.1.

TABLE 19.1

$N^\pi \ldots NH_3^+$ *separations in various conformations of the histamine monocation*

Torsion angles (°)		Separation distances (Å)
θ_1	θ_2	$N^\pi \ldots NH_3^+$
60	180	4.6
60	300	3.6
90	180	4.5
0	180	5.1

Each antagonist has five sites that could participate in H-bonding. The protonated nitrogen atom of the imidazole and the two amidine N—H groups as H-donors and the unprotonated imidazole nitrogen and the terminal electronegative group S, NO_2, or N—CN as hydrogen acceptors. With the exception of one N—H group in cimetidine hydrate all these groups were found to be involved in H-bonding in every crystal studied. The functional parts of the antagonist molecules appear to be the imidazole ring and the terminal amidine residue. The side-chain is thought to adjust the electronic properties of the imidazole group and to determine the separation distance between the functional groups.

In each of the H_2-receptor antagonists the imidazole hydrogen has a specific location on one or other of the imidazole nitrogen atoms. There is no disorder, nor any ordered mixture of tautomers.

The dimensions of the imidazole ring in antagonists and other imidazole structures show that a change in tautomeric form is accompanied by a change in ring geometry. Characteristically, the angle at the unprotonated nitrogen atom is smaller than the angle at the protonated nitrogen, the angles at carbon atoms adjacent to the unprotonated nitrogen atom are the largest in the ring, and the bond lengths are asymmetrical with the C—N bonds involving the unprotonated nitrogen atom invariably the shorter bonds.

The amidine-like terminal group has been observed in the E,Z; Z,E; and Z,Z configurations and cimetidine and its hydrate exist one in the E,Z configuration and the other in Z,E. In this respect all the terminal groups appear sufficiently flexible for it to be impossible to deduce anything definite about the possible configuration at the receptor from the cystallographic studies. One point that emerges from the work on the nitroethene derivative is that it is almost certain that one or other of the end group N—H bonds can be engaged in a strong intramolecular H-bond without affecting activity.

From the nine crystal structure analyses, seven different conformations of the antagonists have emerged, but chronologically the first four analyses revealed only two structural types, and the intramolecular H-bonded form found in thiaburimamide, metiamide, and cimetidine was much favoured as a potentially active form. However, as an active form it suffers from one main disadvantage; the three available H-bonding sites for intermolecular interaction bear no relation to histamine either in conformation or site separation. Indeed the same disadvantages apply to the extended and semi-extended conformations of the tri- and pentamethylene homologues of burimamide, burimamide itself, and the nitroethene antagonist. From the cimetidine hydrate structure perhaps appears the most exciting clue to a possible antagonist conformation at the receptor site. In whatever form and conformation histamine may take at the receptor site the separation distance between the interacting sites at head and the tail of the molecule must lie in the region 3.5 and 5.2 Å, most probably around 4.0–4.5 Å. Models suggest that the *cis* pair of H-bonds from sites at this separation will be parallel. The cimetidine hydrate structure alone offers the possibility of a pair of parallel near coplanar H-bonds from sites about 4 Å apart. These could be either N(1) and N(6) which interact with the water molecule in the crystal structure or N(2) and N(3) which in the hydrate interact with a neighbouring molecule (Figs. 19.16(a) and (b)). The former pair includes the electronegative terminal group which might suggest that the biological activity should be sharply dependent on the nature of the terminal group, but this appears to be contrary to the available evidence. The N(2) and N(3) pair of sites are more histamine-like and if receptor interaction occurs through these atoms then the role of the electronegative group may be to determine the ability of N(3) to donate a proton in the H-bond.

Acknowledgements

We thank Dr S. R. Critchley and Dr P. E. Glover, who as D.Phil. students in the Chemical Crystallography Laboratory were largely responsible for the experimental X-ray diffraction work; and also the SRC CASE scheme for its financial support.

References

ARIENS, K. J. and SIMONIS, A. M. (1966). *Farmaco, Ed. Sci,* **21,** 581.

ASH, A. S. F. and SCHILD, H. O. (1966). *Br. J. Pharmac. Chemother.* **27,** 427.

BENNETT, I., DAVIDSON, A. G. H., HARDING, M. M., and MORELLE, I. (1970). *Acta crystallogr.* **B26,** 1722.

BLACK, J. W. and GANELLIN, C. R. (1974). *Experientia* **30,** 111.

——DURANT, G. J., EMMETT, J. C., and GANELLIN, C. R. (1974). *Nature, Lond.* **248,** 65.

——DUNCAN, W. A. M., DURANT, G. J., GANELLIN, C. R., and PARSONS, M. E. (1972). *Nature, Lond.* **236,** 385.

BONNET, J. J. and IBERS, J. A. (1973). *J. Am. chem. Soc.* **95,** 4829.

——and JEANNIN, Y. (1972), *Acta crystallogr.* **B28,** 1079.

————and LAAOUINI, M. (1975). *Bull. Soc. fr. Minér. Cristallogr.* **98,** 208.

BRIMBLECOMBE, R. W., DUNCAN, W. A. M., DURANT, G. J., EMMETT, J. C., GANELLIN, C. R., LESLIE, G. B., and PARSONS, M. E. (1978). *Gastroenterology* **74,** 339.

COUBEILS, J. L., COURRIERE, P., and PULLMAN, B. (1971). *C. r. hebd. Séanc. Acad. Sci., Paris* **272,** 1813.

CRITCHLEY, S. R. (1979). D.Phil. thesis, Oxford.

DECOU, D. F. (1964). Dissertation No. 64–9987. Universal Microfilms, Ann Arbor.

DURANT, G. J., EMMETT, J. C., and GANELLIN, C. R. (1973). British Patent 1338169.

——GANELLIN, C. R., and PARSONS, M. E. (1975). *J. mednl Chem.* **18,** 905.

————(1977). *Agents Actions* **7,** 39.

——EMMETT, J. C., GANELLIN, C. R., and PRAIN, H. D. (1976). British Patent 1421792.

————and WHITE, G. R. (1973). British Patent 1307539.

——DUNCAN, W. A. M., GANELLIN, C. R., PARSONS, M. E., BLAKEMORE, R. C., and RASMUSSEN, A. C. (1978). *Nature, Lond.* **276,** 403.

——EMMETT, J. C., GANELLIN, C. R., MILES, P. D., PARSONS, M. E., PRAIN, H. D., and WHITE, G. R. (1977). *J. mednl Chem.* **20,** 901.

FARNELL, L., RICHARDS, W. G., and GANELLIN, C. R. (1975). *J. mednl Chem.* **18,** 662.

GANELLIN, C. R. (1973a). *J. mednl Chem.* **16,** 620.

——(1973b). *J. Pharm. Pharmac.* **25,** 787.

——PORT, G. N. J., and RICHARDS, W. G. (1973). *J. mednl Chem.* **16,** 616.

——PEPPER, E. S., PORT, G. N. J., and RICHARDS, W. G. (1973). *J. mednl Chem.* **16,** 610.

GLOVER, P. E. (1979). D.Phil. thesis, Oxford.

GREEN, J. P., KANG, S., and MARGOLIS, S. (1971). *Mém. Soc. Endocr.* **19,** 727.

HÄDICKE, E., FRICKEL, F., and FRANKE, A. (1978). *Chem. Ber.* **111,** 3222.

HAM, N. S., CASEY, A. F., and ISON, R. R. (1973). *J. mednl Chem.* **16**, 470.

KAMENAR, B., PROUT, K., and GANELLIN, C. R. (1973). *J. chem. Soc. Perkin II*, 1734.

KANG, S. and CHOU, D. (1975). *Chem. Phys. Letts* **34**, 537.

KIER, L. B. (1968). *J. mednl Chem.* **11**, 441.

KUMBAR, M. (1975). *J. theor. Biol.* **53**, 333.

PROUT, K., CRITCHLEY, S. R., and GANELLIN, C. R. (1974). *Acta crystallogr.* **B30**, 2884.
———————and MITCHELL, R. C. (1977). *J. chem. Soc. Perkin II* 68.

PULLMAN, B. and PORT, J. (1974). *Mol. Pharmac.* **10**, 360.

REYNOLDS, W. F. and TZENG, C. W. (1977). *Can. J. Biochem.* **55**, 576.

RICHARDS, W. G., WALLIS, J., and GANELLIN, C. R. (1979). *Eur. J. mednl Chem.* **14**, 9.

VEIDIS, M. V., PALENIK, G. J., SCHAFFRIN, R., and TROTTER, J. (1969). *J. chem. Soc. A* 2659.

WASYLISHEN, R. E. and TOMLINSON, G. (1977). *Can. J. Biochem*, **55**, 579.

WITIAK, D. J. (1970). *Burger's medicinal chemistry*, Part 2, chapter 65. Interscience, New York.

YAMANE, T., ASHIDA, T., and KAKUDO, M. (1973). *Acta crystallogr.* **B29**, 2884.

20. Gramicidin S: some stages in the determination of its crystal and molecular structure

MARJORIE M. HARDING

Introduction

THE cyclic decapeptide antibiotic, gramicidin S, was isolated by Gause and Brazhnikova (1944); molecular weight and chemical sequence determination (Consden, Gordon, Martin, and Synge 1947; Synge 1948) lead to the formulation

L-pro—L-val—L-orn—L-leu—D-phe

↑   ↓

D-phe—L-leu—L-orn—L-val—L-pro

or

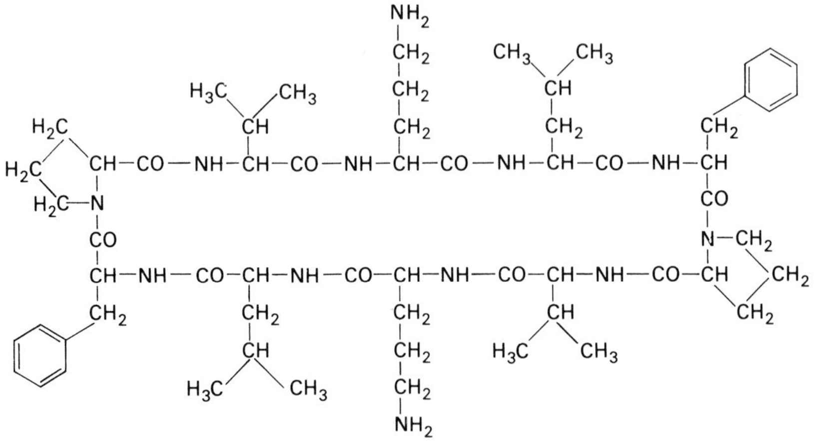

As an antibiotic its application is limited by its toxic properties, but much effort has been given to attempts to determine its crystal structure, and, since this was unsuccessful for so many years, to the prediction of its molecular structure by other means. Crystal structure determination has at last been

successful; Hull, Karlsson, Main, Woolfson, and Dodson (1978) at York
have solved the structure of the trigonal crystals of a gramicidin S urea complex. In this article I wish first to dip into some of the early history, the various
crystalline forms, the attempts at structure determination of the hexagonal
N-acetyl gramicidin S, particularly those parts in which I have been directly
involved, and the model building. Then I shall describe how the knowledge
of the molecular structure in the trigonal form has allowed at least a partial
structure determination of the hexagonal one.

Various crystalline forms

X-ray photographs of a variety of crystalline gramicidin S derivatives (prepared by R. L. M. Synge) were taken by Gerhardt Schmidt, Dorothy Hodgkin, and Beryl Oughton between 1944 and 1956 (Schmidt, Hodgkin, and
Oughton 1957). The initial aim was to establish the molecular complexity,
but the crystallographic evidence was consistent with either a pentapeptide
or a decapeptide having a twofold axis of symmetry; only the latter is consistent with the chemical synthesis and molecular weight. A suggestion could
be made about the shape of the molecule—it should have two 'layers' of
atoms about 4.8 Å apart. From these early studies, too, the N-acetyl, N-
chloroacetyl, and N-iodoacetyl derivatives (acetylated on the δ-amino
group of ornithine) were selected as most suitable for detailed X-ray analysis.
    Years later, in other parts of the world, other derivatives have been studied,
a hydrochloride, chloraurate, and iodo mercurate by Russian crystallographers (Tishchenko, Zikalova, Silantjeva, and Shadiro 1966) and a copper
complex of the bis(salicylaldimate) (Camilletti, de Santis, and Rizzo 1970)
prepared by substitution at both ornithine δ-amino groups; the latter is isomorphous with the N-acetyl derivatives. Eventually, with the great advances
in solving the phase problem by direct methods it was worth while to examine
crystalline forms without heavy atoms; the structure solved by Hull *et al.*
(1978) contains one-half urea and eight water molecules per gramicidin S
molecule.

N-acetyl gramicidin S—crystallographic work on the isomorphous series

Detailed structure analysis was started by Beryl Oughton (now Rimmer) and
Dorothy Hodgkin on the N-acetyl, N-chloroacetyl, and N-iodoacetyl derivatives (AGS, Cl-AGS and I-AGS) and the series was completed later by the
N-bromoacetyl derivative Br-AGS. All have the space group $P6_522$ (only
distinguished from $P6_122$ by the solution of the structure, see p. 203), and
1.5 molecules per asymmetric unit; therefore at least six of the eighteen decapeptide molecules, and possibly all of them, lie on crystallographic twofold
axes. Intensity data for AGS and I-AGS were estimated visually from multiple

Gramicidin S

TABLE 20.1

Some crystal data

	AGS	Cl-AGS	Br-AGS	I-AGS
Unit cell dimensions (Å) a	27.5	27.6	27.6	27.7
c	55.4	54.8	55.0	54.6
Minimum observed plane spacing (Å), along a^*	1.9	1.7	2.2	1.9
c^*	1.6	1.6	1.6	1.9
Number of independent reflections measured	810	950	605	680

AGS $=(C_{32}H_{48}N_6O_6)_2$, MW $=1140$.
Hexagonal, space group $P6_522$, $D_m = 1.165\,\mathrm{g\,cm^{-3}}$, $Z = 18$.
The (air-dried) crystals also contain ethanol and water of crystallization, probably two and three molecules respectively per AGS molecule (Schmidt *et al.* 1957).

film oscillation photographs (taken with a 4-cm radius camera, oscillation ranges of $2°$, and $CuK\alpha$ radiation); later, data for Cl-AGS and Br-AGS were measured by microdensitometer from $(CuK\alpha)$ precession photographs.

In the Patterson series of each compound there is a large peak at $\frac{1}{3}, \frac{2}{3}, \frac{1}{2}$ and a smaller peak at $\frac{1}{3}, \frac{2}{3}, 0$; pseudosymmetry is also very evident from inspection of the photographs. The asymmetric unit contains three chemically identical pentapeptide units and the pseudosymmetry can be accounted for if the relation of the three independent pentapeptides is approximately that of (x, y, z), $(\frac{1}{3}+x, \frac{2}{3}+y, \frac{1}{2}+z)$ and $(\frac{2}{3}+x, \frac{1}{3}+y, \frac{1}{2}+z)$. Table 20.2 gives details of the structure factors of a crystal with exactly this pseudosymmetry; the average value of F^2 will be different for the different classes of reflections and we can predict the ratio of these averages on the assumption that in the

TABLE 20.2

Pseudosymmetry

Class of reflection according to indices				
$h+k =$	$3n \pm 1$	$3n$	$3n \pm 1$	$3n$
$l =$	$2n$	$2n+1$	$2n+1$	$2n$
Structure factor for n atoms	F	F	F	F
Structure factor for $3n$ atoms in pseudosymmetric relation	0	$-F$	$2F$	$3F$
Thus average values of F^2 in perfectly pseudosymmetric structure are in the ratio	$0\quad :$	$1\quad :$	$4\quad :$	9

The average values of the observed F^2s in the four classes, and in ranges according to $\sin^2\theta$, are

$4\sin^2\theta/\lambda^2 = 0\text{--}0.1$	$0.1\text{--}0.2$	$0.2\text{--}0.3$
AGS $1.7:1.2:4:12.3$	$2.5:1.1:4:5.4$	$2.1:1.8:4:6.8$
Cl-AGS $1.5:1.4:4:10.6$	$2.1:1.6:4:6.3$	$1.8:2.3:4:6.3$
Br-AGS $1.5:1.5:4:11.4$	$2.4:1.4:4:6.5$	
I-AGS $1.5:1.1:4:\ 8.4$	$2.2:1.2:4:6.1$	$2.0:2.2:4:6.2$

simple structure the average is the same for each class; the predictions are compared with the observed ratios. The pseudosymmetry is by no means exact and the statistics cannot show how the real structure deviates from the exactly pseudosymmetric one.

A great variety of Patterson series, sharpened Patterson series, and difference series were calculated. The four derivatives should have provided an unambiguous solution for $|F_H|$ for reflections in the centrosymmetric zones and hence for heavy atom positions. Much time was spent trying to establish the best scales and temperature factors for the four data sets and to interpret the Patterson series. But heavy atom positions were never satisfactorily found; although some of the difficulty was no doubt caused by the pseudosymmetry it also appeared that the derivatives were not exactly isomorphous. Some of this work was done in Oxford and some in Edinburgh where I brought the data in 1962.

Models of the gramicidin S molecule

In an appendix to the paper on gramicidin S derivatives (Schmidt *et al*. 1957) Dorothy Hodgkin and Beryl Oughton described four possible molecular models, explaining that 'it may still be years before we can solve the detailed structure'. They were built with the types of conformation which had recently been proposed for α- and β-folded proteins (Pauling and Corey 1951; Pauling, Corey, and Branson 1951). They incorporated the molecular twofold axis and the layers of atoms 4.8 Å apart indicated by the crystal data and also information from the infrared spectra of the crystals of the four models. They favoured one of a β-pleated sheet type illustrated here in Fig. 20.1.

In any such model-building, bond lengths and interbond angles can be regarded as essentially constant but the torsion angles around single bonds may be varied. Several groups (Vanderkooi, Leach, Némethy, Scott, and Scheraga 1966; Liquori, de Santis, Kovacs, and Mazzarella 1966) started the search for conformations of minimum energy; as potential energy functions (and computing power) improved, these conformations were modified, and one of the latest, most favoured versions (de Santis and Liquori 1970) is shown in Fig. 20.2; it is of the same beta type as the early Hodgkin and Oughton model but less regular. Conformations were also deduced from n.m.r. and other physicochemical data (Hardy and Ridge 1973), for example by Ovchinnikov *et al*. (1970) whose set of torsion angles for the polypeptide backbone again corresponds to a rather irregular version of the beta model.

By 1970 all these approaches strongly favoured a beta model, less regular than in Fig. 20.1, but there were substantial differences in detail between the predictions, not only in the side-chain conformations but also in the backbone polypeptide chain. Energy minimization procedures are of course considering the isolated molecule, n.m.r. results refer to the molecule in a

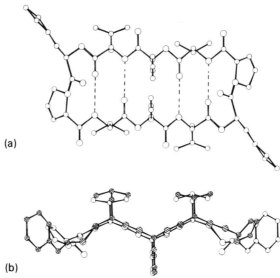

(a)

(b)

FIG. 20.1. Hodgkin and Oughton model of gramicidin S. Hydrogen atoms are omitted. Oxygen atoms are largest, then nitrogen, then carbon; (a) view along the molecular twofold axis and (b) view nearly normal to this axis. In (b) the atoms in the upper part of the molecular are shaded. (Redrawn from Schmidt *et al.* (1957).)

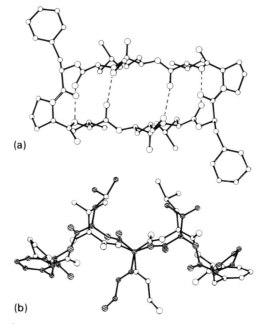

(a)

(b)

FIG. 20.2 De Santis and Liquori model of gramicidin S. (a) View along the molecular twofold axis and (b) view normal to this axis. (Redrawn from de Santis and Liquori (1970).)

particular solvent, and the crystallographers can establish the conformation in a particular crystalline environment; it is my guess that there may be some differences in side-chain arrangements but not in the backbone in these different situations. The detailed stereochemistry is undoubtedly of importance in its antibiotic activity; cyclic penta-, hexa-, and heptapeptides containing the correct amino-acid sequence are inactive as is a cyclic analogue in which the decapeptide ring has been enlarged by two carbon atoms (Hardy and Ridge 1973).

Gramicidin S in the crystalline urea complex

In 1978 the crystal structure of a hydrated gramicidin S urea complex was solved at York (Hull *et al.* 1978) using the multiple solution direct methods program MULTAN (Main 1978). This is one of the largest structures solved by the method (92 atoms other than hydrogen in the asymmetric unit) and illustrates its increasing power. There is one gramicidin S molecule per asymmetric unit and the space group, $P3_12_1$, does not require it to have any molecular symmetry. The polypeptide backbone does have twofold symmetry (to better than 0.5 Å), but the side-chains, especially ornithine and leucine, deviate substantially from it. The molecular conformation (Fig. 20.3)

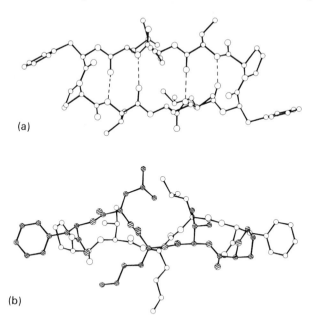

(a)

(b)

FIG. 20.3. Experimentally determined conformation of gramicidin S in trigonal crystals (Hull *et al.* 1978); (a) view along the approximate molecular twofold axis and (b) view normal to this axis.

is a distinctly distorted version of the Hodgkin and Oughton beta model, rather more distorted in the backbone chain than any of the predictions would have suggested.

N-acetyl gramicidin S—model fitting and a partial structure solution

In 1966 when isomorphous replacement failed to yield a structure and new molecular models had just been published, Edmund Komorowski and I attempted the fitting of two of these (Vanderkooi *et al.* 1966; Liquori *et al.* 1966) to the crystal data. Assuming that the pseudosymmetry is exact, if the model molecule is arranged with its diad on a crystallographic diad there remain two degrees of freedom, rotation about this diad, ϕ, and translation of the molecular centre along it, D. Initially steric maps were calculated, to represent in terms of ϕ and D the interactions between the main chain atoms (N, C_α, C_β, C, O) of adjacent molecules related by the sixfold screw axis. The molecular twofold axis may be placed either along $[10\bar{1}0]$ or along $[11\bar{2}0]$; the absolute configuration of the crystal was not known ($P6_122$ or $P6_522$) so it was necessary to explore both 6_1 and 6_5 screw axes, as well as ϕ in the range 0–180° and positive and negative values of D. Large regions in these maps indicated arrangements with unacceptably close van der Waals contacts. The acceptable, or possibly acceptable, regions were explored by structure factor calculation for a selected group of reflections, and Fourier refinement attempted on the best one. The results were inconclusive and it is now clear that the models, roughly beta and alpha in type respectively, were too different from the real structure. Five years later, with help from Robert Gould, the same procedures were used on the model published by Ovchinnikov *et al.* (1970). This time, with an R factor for the 80 selected reflections of 0.50, the molecular position close to the sixfold screw axis and the tilt of the beta-type chains were correctly indicated but refinement was still not achieved.

Eventually, when the York group had solved their crystal structure they kindly provided me with a set of atomic co-ordinates and this molecular backbone was used as a model. It could be fitted well and in a position very similar to that found above, but in that trial the sense of the polypeptide chain had been wrongly chosen (C_α—CO—N—C_α in place of C_α—N—CO—C_α) in relation to the crystal symmetry elements. Electron-density maps, phased on this molecular backbone, have indicated the regions occupied by side-chains; ornithine and phenylalanine are particularly clear. The R factor is still high, 0.40 for all observed reflections in AGS. Refinement will be continued, making as much use as possible of constraints to bond lengths and angles. The pseudosymmetry and the limited data will make it slow and there are hints of disorder in some regions, particularly the acetyl groups on ornithine.

In conclusion we can describe the structure of the hexagonal crystals as

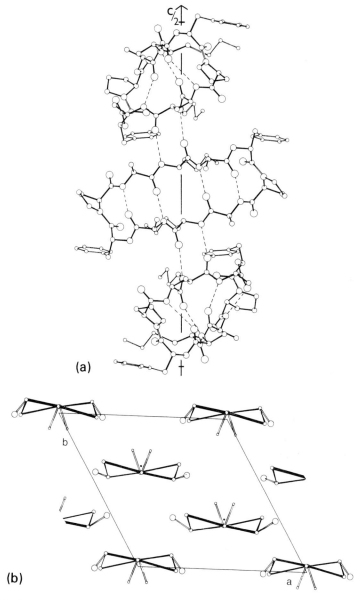

FIG. 20.4. (a) Hexagonal *N*-acetyl gramicidin S: part of a column of molecules up a 6_5 screw axis, between 0 and $c/2$. The centre molecule is viewed along its twofold axis (at $c/4$). Valine and leucine are omitted. (b) Hexagonal *N*-acetyl gramicidin S: one 'layer' in projection along c, showing all the molecules with molecular twofold axes at $c/4$. Each molecule is represented by the C_α atoms of phenylalanine, proline, and ornithine, the phenyl group (large circle), and one atom of the ornithine side-chain.

follows and it is illustrated in Fig. 20.4. The space group is $P6_5 22$. The molecular twofold axis is parallel to $[10\bar{1}0]$, and the molecules are stacked up the sixfold and threefold screw axes (the latter would be locally sixfold screw if the pseudosymmetry were exact); the beta chains are not normal to c but are tilted at 70–75° to c. There are hydrogen bonds of 2.8 Å between ornithine NH and CO of neighbouring sixfold screw related molecules. The general character of the structure, stacks of molecules tightly hydrogen bonded along c with looser association between the stacks, accounts for the marked anisotropy of the intensity data. There is at present no reason to suggest that the molecular backbone is different from that determined at York; the side-chains have not yet been very precisely located, but it is already evident that there are some differences between those of the three pseudosymmetry related molecules.

Acknowledgements

Many people have been associated with the work on gramicidin S at different times: Maureen Innes and David Dale in Oxford and Elizabeth Brackett and Christopher Clayton in Edinburgh should be mentioned in addition to those who appear earlier in the text. I am particularly grateful to Robert Gould for many useful discussions and suggestions. I also thank Peter Main and Stephen Hull for sending me their atomic co-ordinates, Sam Motherwell for the program PLUTO which produced the drawings, and the Edinburgh Regional Computing Centre for their co-operation.

References

CAMILLETTI, G., SANTIS, P. DE, and RIZZO, R. (1970). *Chem. Commun.* **17**, 1073.
CONSDEN, R. J., GORDON, A. H., MARTIN, A. J. P., and SYNGE, R. L. M. (1947). *Biochem. J.* **41**, 596.
GAUSE, G. F. and BRAZHNIKOVA, M. G. (1944). *Nature, Lond.* **154**, 703.
HARDY, P. M. and RIDGE, B. (1973). *Prog. Org. Chem.* **8**, 129.
HULL, S. E., KARLSSON, R., MAIN, P., WOOLFSON, M. M., and DODSON, E. J. (1978). *Nature, Lond.* **275**, 206.
LIQUORI, A. M., SANTIS, P. DE, KOVACS, A., and MAZZARELLA, L. (1966). *Nature, Lond.* **211**, 1039.
MAIN, P. (1978). *MULTAN 78—a program for the automatic solution of crystal structures from X-ray diffraction data.* University of York.
OVCHINNIKOV, YU. A., IVANOV, V. T., BYSTROV, V. F., MIROSHNIKOV, A. I., SHEPEL, E. N., ABDULLAEV, N. D., EFREMOV, E. S., and SENYAVINA, L. B. (1970). *Biochem. biophys. Res. Commun.* **39**, 217.
PAULING, L. and COREY, R. B. (1951). *Proc. natn. Acad. Sci. U.S.A.* **37**, 729.
——and BRANSON, H. R. (1951). *Proc. natn. Acad. Sci. U.S.A.* **37**, 205.
SANTIS, P. DE and LIQUORI, A. M. (1970). *Biopolymers* **10**, 699.
SCHMIDT, G. M. J., HODGKIN, D. C., and OUGHTON, B. M. (1957). *Biochem. J.* **65**, 744.
SYNGE, R. L. M. (1948). *Biochem. J.* **42**, 99.
TISHCHENKO, G. N., ZIKALOVA, K. A., SILANTJEVA, I. A., and SHADIRO, L. L. (1966). *Acta crystallogr.* **21**, A165.
VANDERKOOI, G., LEACH, S. J., NÉMETHY, G., SCOTT, R. A., and SCHERAGA, H. A. (1966). *Biochemistry* **5**, 2991.

21. Connections in the crystal structures of mercury compounds

DRAGO GRDENIĆ

ACCORDING to the connection principle, the links of a given atom with other atoms are selected in order to find out the building pattern of the structure, which may be composed of isolated groups or islands, chains, ribbons, columns, nets, layers, or lattices. The best example of such an approach is the structure of silicates in which the Si—O bond as the shortest link between atoms in the structure is used to characterize the structure. This shortest-bond criterion had been tacitly applied in the past, but it was explicitly defined for the first time by F. Laves (1967): 'If all links', he said, 'except the shortest are dropped, those atoms that are still connected with each other form a connection.' The description of crystal structure in terms of connection is particularly convenient because it differentiates the nature of bonding in the structure and makes it possible to classify different structures according to the same principle.

When I was about to write a review of the structural chemistry of mercury the first problem I faced was the classification (Grdenić 1965). I found it convenient to divide all observed co-ordinations in which mercury occurs in crystal structures of its compounds into two groups: the characteristic and the effective co-ordination. Later on, when it was proposed that I should write a short review on the crystal chemistry of mercury (Grdenić 1969), I used the connection principle to classify various structures of mercury compounds and minerals. Here I am going to show that the connection principle can be used for classification of all mercury compounds and that the characteristic and effective co-ordinations are only special cases of the connection principle.

Characteristic co-ordination

For the stereochemistry of mercury, the typical pattern is molecules or complexes with two collinear, three co-planar, or four tetrahedral bonds at the mercury atom formed by means of the sp, sp^2, or sp^3 hybrid orbitals. Few examples of mercury in a regular octahedral co-ordination are known, while only one structure with mercury in a regular trigonal bipyramidal co-ordination has been observed so far. In terms of the valence-bond method the geometry of these last two co-ordinations is likely based upon the sp^3d^2 and sp^3d hybrid orbitals respectively.

To these five characteristic co-ordinations of mercury, a characteristic co-ordination number m is assigned, which by its value of $m = 2$, 3, 4, 5, and 6 defines not only the number of the ligand atoms L in HgL_m but also the geometry, i.e. the digonal collinear, trigonal co-planar, tetrahedral, trigonal bipyramidal, and octahedral co-ordination.

For each value of m a corresponding covalent radius of mercury $r_m(Hg)$ can be determined with the condition that the additivity rule is satisfied, i.e. that the sum $r_m(Hg) + r(L)$ is in reasonable agreement with the observed value of the Hg—L bond length. The most reliable are the $r_2(Hg)$ values of 1.34 Å proposed by Aurivillius (1965) and 1.30 Å independently by myself (1965), as well as the $r_4(Hg)$ value of 1.48 Å (Pauling and Huggins 1934). Less reliable are the $r_m(Hg)$ values for other co-ordinations. By using observed values for Hg—L bond lengths, preferentially those with small electronegativity difference between mercury and the ligand atom, I completed the set of $r_m(Hg)$ radii and recommended the values 1.30 (or 1.34), 1.39, 1.49, 1.60, and 1.68 Å

TABLE 21.1

Molecular and complex ionic species as finite connections (HgL_m) with mercury atom in the characteristic co-ordination (m)

HgL_m	Geometry of co-ordination	Selected† examples and references
HgL_2	Digonal collinear	$HgCl_2$ (Braekken and Sholten 1934; Grdenić 1950), CH_3HgCl (Grdenić and Kitaigorodsky 1949), HgO_2^{2-} (Hoppe and Röhrborn 1964), $Hg(SCN)_2$ (Beauchamp and Goutier 1972), $HgPh_2$ (Grdenić, Kamenar, and Nagl 1977), $Hg(CF_3)_2$ (Brauer, Bürger, and Eujen 1977), Hg(2-thienyl)$_2$ (Grdenić, Kamenar, and Žeželj 1979)
HgL_3	Trigonal planar	$HgBr_3^-$ (Brodersen 1955), HgI_3^- (Fenn 1966a; Grdenić, Sikirica, and Vicković 1977), $HgCl_3^-$ (Biscarini, Fusina, Nivellini, and Pelizzi 1977; Sandström and Liem 1978), $Hg(TePh)_3^-$ (Behrens, Hoffman, and Klar 1977)
HgL_4	Tetrahedral	HgI_4^- (Fenn 1966b), $Hg(SCN)^{2-}$ (Jeffery and Rose 1968; Sakhri and Beauchamp 1975), $HgCl_4^{2-}$ (Mason, Robertson, and Rusholm 1974), $HgBr_4^{2-}$ Kamenar and Nagl 1976), HgS_4^{6-} (Sommer, Hoppe, and Jansen 1976)
HgL_5	Trigonal-bipyramidal	$HgCl_5^{3-}$ (Clegg, Greenhalgh, and Straughan 1975)
HgL_6	Octahedral	Hg(pyridine 1-oxide)$_6^{2+}$ (Kepert, Taylor, and White 1973), $Hg(DMSO)_6^{2+}$ (Sandström and Persson 1978), $Hg(OH_2)_6^{2+}$ (Johansson and Sanström 1978)

† The structures with the best agreement between the observed and ideal co-ordination geometry are given as examples.

for $m=2$, 3, 4, 5, and 6 respectively. It is interesting that these values of $r_m(\text{Hg})$, when plotted against the characteristic co-ordination number m, lie closely on a straight line given by the equation: $r_m(\text{Hg})=0.095m+1.110$ (Grdenić 1979).

The knowledge of the covalent radius of mercury for different co-ordination number m allows us the possibility of distinguishing the HgL polyhedra in the structure. All Hg—L bonds with length equal or close to the sum of covalent radii, r_m (Hg)+r(L), belong to a characteristic molecular or ionic species HgL_m which we consider now as a finite connection, as an 'island' in the structure. Selected examples of the molecular and ionic species with HgL_m as finite connections are given in Table 21.1.

Effective co-ordination

However, the actual co-ordination in which mercury frequently occurs in the crystal structure, and which I have called effective co-ordination, is irregular and can be described only in terms of deformed polyhedra, mostly tetrahedra, pyramids, or octahedra. The bond lengths and angles at the mercury atom are within a wide range of values which do not obey the rules either for the additivity of covalent radii or for the orientation of bonds. The effective co-ordination is a secondary formation, derived from the characteristic co-ordination by the interaction of the mercury atom with additional ligand atoms. Therefore, the group defined with HgL_m appears in the crystal structure as $\text{HgL}_m\text{L}'_{n-m}$ if n is the effective co-ordination number and L' any additional ligand atom. The value of the number n is determined by the mercury co-ordination sphere which includes all Hg - - - L and Hg - - - L' interactions sufficiently strong to be taken into account. Accordingly, the distances within the effective co-ordination polyhedron should be less than the sum of the corresponding van der Waals radii, i.e. less than $R_W(\text{Hg})+R_W(\text{L}')$. The value of 1.50 Å for $R_W(\text{Hg})$, which I recommended on the basis of few available data, has been improved recently (Grdenić 1977) to 1.54 Å thanks to many data from structural analyses. It is practically equal to the value of 1.55 Å recommended by A. Bondi (1964).

The effective co-ordination of the mercury atom is based mostly upon two collinear or three co-planar bonds. In both these cases the mercury atom is open to other ligands, either from the medium where the compound was prepared, from the same molecule, or from the adjacent atoms or groups in the crystal structure. The resulting co-ordinations are given by the general formulae $\text{HgL}_2\text{L}'$, $\text{HgL}_2\text{L}'_2$, $\text{HgL}_2\text{L}'_4$ for $m=2$, or $\text{HgL}_3\text{L}'$ and $\text{HgL}_3\text{L}'_2$ for $m=3$. In the former case the collinearity is more or less disturbed, the previous Hg—L bonds lengthened according to the intensity of the interaction with the L′ atoms of the additional ligands. Analogous deformations are observed in the latter case.

The criterion for the effective co-ordination gives another possibility to distinguish groups of atoms as finite connections in the structure. It is used for the explanation of the mercury co-ordination sphere in the crystal structure. We are not going to use it here. It focuses our attention on one mercury atom and its co-ordination polyhedron but does not tell us anything about the bonds of the ligands with the neighbouring mercury atoms. We are going to use a more restrictive criterion, the Hg—L bond lengths with value close to the sum of the covalent radii. In other words, in defining connection we use the characteristic co-ordination of mercury.

Infinite connections

When the valency or donating capacity of a ligand is saturated by one Hg—L bond ($k=1$) the HgL_m connection is finite. But, because mercury, although divalent, does not form double bonds, all ligands or ligand atoms with $k>1$ are free to link another mercury atom (or atoms) giving rise to an infinite connection. When the remaining valency is used for the ionic bond the result is a finite connection, e.g. HgO_2^{2-} or HgS_4^{6-} (Table 21.1).

One expects that the most stable characteristic co-ordination is likely to form infinite connections. That is actually what happens: all observed infinite connections with mercury are based upon the characteristic co-ordination with $m=2$ or 4, i.e. with mercury in collinear and tetrahedral co-ordination. What is unexpected is the ability of mercury to use two collinear bonds for infinite connection: the bond angle of 180° and two relatively long bonds (2.05 Å for Hg—O) seems to be not favourable for satisfactory space filling. But the effective co-ordination with the ligands from the adjacent atoms or the anions, in the case of a cationic connection, compensates that stereochemical disadvantage.

Therefore, there are two groups of infinite connections with mercury, the first one derived from mercury in collinear ($m=2$) and the second from the mercury in tetrahedral ($m=4$) characteristic co-ordination.

An additional subdivision in both groups is necessary according to the valency k of the ligand atom ($k=2$, 3, or 4) because there are m/k ligand atoms per one mercury atom in the connection based upon the HgL_m co-ordination. Consequently, in the subgroup with $m=2$, the possible connections are $HgL_{2/2}$ $HgL_{2/3}$, and $HgL_{2/4}$. Thus, in $HgL_{2/2}$ each ligand atom belongs to two mercury atoms, in $HgL_{2/3}$ to three mercury atoms, and in $HgL_{2/4}$ to four mercury atoms, and the connections are chains (1D), ribbons or nets (2D), or three-dimensional lattices (3D) respectively. The examples of structures containing these connections are given in Table 21.2.

TABLE 21.2

Infinite connections $(HgL_{m/k})$ *with collinear* $(m = 2)$ *Hg—L bonds and the ligand atoms L of valency* $k = 2, 3,$ *and 4*

$HgL_{m/k}$	Connection type and composition		Compound and references
$HgL_{2/2}$	Planar zigzag $(+-)$ chain	HgO	in the orthorhombic HgO (Aurivillius 1956, 1964)
		$Hg(OH)^+$	in the bromate (Weiss, Lyng, and Weiss 1960: Björnlund 1971), in the fluoride (Grdenić and Sikirica 1972; Nozik, Fykin, Bukin, and Lantash 1979; Stålhandske 1979), in the nitrate (Matković, Ribar, Prelesnik, and Herak 1974)
		$Hg(NH_2)^+$	in the chloride and bromide (Lipscomb 1951; Nijsen and Lipscomb 1952; Rüdorff and Brodersen 1952; Brodersen and Rüdorff 1954)
		HgI^+	in the hexafluorotitanate (Breitinger and Köhler 1976)
	Planar $(++--)$ chain	HgO	in $Hg(OHg)_4Br_2$ (Aurivillius 1965, 1968)
	Helix	HgO	in hexagonal HgO (Aurivillius and Carlsson 1958)
		HgS	in hexagonal HgS (Aurivillius 1950)
$HgL_{2/3}$	Net	$Hg_3O_2^{2+}$	in the sulphate (Nagorsen, Lyng, Weiss, and Weiss 1962; Bonefačić 1963)
		$Hg_3(NH)_2^{2+}$	in 'imidobromide' (Brodersen 1955)
		$Hg_3S_2^{2+}$	in γ-$Hg_3S_2Cl_2$ (Durovič 1968)
	3D lattice	$Hg_3S_2^{2+}$	in α-$Hg_3S_2Cl_2$ (Puff and Küster 1962; Aurivillius 1967)
$HgL_{2/4}$	3D lattice	Hg_2N^+	in Millon's base and its salts (Lipscomb 1951; Rüdorff and Brodersen 1953)
		Hg_2O^{2+}	in the pyrochlore type niobiate (Sleight 1968)
		Hg_2S^{2+}	in the pyrochlore type hexafluoronickelate (Champlon, Bernard, Pannetier, and Lucas 1974)

Infinite connections with collinearly linked mercury

In the structure of mercury(II) oxide the infinite $HgO_{2/2}$ connection appears in two conformations: as a planar zigzag chain in the common orthorhombic (red or yellow) modification and as a helix in the less known hexagonal modification. For mercury(II) sulphide only the helical conformation is known in the structure of cinnabar. The infinite HgO chain can also appear in a composite structure. In $(HgO)_2NaI$ (Aurivillius 1964) and in $\alpha Hg_2V_2O_7$ (Quarton, Angenault, and Rimsky 1973) it is like that in orthorhombic HgO, but in $Hg(OHg)_4Br_2$ it appears in a particular planar $(++--)$ conformation which does not correspond to any known modification of mercury(II) oxide. However, the oxygen atoms are bridged over by an extra mercury atom and a double net is formed as the infinite $Hg(OHg)_4^{2+}$ connection.

In the infinite $HgL_{2/2}$ chain L can also be any bivalent atom or atomic group which can provide a convenient Hg—L—Hg bond angle. So, iodine in $(HgI)_2TiF_6$ appears as iodonium in an infinite and planar zigzag HgI^+ cation. The infinite $Hg(OH)^+$ cation in the same conformation occurs in some basic salts of the type $Hg(OH)X$, with X=chlorate, bromate, nitrate, or fluoride. The chain in the basic sulphate $Hg_3(OH)_2(SO_4)_2 \cdot H_2O$ is finite, it consists only of two $Hg(OH)^+$ links with the $—OSO_3^-$ and $—HgOSO_3^-$ groups at the respective ends of the chain (Aurivillius and Stålhandske 1976). Better known is the $Hg(NH_2)^+$ chain in amidomercury(II) halides. Inasmuch as the mercury and nitrogen atoms are concerned the geometry of the chain is analogous to that observed in the orthorhombic mercury(II) oxide.

When oxygen stands for L in $HgL_{2/3}$ the connection is two-dimensional, because three Hg—O bonds form a flattened pyramid characteristic for the oxonium ion. The resulting infinite $Hg_3O_2^{2+}$ cation is a slightly puckered hexagonal net and occurs in the basic sulphate as well as in the isostructural selenate and chromate. The analogous $Hg_3(NH)_2^{2+}$ cation has been observed so far in $Hg_3(NH)_2Br(HgBr_3)$, known previously as mercuric imidobromide, Hg_2NHBr_2.

With the sulphur atom introduced instead of oxygen the stereochemical conditions are changed. The sulphonium pyramid is not flat, the Hg—S—Hg bond angle is about $95°$ and a $Hg_3S_2^{2+}$ net similar to the $Hg_3O_2^{2+}$ net probably does not exist. In α-$Hg_3S_2Cl_2$ the Hg_3S pyramids are connected by common mercury atoms in a highly symmetrical cubic 3D lattice. In a less stable orthorhombic γ-modification the rings of four Hg_3S pyramids are connected in pairs chequerwise arrayed in a puckered layer. Such rings can be connected like stairs in one direction with the common mercury atoms in the *anti* position. The resulting $Hg_3S_2^{2+}$ ribbon was observed in the crystal structure of the acetate, $Hg_3S_2(ac)_2$, by the author and co-workers.

The $HgL_{2/4}$ connection does not offer many varieties. The representative is the structure of Millon's base, $(Hg_2N)OH \cdot 2H_2O$, and its salts. The Hg_4N tetrahedra are connected by common mercury atoms. The 3D lattice of a tetrakismercurioammonium cation, Hg_2N^+, is realized in a cubic and a hexagonal version like the lattice of SiO_2 in its cubic (β-cristobalite) and hexagonal (β-tridymite) modification.

The introduction of an L different from N in $HgL_{2/4}$ is possible but success depends upon the choice of a proper method of preparation. So oxygen in the tetrahedral co-ordination (k=4) offers almost the same sterical conditions as the ammonium nitrogen and only a suitable anion in the cavities is needed to stabilize the $HgO_{2/4}$ connection. Such a connection was actually found as Hg_2O^{2+} cation in mercury(II) niobiate, $Hg_2Nb_2O_7$, and its congeners with the pyrochlore structure. The SiO_2 framework pattern of Hg_2N^+ in Millon's base is replaced by the Cu_2O framework which is spacious enough to accommodate niobiate anions in the form of NbO_6 octahedra sharing corners.

Infinite connections with tetrahedrally linked mercury

In the tetrahedral co-ordination ($m=4$) the chances of the mercury atom for connections, at least in a steric sense, seem to be not less than in the collinear co-ordination; nevertheless such connections are not numerous. Only those with the iodine, sulphur, selenium, and tellurium atom as common tetrahedron corner are known at present (Table 21.3). The structure containing the 2D or 3D connections with the tetrahedral $HgL_{4/2}^{2-}$ unit, where L=chalcogen, is a challenge for preparative chemistry. In an indirect way, over the S—C—N—Co bridges, such a connection is realized in the structure of

TABLE 21.3

Infinite connections ($HgL_{m/k}$) with tetrahedral (m=4) $Hg-L$ bonds and the ligand atoms L of valency k=2 and 4

$HgL_{m/k}$	Connection type and composition		Example of the structure and references
$HgL_{4/2}$	Layer	HgI_2	in the tetragonal (red) HgI_2 (Bijvoet, Classen and Karssen 1926)
$HgL_{4/4}$	3D lattice	HgS	in the black HgS, metacinnabarite (Kolkmeijer, Bijvoet, and Karssen 1924; Aurivillius 1964)
		HgSe	in HgSe, tiemannite (Hartwig 1926; Zachariasen 1926)
		HgTe	in HgTe, coloradoite (Hartwig 1926; Zachariasen 1926)

bispyridine adduct of cobalt(II) mercury(II) tetrathiocyanate (Beauchamp, Pazdernik, and Rivest 1976). By the way, the double net in the structure of $Hg(OHg)_4Br_2$ (Aurivillius 1968) may be also interpreted in terms of connected HgO_4 tetrahedra. In combination with the PS_4 tetrahedron from the P_2S_7 groups the HgS_4 tetrahedron is built in a 3D connection of the $Hg_2P_2S_7$ structure (Jandali, Eulenberger, and Hahn 1978a). The HgS_4 tetrahedron occurs also in the 2D connection of the $Hg_2P_2S_6$ structure (Jandali, Eulenberger, and Hahn 1978b).

Composite connections

When the joining ligand is a group of atoms, instead of one atom, the connection type does not change necessarily, e.g. the structure of mercury(II) peroxide (Vannerberg 1959) is composed of the —Hg—O—O— planar zig-zag chains related to the orthorhombic mercury (II) oxide chains. The oxo-acid anions are the most frequent polyatomic ligands with $k = 2$, 3, or 4. The O—Hg—$O(CrO_2)O$—Hg chain in mercury (II) chromate and its hemihydrate is a typical and simple example of such a connection.

In our notation L stands not only for the ligand atom but also for a group of atoms, so that the connections with the oxo-acid anions might be also included in Table 21.2. The simplest are the O—Hg—$O(CrO_2)O$—Hg chains in mercury(II) chromate (Stålhandske 1978) and its hemihydrate (Aurivillius and Stålhandske 1975). In the former the chain is comparable to that in ortho-rhombic HgO, while in the latter the chromate groups are along the one side of the chain and the characteristic zigzag conformation is lost. The structure of anhydrous mercury sulphate, where the chain is deformed to the O—Hg—O bond angle of $157°$, was described in terms of a three-dimensional frame-work (Aurivillius 1965). Yet it seems that it should be classified as built up from O—Hg—$O(SO_2)O$—Hg chains. In the structure of $HgMoO_4$ the O—Hg—O—Mo—O—Hg chains can be recognized, but the structure is better described in terms of MoO_6-octahedra with shared edges and connected by the mercury atoms (Jeitschko and Sleight 1973). The PO_4^{3-} ion in $Hg_3(PO_4)_2$ behaves like a ligand L with $k = 3$. The $Hg(PO_4)_{2/3}$ nets are best described in terms of rows of $-Hg$—$O(PO_2)O-$ chains, running across to each other, interconnected by additional mercury(II) ions. The nets penetrate each other over centres of symmetry giving a remarkable and unique orthophosphate structure (Aurivillius and Nilson 1975).

In the typical infinite connections (Table 21.2), when L stands for oxygen, the neutral $Hg_{2/2}O$ group, the dimercurated $Hg_{2/2}(OH)^+$, or trimercurated $Hg_{3/2}O^+$ cation appear as structural units. These units combine by giving particular connections. In the last few years several surprisingly sophisticated connections were discovered in the structure of mercury oxohalides and basic salts.

For the structure of $Hg_3O_2Cl_2$ there are two descriptions, the first one based upon two-dimensional photographic data (Šćavničar 1955) and the second solved from the diffractometric data by modern methods (Aurivillius and Stålhandske 1974). It seems that the structure is best described in terms of hexagonal $Hg_3O_2^{2+}$ nets parallel to ($10\bar{1}$) plane. The nets are deformed owing to close Hg---O contacts between them and an apparent irregular trigonal co-ordination about one half of the mercury atoms.

In another oxychloride of the same composition the $(+ + - -)$ HgO chains, in the form of an $(HgO)_4$ repeating unit with one oxygen joined to the $-HgCl$ group, are connected by two mercury atoms per unit. The result-ing $Hg_6O_4Cl^{3+}$ connection has three $Hg_{3/2}O^+$ cations and the oxychloride has to be formulated as $Hg_6O_4Cl_4$ (Aurivillius and Stålhandske 1978). The structure was solved also by other authors (Neuman, Petersen, and Lo 1976).

In basic mercury(II) perchlorates new remarkable connections were found. In $2Hg(ClO_4)_2 \cdot 3HgO \cdot xH_2O$ the cation $[Hg_5O_2(OH)_2]^{4+}$ is a flat ribbon composed of hexagons formed by two zigzag Hg—O—Hg(OH)— chains linked together by mercury(II) ions (Johansson 1971a). In HgO·Hg(OH)ClO$_4$ the mercury–oxygen connection is a unique ribbon composed of a planar zigzag HgO chain to every oxygen of which one HgO(HgOH)$_2$ group is attached. In the Hg$_4$O$_2$(OH)$_2^+$ connection the carriers of the positive charge are again Hg$_{3/2}$O$^+$ cations (Johansson and Hansen 1972). In Hg$_7$(OH)$_2$O$_4$(ClO$_4$)·H$_2$O, as defined by Johansson (1971b), the cation is a column composed of two parallel Hg$_3$O$_2^+$ chains crosslinked alternately from both sides by the Hg4(OH)$_2$O$_2^+$ bridges. Again, the structure units are mercurated oxonium cations but connected in a particular way observed so far only in the structure of this basic perchlorate.

Connections with the mercury atom pairs

The pair of mercury atoms in the mercury (I) compounds prefer to form two covalent bonds with the ligand atoms L by giving the finite L—Hg—Hg—L chain connection with collinear or essentially collinear bonds. The crystal structure of a mercury(I) compound with definitely ionic character is not known. The crystal structure of the phosphate, AgHg$_2$PO$_4$, as well as the isostructural arsenate (Masse, Guitel, and Durif 1978) seems to be predominantly ionic. It is a packing of Hg$_2$ pairs between layers of PO$_4$ tetrahedra in which each mercury atom is in contact with three oxygen atoms in a deformed pyramid. A relatively long Hg—Hg bond distance of 2.608 Å also speaks in favour of the ionic nature of the structure.

In the structure of hydrated salts the [H$_2$O—Hg—Hg—OH$_2$]$^{2+}$ aqua-complex is always present. In other complexes water is replaced by a moderately basic ligand whose donor atom takes over the positive charge from the mercury atom.

The Hg$_2$L$_2$ and Hg$_2$L$_2^{2+}$ species are mostly symmetric in respect of the ligands (Table 21.4) and in the crystal structure they retain symmetry of 1 or 2, even 2/m, as in the acridine complex. Owing to polar interaction or packing condition the Hg—Hg—L bond collinearity may be slightly lost in the crystal structure, the greatest departure among the examples listed being observed for the chinoline complex where the Hg—Hg—N bond angle is 164.5°. In Hg$_2$(phen)(NO$_3$)$_2$ the ligands are different, the bidentate phen-ligand at one mercury atom and the unidentate nitrate at the other (Elder, Halpern, and Pond 1967).

Several complexes with more than one ligand per mercury atom in the Hg$_2$ pair have been recently prepared and studied. In most of them mercury has a fairly regular co-ordination, particularly in diphenylselenium and triphenylphosphinoxide complexes. This is not the case in Hg$_2$(pyridine 1-oxide)$_4$(ClO$_4$)$_2$ where the co-ordination is neither regular nor identical for both mercury atoms (Kepert, Taylor, and White 1973c).

T ABLE 21.4

Finite connections Hg_2L_{2n} (n = 1, 2, and 3) with Hg_2 pairs in the structure of mercury(I) compounds

Hg_2L_{2n}	Co-ordination type and composition	References
Hg_2L_2	Collinear	
	Hg_2X_2 (X = F, Cl, Br, and I)	Dorm (1971a)
	$Hg_2(BrO_3)_2$	Dorm (1967)
	$Hg_2(OCOCF_3)_2$	Sikirica and Grdenić (1974)
	$Hg_2(H_2PO_4)_2$	Nilson (1975)
$Hg_2L_2^{2+}$	$Hg_2(OH_2)_2(NO_3)_2$	Grdenić (1956); Grdenić, Sikirica, and Vicković (1975)
	$Hg_2(OH_2)_2(ClO_4)_2 \cdot 2H_2O$	Johansson (1966)
	$Hg_2(OH_2)_2(SiF_6)_2$	Dorm (1971b)
	$Hg_2(4\text{-cyanopyridine})_2(ClO_4)_2$	Kepert, Taylor, and White (1972)
	$Hg_2(3\text{-chloropyridine})_2(ClO_4)_2$	Keppert, Taylor, and White (1973a)
	$Hg_2(\text{chinoline})_2(NO_3)_2$	Brodersen, Hacke, and Liehr (1975)
	$Hg_2(1,8\text{-naphthyridine})_2(ClO_4)_2$	Dewan, Kepert, and White (1975)
	$Hg_2(\text{acridine})_2(ClO_4)_2$	Taylor (1976)
$Hg_2L_4^{2+}$	Trigonal planar	
	$Hg_2(4\text{-benzylpyridine})_4(ClO_4)_2$	Taylor (1977)
	$Hg_2(SePh_2)_4(ClO_4)_2$	Brodersen, Liehr, Rosenthal, and Thiele (1978)
$Hg_2L_6^{2+}$	Deformed tetrahedral	
	$Hg_2(OPPh_3)_6(ClO_4)_2$	Kepert, Taylor, and White 1973b)

When the valency or the donating capacity of the ligand is greater than one ($k > 1$) connections with the Hg_2 pairs like those with the Hg atoms are formed. Thus, in respect to stereochemistry, both units, —Hg— and —Hg—Hg—, resemble each other so that some Hg_2 connections may be considered as homologues of the corresponding Hg connection. The $Hg_2L_{2/2}$ chain corresponds to the $HgL_{2/2}$ zigzag chain, and the structure of mercury(I) sulphate (Dorm 1969) contains infinite $-Hg-Hg-O(SO_2)O-$ connections comparable to those found in mercury (II) chromate. In the structure of mercury(I) *o*-phthalate, according to the same model, the Hg_2 pairs are connected by sharing one phthalate ion but dividing carboxyls (Lindh 1967). In the polymeric complexes formed by 1,4-diazine (Brodersen, Hacke, and Liehr 1974) and 1,3-dithiane (Brodersen, Liehr, and Rölz 1977), the $Hg_2L_2^{2+}$ connection is almost linear, with the Hg—Hg—L bond angle of 168 and 174°, and has not a proper analogue among mercury(II) 1D connections. In mercury(I) phosphate-nitrate-hydrate the PO_4^{2-} ion uses only two of its three valencies to connect Hg_2 pairs into a chain which is again a zigzag of the Hg_2L_2 type. The third oxygen is used to connect a terminal —Hg—Hg—OH$_2^+$ branch, and the compound is the nitrate of the polymeric $[Hg_2(PO_4)Hg_2(OH_2)]^+$ cation which can be classified as a ribbon (Durif, Tordjman, Masse, and

Guitel 1978). The connections of which the crystal structure of mercury(I) arsenate is built up are also unique. The $AsO_4(Hg_2)_2AsO_4$ rings are connected by the Hg_2 pairs to an endless ribbon of the composition $-Hg_2-OAs(O)O_2(Hg_2)_2O_2As(O)O-$ (Kamenar and Kaitner 1973).

The only 3D connection with Hg_2 pairs was found in the structure of eglestonite, a rare mineral with the chemical composition $(Hg_2)_3O_2HCl_3$ as formulated recently (Mereiter and Zemann 1976). The $(Hg_2)_3O_2^{2+}$ polymeric cation is at once recognized as a homologue of $Hg_3O_2^{2+}$. Actually, it is a 3D connection and can be compared with the $Hg_3S_2^{2+}$ connection in the cubic $Hg_3S_2Cl_2$ (Table 21.2). Four such interpenetrating $(Hg_2)_3O_2^{2+}$ frameworks are contained in the unit cell, and are linked in pairs by hydrogen bridges. The $Hg_2(OH_2)^{2+}$ in hydrated salts and the $(Hg_2)_{3/2}O^+$ in eglestonite are the only oxonium cations observed so far in the mercury(I) series.

Connections of catenated mercury atoms

The relative stability of the Hg_2 pair and its preference for the covalent rather than ionic bonds with ligands indicates a distinct property of mercury for catenation. The doublet can be extended to a triplet or quadruplet in the same way as the L—Hg—Hg—L complexes are formed from Hg_2^{2+}. Actually, the compounds containing Hg_n chains, with $n=3$, 4, and ∞, were discovered quite recently. They are listed in Table 21.5 according to the data from a review article (Cutforth, Gillespie, and Ummat 1976).

The Hg_3 chain in chloroaluminate is almost linear, but the Hg---Cl contact of only 2.562 Å, collinear with the Hg—Hg bonds, is notable. In the fluoroarsenate the Hg_3 group is linear, centrosymmetric, but with ionic

TABLE 21.5

Hg_n connections with n > 2

Connection type	Compound and references
Finite connections	
Hg_3^{2+} linear	$Hg_3(AlCl_4)_2$, Ellison, Levy, and Fung (1972)
	$Hg_3(AsF_6)_2$, Cutforth, Davies, Dean, Gillespie, Ireland, and Ummat (1973)
	$Hg_3(Sb_2F_{11})_2$, Cutforth *et al.* (1973)
Hg_4^{2+} linear	$Hg_4(AsF_6)_2$, Cutforth, Gillespie, and Ireland (1973)
Infinite connections	
$Hg_n^{0.35n+}$	$Hg_{2.86}AsF_6$, Brown, Cutforth, Davies, Gillespie, Ireland, and Vekris (1974)
$Hg_n^{0.34n+}$	$Hg_{2.91}SbF_6$, Cutforth, Gillespie and Ummat (1976)

character of the Hg---F contact. The Hg_4 group is nearly linear, centrosymmetric, in the *trans* conformation.

The culmination has been the discovery of the infinite linear $Hg_n^{0.35n+}$ cation in a fluoroarsenate of composition $Hg_{2.86}AsF_6$. The tetragonal unit cell is defined by packing of the AsF_6 anions, and the infinite linear cations, running through the channels in two mutually perpendicular directions, do not match the cell dimensions. The Hg—Hg distance of 2.64 Å is larger than that in dimer, trimer, and tetramer.

Are mercury clusters possible?

The formation of clusters is another version of catenation, yet the collinearity of the Hg—Hg bonds has to be given up. Until now only one structure is known with three mercury atoms linked in an equilateral triangle with Hg—Hg sides of 2.71 Å. The compound known as $Hg_2Cl_2 \cdot 2HgO$ and as mineral terlinguaite has a layer structure with $Hg_4O_2Cl_2$ as the repeating unit (Ščavničar 1956). The characteristic geometry of the HgO chain in the structure is lost just in favour of the triangular Hg_3 group (Aurivillius and Folkmarson 1968). The connections of the mercury atoms in the structure of amalgams with Hg—Hg distances less than 3.0 Å raise the question of the possibility of such connections in non-metallic structures.

When the opportunity was offered to me to write an article for this book in Dorothy's honour I was very happy, although I was aware that my structural studies were not of biological interest, except those I did with Dorothy on ferroverdin in Oxford 1955–6. At the time it was not feasible to begin X-ray analysis on biological structures in Zagreb, and I went on with my studies on the mercury compounds, but greatly stimulated by my experience in Dorothy's lab. The knowledge I gained in this field and used in writing this article, however modest it may be, is the best I could dedicate to Dorothy.

References

AURIVILLIUS, K. (1950). *Acta chem. scand.* **4**, 1413.
——(1956). *Acta chem. scand.* **10**, 852.
——(1964). *Acta chem. scand.* **13**, 1305.
——(1964). *Acta chem. scand.* **18**, 1552.
——(1965). *Ark. Kemi* **23**, 205.
——(1965). *Ark. Kemi* **23**, 469.
——(1965). *Ark. Kemi* **24**, 151.
——(1967). *Ark. Kemi* **26**, 497.
——(1968). *Ark. Kemi* **28**, 279.
——and CARLSSON, I.-B. (1958). *Acta chem. scand.* **12**, 1297.
——and FOLKMARSON, L. (1968). *Acta chem. scand.* **22**, 2529.
——and NILSON, B. A. (1975). *Z. Kristallogr. Kristallgeom.* **141**, 1.

——and STÅLHANDSKE, C. (1974). *Acta crystallogr.* **B30**, 1907.
————(1975). *Z. Kristallogr. Kristallgeom.* **142**, 129.
————(1976). *Z. Kristallogr. Kristallgeom.* **144**, 1.
————(1978). *Acta crystallogr.* **B34**, 79.
BEAUCHAMP, A. L. and GOUTIER, D. (1972). *Can. J. Chem.* **50**, 977.
——PAZDERNIK, L., and RIVEST, R. (1976). *Acta crystallogr.* **B32**, 650.
BEHRENS, U., HOFFMANN, K., and KLAR, G. (1977). *Chem. Ber.* **110**, 650.
BIJVOET, J. M., CLASSEN, A., and KARSSEN, A. (1926). *Proc. Acad. Sci. Amsterdam* **29**, 529.
BISCARINI, P., FUSINA, L., NIVELLINI, G., and PELIZZI, G. (1977). *J. chem. soc., Dalton* 664.
BJÖRNLUND, G. (1971). *Acta chem. scand.* **25**, 1645.
BONDI, A. (1964). *J. Phys. Chem.* **68**, 441.
BONEFAČIĆ, A. (1963). *Acta crystallogr.* **16**, A30.
BRAEKKEN, H. and SCHOLTEN, W. *Z. Kristallogr. Kristallgeom.* **89**, 488.
BRAUER, D. J., BÜRGER, H., and EUJEN, R. (1977). *J. organometal. Chem.* **135**, 281.
BREITINGER, D. and KÖHLER, K. (1976). *Z. anorg. allg. Chem.* **421**, 151.
BRODERSEN, K. (1955). *Acta crystallogr.* **8**, 723.
——and RÜDORFF, W. (1954). *Z. anorg. allg. Chem.* **275**, 141.
————(1974). *Z. anorg. allg. Chem.* **409**, 1.
————(1975). *Z. anorg. allg. Chem.* **414**, 1.
————(1977). *Z. anorg. allg. Chem.* **428**, 166.
————and THIELE, G. (1978). *Z. Naturforsch.* **33b**, 1227.
BROWN, I. D., CUTFORTH, B. D., DAVIES, C. G., GILLESPIE, R. J., IRELAND, P. R., and VEKRIS, J. E. (1974). *Can. J. Chem.* **52**, 791.
CHAMPLON, F., BERNARD, D., PANNETIER, J., and LUCAS, J. (1974). *C.r. hebd. Acad. Séanc. Paris* **278**, 1185.
CLEGG, W., GREENHALGH, A., and STRAUGHAN, B. P. (1975). *J. chem. Soc. Dalton* 2591.
CUTFORTH, B. D., GILLESPIE, R. J., and IRELAND, P. R. (1973). *Chem. Commun.* 723.
————and UMMAT, P. K. (1976). *Rev. Chim. Miner.* **13**, 119.
——DAVIES, C. G., DEAN, P. A. W., GILLESPIE, R. J., IRELAND, P. R., and UMMAT, P. K. (1973). *Inorg. Chem.* **12**, 1343.
DEWAN, J. C., KEPERT, D. L., and WHITE, A. H. (1975). *J. Chem. Soc. Dalton* 490.
DORM, E. (1967). *Acta chem. scand.* **21**, 2834.
——(1969). *Acta chem. scand.* **23**, 1607.
——(1971*a*). *Chem. Commun.* 466.
——(1971*b*). *Acta chem. scand.* **25**, 1655.
DURIF, A., TORDJMAN, I., MASSE, R., and GUITEL, J. C. (1978). *J. solid state Chem.* **24**, 101.
DUROVIČ, S. (1968). *Acta crystallogr.* **B24**, 1661.
ELDER, R. C., HALPERN, J., and POND, J. S. (1967). *J. Am. chem. Soc.* **89**, 6877.
ELLISON, R. D., LEVY, H. A., and FUNG, K. W. (1972). *Inorg. Chem.* **11**, 833.
FENN, R. H. (1966*a*). *Acta crystallogr.* **20**, 20.
——(1966*b*). *Acta crystallogr.* **20**, 24.
GRDENIĆ, D. (1950). *Arhiv za Kemiju*, **22**, 14.
——(1956). *J. chem. Soc.* p. 1312.
——(1965). *Quart. Rev. Chem. Soc. Lond.* **19**, 303.
——(1969). In *Handbook of geochemistry*, Vol. II/1 (ed. K. H. Wedpohl), p. 80 A 1. Springer-Verlag, Berlin.
——(1977). *Izvj. Jugoslav. centr krist.* (*Zagreb*) **12**, 5 (in English).

220   *Connections in the crystal structures of mercury compounds*

——(1979). In *III Congresso Italo-Iugoslavo, Parma 29th May–1st June 1979, Abstracts of Papers*, p. B1. Associazone Italiana di Cristallografia, Parma.
—— and KITAIGORODSKY, A. I. (1949). *Zhur. fiz. Khim.* **23**, 1162.
—— and SIKIRICA, M. (1972). *Inorg. Chem.* **12**, 544.
—— KAMENAR, B., and NAGL, A. (1977). *Acta. crystallogr.* **B33**, 587.
—————— and ŽEŽELJ, V. (1979). *Acta crystallogr.* **B35**, 1889.
—— SIKIRICA, M., and VICKOVIĆ, I. (1975). *Acta crystallogr.* **B31**, 2174.
—————— (1977). *Acta crystallogr.* **B33**, 1630.
HARTWIG, W. (1926). *Sitz.-Ber. Preuss. Akad. Wiss.* **10**, 1926.
HOPPE, R. and RÖHRBORN, H. J. (1964). *Z. anorg. allg. Chem.* **329**, 110.
JANDALI, M. Z., EULENBERGER, G., and HAHN, H. (1978a). *Z anorg. allg. Chem.* **445**, 184.
—————— (1978b). *Z. anorg. allg. Chem.* **447**, 105.
JEFFERY, J. W. and ROSE, K. M. (1968). *Acta crystallogr.* **B24**, 653.
JEITSCHKO, W. and SLEIGHT, A. W. (1973). *Acta crystallogr.* **B29**, 869.
JOHANSSON, G. (1966). *Acta chem. scand.* **20**, 553.
—— (1971a). *Acta chem. scand.* **25**, 1905.
—— (1971b). *Acta chem. scand.* **25**, 2799.
—— and HANSEN, E. (1972). *Acta chem. scand.* **26**, 796.
—— and SANDSTRÖM, M. (1978). *Acta chem. scand.* **A32**, 109.
KAMENAR, B. and KAITNER, B. (1973). *Acta crystallogr.* **B29**, 1666.
—— and NAGL, A. (1976). *Acta crystallogr.* **B32**, 1414.
KEPERT, D. L., TAYLOR, D. and WHITE, A. H. (1972). *Inorg. Chem.* **11**, 1639.
—————— (1973a). *J. Chem. Soc., Dalton* 893.
—————— (1973b). *J. Chem. Soc., Dalton* 1658.
—————— (1973c). *J. Chem. Soc., Dalton* 392.
—————— (1973d). *J. Chem. Soc., Dalton* 670.
KOLKMEIJER, N. H., BIJVOET, J. M., and KARSSEN, A. (1924). *Rec. Trav. Chim.* **43**, 894.
LAVES, F. (1967). In *Intermetallic compounds* (ed. J. H. Westbrook) p. 129. John Wiley, New York.
LIPSCOMB, W. N. (1951). *Acta crystallogr.* **4**, 156.
—— (1951). *Acta crystallogr.* **4**, 266.
LINDH, B. (1967). *Acta chem. scand.* **21**, 2743.
MASON, R., ROBERTSON, G. B., and RUSHOLM, G. A. (1974). *Acta crystallogr.* **B30**, 894.
MASSE, R., GUITEL, J.-C., and DURIF, A. (1978). *J. solid state Chem.* **23**, 369.
MATKOVIĆ, B., RIBAR, B., PRELESNIK, B., and HERAK, R. (1974). *Inorg. Chem.* **13**, 3006.
MEREITER, K. and ZEMANN, J. (1976). *Tschermaks Min. Petrogr. Mitt.* **23**, 105.
NAGORSEN, G., LYNG, S., WEISS, A., and WEISS, A. (1962). *Angew. Chem.* **74**, 119.
NEUMAN, M. A., PETERSEN, D. R., and LO, G. Y. S. (1976). **6**, 177.
NIJSEN, L. and LIPSCOMB, W. N. (1952). *Acta crystallogr.* **5**, 604.
NILSON, B. A. (1975). *Z. Kristallogr. Kristallgeom.* **141**, 321.
NOZIK, YU. Z., FYKIN, L. E., BUKIN, V. I., and LANTASH, N. M. (1979). *Koordinatsionnaya khimiya* **5**, 276.
PAULING, L. and HUGGINS, M. L. (1934). *Z. Kristallogr. Kristallgeom.* **37**, 205.
PUFF, H. and KÜSTER, J. (1962). *Naturwissenschaften* **49**, 299.
QUARTON, M., ANGENAULT, J., and RIMSKY, A. (1973). *Acta crystallogr.* **B29**, 567.
RÜDORFF, W. and BRODERSEN, K. (1952). *Z. Naturforsch.* **7b**, 56.
—————— (1953). *Z. anorg. allg. Chem.* **274**, 323.
SAKHRI, A. and BEAUCHAMP, A. L. (1975). *Acta crystallogr.* **B31**, 409.
SANDSTRÖM, M. and LIEM, D. H. (1978). *Acta chem. scand.* **A32**, 509.
—— and PERSSON, I. (1978). *Acta chem. scand.* **A32**, 95.

ŠĆAVNIČAR, S. (1955). *Acta crystallogr.* **8**, 379.
——(1956). *Acta crystallogr.* **9**, 956.
SIKIRICA, M., and GRDENIĆ, D. (1974). *Acta crystallogr.* **B30**, 144.
SLEIGHT, A. W. (1968). *Inorg. Chem.* **7**, 1704.
SOMMER, H., HOPPE, R. and JANSEN, M. (1976). *Naturwissenschaften* **63**, 194.
STÅLHANDSKE, C. (1978). *Acta crystallogr.* **B34**, 1968.
——(1979). *Acta crystallogr.* **B35**, 949.
TAYLOR, D. (1976). *Aust. J. Chem.* **29**, 723.
——(1977). *Aust. J. Chem.* **30**, 2647.
VANNERBERG, N.-G. (1959). *Ark. Kemi* **13**, 515.
WEISS, A., LYNG, S., and WEISS, A. (1960). *Z. Naturforsch.* **15b**, 678.
ZACHARIASEN, W. H. (1926). *Z. Phys. Chem.* **124**, 436.

III Methods in crystallography

22. Data collection methods for large molecules

ROBERT A. SPARKS

THE intensity data for the X-ray structure determination of the hexacarbo-xylic acid derivative of vitamin B_{12} were collected on Weissenberg films (Hodgkin *et al.* 1959). Integrated intensities were estimated by visual comparison with an intensity strip prepared by exposing to X-rays given reflections for different amounts of time. This procedure was common practice in the 1940s and 1950s and was very time-consuming. With the advent of the computer-controlled four-circle X-ray diffractometer, data collection for structures as large as the hexacarboxylic acid derivative of vitamin B_{12} is now completely automatic. For molecules this size (73 non-hydrogen atoms) and smaller, it is now possible to mount a single crystal and centre it within the instrument with a telescope so that the crystal will always be bathed in a uniform part of the X-ray incident beam. During a period of about one to two hours the computer will then centre on a number of reflections 6–12) and determine the Bravais lattice type, unit cell dimensions, and the orientation of the crystallographic axes with respect to the diffractometer axes. Data collection can now be started, and within 24 hours to a few days a complete set of data will be available on magnetic tape. On the same computer that collected the data it is possible to make absorption and Lorentz-polarization corrections, calculate normalized structure factors, solve the phase problem (with programs like MULTAN), refine the parameters with a least-squares refinement program, calculate an electron-density map, calculate distances and angles, and finally plot stereo pairs with the ORTEP program (Sparks 1976). Indeed, the probability of being able to do all of this successfully for a given substance is now primarily limited by the researcher's ability to obtain a single crystal of the appropriate size.

Unfortunately the collection of intensity data is not so routine for biologically important structures much larger than the hexacarboxylic acid derivative of vitamin B_{12}. The reasons for this are the following.

(1) The unit-cell dimensions are large and thus the reciprocal lattice points

are close together. Thus care must be exercised so that intensity from only one reflection is measured at a time.

(2) X-ray photons cause damage to most of the large biologically important molecules, and hence the intensities of the reflections will change as a function of the time the crystal is exposed to X-rays.

(3) The intensities of most reflections are small, and long counting times are necessary to achieve measurements which are significant compared to background.

(4) The crystals will disintegrate if they are removed from the mother liquor from which they were crystallized. Therefore they are usually placed in a capillary with a small amount of mother liquor present. The capillary is then mounted on the goniometer of the diffractometer. For each reflection the diffractometer must position the crystal with respect to the incident X-ray beam so that the Bragg condition takes place. Because of the resulting large movements of the goniometer and because the crystal is not firmly attached to the goniometer the crystal can translate or rotate with respect to the goniometer. Unless this movement is detected and corrected for, the computer-controlled diffractometer will measure subsequent intensities at the wrong positions in reciprocal space.

(5) Because the irregularly shaped capillary, mother liquor, and crystal all absorb X-rays, analytical algorithms for absorption corrections cannot be used.

(6) Finally, to solve the phase problem, many heavy-atom derivatives must be screened to determine whether they are good candidates for being used in the determination of phases by the isomorphous replacement method. Data sets must be collected for those derivatives which pass this test.

Crystal orientation

Unlike small-molecule crystallographers the protein crystallographer usually knows something about the lattice parameters and orientation of his crystal before he mounts it on an X-ray diffractometer. Nevertheless, it is necessary to find a few reflections and determine their centres accurately in order to obtain a good orientation matrix. A full-rotation Polaroid photograph cannot be used because the reflections are usually so weak that they cannot be seen above the background. A slow rotation of one degree or less is usually sufficient to find some reflections on a Polaroid photograph. To centre on weak reflections, the centring program on the Syntex diffractometers had to be slowed down to obtain good counting statistics.

Background measurements

The most striking difference between a diffractometer collecting data from a small-molecule crystal and one collecting data from a protein crystal is that the detector is placed a long way from the sample in the case of proteins. A distance of 35–40 cm is typical. A helium-filled tube can be placed between the detector and the sample. Scattered radiation, the major source of background, is spread uniformly along the surface of a cone for a given 2θ value. The diffracted Bragg reflection, however, is almost parallel because of the small mosaic spread ($0.1°$ typical for most protein crystals). Helium is used so that the diffracted beam will be minimally absorbed. Thus peak-to-background ratio can be enhanced by this technique. Apertures at the front and rear of the tube are adjusted so that all of the crystal can be seen by the detector, but as little as possible of the scattering volume can be seen. Krieger and Stroud (1976) have shown that background radiation can be further reduced by placing the crystal and capillary inside a volume of helium surrounded by a thin (0.001 in.) Mylar cylinder, so that no air is in the scattering volume.

The incident-beam collimator is sometimes also filled with helium. Peak-to-background ratio is not affected by this technique, but of course the incident beam intensity is increased.

In small-volume crystallography normally as much time is spent in collecting background data as is spent collecting integrated intensity for each reflection. Normally backgrounds are collected on the low-2θ and high-2θ sides of the reflection. For crystals which decay owing to X-ray damage, it is desirable to spend most of the time where most of the information is, namely at the peaks of the reflection. Background measurements are made either before or after data collection on a given crystal. Measurements are made at intervals of 2θ, ϕ, and χ. A smooth analytical function is then fitted by least squares to these background measurements (Krieger, Chambers, Christoph, Stroud, and Trus 1974). This analytical expression is then used in calculating the background for each reflection. It is assumed, of course, that background intensity does not change significantly with time. Krieger and Stroud (1976) have shown that very little of the background comes from the decaying crystal and therefore this assumption is justified.

Efficiency of data collection

For small-molecule crystallography the $2\theta{:}\theta$ scan method is chosen. In this method the detector and crystal move simultaneously (the ω-axis drives at half the speed of the 2θ-axis). Usually, the entire peak is scanned from the background on the low-2θ side to the background on the high-2θ side. For crystals with large unit cells there is the possibility that the reflections will overlap if this technique is used. Instead, data collection for large molecules is done with the ω-scan technique (Wyckoff *et al*. 1970). In this method the detector is held fixed and the ω-axis is stepped over the range corresponding to approximately full-width half-maximum. The method used with the Syntex single-crystal diffractometers (Sparks 1976) is the following: an odd number (e.g. seven) of intensity measurements are made at equally spaced ω- values. The maximum of these measurements is recorded and a smaller number of these points (e.g. five) are chosen so that the maximum is the central point. The integrated intensity is then the sum of the intensities of the subset. If the maximum occurs at one end of the range, additional steps will be made until enough points have been recorded on each side of the maximum. Along a given reciprocal lattice line, the computer remembers where the central point of the previous reflection was and uses this information to predict where the centre will be for subsequent reflections.

All of the data are stored as a profile on magnetic tape. In post-processing, Chambers and Stroud (1979) use this profile data to calculate a peak centre position and integrated intensity using a moving-average method which results in better intensity values than those based on a centre determined by the method described above. Diamond (1969) has shown that it is possible to use the collected profile data from many reflections that are in a given region of reciprocal space to obtain a best least-squares profile shape. Then from this profile shape it is possible to obtain a best least-squares integrated intensity for each reflection. This method is especially useful in obtaining accurate intensities for the weakest reflections. Rossmann (1979) uses two-dimensional profile information to calculate intensities from data obtained from an Arndt–Wonacott rotation camera.

To determine whether a given crystal is a good heavy-atom isomorphous replacement derivative selected reflections with given indices are measured. In the Syntex diffractometers (Sparks 1976) this list of indices is provided to the computer from paper tape or magnetic tape.

All of these techniques are meant to use the incident X-rays as efficiently as possible by collecting data only at points where the most information can be obtained.

Sample movement

During data collection it is possible that the sample can move. Vandlen and Tulinsky (1971) describe a program which will sample certain check reflections periodically, and if the intensity of any of them falls below a preset value, the computer will centre on a number of reflections and recalculate a new orientation matrix. In the Syntex system the ω-scan routine has a feature that if too many extra steps are needed to find a peak maximum, then the recentring procedure is initiated.

Small movements of the crystal can be detected and corrected for by such methods. A displacement large enough to cause the crystal to move out of the uniform part of the incident beam or to cause the crystal to no longer be seen completely by the detector through the detector apertures can be detected but cannot be corrected for by the computer.

Sample decay

Because of sample decay, data are usually collected on many different samples. It is often desirable to collect data in shells from a minimum 2θ to a maximum 2θ value. In any case, the total amount of time a crystal has been exposed to X-rays is recorded with each measurement. Check reflections are measured periodically and during post-processing the entire data set is scaled to correct for the change in intensities observed with time for these check reflections.

If anomalous scattering information is to be obtained, Friedel pairs must be collected. Because the intensity differences are usually small, it is important to collect a reflection and its Friedel pair close together in time. To avoid large slewing times between measurements, the Hilger–Watts diffractometer can move the detector arm at a speed of $1200°$/min (Hilger and Watts Ltd 1964). Slower instruments (the Syntex $P2_1$ detector arm moves at $234°$/min) collect all those reflections on a given reciprocal lattice line and then make one long drive to collect the corresponding Friedel pairs.

Sample decay due to X-ray damage can be minimized by collecting data at low temperatures. With protein crystals that have a large water content, data are usually collected just above the freezing point of water. Petsko (1975) has managed essentially to eliminate sample decay by replacing the water with various alcohols and collecting data at much lower temperatures.

Absorption measurements

Empirical absorption corrections are made by measuring several reflections at different ψ-values. The ψ-axis is defined as that axis lying along the diffraction vector. Intensity variations observed by rotating about ψ must be due to different amounts of absorption in the incident and diffracted beams. An empirical absorption correction curve as a function of the orientation of the crystal is derived and then applied to each measured intensity in the data set.

No commercially available diffractometer has a ψ-axis. Therefore, it is necessary to drive ϕ, χ, and ω to achieve an effective ψ rotation. Most data-collection routines have the capability of achieving this complex motion.

Intensity enhancement

Because of the simplicity of use and because of the increased usable intensity, most modern diffractometers use graphite monochromators to eliminate all radiation except that due to the characteristic $K\alpha_1$ $K\alpha_2$ doublet. About 80 per cent of the characteristic radiation is diffracted with such monochromators. Care must be taken when using a monochromator because the monochromated beam is not of uniform intensity in the plane defined by the incident and monochromated beams. Thus, the sample cannot be allowed to move very much in this plane. The monochromated beam is polarized by an amount dependent on the mosaic character of the graphite crystal. Therefore, this polarization effect must be measured and then the measured intensities must be corrected for this polarization effect.

The maximum power that can be achieved with a commercially available stationary-target X-ray tube is 2.4 kW. Therefore, a number of protein crystallographers are now using rotating anode tubes which can achieve powers of 6 kW and higher. This enhanced power allows faster data collection. Of course, the greater incident X-ray intensity also means that the samples decay faster.

The greatest source of X-ray intensity available today is from the various synchrotron radiation sources. Typical intensities at a wavelength of 1.54 Å would be about 170 times that which could be achieved with a conventional stationary copper target X-ray tube. Synchrotron radiation is pulsed, is 100 per cent polarized, and has a continuous spectrum of wavelengths. All of these properties have significant impact on data-collection techniques for protein crystallography (Winick and Brown 1978).

Multiple detectors

For large unit cells many reflections will go through the diffraction condition with a small rotation of one of the diffractometer axes. With the single-detector four-circle diffractometer only one of these reflections can be measured at a time. Many investigators have developed instruments which measure more than one reflection at a time. The simplest of these devices is the Arndt–Wonacott rotation camera (Arndt, Gilmore, and Wonacott 1977). The sample is mounted on a verticle ω-axis. A flat film cassette is positioned behind the sample normal to the horizontal incident X-ray beam. The sample is slowly rotated about the ω-axis. The amount of rotation is limited so that no two reflections will overlap one another on the film. The first film cassette is moved out of position and a second film cassette is moved into the position which was occupied by the first. Again the sample is rotated about the ω-axis. This procedure is continued until all of the data (except for those reflections whose reciprocal lattice points lie near the axis of rotation) are collected.

The films are two-dimensional position-sensitive detectors. The positions of the reflections and their intensities are measured by a film reader. One disadvantage of film is that background accumulates at each point on the film during the whole time that the film is in front of the X-ray beam, whereas the X-rays from a reflection impinge on the film only during the short period of time that the reciprocal lattice point moves through the Ewald sphere.

Banner and co-workers (Banner, Evans, Marsh, and Phillips 1977) have modified a Hilger–Watts four-circle instrument by adding a fifth (σ) axis parallel to the 2θ-detector arm. Five detectors are mounted on this axis so that their apertures form a straight line perpendicular to the 2θ arm. By positioning 2θ, ω, ϕ, and χ so that the central detector is measuring a reflection and then by rotating the σ-axis, the other four detectors can also be brought into position to measure four other reflections. A small rotation about the ω-axis will ensure that all five reflections are measured.

More recently, Prince and co-workers (Prince, Wlodawer, and Santoro 1978) have added an electronic linear position-sensitive detector to a four-circle neutron diffractometer. Because this detector can measure a reflection anywhere along its length instead of at just selected apertures, no σ-axis motion is required. In this diffractometer reflections are measured layer by layer as is done with a traditional Weissenberg camera. For a cubic crystal with lattice parameters of 50 Å about ten reflections at 1.5 Å resolution can be measured in a $2°$ rotation of the crystal.

With a two-dimensional detector the number of reflections which can be measured at one time for the example above would be about 400. There are two types of two-dimensional electronic position-sensitive detectors. The first is the multiwire proportional chamber (MWPC) used by Xuong and his co-workers (Cork *et al.* 1974), and the second is the television area detector. It is beyond the scope of this chapter to discuss the two-dimensional detectors in detail. The excellent book, *The rotation method in crystallography*, edited by Arndt and Wonacott (1977) describes both types of detector and the Arndt–Wonacott rotation camera in great detail. At the present time the rotation camera and the traditional one-detector four-circle X-ray diffractometer are the only instruments which are commercially available.

Programming language

Many of the early computer-controlled four-circle diffractometers were programmed in assembly language. As the data-collection problem became more complex, it became apparent that the flexibility of a higher-level language was important. This was especially true for data-collection routines for large molecules. Each new crystal structure seems to require its own data-collection algorithm. Therefore it is necessary that the crystallographer should be able to make modifications to standard data-collection routines. The language best known by crystallographers is FORTRAN, and it is this language which is used on the commercially available diffractometers of Enraf–Nonius and Syntex.

Almost all of the data-collection techniques (both hardware and software) which have been developed in the last decade have arisen because of the stringent requirements of large-molecule crystallography. It can be expected that this trend will continue, and as soon as new detector systems and powerful X-ray sources become more readily available, new instruments will be built to help solve these difficult data-collection problems.

References

ARNDT, U. W. and WONNACOTT, A. J. (ed.) (1977). *The rotation method in crystallography*. North-Holland, Amsterdam.

——GILMORE, D. J., and WONACOTT, A. J. (1977). In *The rotation method in crystallography* (ed. U. W. Arndt and A. J. Wonacott), Chapter 14. North-Holland, Amsterdam.

BANNER, D. W., EVANS, P. R., MARSH, D. J., and PHILLIPS, D. C. (1977). *J. appl. Crystallogr.* **10**, 45–51.

CHAMBERS, J. L. and STROUD, R. M. (1979). Private communication.

CORK, C., FEHR, D., HAMLIN, R., VERNON, W., XUONG, NG. H., and PEREZ-MENDEZ, V. (1974). *J. appl. Crystallogr.* **7**, 319.

DIAMOND, R. (1969). *Acta crystallogr.* **A25**, 43.

HILGER and WATTS LTD (1964). *Four-circle single crystal X-ray diffractometer*. Hilger and Watts Ltd, London.

HODGKIN, D. C., PICKWORTH, J., ROBERTSON, J. H., PROSEN, R. J., SPARKS, R. A., and TRUEBLOOD, K. N. (1959). *Proc. R. Soc.* **A251**, 306–52.

KRIEGER, M. and STROUD, R. M. (1976). *Acta crystallogr.* **A32**, 653–6.

——CHAMBERS, J. L., CHRISTOPH, G. G., STROUD, L. M., and TRUS, B. L. (1974). *Acta crystallogr.* **A30**, 740–8.

PETSKO, G. A. (1975). *J. molec. Biol.* **96**, 381–92.

PRINCE, E., WLODAWER, A., and SANTORO, A. (1978). *J. appl. Crystallogr.* **11**, 173–8.

ROSSMAN, M. G. (1979). *J. appl. Crystallogr.* **12**, 225–38.

SPARKS, R. A. (1976). *Crystallographic computing techniques* (ed. F. R. Ahmed, K., Huml, and B. Sedlacek). Munksgaard, Copenhagen.

VANDLEN, R. L. and TULINSKY, A. (1971). *Acta crystallogr.* **B27**, 437.

WINICK, H. and BROWN, G. (1978). Workshop on X-ray Instrumentation for Synchrotron Radiation Research. Stanford Synchrotron Radiation Laboratory Report No. 78/04. Stanford, California.

WYCKOFF, H. W., TSERNOGLOU, D., HANSON, A. W., KNOX, J. R., LEE, B., and RICHARDS, F. M. (1970). *J. biol. Chem.* **245**, 305–28.

23. On anomalous scattering and the multiple wavelength method

S. RAMASESHAN AND RAMESH NARAYAN

Introduction

WHEN the wavelength of the incident radiation approaches an absorption edge of an atom, the atomic scattering factor becomes complex and may be expressed in the form

$$f = f_0 + f' + if''. \tag{23.1}$$

In the case of X-rays, the imaginary part f'' exists only on the shorter wavelength side of the absorption edge. If an absorbing atom is present in a noncentrosymmetric crystal the complex scattering factor leads to the failure of Friedel's Law, i.e. $I(hkl) \neq I(\overline{hkl})$. This failure was first demonstrated in X-rays by Nishikawa and Matukawa (1928) and later by Coster, Knol, and Prins (1930). After almost two decades, Bijvoet (1949) showed that the breakdown of Friedel's Law could be used to determine the absolute configuration of molecules, as could the phases of X-ray reflections. Quantitative expressions relating phases and observed intensities of Bijvoet pairs were given by Ramachandran and Raman (1956). Patterson techniques involving the sum and difference of the intensities of Bijvoet pairs were evolved by Pepinsky and Okaya (1956) to solve crystal structures. It was also shown at that time that by using multiple wavelengths (by which the magnitudes of the real and imaginary parts of the scattering factor could be changed by varying the wavelength of the incident radiation), the phase problem could be solved using techniques similar to isomorphous replacement (Ramaseshan 1962). Since then the anomalous scattering technique has played an increasingly important role in crystal structure analysis and now forms a standard tool for resolving the phase problem (Ramaseshan and Abrahams 1975).

Anomalous scattering is also present in (i) neutron scattering, (ii) recoilless γ-ray scattering, and (iii) electron scattering. Therefore, in theory, these methods could also be used for phasing. Unfortunately, dynamical effects predominate in electron diffraction and these may lead to the violation of Friedel's Law even in the absence of resonance effects (Cowley 1975).

In this article we shall examine the merits of these different techniques. The multiple wavelength method seems to hold high promise owing to the availability of intense synchrotron X-radiation from storage rings tunable to different wavelengths. We shall consider optimum methods of using this technique.

Neutron anomalous scattering

The breakdown of Friedel's Law for neutron diffraction was first reported by Petersen and Smith (1961) in a crystal of α-CdS containing ^{113}Cd isotope which exhibits resonance absorption in the thermal neutron region. The resonant scattering of neutrons is rather large (as compared to X-rays) and the dispersion terms are almost an order of magnitude greater than the normal scattering amplitudes. The scattering length of neutrons in the resonance region is well represented by a one-term Breit–Wigner formula. The imaginary part f'' always has the same sign while the real part changes sign, being negative in the longer wavelength side of the resonance absorption and positive in the shorter wavelength side.

It was proposed, from Dorothy Hodgkin's laboratory in Oxford, that the large anomalous dispersion in nuclei like ^{113}Cd, ^{149}Sm, ^{151}Eu, and ^{157}Gd could be used for solving the structures of large molecules like proteins (Ramaseshan 1966). The following were some of the suggestions that were made in that note:

(i) It was originally thought that heavy atom methods could not be used in neutron diffraction. However, if a suitable incident wavelength is used when an anomalous scatterer is present, one can obtain changes in the real part of the scattering factor by as much as 5 to 10 times the normal value. Hence this procedure is equivalent to collecting data in a structure having a heavy atom. All the techniques of phasing associated with the 'heavy atom' method can be directly used.

(ii) In a non-centrosymmetric crystal, (a) if the position of the anomalous scatterer is determined by the above method and (b) if the intensities of Bijvoet pairs are measured, the phases of the reflections could be determined (with a residual twofold ambiguity).

(iii) If the measurements are made at two wavelengths (preferably on either side of the resonant wavelength), it is equivalent to collecting data on two perfectly isomorphous crystals.

(iv) By collecting data at three wavelengths, both the positions of the anomalous scatterers as well as the phases of the reflections can be uniquely determined without any ambiguity.

(v) Because of the peculiar shapes of the f' and f'' versus wavelength curves, two wavelengths λ_1 and λ_2 can be chosen such that the real parts of the form factor or the imaginary parts are equal. Such choices have some advantages.

Many techniques and procedures have since been suggested which are variations of the above. Expressions have been derived for determining the positions of the anomalous scatterers as well as for the calculation of the phases when data is collected at two wavelengths (Singh and Ramaseshan 1968). This method has actually been used to solve the structure of $NaSm(EDTA) \cdot 8H_2O$. (Koetzle and Hamilton 1975). Modifications of (v) have been suggested by Sikka (1969) and Bartunik (1978).

Sikka (1973) has made an elegant proposal—which has proved of real practical use—that the tangent formula can be used to resolve the bimodal phase ambiguity in the neutron anomalous scattering method.

Given below is a list of almost all the structures that have been solved using neutron anomalous scattering methods:

$Cd(NO_3)_2 \cdot 4D_2O$	MacDonald and Sikka (1969)
$Sm(BrO_3)_3 \cdot 9H_2O$	Sikka (1969)
$NaSm(EDTA) \cdot 8H_2O$	Koetzle and Hamilton (1975)
$Cd(tartrate) \cdot 5H_2O$	Sikka and Rajagopal (1975)
Cd-histidine$\cdot 2H_2O$	Bartunik and Fuess (1975)
Aqua (L-glutamate) $Cd(II)H_2O$	Flook, Freeman, and Scudder (1977)

Proteins have also given rather unsatisfactory results. The following quote is taken from an unpublished manuscript of S. A. Mason:

Following the suggestion of Ramaseshan in 1964, work was begun at Harwell on insulin crystals by Moore and MacDonald (1970) and later by the present author (Hodgkin, Willis, Fuess, and Mason 1973). In the beginning it was thought that the potential power of the anomalous dispersion of neutrons would contribute to the solution of the then unknown crystal structure of insulin. However, limitations of flux did not permit this objective to be reached.

The only protein structure in which successful phasing has been achieved is cadmium myoglobin (Schoenborn 1975). Sperm whale metmyoglobin crystals in space group $P2_1$ were grown from deuterated ammonium sulphate solutions soaked in ^{113}Cd acetate. Data for this derivative were collected at 0.8 and 1.24 Å, both on the low-energy side of the anomalous peak. At both wavelengths about 5000 of the stronger reflections in the 2 Å sphere were measured yielding 1200 independent reflections. The bimodal phase ambiguity was solved by the tangent formula method (Sikka 1973). The resultant phases (calculated by the method of Blow and Crick 1959) were used to determine and refine the Cd parameters, which were very close to those determined by the X-ray method.

This exercise proved that it is possible to determine the phases in a protein crystal using neutron anomalous scattering. But the collection of data was very time-consuming. Improvements in technology, such as increasing the neutron flux, installing position-sensitive detectors and multicounter systems, are required if this method is to be used in a big way.

For some time to come, neutron anomalous scattering will probably be used in fields other than in the crystallography of large molecules, such as solid-state and liquid-state physics (Ramaseshan, Ramesh, and Ranganath 1975).

Anomalous dispersion of gamma radiation

The application of resonant recoil-less γ-ray scattering to the determination of phases of Bragg scattered waves has been suggested by many authors (e.g. Moon 1961; Raghavan 1961) and many experimental studies have been pursued (Black 1965; Parak, Mossbauer, Hoppe, Thomanek, and Bade 1976; Mossbauer 1975). Since nuclear energy levels are much sharper, the anomalous scattering effects at resonance are much larger—almost an order of magnitude greater than even the neutron case. This is why the method has attracted interest. In fact, when this method was suggested, many other advantages were predicted. The wavelength of the radiation can be tuned to various points in the resonance curve by means of linear Doppler effect—velocities as low as $0.2\,\mathrm{mm\,s^{-1}}$ can tune the wavelength right across resonance. γ-ray sources have very stable intensity outputs—far superior to X-ray or neutron sources. One can get accurate phase determination with much less measured intensity than with other methods. This is useful for crystals which undergo rapid decomposition when irradiated.

Unfortunately, all these advantages are completely offset by the very low intensity of the sources. While the anomalous dispersion effects dramatically increase in going from X-rays to neutrons to γ-ray resonant scattering, the available radiation sources decrease abysmally in intensity. Mossbauer (1975) estimates that the incident intensity for X-rays, neutrons, and resonant γ-rays is respectively 2×10^9, 10^7, and 8×10^4 counts $\mathrm{cm^{-2}\,s^{-1}}$.

Because of this severe intensity restriction, the method seems to have little future as a routine technique in crystal structure analysis. It can possibly be used in very special applications.

Synchrotron radiation

It is clear from the discussion above that X-ray methods appear to be the most rapid, the larger intensity more than compensating for the lower values of the dispersion terms. The number 2×10^9 counts $\mathrm{cm^{-2}\,s^{-1}}$ quoted above is for an ordinary laboratory X-ray generator (Mo Kα radiation with β filter operated at 50 kV and 10 mA). Synchrotron X-radiation from storage rings have become available and these have two to four orders of magnitude higher intensity! Further, with a suitable monochromator, any wavelength of inter-

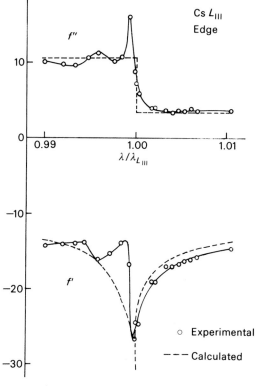

Fig. 23.1.

est can be tuned quite easily. One can, therefore, go close to the atomic resonance edges and obtain larger anomalous scattering factors. In addition, dispersion effects much larger than predicted have been observed close to the L_{III} and L_{II} absorption edges of caesium in caesium tartrate (see Fig. 23.1) with f' and f'' values being as high as 26.7 and 16.1 electrons respectively. This is an effect important to the crystallographer, but it will not be discussed further here (see Phillips 1978). It is almost certain that this effect would also be observable in other crystals. This opens up immense possibilities for the crystal structure analysis of large molecules.

The multiple wavelength method and crystal structure analysis

With the availability of intense tunable synchrotron X-radiation sources, the multiwavelength method may come possibly into its own and become a rather effective tool in crystal structure analysis. The recent work at Stanford (Phillips 1978) is definitely a pointer in this direction.

Traditionally two methods have been employed either separately or in combination for solving the phase problem in structure analysis:

(i) Isomorphous replacement.
(ii) Anomalous scattering.

In both the techniques the crystal has two sets of atoms.

(a) A special set of atoms which we call the 'heavy atoms' (designated by the subscript H).
(b) The rest of the crystal which we call the 'protein' (designated by the subscript P).

In isomorphous replacement one has different crystals with different heavy atoms while in anomalous scattering one has only one crystal but the 'heavy atoms' are the anomalous scatterers. The scattering factor of these 'heavy atoms' can be altered by changing the wavelength of the incident radiation.

The principles of the two techniques are illustrated in Fig. 23.2 for a simple case. The origin in the complex plane is O and the various vectors are the complex structure factors. **PH** is the scattering vector F_H of the heavy atoms alone. In the isomorphous technique one measures the length $|\mathbf{OP}|$ $(=|F_{HP}|)$ of the heavy-atom derivative. If the heavy-atom positions are known (i.e. **PH** is known), then from the experimentally measured values of $|F_P|$ and $|F_{HP}|$ one can get the origin O and hence the phases of **OP** and **OH**. One should note the inherent ambiguity in the solution. The origin could equally well be O' which is also consistent with the available information. It cannot be ruled out unless there are some additional data with which to do so.

In the case of anomalous scattering, the 'heavy atoms' are anomalous scatterers whose atomic scattering factors are of the form $f_0 + f' + if''$. This implies that the vector **PH** is replaced by the vectors **PA** and **PA'** (Fig. 23.2). The

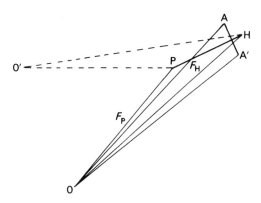

FIG. 23.2.

measured intensities of these two Friedel reflections are given by the lengths $|\mathbf{OA}|$ and $|\mathbf{OA}'|$ which are in general different. We can again solve for the phase of F_p, but once again there is an ambiguity (not shown in Fig. 23.2). It must be pointed out that if $|\mathbf{OP}|$, $|\mathbf{OA}|$, and $|\mathbf{OA}'|$ are all known then this ambiguity can be resolved. This corresponds to the standard technique of combining the isomorphous and anomalous data.

The accuracy of the phase determination depends on the length of $|\mathbf{PH}|$ in the isomorphous case and $|\mathbf{AA}'|$ in the anomalous case. $|\mathbf{PH}|$ is proportional to f_H the atomic form factor of the heavy atom which can be as high as 92 electrons, while $|\mathbf{AA}'|$ is proportional to $2f''$ which is usually less than 10 electrons. The isomorphous technique should therefore be the more effective one. Unfortunately, strict isomorphism is never achieved and this can introduce significant errors. In the anomalous case since the *same crystal* is used to measure $|\mathbf{OA}|$ and $|\mathbf{OA}'|$, one has perfect isomorphism and such problems are not present. With the availability of high-intensity tunable synchrotron radiation and with the possibilities of $|\mathbf{AA}'|$ being quite large (as much as 25 to 30 electrons), it seems that the anomalous scattering method will become more and more important. We shall deal exclusively with anomalous scattering methods in the rest of this chapter.

There are two stages in the use of anomalous scattering in solving the structure of a protein:

(i) Locating the anomalous scatterers (which we call the heavy atoms).
(ii) Solving for the phases of the reflections.

Using measurements at two wavelengths (Singh and Ramaseshan 1968), it is possible to solve the positions of the heavy atoms. The same data can then be used to determine the phase angles.

Before we discuss this method it is convenient to introduce a useful pictorial representation of anomalous scattering effects (see, for example, Hosoya 1975). Eqn (23.1) for the scattering length of a neutron from a resonant nucleus is given by a single-term Breit–Wigner formula

$$f = f_0 + C\left[\frac{E}{E^2 + \gamma^2} + \frac{i\gamma}{E^2 + \gamma^2}\right] \tag{23.2}$$

where C is a constant, E is the energy difference between the incident neutrons and the nuclear resonance energy, and γ is the half width of the resonance. It is easy to show that

$$\left|f - \left(f_0 + \frac{i}{2\gamma}\right)\right|^2 = \frac{1}{4\gamma^2} = \text{constant.} \tag{23.3}$$

In the Argand diagram, as E (i.e. λ) varies, f moves in a circle (Fig. 23.3(a)), there being two circles corresponding to the Friedel pairs of reflections. In the X-ray example as λ varies, f moves along two symmetric curves

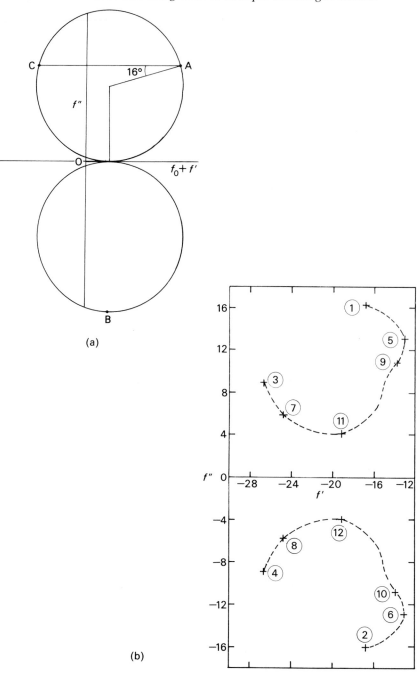

(a)

(b)

FIG. 23.3.

whose shapes, however, are not circles. Fig. 23.3(b) represents the values for the Cs L_{III} edge in caesium tartrate (Phillips 1978). By choosing suitable wavelengths, the heavy-atom form-factor vector f_H can be made to go from O (in Fig. 23.3(a)) to *any* point on the two circles. This is the extra freedom that the anomalous scattering phenomenon coupled with the tunability of the incident radiation allows us.

We shall now consider the technique of locating the anomalous scatters (heavy atoms). The structure factor of the heavy atoms F_H can be written as

$$F_H = S_H f_H, \tag{23.4}$$

where S_H is the scattering function depending on the position of the heavy atoms. At the beginning of the problem, S_H is not known, i.e. in Fig. 23.2, neither the length nor the orientation of **AA′** is known. However, if now we obtain data at a second wavelength, we have a second vector **BB′** which is shifted parallel to **AA′** with a different length (Fig. 23.4). All the four points A, A′, B, and B′ must lie on curves similar to those given in Fig. 23.3. We thus have four points in the complex plane whose internal geometry is known except for a scale $|S_H|$. Experimentally one has the distances of the origin from these four points. For any arbitrary choice of $|S_H|$, the four lengths will not be consistent. However, insisting on consistency, $|S_H|$ can be solved (Singh and Ramaseshan 1968). It may be mentioned that although two Friedel pairs give four lengths, one of them is redundant in the sense that it is uniquely determined by the other three. Hence one generally has to measure only three intensities though the fourth may also be measured to reduce errors and to act as a consistency check.

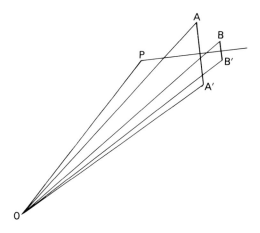

FIG. 23.4.

At present the standard technique to locate the heavy atoms employs a combination of isomorphous and anomalous techniques. Here one measures the intensity for the pure protein and also the Friedel pairs of the heavy anomalous scatterer derivative. In Fig. 23.4 one thus experimentally obtains the lengths $|\mathbf{OP}|$, $|\mathbf{OA}|$, and $|\mathbf{OA'}|$. It is clear from the discussion in the previous paragraph that this method is, in principle, identical to the two-wavelength method—all that one needs are the distances from the origin of three different points in the complex plane.

As in the combined isomorphous–anomalous method, in the two-wavelength method also there is in general a twofold ambiguity in $|S_H|$. In certain cases, it may be possible to eliminate one of the solutions, as when $|S_H| > n_H$,

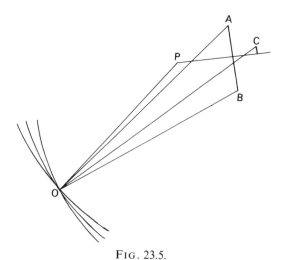

FIG. 23.5.

the number of heavy atoms. But, in all other cases, one has to select one of the two solutions. The standard approach is to select the smaller of the two values (Harding 1962). However, we note that each $|S_H|$ predicts a value of $|F_P|$, the protein structure factor. Statistical arguments can well be used to select the more probable pair of $|F_P|$ and $|F_H|$. Another possibility is to measure data at a third wavelength which will obviously resolve the ambiguity in $|S_H|$.

Once $|S_H|$ has been obtained for a number of reflections, the Patterson and other methods can be used to solve the heavy-atom structure. At the end of this stage, the heavy-atom vector F_H for all the reflections will be known and one can proceed to the next stage of obtaining the phases of the protein reflections.

Our earlier discussion (Fig. 23.2) showed that, given a single Friedel pair of reflections, there is a twofold ambiguity in the phase ϕ_P of F_P. Thus, one needs data at least at two wavelengths to solve the problem. In other words, one needs the distances from the origin of at least three points A, B, C which are not all collinear (Fig. 23.5). Once the lengths **OA** , **OB** , and **OC** are known, circles with these radii are drawn around the respective points and their intersection gives the origin and hence ϕ_P (Fig. 23.5).

In an ideal experiment, the three circles will of course intersect at a point. However, in an actual case, they will never intersect exactly. One then has to make the best estimate for the direction of O. The standard technique for doing this is to use the method of Blow and Crick (1959) and its many modifications (North 1965; Matthews 1965, 1966). Essentially, in this method, different radius vectors are drawn for different choices of the direction of O from ABC. In each case, the intersections with the three circles are computed and the spread in these is measured as a variance around the mean. The direction corresponding to the minimum in the variance is taken to be the correct solution. It is also possible to estimate the r.m.s. error in the computed ϕ_P which helps in selecting weights for the reflections in further calculations.

The optimum choice of wavelengths

If one is to use the multiwavelength method, an important question is how to select the wavelengths of the incident radiation. In the previous discussion (Fig. 23.5), of the three points A, B, and C, two are Friedel pairs (so that measurements are made at only two wavelengths). However, this restriction is not necessary since the ease of tuning synchrotron radiation permits us to use three different wavelengths for the three points. Hence, the question is how to select three points A, B, and C, lying on a given pair of symmetrical curves such as in Fig. 23.3, so as to minimize the r.m.s. error in ϕ_P averaged over all directions of F_p. Phillips (1978) has developed a computer simulation approach to this problem. However, we have investigated the question analytically (Narayan and Ramaseshan 1980) and the solution we obtain is that the area of the triangle ABC must be maximized subject to the constraints on the points A, B, and C. For the neutron case, the positions of the points are shown in Fig. 23.3(a). For the X-ray case, the constraints are not as simple as in the neutron case (Fig. 23.3(b)) and the optimum triangle ABC has to be numerically calculated. However, we note the useful fact that the optimum wavelengths have to be calculated only once for each crystal and can then be used for any reflection.

Protein crystallographers always try to produce a large number of isomorphous crystals so that the data can be extended to more than three 'centres'. Unfortunately, protein derivatives are not very easy to prepare and they are rarely perfectly isomorphous. In the multiwavelength method, however, any number of 'isomorphous crystals' can in effect be obtained by just tuning the incident radiation to different wavelengths. When there are more than three centres, the criterion to be followed to minimize the average r.m.s. error in ϕ_P is to maximize D given by

$$D = (\Sigma x_i^2)(\Sigma y_i^2) - (\Sigma x_i y_i)^2 \tag{23.5}$$

where (x_i, y_i) are the co-ordinates in the complex plane of the ith centre measured from the 'centre of mass' of the points as origin. We note again that (x_i, y_i) are arbitrary, subject only to the constraint that they should lie on permitted curves such as Fig. 23.3. The choice of the set of points which maximizes D will permit the best choice of wavelengths to be made. It will be noticed that D is the determinant of the 'moment of inertia' ellipse of the two-dimensional set of points, i.e. it is the product of the principal moments of inertia. This would seem to be a natural generalization to n points of the area of the triangle encountered earlier in the three-point case.

All the methods discussed require the absolute values of the radius vectors in the complex plane. Unangst, Muller, Muller, and Keinert (1967) and Bartunik (1978) have suggested that, since this requires absolute scaling and is therefore susceptible to error, it may be better to use Bijvoet ratios. With the multiwavelength method, one uses the same crystal for all the experiments and therefore scaling may not be a problem. Hence, it is preferable not to ignore valuable information by using only Bijvoet ratios.

References

BARTUNIK, H. D. (1978). *Acta crystallogr.* **A34**, 747.
——and FUESS, H. (1975). *Proc. Neutron Diffraction Conf.* Petten P. 527. Institut Laue-Langevin Internal Report 75D383.
BIJVOET, J. M. (1949). *Proc. Acad. Sci. Amst.* **52**, 313.
BLACK, P. J. (1965). *Nature, Lond.* **206**, 1223.
BLOW, D. M. and CRICK, F. H. C. (1959). *Acta crystallogr.* **12**, 794.
COSTER, D., KNOL, K. S., and PRINS, J. A. (1930). *Z. Phys.* **63**, 345.
COWLEY, J. M. (1975). In *Anomalous scattering* (ed. S. Ramaseshan and S. C. Abrahams) p. 113. Munskgaard, Copenhagen.
FLOOK, R. J., FREEMAN, H. C., and SCUDDER, M. L. (1977). *Acta crystallogr.* **B33**, 801.
HARDING, M. M. (1962). D.Phil. thesis, University of Oxford.
HODGKIN, D. C., WILLIS, B. T. M., FUESS, H., and MASON, S. A. (1973). *Scientific Research Council Annual Report*, 16A–16B.
HOSOYA, S. (1975). In *Anomalous scattering* (ed. S. Ramaseshan and S. C. Abrahams) p. 275. Munskgaard, Copenhagen.
KARTHA, G. and PARTHASARATHY, R. (1965). *Acta crystallogr.* **18**, 745.
KOETZLE, T. F. and HAMILTON, W. C. (1975). In *Anomalous scattering* (ed. S. Ramaseshan and S. C. Abrahams) p. 489. Munskgaard, Copenhagen.
MACDONALD, A. C. and SIKKA, S. K. (1969). *Acta crystallogr.* **B25**, 1804.
MATTHEWS, B. W. (1966). *Acta crystallogr.* **20**, 82.
MOON, P. B. (1961). *Nature, Lond.* **185**, 427.
MOORE and MACDONALD, A. C. (1970) Unpublished work.
MOSSBAUER, R. L. (1975). In *Anomalous scattering* (ed. S. Ramaseshan and S. C. Abrahams) p. 463. Munskgaard, Copenhagen.

NARAYAN, R. and RAMASESHAN, S. (1980). To be published.

NISHIKAWA, S. and MATUKAWA, K. (1928). *Proc. imp. Acad. Japan* **4**, 97.

NORTH, A. C. T. (1965). *Acta crystallogr.* **18**, 212.

PARAK, F., MOSSBAUER, R. L., HOPPE, W., THOMANEK, U. F., and BADE, D. (1976). *J. Phys. Radium, Paris* **37**, C6–703.

PEPINSKY, R. and OKAYA, Y. (1956). *Proc. natn. Acad. Sci. U.S.A.* **42**, 286.

PETERSON, S. W. and SMITH, H. G. (1961). *Phys. Rev.* **6**, 7.

PHILLIPS, J. C. (1978). Ph.D. thesis, Stanford University.

RAGHAVAN, R. S. (1961). *Proc. Indian Acad. Sci.* **A53**, 265.

RAMACHANDRAN, G. N. and RAMAN, S. (1956). *Curr. Sci.* **25**, 348.

RAMASESHAN, S. (1962). In *Advanced methods of crystallography* (ed. G. N. Ramachandran) p. 67. Academic Press, London.

——(1966). *Curr. Sci.* **35**, 87.

——and ABRAHAMS, S. C. (Eds.) (1975). *Anomalous scattering.* Munskgaard, Copenhagen.

——RAMESH, T. G., and RANGANATH, G. S. (1975). In *Anomalous scattering* (ed. S. Ramaseshan and S. C. Abrahams) p. 139. Munskgaard, Copenhagen.

SCHOENBORN, B. P. (1975). In *Anomalous scattering* (ed. S. Ramaseshan and S. C. Abrahams) p. 407. Munskgaard, Copenhagen.

SIKKA, S. K. (1969a). *Acta crystallogr.* **A25**, 396.

——(1969b). *Acta crystallogr.* **A25**, 539.

——(1973). *Acta crystallogr.* **A29**, 211.

——and RAJAGOPAL, H. (1975). In *Anomalous scattering* (ed. S. Ramaseshan and S. C. Abrahams) p. 503. Munskgaard, Copenhagen.

SINGH, A. K. and RAMASESHAN, S. (1968). *Acta crystallogr.* **B24**, 35.

UNANGST, D., MULLER, E., MULLER, J., and KEINERT, B. (1967). *Acta crystallogr.* **23**, 898.

24. The crystal structure of triclinic lysozyme by neutron diffraction

G. A. BENTLEY AND S. A. MASON

BECAUSE intense neutron sources have been, and will remain, more expensive than X-ray sources, few attempts have been made to study the crystal structures of large molecules by neutron diffraction. Notable among the successes are the work on a monocarboxylic acid of vitamin B_{12} by Dorothy Hodgkin and co-workers (Moore, Willis, and Hodgkin 1967), and studies of derivatives of myoglobin by Schoenborn and co-workers (Schoenborn 1969, 1971; Schoenborn and Diamond 1976; Schoenborn and Norvell 1976; Norvell, Nunes, and Schoenborn 1975). Very recently, new experiments have been started on vitamin B_{12} itself by Finney, Lindley, and Timmins (personal communication), and very interesting measurements are under way on single crystals of tomato bushy stunt virus (Bentley, Lewit-Bentley, and Timmins 1978) and the nucleosome core particle (Finch, Lewit-Bentley, Roth, Timmins, and Bentley, personal communication).

In this chapter, we summarize our progress in refining the structure of hen-egg-white lysozyme in the triclinic form to high resolution using neutron-diffraction data.

X-ray and neutron diffraction from proteins

The most important difference between X-ray and neutron diffraction arises from the different character of the coherent scattering factors. The X-ray scattering factor is proportional to the atomic number of the atom and decreases significantly at high scattering angles. The neutron scattering factor, in contrast, does not vary regularly from atom to atom and is constant as the scattering angle changes. An important advantage of neutron diffraction over X-ray diffraction for protein structure analysis is clear from the scattering factors, which are compared in Table 24.1. Hydrogen contributes more strongly to the diffracted intensities in the neutron experiment, particularly if it is replaced by deuterium. Consequently, these atoms can be more easily

TABLE 24.1

Neutron and X-ray scattering
factors (Bacon 1975)

	Scattering factor $(10^{-12}$ cm)	
Element	Neutron	X-ray $(\sin \theta = 0)$
H	−0.37	0.28
D	0.67	0.28
C	0.67	1.69
N	0.94	1.97
O	0.58	2.25
S	0.28	4.5

located, compared to an X-ray structure analysis. Neutron diffraction there-fore has the potential to give a more detailed picture of a protein molecule.

Exchanging the mother liquor of the protein crystal for one prepared in D_2O offers the following advantages in structure analysis by neutron diffrac-tion.

(i) Identification of those potentially exchangeable hydrogens which have been replaced by deuterium will give some indication of the flexi-bility of the molecule.

(ii) Heavy water diffracts neutrons more strongly than ordinary water; the description of the ordered water structure around the protein will be correspondingly more accurate.

(iii) Hydrogen, unlike deuterium, has a very large incoherent scattering cross-section; consequently protein crystals produce a high back-ground when placed in the neutron beam. Therefore, by introducing D_2O mother liquor, less time is needed to measure the coherent dif-fraction intensities to acceptable statistical precision.

The intensity data collection with X-rays and neutrons differs in the follow-ing ways.

(i) Protein crystals suffer radiation damage in the X-ray beam and their useful life-time for data collection is limited. Several crystals must be used for high resolution data collection and their deterioration in the beam sets a limit on the quality of the structure analysis. Neutrons, however, do not damage the protein and a complete data set may be obtained from one crystal. Therefore there are no errors from radiation damage and the scaling together of data from different crystals.

(ii) Available neutron fluxes are smaller than in the case of X-rays and the crystal volumes must be several mm³ if the neutron intensities are to be collected in an acceptable time. This poses problems in crystal growth and also requires that accurate absorption corrections be applied to the intensity data.

Triclinic lysozyme

Hen egg-white lysozyme in the tetragonal crystal form has been extensively studied by X-ray diffraction (Blake *et al.* 1967). A detailed examination of this molecule, together with its substrate complexes, has led to a hypothesis on the molecular mechanism of the enzyme reaction. One important feature of the proposed mechanism depends crucially on the state of protonation of the acidic residues Glu 35 and Asp 52, for which the pK_as have been assigned on indirect evidence (Imoto, Johnson, North, Phillips, and Rupley 1972).

The triclinic modification of lysozyme was first studied structurally by Joynson and co-workers (Joynson, North, Sarma, Dickerson, and Steinrauf 1970) who used the rotation function to relate it to the tetragonal form. A more detailed X-ray structure analysis was subsequently performed by Jensen and co-workers (Kurachi, Jensen, and Sieker 1976; Jensen, Hodsdon, and Sieker, personal communication) to a resolution of 1.5 Å, and independently, by Moult *et al.* (1976) to 2.5 Å resolution.

Triclinic lysozyme is well suited to a neutron diffraction study. The unit cell contains only one molecule, the mean temperature factor is small (a value of 9.5 Å² was obtained from our neutron intensity data) and large crystals are readily available; the intensities are therefore relatively large even at high Bragg angles. Furthermore, the unit-cell dimensions are small (see Table 24.2) so that the resolution of individual diffracted intensities at the detector is not a severe problem. From the point of view of the structural analysis, the results would complement the wealth of information from X-ray diffrac-

TABLE 24.2

Unit-cell dimensions for triclinic lysozyme

	a (Å)	b	c	α	β	γ
Undeuterated						
Jensen *et al.* (1976)	27.283 (10)	31.980 (10)	34.291 (13)	88.53 (5)	108.54 (3)	111.85 (.
2a	27.22	31.91	34.22	88.50	108.50	112.00
Partially deuterated						
2b	27.17	31.87	34.22	88.64	108.43	111.97
3	27.18 (2)	31.87 (2)	34.21 (2)	88.60 (2)	108.43 (2)	111.96 (2

The values of Jensen *et al.* (1976) are an average over eight crystals. The values in rows 2a and 2b we measured using X-rays on the same small crystal before and after deuteration. The values in row 3 a from a large crystal and were measured on the neutron diffractometer.

The figures in parentheses are the standard deviations in the unit-cell parameters.

tion and other chemical and physical data. A special point of interest lies in the highly ordered water structure in the triclinic form (Jensen, personal communication).

The collection of the neutron intensity data

Crystals of triclinic lysozyme were prepared by the procedure outlined by Steinrauf (1959), but with conditions optimized to encourage the growth of large, well-formed specimens. Clusters of needles, or sometimes large mono-clinic crystals, would often appear in the crystallizing tubes, but the large triclinic crystals were very recognizable. These large crystals had many inter-nal cracks and most proved to be multiple with individual components slightly misaligned with respect to each other. However, we found a good single crystal of volume approximately 20 mm³ and having a small mosaic spread. This crystal was soaked for several months in a large excess of D_2O buffer of pD 4.6 before being mounted in a fine-walled quartz capillary. A packing of quartz wool secured the crystal in a stable orientation inside the capillary. Unit-cell dimensions were only slightly altered upon deuteration (see Table 24.2).

The measurement of the neutron intensities was made on the D8 four-circle diffractometer which is installed on a thermal neutron beam tube at the high-flux reactor at the Institut Laue-Langevin, Grenoble. We chose the relatively long wavelength of 1.68 Å as a compromise between having strong diffracted intensities and minimal overlap of reflections on one hand, without introducing a large half-wavelength component from the thermal neutron spectrum on the other. This wavelength was obtained from the (002) planes of a pyrolytic graphite monochromator in reflection. A neutron flux of $2 \times 10^8 \, \mathrm{n \, cm^{-2} \, s^{-1}}$ was measured by the activation of a gold foil at the sample position. To reduce the effect of the large incoherent background, we used a detector aperture smaller than that needed to collect the whole of the dif-fracted beam; an empirical correction was then made for the loss of intensity at higher Bragg angles.

The intensity data to 1.85 Å d-spacing were recorded with 31-step omega scans, counting for 2.1 s at each step. The counting time was tripled for the data between 1.85 Å and 1.7 Å. Ten months later, we used the same crystal under identical measuring conditions to extend the resolution of the data to 1.5 Å. The counting time per step was 9.7 s for $1.72 \, \text{Å} \geqslant d \geqslant 1.6 \, \text{Å}$ and 6.7 s for $1.6 \, \text{Å} \geqslant d \geqslant 1.4 \, \text{Å}$. Between 1.5 Å and 1.4 Å, a preliminary four-step scan was used to select the 2000 strongest reflections for measurement. Resolution on D8 is poor at high Bragg angles ($2\theta_{max}$ was 74° compared with $2\theta_{monochromator}$ of 29°) and overlap may be significant at high resolution (i.e. $d < 1.5 \, \text{Å}$). In total, 17 889 unique reflections were measured. The data were corrected for the Lorentz factor, the cutting effect of the small detector aperture, the monitor

TABLE 24.3

Statistical precision of the intensity data

Range of d-spacings (Å)	No. of measured reflections	No. with I > 2σ	% with I > 2σ
d > 4.5	585	581	99
2.5→4.5	2872	2753	96
2.0→2.5	3275	2851	87
1.7→2.0	4229	3308	78
1.5→1.7	4930	3675	75
1.4→1.5†	1877	1567	83†

† Only the strong reflections were measured—if all reflections in this range were considered, only 44 per cent would have I > 2σ.

and detector dead-time, absorption and the drift in three standard intensities. Corrections for the half-wavelength component, which are about 2 per cent, have not been applied yet. The statistical precision of the data is summarized in Table 24.3. From the 2373 non-zero equivalent reflections measured the following residuals were calculated:

$$R(\Sigma \mid F_1^2 - F_2^2 \mid /\Sigma F_1^2) = 0.08,$$
$$RS(\Sigma \sigma(F_1^2)/\Sigma F_1^2) = 0.06.$$

The method of refinement

There are three main approaches currently used to refine protein structures. These are difference Fourier techniques, crystallographic least-squares refinement and the real-space refinement method of Diamond (1971). To date, only one protein structure has been refined from neutron diffraction data; this is carbon monoxide myoglobin (Schoenborn and Diamond 1976) which has been refined by Diamond's real-space procedure with data extending to 1.8 Å d-spacing. Although the details of this refinement have not been published extensively, they show that some useful information can be gained on the hydrogen parameters. In all methods of refinement, the geometry of the molecule must be constrained to sensible bond lengths and angles to prevent convergence to false minima. This is because of the difficulty of collecting data beyond about 1.5 Å d-spacing where the intensities are weak; consequently the ratio of observations to the number of parameters being refined is small. The situation is worse in a neutron structure determination since we must consider the hydrogens which account for approximately half the number of atoms in a protein (see Table 24.4).

To refine the structure of triclinic lysozyme using neutron diffraction data, we have chosen the least-squares technique of Agarwal (1978). This method

TABLE 24.4

| | | Ratio† | |
Resolution (Å)	No. of data	X-ray (1001) atoms)	Neutron (1950 atoms)
2.0	6770	1.7	0.9
1.7	10 500	2.6	1.3
1.4	17 900	4.5	2.3

† Ratio of the number of intensity observations to the number of parameters refined.

of refinement is ideal for large molecules since the demands on the computer with respect to time and core memory are acceptable. We constrain the shifted atom co-ordinates to an acceptable geometry using the procedure developed by Dodson, Isaacs, and Rollet (1976).

The progress of the refinement

We took as a starting model a set of co-ordinates given to us by Jensen, Hodson, and Sieker. These co-ordinates came from an intermediate stage of their X-ray analysis ($R=0.25$ with 2.0 Å X-ray data, including 170 water molecules). Initially we attempted refining with these co-ordinates (but excluding the solvent molecules), introducing most of the hydrogen atoms at calculated positions (1914 atoms). Those hydrogens which were potentially exchangeable were assigned as deuterium in the first instance. The neutron intensity data to 1.7 Å d-spacing only were included in the refinement. The convergence was poor, but a sequence of $(2F_o-F_c)$ Fourier maps calculated with data to 1.4 Å failed to give any clear indication of the reasons. In these maps, 10–15 per cent of the protein was successively left out of the phasing to give an unbiased view of that region of the structure. The hydrogen positions were very ambiguous in the maps, no doubt due to the poor phasing at this stage of the refinement and the low resolution of the data compared to the covalent bond length of hydrogen. In particular, the hydrogens which scatter negatively were obscured by the first trough of the diffraction ripple emanating from the atoms to which they are bonded.

We subsequently tried the more cautious approach of removing all hydrogen and deuterium atoms from the calculation. Although these atoms account for about 30 per cent of the protein contribution to the structure factor, we felt that the remaining atoms should be refined to a greater precision, if possible, before reintroducing the hydrogens at calculated positions. Continuing from this point, we have, at present, calculated 30 cycles of refinement, the current R-factor being 0.282 (after idealization of geometry) with the data extending to 1.5 Å d-spacing. So far 1730 atoms have been placed; these include 165 solvent positions, and 623 hydrogen and deuterium atoms which were introduced at various stages. Of the original 1001 atoms from the X-ray crystal structure only 59 atoms did not refine well; we have not yet been able to place these with confidence in difference maps. These atoms are on the surface of the molecule and have high thermal parameters.

The hydrogens on the amino-acid side-chains are refining with difficulty, but most of those on the peptide backbone appear to have converged giving acceptable bond parameters and temperature factors. We can say that a large fraction of the hydrogens on the peptide nitrogens appear to have undergone substantial exchange with the D_2O mother liquor, even those involved in secondary structure. However, we must await a more complete refinement for a better quantitive evaluation of the exchange.

Conclusions

Although the refinement is far from complete, some aspects of high resolution refinement with neutron intensities have emerged. A major problem is the low ratio of observations to parameters refined (Table 24.4). Collecting data with a single-counter diffractometer to a resolution higher than that which we have now would make inefficient use of the neutron beam, and future work in this direction awaits the implementation of the multidetector, D19, at the Institut Laue-Langevin, Grenoble.

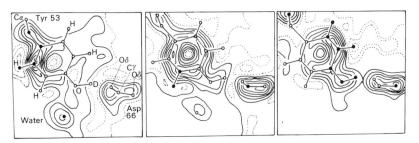

FIG. 24.1. Successive sections of a $(2F_O - F_C)$ Fourier in the region of Tyrosine 53. The tyrosine has been excluded from the phase calculation. Atoms which lie in given section are in full circles; the sections are separated by 0.5 Å.

We have found the appearance of hydrogens in the Fourier maps extremely ambiguous and have so far relied entirely on using the positions predicted from the geometry of the other atoms. It is therefore important to start with an accurate model from which good hydrogen co-ordinates may be calculated. Only in regions of the molecule which appeared to refine correctly did we begin to place the hydrogens at calculated positions. Figure 24.1 shows the contours in a $(2F_O - F_C)$ Fourier in the region of Tyr 53. (This residue, which has refined well, was not used to phase the map.) Although the positions of the hydrogens are certain from the geometry of the aromatic ring, they do not all correspond to well-defined minima in the map. It is interesting to note, in contrast, the strong evidence for a deuterium on the phenol group forming a hydrogen bond to Asp 66.

We have experienced much difficulty in using least-squares refinement with neutron diffraction data because of strong interaction between the parameters of neighbouring atoms in the least-squares normal matrix. The refinement programme uses the block-diagonal matrix approximation, accounting only for the interaction between x, y, and z on the same atom. A particularly strong interaction occurs between hydrogens and their neighbouring atoms; the effect is to displace a hydrogen away from the atom to which it is bonded. This seems to be a consequence of its negative scattering factor rather than the short covalent bond length since the effect does not occur with the positively scattering deuterium atom. It is of interest to note that the perturbation decreases as the hydrogen temperature factor increases. We have tried to overcome this interaction by idealizing the geometry after each round of co-ordinate refinement, placing a low weight on the hydrogen co-ordinates. However, the rate of convergence has been very slow and the high R-factor suggests that we may be refining to a false minimum. Simultaneous crystallographic and geometric refinement may perhaps help here (Konnert 1976). Extending the resolution of the data may also reduce the parameter interaction since the block-diagonal approximation assumes that the atoms are resolved (Srinivasan 1961). The most satisfactory way to deal with this problem, however, would be to use the complete normal matrix. Agarwal (1978) has outlined a procedure to calculate its elements using the Fast Fourier Transform. This entails considerable modification of the program and will require increased computer time and core memory.

The scale factor and temperature factors are highly correlated in least-squares refinement. In the block-diagonal matrix approximation, this interaction is considered by refining the scale and mean temperature factor in a 2×2 matrix. From this a correction may be calculated and applied to each individual temperature factor (Rollett 1965). We have observed a particularly strong correlation between the scale and temperature factors, which may point to the poor quality of our starting model (i.e. the incomplete description of the hydrogen atoms and solvent structure).

In spite of the difficulty we have experienced with the refinement of triclinic lysozyme by neutron diffraction, we are optimistic that the final result will yield new information not accessible to an X-ray structure analysis. The present stage of our work shows that we shall soon have a good quantitative measure of the H/D exchange on the peptide backbone; in the final stages of the refinement we hope to determine the state of protonation of the residues involved in the enzyme mechanism. This will determine the course of future work on the substrate complexes of lysozyme.

Acknowledgements

We would like to thank many colleagues for their support: E. Duée for providing the crystals; A. C. Nunes for his vital participation in the experimental work; L. H. Jensen and L. Sieker for providing the starting model for the refinement and for helpful discussions; and Eleanor Dodson for her generous help in preparing the least-squares refinement programme.

Appendix: new experimental work

Rather than extending our data set on a partially deuterated crystal beyond 1.4 Å *d*-spacing, we have measured recently all data to 1.8 Å on a crystal prepared from H_2O buffer. Measurement conditions were similar to those for the deuterated crystal. As the incoherent background is higher the data have lower statistical precision than those from the deuterated crystal. We changed drastically one important experimental parameter: the sample temperature was 5 °C. Jensen (personal communication) had suggested cooling triclinic lysozyme, which could almost be considered as a crystal hydrate in view of the highly ordered water structure, to just above the freezing point of water.

During this experiment, the crystal was cooled several times to very low temperatures (as low as −210 °C). The quality of the diffraction pattern at 5 °C was apparently not affected by this cooling. There are some unexplained reversible phenomena at temperatures well below 0 °C suggestive of a non-destructive phase transition. Further experiments are in progress.

References

AGARWAL, R. C. (1978). *Acta crystallogr.* **A34**, 791.

BACON, G. E. (1975). *Neutron diffraction*, p. 39. Oxford, University Press.

BENTLEY, G. A., LEWIT-BENTLEY, A., and TIMMINS, P. A. (1978). *Abstracts, 11th Int. Congr. Crystallogr.*

BLAKE, C. C. F., JOHNSON, L. N., MAIR, G. A., NORTH, A. C. T., PHILLIPS, D. C., and SARMA, V. R. (1967). *Proc. R. Soc.* **67**, 378.

DIAMOND, R. (1971). *Acta crystallogr.* **A27**, 436.

DODSON, E. J., ISAACS, N. W., and ROLLETT, J. S. (1976). *Acta crystallogr.* **A32**, 311.

IMOTO, T., JOHNSON, L. N., NORTH, A. C. T., PHILLIPS, D. C., and RUPLEY, J. A. (1972). *The enzymes* (ed. P. D. Boyer), 7, 665.

JOYNSON, M. A., NORTH, A. C. T., SARMA, V. R., DICKERSON, R. E., and STEINRAUF, L. K. (1970). *J. molec. Biol.* **50**, 137.

KONNERT, J. H. (1976). *Acta crystallogr.* **A32**, 614.

KURACHI, K., JENSEN, L. H., and SIEKER, L. C. (1976). *J. molec. Biol.* **101**, 11.

MOORE, F. H., WILLIS, B. T. M., and HODGKIN, D. C. (1967). *Nature, Lond.* **214**, 130.

MOULT, J., YONATH, A., TRAUB, W., SMILANSKY, A., PODJARNY, A., RABINOVICH, D., and SAYA, A. (1976). *J. molec. Biol.* **100**, 179.

NORVELL, J. C., NUNES, A. C., and SCHOENBORN, B. P. (1975). *Science, N.Y.* **190**, 568.

ROLLETT, J. S. (1965). *Computing methods in crystallography* (ed. J. S. Rollett) p. 47. Pergamon Press, Oxford.

SCHOENBORN, B. P. (1969). *Nature, Lond.* **224**, 143.
——(1971). *Cold Spring Harb. Symp. Quant. Biol.* **36**, 569.
——and DIAMOND, R. (1976). *Brookhaven Symp. Biol.* **27**, II.3.
——and NORVELL, J. C. (1976). *Brookhaven Symp. Biol.* **27**, II.12.
SRINIVASAN, R. (1961). *Acta crystallogr.* **14**, 1613.
STEINRAUF, L. K. (1959). *Acta crystallogr.* **12**, 77.

25. Early history of the convolutional and probabilistic direct-method relations

D. SAYRE

IN this note I shall describe the discovery of the squaring-method convolutional equations and the probabilistic triplet phase relations. These events took place in 1950 while I was a graduate student of Dorothy's at Oxford.

My introduction to crystal structure analysis came in 1947, when I came across a review article by J. M. Robertson on Fourier methods in X-ray analysis (Robertson 1937). I was studying biochemistry under George Wald at Harvard at the time. I was intensely excited by the map of nickel phthalocyanine which Robertson included in that review, and by the discovery that a complex molecule could be visualized so beautifully and exactly by a physical method. I immediately decided that this was the field I wanted to work in, and not long thereafter I joined Ray Pepinsky's group in X-ray analysis at the Alabama Polytechnic Institute in Auburn, Alabama. Incidentally, it was in Auburn in 1948 that I first met Dorothy when she paid a visit to our laboratory there.

At Auburn I tried to relate structure-factor phases to structure-factor amplitudes in a manner analogous to the way Bode had related the phase and amplitude characteristics of electrical networks (Bode 1945). The idea is still of some interest, and Burge and his colleagues have recently published papers on this approach (Burge, Fiddy, Greenaway, and Ross 1976).

When I came to England to be Dorothy's student in 1949 I met Vittorio Luzzati in Paris and Bill Cochran in Cambridge, and discovered that they, like me, were wondering whether the structure-factor phases might be implicit in the amplitudes. The Harker–Kasper inequalities had appeared not long before, and they showed that the values which the phases could assume were limited by the amplitudes, but our hope was that they might actually be determined by the amplitudes. I had several long conversations with Vittorio and with Bill in which we tried to utilize the theorems concerning Fourier transforms to obtain such a result, but without success.

One day I was sitting in the Radcliffe Science Library in Oxford, reading a discussion of the convolution theorem (Franklin 1940). I knew the theorem, but on this occasion I found myself somehow in a concentrated mood. With convolution in direct space and multiplication in reciprocal space, one had the familiar argument leading to the Patterson function. Now, however, I found myself thinking about multiplication in direct space and convolution in reciprocal space. But what to multiply? $d \times d$ seemed the only thing. Then suddenly I saw it: with equal resolved atoms, multiplying d by d would only sharpen d. But then convoluting F by F must do no more to F than multiply it by a sharpening factor. And if this were true, it would be a new and unexpected property of the Fs.

The whole thought-process, I think, must have lasted only a second or two. I went back to the crystallography laboratory to see by numerical test whether convoluting F by F would in fact merely modify F slightly, as predicted. By the end of the afternoon I had worked through a small one-dimensional test case and knew that the answer was yes.

By now I was wondering whether I could use the equations to deduce the phases of the structure factors in the test case. After supper I went back to the laboratory to try. By 2 a.m., when I went home to sleep, I had found some of the phases, and the remainder were found the next day. In this work it was obvious that the phases of $F(a)$, $F(b)$, and $F(a+b)$ are linked if all three magnitudes are large enough. This is the triplet relationship. At this point there was nothing probabilistic in this, however, and in fact all the phases in the test case were deduced rigorously.

Bill Cochran was one of the first persons to whom I described the squaring-method equations, and it was he who asked Joel Zussman to send me the data on hydroxyproline (Zussman 1951) to try as a more realistic test of the method. With hydroxyproline there were no terms large enough to compel any phase relations beyond doubt, and it was necessary to regard the triplet relation as a probabilistic one. Because of this the solving of hydroxyproline proceeded along the now-familiar lines of choosing origin-determining reflections, finding eight basic probable sign sets, extending the phases by use of the triplet relations and a crude form of phase refinement, and finally using the squaring-method equations to evaluate the eight sets of signs and select a best set.

These events concluded with the writing of a paper describing the convolutional equations, the triplet relationship,† and the solution of hydroxyproline and the small test case (Sayre 1952). The work was also described

---

† It would not be fair in this history to leave unmentioned the relationship between the probabilistic triplet relation and the important work of Karle and Hauptman (Karle and Hauptman 1950) on determinantal inequalities, which appeared in 1950 at about the time when my own work was being done. (At the time of doing the work I think I had not yet read the Karle–Hauptman paper, however.) In their paper they placed $F(a+b)$ within a circle centred on $F(a)F(b)/F(0)$, and gave the expression for the circle radius. As I stated in my own paper, the

in my Oxford D.Phil. thesis and at the Stockholm crystallography congress in 1951.

Perhaps it would be in order to state briefly some of the subsequent history of these ideas. The probabilistic triplet phase relation was quickly rederived by methods not requiring the assumption of equal resolved atoms (Cochran 1952; Zachariasen 1952; Hauptman and Karle 1953); actual probability distributions (not just expected values) were calculated (Hauptman and Karle 1953; Cochran and Woolfson 1955; Cochran 1955; Klug 1958); and the probabilistic theory was integrated into the theory of structure invariants and semi-invariants (Hauptman and Karle 1953). During this period the triplet relation was adopted as the main initial phasing device in direct-method systems. Recently the probabilistic theory has been extended to higher-order invariants and semi-invariants (Giacovazzo 1975; Hauptman 1975), supplementing the triplet relation with important additional relations based on quartets, quintets, etc.

The squaring-method equations were rederived probabilistically, allowing weaker assumptions concerning atom equality and data completeness to be used (Hughes 1953; Karle and Karle 1966; Hauptman 1971). Tangent-formula refinement (Karle and Hauptman 1956), an iterative technique for solving the equations, became the main method for the phase-extension and phase-refinement portions of direct-method systems. (Density modification, an iterative technique for solving the equations in direct space, is occasionally used instead (Hoppe and Gassmann 1968).) Recent developments include a more stable iterative solution method using least-squares (Sayre 1972, 1974), and a system of exact convolutional equations for the unequal-atom case (Rothbauer 1976).

The remaining elements of the work (definition of starting sets; extension, refinement, and evaluation of a phase set), which had been improvised for

---

recognition that $F(a+b)$ is linked to $F(a)$ and $F(b)$ in magnitude and phase is therefore unquestionably theirs. Their result, however, seems to have been advanced, and was certainly read, as a result in the same spirit as the Harker–Kasper inequalities, i.e. simply as a limitation on the values that phases can assume, and the quantitative situation is unfortunately such that in the great majority of cases the limitation does not come into play. (It was not until some years later that Tsoucaris showed that in these cases the Karle–Hauptman result may still be probabilistically significant (Tsoucaris 1970).) In contrast to this, the probabilistic triplet relation could always be formed and tried; and even if any individual relation might give a false indication, there was a feeling that the indications could still be correctly sorted out in many cases; as experience with the triplet relation increased, this proved to be true. On such small differences do events sometimes turn.

The determinantal inequalities, convolutional equations, and probabilistic phase relations constitute the three major categories of direct-method relations which we possess today. (The molecular replacement equations, due to Rossmann and Blow (Rossmann and Blow 1963), make up a fourth category of more specialized relations of great importance in protein and virus crystallography). The determinantal inequalities have thus far not played as large a role in practical direct-method systems as the probabilistic and convolutional relations, but with the Tsoucaris result there appears to be no reason why they should not do so.

the solution of hydroxyproline, were also more carefully defined and improved (Zachariasen 1952; Karle and Karle 1963, 1966). Finally, with the arrival of the computer starting in the mid-1950s, the partner for which direct methods had been waiting was at hand. The methods began to be programmed for execution by computer (Cochran and Douglas 1955; Germain and Woolfson 1968; Germain, Main, and Woolfson 1971), and the use of direct methods on a large scale could now begin.

I must say that I was never quite sure what Dorothy's attitude toward my involvement with direct methods was. I think she was happy for my sake that a significant result had emerged, but I also think it was a slight relief to her to know that she herself would probably always be working on structures of a complexity that direct methods had not yet learned to deal with. Just a few years ago she surprised and very much pleased me one day by asking me about some of the events I have described here. It was her questions that day that decided me to write this article for her book.

References

BODE, H. W. (1945). *Network analysis and feedback amplifier design*. D. van Nostrand, New York.
BURGE, R. E., FIDDY, M. A., GREENAWAY, and ROSS, G. (1976). *Proc. R. Soc.* **A350**, 191.
COCHRAN, W. (1952). *Acta crystallogr.* **5**, 65.
——(1955). *Acta crystallogr.* **8**, 473.
——and DOUGLAS, A. S. (1955). *Proc. R. Soc.* **A227**, 486.
——and WOOLFSON, M. M. (1955). *Acta crystallogr.* **8**, 1.
FRANKLIN, P. (1940). *Treatise on advanced calculus*. Wiley, New York.
GERMAIN, G. and WOOLFSON, M. M. (1968). *Acta crystallogr.* **B24**, 91.
——MAIN, P., and WOOLFSON, (1971). *Acta crystallogr.* **A27**, 368.
GIACOVAZZO, G. (1975). *Acta crystallogr.* **A31**, 252.
HAUPTMAN, H. (1971). Abstract H3, ACA Winter Meeting, Columbia, South Carolina.
——(1975). *Acta crystallogr.* **A31**, 680.
——and KARLE, J. (1953). *Solution of the phase problem. I. The centrosymmetric crystal*. ACA Monograph No. 3.
HOPPE, W. and GASSMAN, J. (1968), *Acta crystallogr.* **B24**, 97.
HUGHES, E. W. (1953). *Acta crystallogr.* **6**, 871.
KARLE, I. L. and KARLE, J. (1963). *Acta crystallogr.* **16**, 969.
KARLE, J. and HAUPTMAN, H. (1950). *Acta crystallogr.* **3**, 181.
————(1956). *Acta crystallogr.* **9**, 635.
——and KARLE, I. L. (1966). *Acta crystallogr.* **21**, 849.
KLUG, A. (1958). *Acta crystallogr.* **11**, 515.
ROBERTSON, J. M. (1937). *Rep. Prog. Phys.* **4**, 332.
ROSSMANN, M. G. and BLOW, D. M. (1963). *Acta crystallogr.* **16**, 39.
ROTHBAUER, R. (1976). *Acta crystallogr.* **A32**, 169.
SAYRE, D. (1952). *Acta crystallogr.* **5**, 60.
——(1972). *Acta crystallogr.* **A28**, 210.
——(1974). *Acta crystallogr.* **A30**, 180.
TSOUCARIS, G. (1970). *Acta crystallogr.* **A26**, 492.
ZACHARIASEN, W. H. (1952). *Acta crystallogr.* **5**, 68.
ZUSSMAN, J. (1951). *Acta crystallogr.* **4**, 72.

26. X-ray analysis of 2Zn insulin: some crystallographic problems

ALTHOUGH the X-ray diffraction pattern from 2Zn insulin crystals was recorded as early as 1935 by Dorothy Hodgkin, the solution of the crystal structure was achieved in her laboratory only by the late sixties. The structure was subsequently refined to a high degree of accuracy in the seventies. Because of the difficulties encountered, the X-ray analysis of 2Zn insulin crystals had many interesting features. I have had the privilege and the good fortune to be associated with Dorothy Hodgkin in the solution of the structure from the isomorphously and anomalously phased map as well as its subsequent refinement at high resolution. I have been asked to discuss in this article some of the crystallographic problems encountered in the course of the structure analysis. This contribution is mainly concerned with the analysis using the isomorphous replacement and the anomalous dispersion methods. The problems associated with the high-resolution refinement of the structure have also been very interesting, but I shall confine myself to a few general remarks on them. I would like to emphasize that the ideas and the procedures discussed here arose out of the combined efforts of many workers, notably, Margaret Adams, Tom Blundell, Eleanor Dodson, and Guy Dodson in addition to Dorothy Hodgkin herself.

Background

2Zn insulin crystallizes in the hexagonal (rhombohedral) space group $R3$ ($a = b = 82.5$, $c = 34.0$ Å) with 18 insulin molecules and six zinc ions in a triply primitive unit cell (Harding, Hodgkin, Kennedy, O'Connor, and Weitzman 1966). The insulin molecules exist as hexamers in the structure. Each hexamer is associated with two zinc ions. A schematic representation of the insulin hexamer is shown in Fig. 26.1. It would appear that the insulin molecules first associate into dimers and the dimers in turn associate into hexamers in the presence of zinc ions. The three dimers in the hexamer are related to one another by the crystallographic threefold axis. The two molecules in each

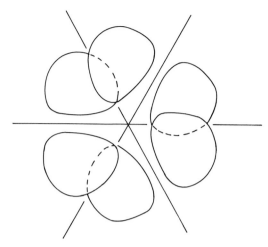

FIG. 26.1. A schematic representation of the 2Zn insulin hexamer in the crystal.

dimer are related to each other by an approximate non-crystallographic two-fold axis perpendicular to and passing through the threefold axis (Dodson, Harding, Hodgkin, and Rossmann 1966). Each dimer is also then related to the adjacent dimers by similar twofold axes displaced by 60° and −60° from the one that relates the two molecules in the dimer. The two zinc ions lie on the threefold axis. Taking the point of intersection of the twofold axis and the threefold axis as the origin, the zinc ions are located approximately at $c/4$ and $-c/4$.

A major problem encountered in the structure analysis of 2Zn insulin crystals was the difficulty in preparing suitable isomorphous heavy-atom derivatives. The chief reason for this was the tight packing of insulin molecules in the hexamer and that of the hexamers in the crystal. The derivatives ultimately used had complicated heavy-atom substitution patterns. Also, the major derivatives were not truly independent of one another.

Determination and refinement of heavy-atom parameters

The structure factor of the native protein crystal $\overline{F}_{P}$, that of the heavy-atom derivative $\overline{F}_{PH}$, and the heavy-atom contribution $\overline{F}_{H}$ are all real numbers when the reflections are centric. The magnitude of the heavy-atom contribution, F_{H}, can then be estimated as the difference between the magnitudes of $\overline{F}_{PH}$ and $\overline{F}_{P}$. Thus, a difference Patterson map with $|F_{PH} - F_{P}|^{2}$ as coefficients and a difference Fourier map with $(F_{PH} - F_{P}) \exp(i\alpha_{P})$ as coefficients (when approximate protein phases, α'_{P}s, are known) are good approximations to the vector-density and the electron-density distributions respectively of the heavy atoms when centric reflections are used. Also, realistic estimates of F_H are of great value in the least-squares refinement of heavy-atom parameters, especially the occupancy factors. Therefore, centric zones, when present, are used extensively in the determination and refinement of heavy-atom parameters in the X-ray analysis of proteins.

The diffraction pattern of 2Zn insulin crystals, with space group $R3$, has no centric zones. Some limitations are introduced in the effectiveness of difference Patterson and difference Fourier syntheses when all the reflections are acentric. The nature and extent of these limitations have been discussed by Dodson and Vijayan (1971). In the analysis of 2Zn insulin crystals, the situation was further complicated, when refining the heavy-atom parameters, by the existence of common features in the heavy-atom substitution patterns in the major derivatives. These problems were overcome by the extensive use of anomalous scattering data.

Harding (1962), Kartha and Parthasarathy (1965), Matthews (1966), and Singh and Ramaseshan (1966) have shown that the isomorphous difference (difference between the magnitudes of the structure factors of the derivative and the native protein) and the anomalous difference (the difference between the magnitudes of the structure factors of the Friedel pairs in the data from the derivatives) can be combined to provide an experimental estimate of the heavy-atom contribution to each reflection. According to the expression given by Singh and Ramaseshan

$$F_H^2 = F_{PH}^2 + F_{PH}^2 - 2F_{PH}F_P \cos(\alpha_P - \alpha_{PH})$$
$$= F_{PH}^2 + F_P^2 \pm 2F_{PH}F_P[1 - (k[F_{PH}(+) - F_{PH}(-)]/2F_P)^2]^{1/2},$$

where $k = (f_{0H} + f_H')/f_H''$ and $F_{PH} = (F_{PH}(+) + F_{PH}(-))/2$. Here f_{0H} is the scattering factor of the heavy atom in the absence of anomalous scattering, and f_H' and f_H'' are the real and the imaginary components of the dispersion correction. On account of the ambiguity in the third term in the above expression, there are two possible estimates of F_H. The upper estimate is called F_{HUE} and the lower estimate F_{HLE}. Most often F_{HLE} corresponds to the correct estimate (Harding 1962; Dodson and Vijayan 1971).

The heavy-atom vector distribution in each derivative was determined from a Patterson synthesis with F_{HLE}^2 as coefficients. As is to be expected, this synthesis was clearly superior in most cases to those with $|F_{PH} - F_P|^2$ and $|F_{PH}(+) - F_{PH}(-)|^2$ as coefficients. This can be clearly seen from the Harker sections of the three syntheses, given in Fig. 26.2, calculated using the data from the crystals of zinc-free insulin lead acetate and zinc-free insulin.† However, one cannot always expect the Patterson synthesis with F_{HLE}^2

† Historically the first, and in phasing the most useful, derivative was zinc-free insulin lead acetate. This derivative was prepared by soaking zinc-free insulin crystals, obtained by the removal of zinc ions from 2Zn insulin crystals by treatment with EDTA solution, in 0.01M lead acetate solution. The lead ions enter into not only the two axial positions formerly occupied by the zinc ions, but also into three other sites with varying occupancies. The data from zinc-

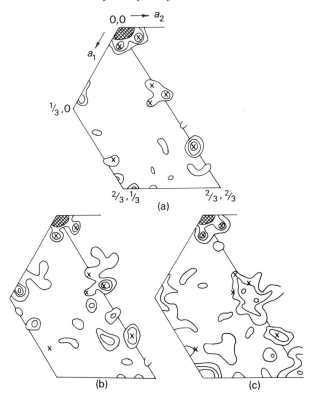

FIG. 26.2. Harker sections of the Patterson syntheses using (a) F_{HLE}^2, (b) $| F_{PH} - F_P |^2$, and (c) $| F_{PH}(+) - F_{PH}(-) |^2$ as coefficients calculated with data from zinc-free insulin lead acetate and zinc-free insulin. The contours in each map are drawn at equal but arbitrary intervals. The expected peak positions are indicated by crosses.

as coefficients to be superior to the conventional difference Patterson synthesis. When the substitution of heavy atoms is low, the contribution of the heavy atoms to the structure factors is small and consequently the anomalous differences are also small. In such a situation, the experimental error in the measurement of intensities is often comparable to, or even greater than, the anomalous difference and hence F_{HLE} could be in great error. For example, 2Zn insulin mercury benzaldehyde is a weakly substituted derivative. Harker sections of the heavy-atom Patterson syntheses for this derivative are given in Fig. 26.3. Neither of the two maps is really good; the conventional dif-

free insulin crystals were used as the native series in the analysis of the heavy-atom substitution pattern in zinc-free insulin lead acetate. All other derivatives used in the structure solution at 2.8 Å resolution were prepared by soaking 2Zn insulin crystals in the appropriate heavy-atom-containing solutions (Adams *et al.* 1969) and hence the data from 2Zn insulin crystals were used as the native series in analysing the heavy-atom distribution in them.

ference Patterson is, however, perhaps marginally better than the map with F^2_{HLE} as coefficients.

The heavy-atom substitution patterns in the derivatives determined from Patterson and difference Fourier syntheses at 4.5 Å resolution contained some disturbing features. First, in none of the derivatives did the heavy-atom positions and their occupancies obey the non-crystallographic twofold symmetry exhibited by the insulin monomers in the dimer. This feature, though worrying, did not cause any formal crystallographic difficulty. Indeed, the

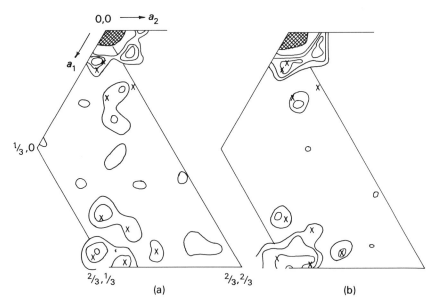

FIG. 26.3 Harker sections of the Patterson syntheses using (a) F^2_{HLE}, (b) $|F_{\mathrm{PH}} - F_{\mathrm{P}}|^2$ as coefficients calculated with data from 2Zn insulin mercury benzaldehyde and 2Zn insulin. See the underline for Fig. 26.2 for other information.

absence of symmetry in heavy-atom substitution was a reflection of the departure of the insulin molecules in the asymmetric unit from twofold symmetry. The second worrying feature of the substitution patterns was the presence of heavy-atom sites common to different derivatives. Of the five derivatives used in the 2.8 Å analysis, only two, namely, 2Zn insulin mercury benzaldehyde and 2Zn insulin uranyl fluoride, could be considered as completely independent of the rest of the derivatives as well as of each other. Of these, the heavy-atom substitution in 2Zn insulin mercury benzaldehyde was very low, thus making it comparatively ineffective. 2Zn insulin uranyl fluoride was not useful beyond a resolution of 4.5 Å as the heavy atoms had very high 'temperature factors'. The three major and highly substituted derivatives, namely zinc-free insulin lead acetate, 2Zn insulin lead acetate, and 2Zn insulin uranyl acetate, were related to one another through two common heavy-atom sites of which one was the major site in each of them. In addition, there was yet another site common to zinc-free insulin lead acetate and 2Zn insulin lead acetate. Thus, the essential differences between the three major derivatives lay not so much in the positions of the heavy atoms but in the relative occupancies of mostly common sites. The 'relatedness' of the major derivatives could be demonstrated in reciprocal space as well. If the phase angles of the heavy-atom contributions from derivatives 1 and 2 are α_{H1} and α_{H2} respectively and if the phase angles are adjusted to be between $-180°$ and $180°$, the average value of $||\alpha_{H1}| - |\alpha_{H2}||$ should be $90°$ when the two derivatives are completely independent. However, this quantity at $4.5\,\text{Å}$ resolution for zinc-free insulin lead acetate and 2Zn insulin lead acetate was $43°$, and for zinc-free insulin lead acetate and 2Zn insulin uranyl acetate $57°$.

The greatest value of F_{HLE}s was in the refinement of heavy-atom parameters, especially occupancy factors. The analysis of 2Zn insulin was perhaps the first occasion on which the estimated values of F_H were systematically and methodically used to refine heavy-atom parameters. The refinement procedure consists essentially in treating the F_{HLE}s as the observed values of the heavy-atom contribution and using

$$\Sigma W | F_{HLE} - F_{Hcalc}|^2,$$

where F_{Hcalc} is the heavy-atom contribution calculated from unrefined parameters and W is the weight factor, as the minimization function in the conventional least-squares analysis. The use of F_{HLE}s in refinement procedures was necessitated primarily because of the presence of common sites among major derivatives. The results of the conventional refinement using the minimization of the sum of squares of the lack-of-closure errors (Dickerson, Kendrew, and Strandberg 1961; Muirhead, Cox, Mazzarella, and Perutz 1967; Dickerson, Weinzierl, and Palmer 1968) led one to suspect that the occupancies of the common sites were being overestimated at the expense of those of the other sites. Calculations on simulated model structures (Dodson and Vijayan 1971) indicated that this was indeed a possibility. Therefore, it was important to make sure that the heavy-atom parameters in each derivative were refined independently of the information obtained from those in the other derivatives. This is readily achieved when F_{HLE}s are used in the least-squares refinement as the observed heavy-atom contributions.

The problems associated with refinement using F_{HLE}s have been discussed in some detail by Dodson and Vijayan (1971) with the aid of model calculations. Reflections for which F_{HUE} was less than the maximum expected value of F_H (for the given derivative) were excluded from least-squares calculations. This procedure ensured that F_{HLE} corresponded to the correct estimate of F_H for all reflections used in the calculations. The other problem associated with F_{HLE} refinement is concerned with the effects of errors in the data. Following Kartha and Parthasarathy (1965), one can write

$$F_{HLE}^2 \simeq |F_{PH} - F_P|^2 + (k/2)^2 |F_{PH}(+) - F_{PH}(-)|^2.$$

$|F_{PH}(+) - F_{PH}(-)|$ is a small difference between two large quantities and its value is often comparable to that of the statistical errors in the data. In such a situation, it has been shown that the errors in the data, even when random, result, on average, in the systematic overestimation of $|F_{PH}(+) - F_{PH}(-)|$. Thus, anomalous differences tend to be overestimated, resulting in the spurious overestimation of F_{HLE}s and, consequently, the occupancy factors. On the basis of theoretical considerations and model calculations, Dodson and Vijayan (1971) suggested that the situation could be improved by using empirically evaluated values of k (k_{emp}) instead of those calculated from form factors and dispersion correction (k_{theo}). According to Kartha (1965) and Matthews (1966), k_{emp} can be evaluated as a function of $\sin \theta / \lambda$ using the relation

$$k = 2|F_{PH} - F_P|/F_{PH}(+) - F_{PH}(-)|.$$

For each derivative, two sets of F_{HLE}s, one using k_{emp} and the other using k_{theo}, were estimated for the F_{HLE} refinement of heavy-atom parameters. The empirically evaluated values of k for the three major derivatives are shown in Fig. 26.4. The corresponding theoretical values are also given in Fig. 26.4. Two features of the curves given in Fig. 26.4 merit some discussion. One is the abnormal increase in the values of k_{emp} at low angles. It is very likely that at least some of the heavy-atom sites were occupied not by lead or uranyl ions as such, but by lead acetate or uranyl acetate. The acetate group, at

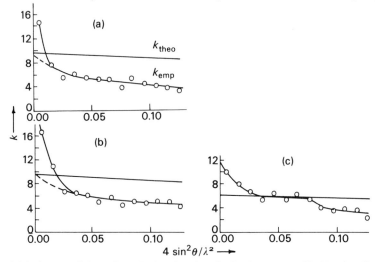

FIG. 26.4. k_{emp} and k_{theo} for (a) zinc-free insulin lead acetate, (b) 2Zn insulin lead acetate and (c) 2Zn insulin uranyl acetate.

low angles, would contribute substantially to the real part of the scattering amplitude, but not to the imaginary part. Also, changes in the structure resulting from the introduction of heavy-atom groups would contribute to the real part, but not to the imaginary part. Hence, when evaluating k_{emp} at low angles, one is in fact dividing the isomorphous difference resulting from the real part of the heavy-atom scattering factor, the scattering factor of the acetate group, and the contribution from gross changes in the structure, solely by the anomalous difference resulting from the imaginary part of the heavy-atom scattering factor. That would necessarily lead to a high value of k_{emp} at low angle. A careful examination of the phases after the solution of the structure showed that the protein phases and the heavy-atom phases were somewhat related to each other at low angle. $\overline{F}_P$ and $\overline{F}_H$ were not randomly oriented with respect to each other; they tended to be parallel or antiparallel. This again tended to increase isomorphous differences at the expense of anomalous differences. The second interesting feature of the k curves is the decrease in the values at high angles. This appears to be related to the effect of errors in the data on anomalous differences. In general, intensities are weak at high angles and are subject to higher percentage errors. This, in turn, leads to spuriously high anomalous differences and, hence, low values of k_{emp}. The abrupt decrease in the value of k_{emp} for 2Zn insulin uranyl acetate at about 3.6 Å resolution is interesting as it was known that the data from this derivative were rather inaccurate from this resolution onwards on account of some instrumental errors in the diffractometer.

It is instructive to compare the results obtained from the refinement using the F_{HLE}s estimated with k_{theo} and the F_{HLE}s obtained with k_{emp}. The positional parameters were refined in the usual manner, but the occupancy factors were refined in three shells of increasing Bragg angle. As the occupancy factors were allowed to vary from shell to shell, there was no need to introduce separate temperature factors. Refinement calculations using the two sets of F_{HLE}s led to nearly the same positional parameters. There were, however, systematic differences in the occupancy factors obtained from the two sets of calculations. As an example, occupancy factors of the heavy-atom sites in zinc-free insulin lead acetate obtained from the two sets of unweighted refinements are shown in Fig. 26.5. Similar results were obtained for the other major derivatives as well. In general, calculations based on k_{emp} led to occupancy factors lower than those obtained from calculations based on k_{theo}. On the basis of theoretical considerations and experience with model calculations, the former were accepted as the true occupancy factors.

Calculation of phase angles

One crystallographic aspect which deserves special mention in relation to the calculation of phase angles in the analysis of 2Zn insulin crystals is the

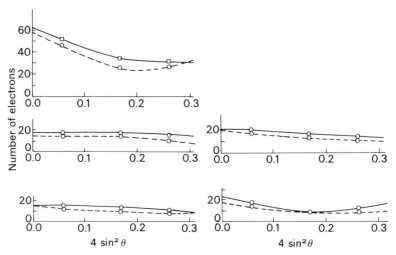

FIG. 26.5. Unweighted refinement of the occupancy factors of the five lead sites in zinc-free insulin lead acetate. The unbroken and broken lines represent the values obtained from F_{HLE}s estimated using k_{theo} and k_{emp} respectively.

evaluation of the root mean square lack-of-closure error, usually denoted by E (Blow and Crick 1959). When centric zones are present in the diffraction pattern, the values of E for different derivatives are evaluated from reflections in such zones using the expression

$$E^2 = \sum_n (|\overline{F}_{\text{PH}} \pm \overline{F}_{\text{P}}| - F_{\text{H}})^2/n.$$

The values of E thus obtained are used for the entire data. In the absence of centric zones in the diffraction pattern, the values of E were evaluated by two different methods (Adams 1968) in the analysis at 4.5 Å resolution. Both the methods are based on combining the information obtained from isomorphous and anomalous differences.

In the first method of evaluating E, the following well-known relations (Ramachandran and Raman 1956) were used:

$$-\cos\alpha = \frac{F_{\text{P}}^2 - F_{\text{PH}}^2 - F_{\text{H}}^2}{2F_{\text{PH}}F_{\text{H}}}, \tag{26.1}$$

$$\sin\alpha = \frac{F_{\text{PH}}(+)^2 - F_{\text{PH}}(-)^2}{4F_{\text{PH}}F_{\text{H}}''}, \tag{26.2}$$

where $\alpha = \alpha_{\text{PH}} - \alpha_{\text{H}}$ and F_{H}'' is the imaginary component of the heavy-atom contribution. The magnitude of α can be determined from eqn (26.1) and the quadrant from eqn (26.2). The resultant angle may be called α_{iso}. Alternatively, the magnitude can be determined from eqn (26.2) and the quadrant from eqn (26.1), resulting in what may be called α_{ano}. Ideally α_{iso} and α_{ano}

should be the same. However, in practice, they differ on account of the various errors present in the data. Rearranging eqn (26.1), we can write

$$F_P^2 = F_{PH}^2 + F_H^2 - 2F_{PH}F_H \cos \alpha. \tag{26.3}$$

Using α_{ano} in eqn (26.3) we can obtain what may be considered as the calculated values of F_P (F_{Pcalc}). Assuming that all errors lie in F_P, the values of E were calculated from the expression

$$E^2 = \sum_n |F_P - F_{Pcalc}|^2 / n. \tag{26.4}$$

Es were also estimated using the expression

$$E^2 = \sum_n |F_{HLE} - F_{Hcalc}|^2 / n, \tag{26.5}$$

where F_{Hcalc} is the calculated value of the heavy-atom contribution. It turned out that the values obtained from eqns (26.4) and 26.5) were nearly the same. Therefore, only eqn (26.5) was used to evaluate Es in the high-resolution analysis.

The root mean square anomalous lack-of-closure error, the so-called E' (North 1965), could also be calculated from eqns (26.1) and (26.2). Assuming F_{PH} to be equal to $(F_{PH}(+) + F_{PH}(-))/2$, eqn (26.2) can be written as

$$F_{PH}(+) - F_{PH}(-) = 2F_H'' \sin \alpha. \tag{26.6}$$

The value obtained by using α_{iso} in eqn (26.6) may be considered as the calculated value of the anomalous difference (ΔF_{PHcalc}). Then the values of E' could be calculated from the expression

$$E' = \sum_n |\Delta F_{PH} - \Delta F_{PHcalc}|^2 / n, \tag{26.7}$$

where $\Delta F_{PH} = |F_{PH}(+) - F_{PH}(-)|$. The values of E' calculated from eqn (26.7) for data in the 4.5 Å sphere turned out to be a third of the corresponding E's. The values of E' for the 2.8 Å analysis were, therefore, assumed to be a third of the corresponding E's as suggested by North (1965).

High-resolution refinement

Isomorphously and anomalously phased maps were calculated first at 2.8 Å resolution and later at 1.9 Å resolution. The atomic co-ordinates obtained from the latter were subjected to refinement by more than one technique using 1.5 Å resolution data. These calculations were aimed at refining the known co-ordinates, correcting them wherever necessary, and locating atoms with unknown positions, especially those belonging to the solvent molecules. A description of these calculations is beyond the scope of this paper. I shall confine myself here to a brief discussion of one particular aspect of high-resolution refinement which is perhaps of fundamental importance and general interest.

The difficulties encountered during the refinement suggested that the Fourier and the difference Fourier maps are, on occasions, not as effective as one would normally expect them to be in indicating the errors in the position of the atoms used for phasing and in giving the correct positions of the unknown atoms. Consequently, it was possible, as indeed happened in the course of the refinement of the structure, to arrive at a crystallographically acceptable yet erroneous structure even when various kinds of Fourier maps had been used at different stages to check the correctness of the atomic positions. This disturbing feature was discovered mainly by comparing the results obtained from independent refinements employing different semi-automatic procedures but using the same data. The discrepancies in the results were subsequently resolved and the correct positions arrived at through a detailed examination of different types of Fourier maps, including those used in the earlier stages of refinement, coupled with geometrical and chemical considerations. When examining Fourier and difference Fourier maps, it appeared that some atoms 'remembered' their previous history even when they were not included in the calculation of phase angles. One was also led to suspect— and it appears to be obvious in retrospect—that the positions obtained from Fourier maps, whether they be the corrected positions or the new ones, are affected in a systematic manner by the errors that might be present in the positions of the input atoms. This prompted us to undertake a theoretical analysis of Fourier refinement of protein structures with special reference to the effect of errors in the input atomic positions.

Typically, at any stage of the refinement of a protein structure, the inaccurate positions of P out of a total of N atoms (which include solvent molecules as well) are known and those of the remaining Q atoms are unknown. The structure factor of any given reflection may be denoted by $F_N \exp(i\alpha_N)$. The correct and the inaccurate positions of the P known atoms may be denoted by $\bar{r}_{Pj}$ and $\bar{r}'_{Pj}$, and the corresponding calculated structure factors by $F_P \exp(i\alpha_P)$ and $F'_P \exp(i\alpha'_P)$. The positions of the Q unknown atoms may likewise be denoted by $\bar{r}_{Qj}$. Then, following the formulation of Ramachandran and Srinivasan (1970), it can be shown (Vijayan 1980) that the peaks in a difference Fourier synthesis with coefficients

$$(F_N - F'_P)\exp(i\alpha'_P)$$

have the following positions and strengths:

$$\bar{r}'_{Pj}, \qquad \frac{S_N - S_P}{S_P} f_{Pj},$$

$$\bar{r}'_{Pj} + \bar{r}'_{Pk} - \bar{r}'_{Pl}, \qquad -\frac{S_N}{2S_P^3} f_{Pj} f_{Pk} f_{Pl},$$
$$(j \neq l)$$

$$\bar{r}'_{Pj}+\bar{r}_{Qk}-\bar{r}_{Ql}, \qquad\qquad \frac{1}{2S_N S_P}f_{Pj}f_{Qk}f_{Ql},$$
$$(k\neq l)$$

$$\bar{r}'_{Pj}+\bar{r}_{Pk}-\bar{r}_{Ql}, \qquad\qquad \frac{1}{2S_N S_P}f_{Pj}f_{Pk}f_{Ql},$$

$$\bar{r}_{Pj}+\bar{r}'_{Pk}-\bar{r}_{Pl} \qquad\qquad \frac{1}{2S_N S_P}f_{Pj}f_{Pk}f_{Pl},$$
$$(j\neq l)$$

$$\bar{r}_{Qj}+\bar{r}'_{Pk}-\bar{r}_{Pl}, \qquad\qquad \frac{1}{2S_N S_P}f_{Qj}f_{Pk}f_{Pl},$$

where $S_N^2=\Sigma f_{Nj}^2$, $S_P^2=\Sigma f_{Pj}^2$ and $S_Q^2=\Sigma f_{Qj}^2$. Thus the features in a difference Fourier map can be considered as consisting primarily of shifted vector sets centred around different atomic positions, some sets with origin peaks and some without. When the errors in the known positions are small and random, the terms listed above can be rearranged in a simple manner and divided into two types: those contributing to density around the atomic positions $\bar{r}'_{Pj}, \bar{r}_{Pj}$, and $\bar{r}_{Qj}$, and those contributing to the background. In such a situation, and assuming N and P to be large, the following peak strengths result at different atomic positions:

$$\bar{r}'_{Pj}, \qquad\qquad \frac{S_N-2S_P}{2S_P}f_{Pj},$$

$$\bar{r}_{Pj}+\langle (\bar{r}'_{Pk}-\bar{r}_{Pk})\rangle, \qquad\qquad \frac{S_P}{2S_N}f_{Pj},$$
$$j\neq k$$

$$\bar{r}_{Qj}+\langle (\bar{r}'_{Pk}-\bar{r}_{Pk})\rangle, \qquad\qquad \frac{S_P}{2S_N}f_{Qj}.$$

The remaining terms contribute to general background. All the known properties of the difference Fourier synthesis follow from these three expressions. Also, as can be readily seen, refinement proceeds smoothly and automatically. 'Mixed' synthesis with coefficients

$$(mF_N-nF'_P)\exp(i\alpha'_P)$$

was also similarly analysed. In this case also, the expected features in the map can be divided in a simple manner into those contributing to peaks at the atomic positions and those contributing to general background, when the errors in the input positions are random and small. The optimum values of the parameters m and n can then be derived in terms of S_N and S_P, and the synthesis used for automatic refinement of protein structures.

The situation becomes more complex when the errors in the positions of the input atoms are systematic, which is often the case in proteins. It can be readily shown that non-random errors lead to shifts in peak positions from $\bar{r}_{P_j}$ and $\bar{r}_{Q_j}$. The features in a difference Fourier or a mixed synthesis can no longer be divided in a simple manner into those contributing to peaks at the atomic positions and those contributing to the background. What was considered earlier as background is also then likely to be important in determining peak positions. The analysis also shows that there is no inherent mechanism for correcting the positional errors in Fourier maps resulting from the non-random errors in the positions of the input atoms.

To summarize, the theoretical analysis outlined above, and the details of which are given elsewhere (Vijayan 1980), establishes the definite though complicated relationship that exists between the errors in the positions of the known atoms on the one hand, and the errors in the new positions of the known atoms as well as the positions of the unknown atoms derived from Fourier syntheses on the other. This is an aspect of paramount importance in the refinement of protein structures. The effects of random errors in the input positions are easily and automatically taken care of by Fourier syntheses employing carefully chosen coefficients. The situation becomes more complex when the errors are non-random; automatic refinement procedures could then lead to erroneous results. The theory thus offers an explanation, though not simple solutions, to some of the problems encountered during the high-resolution refinement of 2Zn insulin structure. It also emphasizes the need for careful examination of different maps at all stages of refinement to eliminate possible systematic errors.

References

ADAMS, M. J. (1968). D.Phil. thesis, University of Oxford.

——BLUNDELL, T. L., DODSON, E. J., DODSON, G. G., VIJAYAN, M., BAKER, E. N., HARDING, M. M., HODGKIN, D. C., RIMMER, B., and SHEAT, S. (1969). *Nature, Lond.* **224**, 491.

BLOW, D. M. and CRICK, F. H. C. (1959). *Acta crystallogr.* **12**, 794.

DICKERSON, R. E., KENDREW, J. C., and STRANDBERG, B. E. (1961). *Acta crystallogr.* **14**, 1188.

——WEINZIERL, J. E., and PALMER, R. A. (1968). *Acta crystallogr.* **B24**, 997.

DODSON, E. and VIJAYAN, M. (1971). *Acta crystallogr.* **B27**, 2402.

——HARDING, M. M., HODGKIN, D. C., and ROSSMANN, M. G. (1966). *J. molec. Biol.* **16**, 227.

HARDING, M. M. (1962). D.Phil. thesis, University of Oxford.

——HODGKIN, D. C., KENNEDY, A. F., O'CONNOR, A., and WEITZMANN, P. D. J. (1966). *J. molec. Biol.* **16**, 212.

KARTHA, G. (1965). *Acta crystallogr.* **19**, 883.

——and PARTHASARATHY, R. (1965). *Acta crystallogr.* **18**, 745.

MATTHEWS, B. W. (1966). *Acta crystallogr.* **20**, 230.

MUIRHEAD, H., COX, J. M., MAZZARELLA, L., and PERUTZ, M. F. (1967). *J. molec. Biol.* **28**, 117.

NORTH, A. C. T. (1965). *Acta crystallogr.* **18**, 212.

RAMACHANDRAN, G. N. and RAMAN, S. (1956). *Curr. Sci.* **25**, 348.
—— and SRINIVASAN, R. (1970). *Fourier methods in crystallography.* Wiley, New York.
SINGH, A. K. and RAMASESHAN, S. (1966). *Acta crystallogr.* **21**, 279.
VIJAYAN, M. (1980). *Acta crystallogr.* **A36**, 295.

27. The refinement of protein crystal structures by fast-Fourier least-squares

NEIL ISAACS

RECENT results on the refinement of protein crystal structures with high-resolution data (Huber *et al.* 1974; Freer, Alden, Carter, and Kraut 1975; Moews and Kretsinger 1975; Adman, Sieker, and Jensen 1975; Deisenhofer and Steigemann 1975; Bode and Schwager 1975; Chambers and Stroud 1977; Takano 1977; Isaacs and Agarwal 1978) have shown that refinement markedly improves the accuracy of the structure and the quality of the electron-density map. With the exception of rubredoxin (Watenpaugh, Sieker, Herriott, and Jensen 1973) all the refinements listed here prior to 1978 were performed using either the real-space method of Diamond (1971, 1974) or difference Fourier methods (see, for example, Watenpaugh *et al.* 1973). Rubredoxin has been extensively refined with difference-Fourier methods followed by block-diagonal least-squares refinement using a conventional program. The quality of this result, and the extra structural information (particularly the water structure) which results from this refinement, showed the validity and value of refining protein structures by least-squares methods.

The principal obstacles to the routine use of least-squares refinement for protein structures are (i) the paucity of diffraction data relative to that for small molecules, and (ii) the enormous computing cost. The paucity of diffraction data is characteristic of most protein crystals. The intensity of the diffracted radiation is reduced because of the large unit-cell volume, and the extent of the scattering is further reduced by the large amount of disordered solvent in the cell. Few protein crystals have had data measured to a resolution of 1.5 Å. The reliability and accuracy of least-squares refinement reduces as the ratio of the number of observations to the number of parameters reduces. For practical purposes this means that diffraction data to a resolution of ≥ 2 Å is required for a meaningful refinement. The computing cost arises from the number of calculations required for least-squares refinement. For a full matrix least-squares calculation the computing required is proportional to NM^2 where N is the number of reflections and M the number of parameters. For the simplest diagonal least-squares calculation, the requirement is proportional to NM since a derivative of each calculated structure factor has to be computed for each variable parameter.

It is possible to circumvent a lack of data by either reducing the number of parameters or by adding additional observations into the calculation. The number of parameters may be conveniently reduced by treating groups of atoms as 'rigid bodies' with their positions described by three rotational and three translational parameters. Alternatively, the number of observations may be increased by including information in the form of constraints on the known geometry (bond lengths and valence angles) of peptides. Least-squares programs have been written which use one (Konnert 1976) or both (Sussman, Holbrook, Church, and Kim 1977) of these approaches, or use least-squares coupled with potential energy minimization (Jack and Levitt 1978). They have been used with remarkable success, particularly where only low-resolution data are available, but both the methods using constraints or restraints are affected (although less acutely when rigid-body constraints are used) by the rapid increase in the cost of computing with the increase in the size of the structure (Schmidt, Girling, and Amma 1977). The method of Jack and Levitt (1978) uses Agarwal's (1978) fast-Fourier least-squares procedures.

This problem of cost required a drastic new approach to least-squares refinement, and this was provided by Agarwal (1978), who showed that most of the calculations involved could be computed by Fourier transforms. With the availability of fast-Fourier transforms (FFT) (Cooley and Tukey 1965; Ten Eyck 1973) the algorithm is extremely fast, and the computing required is proportional to $N \log N$ where N is the number of reflections.

The method

In the least-squares refinement of atomic parameters the function minimized is

$$P = \tfrac{1}{2} \sum_{hkl} W_{(hkl)} [|F_{c(hkl)}| - |F_{obs(hkl)}|]^2$$

where W_{hkl} is a weighting function. The corrections to the parameters are obtained from the matrix equation

$$\Delta u = -H^{-1} G,$$

where Δu_i is the correction to be applied to the ith parameter. H^{-1} is the inverse of the normal matrix whose general term is

$$H_{ij} = \sum_{r=1}^{N} W_r \frac{\partial |F_{cr}|}{\partial p_i} \frac{\partial |F_{cr}|}{\partial p_j},$$

where N is the number of reflections and W_r is a weighting function. G is the gradient vector of general form

$$G_i = \sum_{r=1}^{N} W_r (\Delta F_r) \frac{\partial |F_{cr}|}{\partial p_i}.$$

The size of the normal matrix is $M \times M$ where M is the number of parameters and the length of the gradient vector is M. The calculation of the gradient vector is proportional to NM and that of the normal matrix is proportional to NM^2.

Calculation of structure factors

Structure factors are calculated by a FFT of a model electron density. This is not a new procedure. The use of the basic method was discussed by Sayre (1951) nearly thirty years ago. However, it is only with the development of FFT that the method has become viable for protein structures. Computationally there are two distinct stages in calculating structure factors. The first stage is to calculate the atom electron density of the structure at each point on a uniform grid parallel to the cell axes. The second is the Fourier inversion of this electron-density map to obtain the structure factor magnitudes and phases. The computation required for the first stage depends on the fineness of the sampling interval throughout the cell, or the number of grid points, and the distance from the centre of each atom for which the electron density is to be computed.

Sayre (1951) discussed how too coarse a grid for sampling the electron density can produce overlap errors in the Fourier transform owing to the periodicity of the crystal structure. He showed how this can be prevented by using hypothetical atoms with zero scattering beyond a distance of $2/\lambda$ in reciprocal space and a sampling lattice with a spacing of at least $\lambda/4 \text{ Å}$ in real space. An alternative method, which he did not consider practical at that time because of the inaccuracy of the method of calculation, was to use atoms with large temperature factors. Broadening the atoms in real space sharpens their transform in reciprocal space and allows for a coarser grid to be used. Ten Eyck (1977) has shown that the accuracy of the FFT is sufficient to allow this approach to be used. In this procecure, an artificial temperature factor is added to each of the atoms, and after the transform the effect of the temperature factor is removed. The values of the additional temperature factor, and the sampling interval of the cell, are related as shown by Ten Eyck (1977). In the refinement of insulin an additional B of 15.0 Å^2 was added to each atom. This allowed the transform to be computed on a grid of about 0.6 Å separation with data to a resolution of 1.5 Å, resulting in a saving of about 20–25 per cent of the transform time compared with a normal spacing of 0.5 Å.

The most expensive part of the structure factor calculation is setting up the model atom electron density which is the Fourier transform of the atom scattering factor curve corrected for thermal motion. The isotropic thermal motion of the atoms is represented as a Gaussian function $\exp(-B_m s^2/4)$ where s is $2\sin\theta/\lambda$. If the atom scattering factor curve is also defined as a Gaussian function then the product of these two Gaussian functions is another Gaussian function whose Fourier transform is also Gaussian. Both Agarwal (1978) and Ten Eyck (1977) have given the formulae to calculate the atom electron density. The speed of this computation depends on the number of terms in the Gaussian approximation to the atom scattering factor curve and on the number of grid points for which the density is to be calculated for each atom. Coefficients for analytical approximations of four Gaussian terms are given in *International tables for X-ray crystallography*, Vol. IV (1974). These give an accurate fit to the scattering factor curve over a wide range of $\sin\theta/\lambda$, but for most protein structures the range of angle is smaller and the curve can be approximated by a single Gaussian function, for data limited to low resolution, or by the sum of two or three Gaussian terms for higher resolution data (Agarwal 1978; Ten Eyck 1977). Reducing the radius of the atom will exclude a fraction of its density from the calculation. The fraction excluded beyond a given radius depends on the atom type and its temperature factor. Agarwal (1978) has computed this fraction for different limiting radii and temperature factors so that a radius consistent with the required accuracy of the calculation may be chosen.

Calculation of the gradient vector

Agarwal (1978) has derived the following expression for the gradient vector with respect to the x co-ordinate of the mth atom

$$G(x_m) = \sum_s g_m(s)\,(-i2\pi h)\,W(s)\,E(s)\,\exp\{i\phi(s)\}\,\exp(-i2\pi s \cdot r_m),$$

where $g_m(s) = f_m(s)\,\exp(-B_m s^2/4)$,
$\qquad (s) = 2\sin\theta/\lambda$,
$\qquad W(s) = W_{(hkl)}$, a weighting function,
$\qquad E(s) = |F_{\text{calc}(hkl)}| - |F_{\text{obs}(hkl)}|$,
$\qquad s \cdot r_m = hx_m + ky_m + lz_m$.

Similar expressions hold for $G(y_m)$, $G(z_m)$, and $G(B_m)$ with the term $(-i2\pi h)$ replaced by $(-i2\pi k)$, $(-i2\pi l)$, and $(-s^2/4)$ respectively.
$\quad G(x_m)$ may be rewritten as

$$G(x_m) = \sum_s D_x(s)\,g_m(s)\,\exp(-i2\pi s \cdot r_m),$$

where $\qquad\qquad D_x(s) = (-i2\pi h)\,W(s)\,E(s)\,\exp\{i\phi(s)\}.$

$G(x_m)$, then, is the Fourier transform of the product of two functions $D_x(s)$ and $g_m(s)$ evaluated at r_m (the position of the mth atom). According to the convolution theorem, multiplication in reciprocal space is equivalent to convolution in real space. The Fourier transform of $g_m(s)$ is the electron density of the atom $\rho_m(r)$, and the Fourier transform of $D_x(s)$, which we shall call $d_x(r)$, is a modified difference density map. The gradient then is computed by the summation.

$$G(x_m) = \sum_r d_x(r)\, \rho_m(r - r_m).$$

The computation of all the x gradients requires the calculation of the modified difference density map, $d_x(r)$, by FFT followed by an integration of $d_x(r)$ with the electron-density function for each atom. If the atom electron density is assumed to be zero outside a radius rad_m from the atom centre r_m, the summation need only be carried out within this radius for each atom. Separate difference density functions have to be computed for gradients with respect to y, z, and B.

Calculation of the normal matrix

The following expressions for the normal matrix term $H(x_m, x_n)$, corresponding to interactions between x_m and x_n, have been derived (Agarwal 1978).

$$H(x_m, x_n) = H_1(x_m, x_n) + H_2(x_m, x_n),$$

where

$$H_1(x_m, x_n) = \sum_s \tfrac{1}{2} g_m(s)\, g_n(s)\, (4\pi^2 h^2)\, W(s)\, \exp\{i2\pi s \cdot (r_m - r_n)\},$$

$$H_2(x_m, x_n) = \sum_s -\tfrac{1}{2} g_m(s)\, g_n(s)\, (4\pi^2 h_2)\, W(s)$$
$$\times \exp\{i2\phi(s)\}\, \exp\{-i2\pi s \cdot (r_m + r_n)\}.$$

Similar expressions hold for all other elements in the normal matrix, differing only in that the term $(4\pi^2 h^2)$ is replaced by a similar term depending on the type of interaction. For example, it is replaced by $(4\pi^2 k^2)$ for $y_m y_n$ interactions, by $(4\pi^2 hk)$ for $x_m y_n$ interactions, $(s^4/16)$ for $B_m B_n$ interactions, and $(i\pi h s^2)$ for $x_m B_n$ interactions.

If $A_{xx}(s) = 2\pi^2 h^2\, W(s)$, then the $H_1(x_m, x_n)$ terms represent the Fourier transform of $A_{xx}(s) g_m(s) g_n(s)$ evaluated at $(r_n - r_m)$, the vector between the two atoms. $A_{xx}(s) g_m(s) g_n(s)$ is always real and positive. Its Fourier transform has a large origin peak, then drops rapidly and alternates in sign as the distance between the atoms increases. The $H_2(x_m, x_n)$ terms represent the Fourier transform of $-A_{xx}(s) g_m(s) g_n(s) \exp\{i2\phi(s)\}$ evaluated at $(r_m + r_n)$. This involves phase terms so that, unlike $H_1(x_m, x_n)$, these terms will have no major peaks and their magnitude distribution is likely to be the same in all parts of the normal matrix. As the major contribution to the elements of the normal matrix comes from the H_1 terms, neglecting the H_2 contribution will not affect the final result, but only the rate of convergence.

For a diagonal least-squares approximation, $m=n$ and

$$H_1(x_m, x_m) = \sum_s A_{xx}(s) g_m^2(s).$$

This may be computed directly following the procedure described by Agarwal (1978). The computation required is proportional to the number of unique reflections.

Off-diagonal elements can be calculated in a similar manner to the gradients. We may write

$$H_1(x_m, x_n) = \sum_s A_{xx}(s) g_m(s) g_n(s) \exp\{-i2\pi s.(r_n - t_m)\},$$

which is similar to the expression for the gradients except that $D_x(s)$ is replaced by $A_{xx}(s)$, $g_m(s)$ is replaced by $g_m(s)$, $g_n(s)$, and r_m is replaced by $r_n - r_m$. If $a_{xx}(r)$ is the Fourier transform of $A_{xx}(s)$, and $\rho_{mn}(r)$, the joint Gaussian electron-density function of the mth and nth atoms, is the Fourier transform of $g_n(s) g_m(s)$, then by the convolution theorem $H_1(x_m x_n)$ is the convolution of $a_{xx}(r)$ and $\rho_{mn}(r)$ evaluated at $r = r_n - r_m$. This may be expressed as the summation:

$$H_1(x_m, x_n) = \sum_r a_{xx}(r) \rho_{mn}(r - r_n + r_m).$$

The summation is over all the grid points in real space, and $(r - r_n + r_m)$ is the distance of the grid point from the point $(r_n - r_m)$. If the joint electron-density $\rho_{mn}(r - r_n + r_m)$ is assumed to be non-zero only for grid points within some limiting radius from the point $(r_n - r_m)$, then the summation is much simplified. Furthermore, if the off-diagonal terms of the matrix are restricted to interactions between closely related atoms (atoms in the same side chain or the same peptide unit, for instance), then $a_{xx}(r)$ is required over a limited volume of real space about the origin and could be computed directly. Since $A_{xx}(s)$ will change only if the weights change, the function $a_{xx}(r)$ may be used in a number of refinement cycles until this happens. Similarly, the joint Gaussian electron-density function $\rho_{mn}(r - r_n + r_m)$ will change only if $(r_n - r_m)$, B_m, or B_n changes. The calculation of other off-diagonal terms is similar with A_{xx} replaced by the appropriate function as given above.

Programming considerations

Although the algorithm may appear complex, writing a program is relatively straightforward since Agarwal (1978) has so clearly described all the procedures. The two largest computations are the fast-Fourier transforms and the modelling of the atom electron density. In all of these calculations it is

The refinement of protein crystal structures

important that the minimum amount of computation is done, not only to save computer time but also to save on storage.

The fast-Fourier transforms

The nature of FFT means that the transforms are calculated for a complete unit cell, which requires a full set of data. Ten Eyck (1973) has shown, however, that symmetry may be utilized to reduce both the amount of computation and the amount of data required for the FFT and has written an excellent package of programs to perform these calculations. Savings in time and space may also be made when there are systematic absences in general (*hkl*) reflections. For example, in the space group *R*3, indexed on a hexagonal cell, there are systematic absences of the form $-h+k+l=3n\pm 1$. For constant values of *k* and *l* then, only one in three F_{hkl} is non-zero, and the *h* index will be either $3n$, $3n+1$, or $3n+2$. If the Fourier transform is factored by three, then for constant values of *k* and *l* the transform on *h* is given by (Ten Eyck 1973):

$$T_{h(x, k, l)} = \sum_{h=0}^{N/3-1} F_{(3h)} \, \mathrm{e}\,(3hx/N) + \mathrm{e}\,(x/N) \sum_{h=0}^{N/3-1} F_{(3h+1)}$$

$$\mathrm{e}\,(3hx/N) + \mathrm{e}\,(2x/N) \sum_{h=0}^{N/3-1} F_{(3h+2)} \, \mathrm{e}\,(3hx/N),$$

$$T_{h(x+\frac{1}{3}, k, l)} = \sum_{h=0}^{N/3-1} F_{(3h)} \, \mathrm{e}\,(3hx/N) + \mathrm{e}\,(\tfrac{1}{3})\,\mathrm{e}\,(x/N) \sum_{h=0}^{N/3-1}$$

$$F_{(3h+1)} \, \mathrm{e}\,(3hx/N) + \mathrm{e}\,(\tfrac{2}{3})\,\mathrm{e}\,(2x/N) \sum_{h=0}^{N/3-1} F_{(3h+2)} \, \mathrm{e}\,(3hx/N),$$

$$T_{h(x+\frac{2}{3}, k, l)} = \sum_{h=0}^{N/3-1} F_{(3h)} \, \mathrm{e}\,(3hx/N) + \mathrm{e}\,(\tfrac{2}{3})\,\mathrm{e}\,(x/N) \sum_{h=0}^{N/3-1} F_{(3h+1)}$$

$$\mathrm{e}\,(3hx/N) + \mathrm{e}\,(\tfrac{1}{3})\,\mathrm{e}\,(2x/N) \sum_{h=0}^{N/3-1} F_{(3h+2)} \, \mathrm{e}\,(3hx/N),$$

where $F_{(3h)}$, $F_{(3h+1)}$, $F_{(3h+2)}$ means $F_{(h, k, l)}$ with $h=3n$, $3n+1$, $3n+2$ respectively, and e (*x*) means exp ($-\mathrm{i}2\pi x$).

For any fixed values of *k* and *l*, $F_{(h, k, l)}$ will be non-zero for only one of the three terms in each summation, and in the cases where $h=3n+1$ or $3n+2$, the transforms at $x+\frac{1}{3}$ and $x+\frac{2}{3}$ differ from those at *x* by constant factors. These relationships may be used to advantage if, for fixed values of *k* and *l*, the systematically absent data are ignored and the remaining structure factors reindexed on *h* to fit into a cell compressed to a length $N/3$. The transform on *h* is computed as if this were the full cell, then the results are corrected by the appropriate factors depending on whether F_{3h+1} or F_{3h+2} type data were transformed.

Factoring by two leads to a similar result for centred space groups. The transform becomes (Ten Eyck 1973)

$$T_{k(h,y,l)} = \sum_{k=0}^{N/2-1} F_{(h,2k,l)}\, e\,(2ky/N) + e\,(y/N) \sum_{k=0}^{N/2-1} F_{(h,2k+1,l)}\, e\,(2ky/N),$$

$$T_{k(h,y+\frac{1}{2},l)} = \sum_{k=0}^{N/2-1} F_{(h,2k,l)}\, e\,(2ky/N) - e\,(y/N) \sum_{k=0}^{N/2-1} F_{(h,2k+1,l)}\, e\,(2ky/N).$$

For any centred cell the structure factor data may be arranged so that for fixed values of two of the indices, all the third indices are either odd or even for non-absent terms. For a C-centred cell k is always even when h is even and odd when h is odd. For fixed values of h and l only one of the terms in the summation of the transform on k will be non-zero and furthermore the value of the transform at $y+\frac{1}{2}$ may be obtained directly from the value of the transform at y. If the data are reindexed on k to fit into a cell of length $N/2$ then the transform may be computed as if this were the full cell. The values of the transform for rows with odd values of h (i.e. k all odd for the F_{hkl} data) are then corrected by the factor $e\,(y/N)$.

However, as a centred cell always has symmetry, this should be considered together with the centring to gain the maximum reduction in the size of the transform and computation. The argument in this case is different: consider, for example, the space group $C2$ where the diad along b and the Friedel symmetry made the Fourier transform along k Hermitian symmetric and allows the transform for the whole axis, which is real, to be obtained from half the structure factors, i.e. $k \geqslant 0$. Now, since the FFT is computed for a row along k when h and l are fixed, for odd values of h this row will contain non-zero structure factors only for odd values of k. Similarly for even values of h the row on k will be non-zero only for even values of k. These systematic absences produce periodicity in the transform such that (Ten Eyck 1973)

$$T_{k(y+\frac{1}{2})} = T_{k(y)} \text{ when } k \text{ odd terms are zero, and}$$
$$T_{k(y+\frac{1}{2})} = -T_{k(y)} \text{ when } k \text{ even terms are zero.}$$

Note that the systematic absences in the structure factor data are lost once a transform is computed, so the relationships which occur because of them are valid only for the transform computed along the first cell edge. Ten Eyck (1973) has suggested combining adjacent sets of data of h odd and h even terms so that those where k is even contain data for the h even row and the k odd terms contain data for the h odd row. If the compressed data are indexed $h'kl$, then the true indices of the data are $2h'kl$ when k is even and

$2h' + 1kl$ when k is odd. The transform on k is the sum of these two transforms.

$$T_{k(h',y,l)} = T_{k(2h',y,l)} + T_{k(2h'+1,y,l)},$$

and we can recover the individual transforms by making use of the periodicity and antiperiodicity in the separate transforms.

$$T_{k(h',y+\frac{1}{2},l)} = T_{k(2h',y+\frac{1}{2},l)} + T_{k(2h'+1,y+\frac{1}{2},l)}$$
$$= T_{k(2h',y,l)} - T_{k(2h'+1,y,l)},$$

so that $\qquad T_{k(h',y,l)} + T_{k(h',y+\frac{1}{2},l)} = 2T_{k(2h',y,l)}$

and $\qquad T_{k(h',y,l)} - T_{k(h',y+\frac{1}{2},l)} = 2T_{k(2h'+1,y,l)}.$

In this way the transform on k need be computed over only half the cell in the x direction. Any centred cell may be treated in a similar manner.

Modelling the atom electron density

This is the most expensive part of the calculation and it is important that care is taken to optimize the programming. The use of single and double Gaussian approximations to the atom electron density has been discussed by Agarwal (1978) and Ten Eyck (1977).

The only difficulty in the programming is to allow correctly for atoms which lie close to the edges of the asymmetric cell unit. This may be approached in two ways. In the first method, which was employed in the original program used for the insulin refinement (space group $R3$), the co-ordinates of each atom are transformed by the space-group symmetry to lie within the cell asymmetric unit required for the transforms. If an atom extends outside this asymmetric unit, then this density is added to the symmetrically equivalent point within the asymmetric unit. This method is appropriate for a computer with a large virtual memory where the whole asymmetric unit may be considered to be held in core. In the case where only a small slab of density can be held in core this method is not efficient, since each atom in the structure has to be moved through each of the symmetry operators to determine if any part of its volume falls within the required slab. In the second procedure the initial atom co-ordinates are transformed through all the symmetry positions and a sorted list of those atoms which will have some density within the required asymmetric unit is retained. For any slab of density only those atoms contributing to the slab are used. The disadvantage of this system is that the atom list is very much extended and the exponential factors for duplicate atoms need to be computed a number of times.

The calculation of the gradients uses a similar routine except that premultiplication of the $W\Delta E$ values by $-ih$, $-ik$, $-il$ may change the symmetry of the modified difference map. In $R3$, for example, premultiplication by $-ih$ or $-ik$ destroys the threefold symmetry around the origin which means that in convoluting the atom electron density with the modified difference map, special care has to be taken with atoms which extend over the edge of the asymmetric unit. This problem was overcome by using an expanded asymmetric unit for the modified difference map, which extended in both positive and negative directions on x and y by a distance greater than the maximum atom radius. This is a cumbersome procedure as it requires additional computer core to hold the section of the map and does not lend itself to producing a space-group general routine. An alternative solution adopted by Eleanor Dodson (Baker and Dodson 1980) is to use a symmetry expanded set of atoms, as for the structure factor calculation, and to compute the convolution product of the atom density and modified difference density only for that portion of the atom which lies within the required unit of the cell. If an atom has been shifted from its original position by a symmetry operation, the gradients are transformed through the inverse operation and the total atom gradients are summed from the individual contributions. In this way the calculation is space-group general, and Dodson has written programs for use in the space groups $P1$, $P2_12_12_1$, $P4_12_12$, $R3$, and $P3_121$.

The requirements of the method

In order to use the FFT least-squares certain conditions with regard to the resolution of the data set, the accuracy of the starting co-ordinates, and the size of the computer have to be met. The accuracy of the final refined structure depends on the resolution of the data used, the higher the resolution the more accurate will be the structure. Generally, diffraction data to a resolution of at least 2 Å is required for a meaningful refinement. However, at the beginning of refinement, when the co-ordinate errors are large, high-resolution terms should not be used and a refinement with data to a resolution of less than 2 Å may produce some improvement in the model structure.

Test calculations made by Agarwal (1978) have indicated that the method is capable of correcting co-ordinate errors with an r.m.s. value of 0.75 Å. Obviously, a protein structure with this degree of error in the co-ordinates would not have the geometry expected for peptide units and it is likely that most model structures fitted to electron-density maps will have smaller r.m.s. errors than this, although some individual atoms could have much larger errors. In both insulin (Isaacs and Agarwal 1978) and actinidin (Baker and Dodson 1980) the refinement was able to correct automatically co-ordinates which were in error by 0.5 Å on average. In actinidin, the starting co-ordinates for the refinement were those read from a model fitted to a 2.8 Å m.i.r. map, whereas for insulin the co-ordinates were read by inspection from a 1.5 Å map, phased by the phase refinement method of Sayre (1972, 1974;

Cutfield, Dodson, Dodson, Hodgkin, Isaacs, Sakabe, and Sakabe 1975). It now appears that if the lower-resolution map is of sufficient quality, there is little to be gained by extending it to a higher resolution in order to improve the co-ordinates prior to refinement.

For a protein crystallographer, the computing requirements of the program are very modest. The program written by Dodson is flexible in its core storage requirements, and for the actinidin refinement (1820 atoms, 24 000 data) required 35K words of store. It does need back-up store, preferably on a disc, but could operate with magnetic-tape files. The cost of the refinement, both in computer time and manpower, is its greatest attraction. The complete refinement of actinidin, from a set of co-ordinates read from a 2.8 Å map to a set of refined co-ordinates with 1.7 Å data, took about fourteen hours of computer time on a DEC 10 and was completed in only three months.

The use of the method

Isaacs and Agarwal (1978) and Baker and Dodson (1980) have recorded their experiences in using the method to refine insulin and actinidin respectively. Generally these experiences and those of Hardman with myoglobin and carbonic anhydrase (personal communication) are similar and it seems that differences in the size of the problem do not influence the nature of the refinement. The major difficulty with the method is the fact that the shifts calculated for geometrically related atoms, such as those forming a peptide unit, destroy the geometry. This behaviour is characteristic of protein refinements where atoms are allowed to move individually. Causes of this might be the large initial errors in the co-ordinates, the relative sparseness of the data with fewer than three observations for each variable parameter, and the neglect of atom–atom interactions in the normal matrix. This loss of geometry may be controlled using a program of the type written by Dodson, Isaacs, and Rollett (1976) or Ten Eyck, Weaver, and Matthews (1976) to correct the gross structural irregularities. Although this method of correcting the geometry every few cycles decreases the rate of convergence, there is no evidence so far to suggest that it adversely affects the accuracy of the final model. Test calculations performed by Stenkamp and Jensen (1976) with simulated protein data support this. A faster rate of convergence could be achieved by incorporating the geometrical constraints in the least-squares equations (Konnert 1976), but an advantage in separating the least-squares refinement and regularization procedures is that large shifts on regularization often indicate gross errors in the structure.

The experience of Isaacs and Agarwal (1978) and Baker and Dodson (1980) with high-resolution data indicates that the gross errors in the model may be corrected by refinement with data to a resolution of 2 Å. The use of the

TABLE 27.1.
Some examples of structures refined by FFT least-squares

Molecule	No. atoms	No. data	d_{min} (Å)	Space group	R factors Initial	R factors Final	No. Cycles xyz	No. Cycles B	C.p.u./cycle†	Computer	Reference
Insulin	1077	11 890	1.5	$R3$	0.28	0.11	43	24	3 min	IBM 370/168	Isaacs and Agarwal (1978)
Insulin	1077	11 890	1.5	$R3$	—	—	—	—	12 min	DEC 10	Isaacs and Agarwal (1978)
6-Acetyldolatriol	26	1041	1.0	$R3$	0.50	0.10	13	—	20 s	IBM 370/168	Agarwal (1978)
Beavuricin Ba salt	370	6667	1.2	$P2_1$	0.21	0.12	14	2	38 s	IBM 370/168	Agarwal (1978)
Metmyoglobin	~1400	~9000	2.0	$P2_1$	0.27	0.16	5	10	68 s	IBM 370/168	Hardman (1978)
Carbonic anhydrase	2050	9451	2.0	$P2_1$	0.42	0.18	9	3	—	IBM 370/168	Hardman (personal communication)
Actinidin	1821	23 390	1.7	$P2_12_12_1$	0.43	0.17	28	5	20 min	DEC 10	Baker and Dodson (1980)
Actinidin	1821	23 390	1.7	$P2_12_12_1$	—	—	—	14	10 min	CYBER 74-16	Baker and Dodson (1980)
Gramicidin S	84	4902	1.0	$P3_12_1$	—	0.19	—	—	—	DEC 10	Hull, Karlsson, Main, Woolfson, and Dodson (1978)

†C.p.u. times for the DEC 10 include a charge for operating costs.

weighting scheme proposed by Agarwal (1978) is important in placing most weight on the low-angle terms for these early cycles. Baker and Dodson (1980) found that the initial seven cycles of co-ordinate refinement with 2 Å data on actinidin produced an average shift of 0.43 Å for main chain atoms. The remaining 21 cycles of co-ordinate refinement with the inclusion of data to 1.7 Å produced an average shift of 0.18 Å for the main chain atoms. Much of the manual labour required for a refinement is spent on the poorly defined regions of the structure and on the solvent structure. The well-ordered solvent should be included in the model structure as soon as possible, but with regard to other solvent and to disordered structures it is wise to proceed with caution. Isaacs and Agarwal (1978) have discussed how incorrectly assigned solvent (water) molecules confused the interpretation of difference-Fourier densities of some side chains, particularly of glutamic acid and arginine residues. The solvent structure can be unravelled (Watenpaugh, Margulis, Sieker, and Jensen 1978), but to do so requires considerable effort.

Conclusion

Table 27.1 lists a number of structures which have been subject to refinement using the FFT least-squares refinement programs. The speed of the new algorithm is evident—the complete refinement of actinidin, starting with co-ordinates from a 2.8 Å m.i.r. phased electron-density map, required about fourteen hours of computer time on a DEC 10 and was completed in only three months. This work also provides a good estimate of the radius of convergence of the method. The average shift in position for main chains was 0.45 Å and for side chain atoms 0.56 Å.

The development of the FFT least-squares has made it possible to refine quite large protein structures at a reasonable cost and to an accuracy which is limited by the quality of the diffraction data and the degree of order of the protein. Routine refinement will lead to more accurate descriptions of protein structures and a better understanding of the nature of those enzyme mechanisms and protein interactions where the conformational changes of the molecule may be quite small. The refinements already completed, particularly the work on rubredoxin, has produced a wealth of new information on the nature of the water structure surrounding protein molecules.

References

ADMAN, E. T., SIEKER, L. C., and JENSEN, L. H. (1975). *Acta crystallogr.* **A31**, S34.
AGARWAL, R. C. (1978). *Acta crystallogr.* **A34**, 791–809.
BAKER, E. N. and DODSON, E. J. (1980). *Acta crystallogr.* In press.
BODE, E. and SCHWAGER, P. (1975). *J. molec. Biol.* **98**, 693–717.
CHAMBERS, J. L. and STROUD, R. M. (1977). *Acta. crystallogr.* **B33**, 1824–37.
COOLEY, J. W. and TUKEY, J. W. (1965). *Maths Comput.* **19**, 297–301.

CUTFIELD, J. F., DODSON, E. J., DODSON, G. G., HODGKIN, D. C., ISAACS, N. W., SAKABE, K., and SAKABE, N. (1975). *Acta crystallogr.* **A31**, S21.

DEISENHOFER, J. and STEIGEMANN, W. (1975). *Acta crystallogr.* **B31**, 238–50.

DIAMOND, R. (1971). *Acta crystallogr.* **A27**, 436–52.

——(1974). *J. molec. Biol.* **82**, 371–91.

DODSON. E. J., ISAACS, N. W., and ROLLETT, J. S. (1976). *Acta crystallogr.* **A32**, 311–15.

FREER, S. T., ALDEN, R. A., CARTER, C. W., and KRAUT, J. (1975). *J. biol. Chem.* **250**, 46–54.

HARDMAN, K. D. (1978). *Acta crystallogr.* **A34**, S65.

HUBER, R., KUKLA, D., BODE, W., SCHWAGER, P. BARTELS, K., DEISENHOFER, J., and STEIGEMANN, W. (1974). *J. molec. Biol.* **89**, 73–101.

HULL, S. E., KARLSSON, R., MAIN, P., WOOLFSON, M. M., and DODSON, E. J. (1978). *Nature Lond.* **275**, 206–7.

International tables for X-ray crystallography (1974). Vol. IV. Kynoch Press, Birmingham.

ISAACS, N. W. and AGARWAL, R. C. (1978). *Acta crystallogr.* **A34**, 782–91.

JACK, A. and LEVITT, M. (1978). *Acta crystallogr.* **A34**, 931–5.

KONNERT, J. H. (1976). *Acta crystallogr.* **A32**, 614–17.

MOEWS, P. C. and KRETSINGER, R. H. (1975). *J. molec. Biol.* **91**, 201–28.

SAYRE, D. (1951). *Acta crystallogr.* **4**, 362–7.

——(1972). *Acta crystallogr.* **A28**, 210–12.

——(1974). *Acta crystallogr.* **A30**, 180–4.

SCHMIDT, W. C., Jr, GIRLING, R. L., and AMMA, E. L. (1977). *Acta crystallogr.* **B33**, 3618–20.

STENKAMP, R. E., and JENSEN, L. H. (1976). *Acta crystallogr.* **A32**, 255–8.

SUSSMAN, J. L., HOLBROOK, S. R., CHURCH, G. M., and KIM, S. H. (1977). *Acta crystallogr.* **A33**, 800–4.

TAKANO, T. (1977). *J. molec. Biol.* **110**, 537–84.

TEN EYCK, L. F. (1973). *Acta crystallogr.* **A29**, 183–91.

——(1977). *Acta crystallogr.* **A33**, 386–92.

——WEAVER, L. H., and MATTHEWS, B. W. (1976). *Acta crystallogr.* **A32**, 349–50.

WATENPAUGH, K. D., MARGULIS, T. N., SIEKER, L. C., and JENSEN, L. H. (1978). *J. molec. Biol.* **122**, 175–90.

——SIEKER, L. C., HERRIOTT, J. R., and JENSEN, L. H. (1973). *Acta crystallogr.* **B29**. 943–56.

IV Protein structure

28. Approaches to the structure and function of ferritin

PAULINE M. HARRISON

I was introduced to ferritin by Dorothy Hodgkin. There must be many others who can say that about 'their protein'. Since first getting involved with ferritin as a post-graduate student in Oxford thirty years ago I have been fascinated by this molecule. As a means of expressing my indebtedness to Dorothy I shall try to trace the history of what we have learned about the protein structure and how it stores iron.

Ferritin was first isolated in pure form from horse spleen by Laufberger (1937), who found that it would yield red–brown, octahedral crystals on addition of $CdSO_4$ (or of $ZnSO_4$, $CuSO_4$, $NiCl_2$, or $Co(NO_3)_2$). Laufberger showed that ferritin contained 20 per cent Fe by weight and considered it to act as a storage depot for this element. The classic studies of Granick, Michaelis, Hahn, and their associates followed (Granick 1946). They showed that ferritin iron was in the form of 'micelles' of ferric oxyhydroxide-phosphate, which remained intact when dissociated from the protein in 1N NaOH, but which would dissolve, releasing Fe(II), under the action of sodium dithionite at pH 4.6, leaving intact the colourless, iron-free protein, apoferritin (Granick and Michaelis 1943). A remarkably accurate estimate of apoferritin's molecular weight as 465 000 was made by Rothen (1944). More recent determinations suggest a weight of only 2–4 per cent less than this (e.g. Bjork and Fish 1971; Bryce and Crichton 1971). Rothen (1944) also showed ferritin to be heterogeneous, apparently containing micelles of different sizes as well as apoferritin. Granick (1951) considered the micelles to be outside the protein, even though Mazur and Shorr (1950) had found that ferritin and apoferritin had the same electrophoretic mobilities. The problem was resolved by Farrant (1954), whose electron micrographs of molecules shadowed from opposite sides clearly showed electron-dense micelles (diameter 55 Å) surrounded by less dense protein shells.

The first X-ray work on ferritin was done by Fankuchen (1943). Only small crystals were available but he was able to index their powder diffraction lines on a face-centred cubic lattice of side 186 Å. Later Granick grew large single

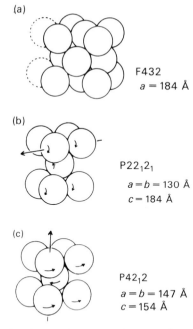

FIG. 28.1. Packing relationships in three crystalline modifications of horse-spleen ferritin or apoferritin: (a) cubic, $a=184$ Å, space-group F432; (b) orthorhombic, $a=b=130$ Å, $c=184$ Å, space group $P22_12_1$; (c) tetragonal, $a=b=147$ Å, $c=154$ Å, space group $P42_12$.

In (a) the approximately spherical molecules are in face-centred cubic close-packing (c is vertical). Packing in orthorhombic crystals, (b), is similar to that in cubic, (a), but the unit cell sides, a and b, lie parallel to cube face-diagonals (the unit cell is approximately body-centred). Tetragonal cells can be derived from orthorhombic by a small expansion along a and b and contraction along c. Intermolecular contacts are made along body-diagonals.

In the cubic modification molecules are arranged such that their symmetry axes (point symmetry 432) are coincident with those of the unit cell. In the orthorhombic cell, molecules are rotated approximately $\pm 6°$ about one of their twofold axes (lying in the a direction) and so no other symmetry axis is used by the lattice. In the tetragonal cell, one of the molecular fourfold axes lies parallel to c, the unique axis, but rotation of $\pm 27.5°$ about this direction again ensures that no other symmetry axis is used in the lattice.

crystals and gave these to Dorothy. As Dorothy's young research assistant, I had the task of photographing these crystals. They were a beautiful golden-brown. Several of them measured at least 0.5 mm across and they were clearly anisotropic and orthorhombic ($a=130$ Å, $b=130$ Å, $c=184$ Å). This was reported by Dorothy in 1950 at a Cold Spring Harbour Symposium and she also remarked that crystals of apoferritin large enough for X-ray work had not yet been obtained, but that comparison of these with ferritin ought to

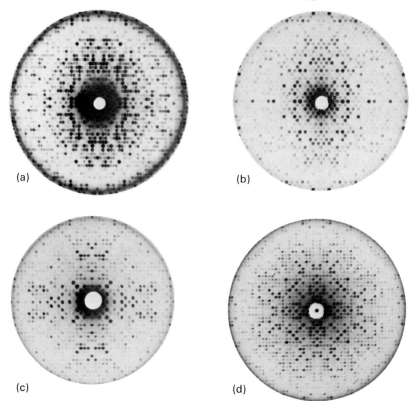

(a)

(b)

(c)

(d)

FIG. 28.2 Precession photographs of three crystalline modifications of ferritin and apo-ferritin (compare with Fig. 28.1).

(a) Tetragonal apoferritin. 8° precession. (*hk*0) zone, [100] horizontal. Two fourfold axes of the 432 molecules lie in the (110) plane, but are rotated about [001] such that they lie approx. $\pm 27.5°$ from [110].

(b) Orthorhombic ferritin. 7.5° precession. (*hk*0) zone with [100] horizontal. One of the twofold axes of the 432 molecules lies parallel to [100], but the molecules are rotated about this direction by approx. $\pm 6°$, such that two fourfold axes lie near to, but not along [110] and [1$\bar{1}$0].

(c) Cubic apoferritin. 8° precession. (*hh*l) zone with [001] horizontal. One of the four-fold axes lies along [001] and a twofold axis along [110].

(d) Orthorhombic ferritin. 8° precession. (0*kl*) zone with [001] horizontal. Molecules are rotated about [100] such that one fourfold axis lies $\pm 6°$ from [001].

be most valuable (Hodgkin 1950). Later, when I came to prepare ferritin and apoferritin myself, I never succeeded in obtaining orthorhombic crystals either of ferritin or of apoferritin, although both grew readily enough in Fan-kuchen's cubic form. For a time a tetragonal modification of apoferritin also appeared, as I shall describe later, but never of pure ferritin. The unit cells and molecular packing relationships for these modifications are indicated in Fig. 28.1 and Fig. 28.2. The X-ray photographs we took between 1949 and 1952 of orthorhombic ferritin were usually 2° oscillation photographs or stills, taken with a flat-plate attachment to a Unicam oscillation camera with film-to-crystal distance usually 10 cm. Exposure times of about 20 h were required with X-rays from a Newton–Victor generator housed in the University Museum. I collected successive oscillation photographs, made an intensity strip from one of the strong reflections and used this to measure (0kl) intensities by eye. From these I calculated a Patterson projection, but since this was down a unit cell side of 130 Å it is not surprising that I found it hard to interpret! It is interesting to note that this method of photography

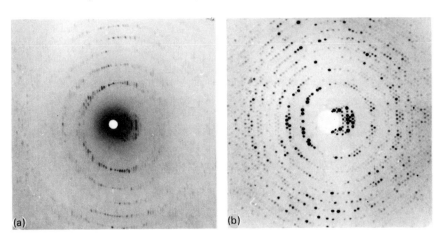

FIG. 28.3. Oscillation and rotation photographs of ferritin and apoferritin crystals. (a) Orthorhombic ferritin, 2° oscillation, $23\frac{1}{2}$ h exposure, taken in 1950 on a Unicam oscillation camera with plate attachment. (b) Cubic apoferritin, 1.5° rotation, 2 h exposure taken in 1978 on an Arndt–Wonacott rotation camera (scale $0.79 \times a$).

(although not of intensity measurement!) is now again in vogue, although the Arndt–Wonacott rotation camera is rather more sophisticated than the camera we used then. A recent rotation photograph of cubic apoferritin (2 h exposure time, Philips fine-focus tube) is compared in Fig. 28.3 with an earlier oscillation photograph of orthorhombic ferritin.

The face-centred cubic packing of ferritin molecules (and a closely related arrangement in the orthorhombic crystals) suggested to us that these molecules may be roughly spherical and composed of submolecules (now usually known as subunits) in a symmetrical array. The larger tomato bushy stunt and turnip yellow mosaic viruses probably also crystallized in cubic space groups, which Dorothy (Hodgkin 1950) pointed out would require them to consist of $12n$ submolecules (unless the apparent cubic symmetry was produced statistically). We thought that both protein and iron-micelles in ferritin might be so subdivided and considered possible arrangements. One idea was that ferritin might resemble phosphotungstic acid. In the $[PW_{12}O_{40}]^{3-}$ ion, the twelve W atoms at the vertices of a cubo-octahedron are linked through twenty-four O atoms to form a closed polyhedral $W_{12}O_{24}$ group. Four of the remaining O atoms are inside this polyhedral shell arranged tetrahedrally around a central P atom (Wells 1945). Farrant's (1954) electronmicrographs showed some iron-micelles as a 'tetrad' of four subparticles (like the O atoms of phosphotungstate?), although others later suggested octahedral and other substructure. The view now is that the micelles are not uniformly subdivided and many of them may be single particles (Haggis 1965; Massover and Cowley 1973).

After a period studying on other molecules I returned to ferritin in 1958. I was now in the Biochemistry Department in Sheffield and most of my preparative work was done in an old cinema, the Scala, which had been converted to a laboratory. I succeeded in producing crystals of both apoferritin and ferritin large enough for single-crystal work, but I had no suitable camera. Fortunately, I was able to spend a few weeks at the Royal Institution, where Tony North taught me how to take precession photographs. Both ferritin and apoferritin crystals turned out to be cubic ($a = 184$ Å, space-group F432) like those examined by Fankuchen. It was exciting to find not only that they were isomorphous, but that, judged by the similarity of their diffraction patterns at high angles, the protein remained unchanged despite a loss of iron component amounting to 300 000 molecular weight (Harrison 1959). At low angles the diffraction patterns were very different. Apoferritin low-angle data could be approximated by a hollow shell ($r_1 = 38$ Å, $r_2 = 62$ Å), see Fig. 28.7(a), whereas the ferritin–apoferritin differences were roughly those expected for a uniform sphere, $r = 35 \pm 3$ Å (the inner radius of the protein shell), and twice the electron density of the protein (Harrison 1963). All these observations strongly supported Farrant's conclusion that ferritin iron-micelles formed the central core of the ferritin molecule and that the protein shell remained intact on removal of the iron. The cubic space group, F432, and the estimated number of molecules in the unit cell, $Z = 4$, indicated that molecules lay on lattice points. Symmetry considerations therefore required that each protein shell consist of $24n$ subunits where n was the number of subunits (or polypeptide chains) in the asymmetric unit. Measurements of crystal density and cell dimensions yielded a molecular weight of 480 000 (assuming the then current value of $\bar{V}$ of 0.747; a more recent estimate of 0.732, based on amino-acid composition, gives 450 000 for the X-ray molecular weight) and it seemed likely that the total number of subunits/molecule should be 24 since this would give a reasonable subunit weight.

I published this preliminary conclusion in the first number of the *Journal of Molecular Biology* (Harrison 1959) with a suggested quaternary structure of 24 subunits at the apices of an Archimedean snub-cube, Fig. 28.9(a). In

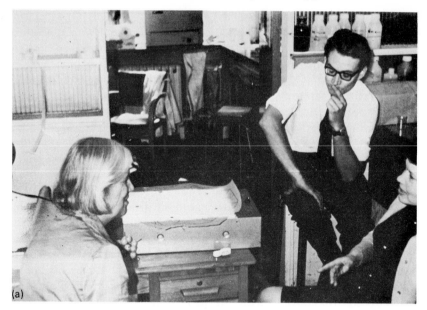

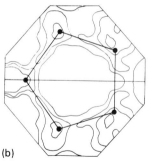

FIG. 28.4. (a) Dorothy Hodgkin, Terry Hoy, and Pauline Harrison in the Biochemistry Department at the University of Sheffield in 1970. On the light box is a Patterson section which appears to show fivefold symmetry in apoferritin. The section is reproduced in (b). It was calculated with data at 14.5 Å resolution at 20 Å from (0 0 0) along a supposed fivefold axis.

addition to very strong intensities along cube axes, this arrangement gives rise to strong diffraction at 30° to the cube axes (Harrison 1959; Hoy, Harrison, and Hoare 1974*a*). Strong intensities are visible in apoferritin diffraction photographs along this direction. However, icosahedral viruses which have fivefold axes lying at 31.7° to their twofold axes also give intensity spikes in similar positions (Caspar and Klug 1962) and Patterson sections of apoferritin perpendicular to this direction gave indications of fivefold symmetry (see Fig. 28.4). Could it be that apoferritin was icosahedral, rather than octahedral, with the observed fourfold symmetry being produced statistically? Ferritin could be said to resemble a virus, with its protein shell surrounding and protecting its functional non-protein core. I remember showing my photographs to Aaron Klug and to Dorothy Hodgkin and discussing ferritin symmetry with them (see Fig. 28.4). Aaron was convinced that ferritin must be icosahedral (Casper and Klug 1962); Dorothy was sceptical. True icosahedral symmetry would demand $60n$ subunits, but a pseudoicosahedral arrangement would be possible with 12, 20, or 30 subunits. It seemed necessary to obtain independent evidence of the number of polypeptide chains so that this could be settled. If the cubic crystal symmetry resulted from several orientations of molecules with lower symmetry giving 'ordered-disorder' then the future for further crystal structure determination was not promising.

I enlisted the help of Dr Theo Hofmann, and we tried several methods of determining subunit size. Unfortunately the subunits turned out to have a blocked N-terminus (Harrison, Hofmann, and Mainwaring 1962) and a clear choice between 20 and 24 subunits could not be made by 'finger printing' (Harrison and Hofmann (1962). Apoferritin also turned out to be more resistant to denaturation than many other proteins, but we found that the freeze-dried protein readily dissolved in sodium dodecyl sulphate (SDS) giving an SDS-subunit complex (Hofmann and Harrison 1963). We measured the molecular weight of this complex in the analytical ultracentrifuge and after allowing for bound SDS we calculated a subunit weight of 25 000 to 27 000. This would give 18–19 subunits per molecule, a number nearer to 20 than to 24. Our cubic crystal symmetry, apparent icosahedral molecular symmetry, and subunit weight could be accounted for by a molecule with 20 subunits at the apices of a pentagonal dodecahedron in twelve orientations (Harrison 1963). The orthorhombic crystals formally required only a single molecular twofold axis, or only one such axis coincident with that of the space-group.

We now know that at that period subunit molecular weights in several other molecules were overestimated, giving erroneously low numbers of subunits, e.g. aldolase was thought to contain three rather than four subunits (Stellwagen and Schachman 1962; Deal, Rutter and Van Holde 1963). With hindsight we should have been more wary of our subunit weight, but the supposed disorder, the formidable task of measuring intensities, and the lack of suitable computing facilities all contributed to a lag in the structure determination. We started to look for other crystalline forms, which might either resolve the symmetry problem or be more suitable for X-ray analysis. A new crystalline form did indeed turn up. This was the tetragonal one referred to above, space group P42$_1$2 $a=b=147.4$ Å, $c=154.4$ Å with a pseudo-body-centred cell containing two molecules (Hoy, Harrison, and Hoare 1974b). This arrangement required the molecules to have at least fourfold axis (along c) and application of the rotation function of Rossmann and Blow (1962) showed that the diffraction patterns could be accounted for by two molecules with 432 symmetry rotated about c such that the remaining fourfold axes were $\pm 17.5°$ from the a axis (Figs. 28.1 and 28.2). It seemed most unlikely that this arrangement could have been produced by a statistical arrangement of molecules with lower symmetry. Further application of the rotation function to the cubic data by Terry Hoy (Hoy *et al.* 1974*a*) showed that the apoferritin rotation function had large peaks only on crystallographic axes, whereas the 20-subunit model gave peaks on non-crystallographic axes, which were not observed with apoferritin. These results together with re-determined subunit molecular weights (Bjork and Fish 1971; Bryce and Crichton 1971) made it quite clear that a 24-subunit model with 432 symmetry was the correct one and this was again shown by the successful

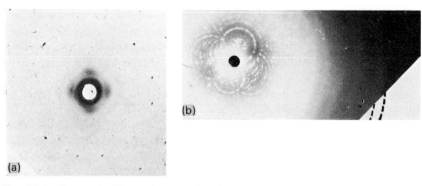

(a)

(b)

FIG. 28.5. Contrasting X-ray photographs of stationary single crystals of cubic ferritin: (a) shows a low-angle diffuse pattern seen with ferritin, but not apoferritin crystals, which represents the shape transform of the ferritin iron-core (the cavity in apoferritin, filled with mineral). Fourfold axes horizontal and vertical. (b) shows both a high-angle powder pattern $d < 2.5\,\text{Å}$) from the mineral component and sharp reflections due to the lattice of protein shells. The atomic structure of the hydrous ferric oxide ('iron-core') inside ferritin molecules is evidently not specifically orientated with respect to that of the apoferritin shell. The outermost powder line visible (dotted lines) is at $d = 1.47\,\text{Å}$.

structure determination, which I shall shortly describe. It is of interest to contrast the apoferritin story with that of the tobacco necrosis virus satellite, which was first thought to be octahedral (Åkervall *et al.* 1971) and then shown more probably to be icosahedral (Klug 1971).

I turn now, as we did in the second half of the sixties, when progress on the protein structure itself was slow, to ferritin 'iron-cores' and their formation. One of the first observations I had made in photographing ferritin crystals was that they, but not apoferritin, gave characteristic low-angle diffuse patterns with maxima along fourfold axes (Fig. 28.5(*a*)). These maxima were close to those predicted for a uniform cube of side 58 Å, except that they were more arced suggesting an object with shape between a cube and a sphere. Ferritin preparations contain some apoferritin, with which it co-crystallizes, and a range of mixed crystals can be made artificially. I concluded that the diffuse patterns arose from variations in electron density from one lattice point to another and thus represented the shape transform of the hole inside the protein, occupied by iron-core (Harrison 1963; Fischbach, Harrison, and Hoy 1969). Electron diffraction patterns of dried iron-cores (Haggis 1965) and X-ray diffraction from wet molecules or isolated moist iron-cores showed that the iron-core particles were small crystals (Harrison, Fischbach, Hoy, and Haggis 1967). We found that the higher the iron-content, the larger the crystallite size, full ferritin apparently containing single crystals filling the protein cavity. We also obtained remarkable X-ray photographs of single ferritin crystals with MoKα radiation (Fig. 28.5(b)), which showed sharp Bragg reflections, due to the protein lattice, giving way to more diffuse powder lines at spacings less than about 2.5 Å (Fischbach *et al.* 1969). Thus the atomic structure of the microcrystalline mineral is not specifically orientated with respect to that of the protein. Taking this in conjunction with the orientated diffuse patterns at first sight it would appear that the Law of Rational Indices is not obeyed by the mineral in ferritin! However, we do not believe that this is the case. The low-angle diffuse patterns represent the shape of the protein cavity and only the *average* shape of the crystalline iron-core which can grow in many directions until it fills the cavity. Ferritin fractions of low iron content separated by density gradient centrifugation do not show the characteristic maxima at low angles, only a diffuse halo. This suggests that iron is not distributed uniformly throughout the cavity, but is in small discrete particles which vary in position and orientation with respect to the protein shell. Electron micrographs of iron-poor ferritin also show smaller particles than those found in iron-rich fractions (Haggis 1966). Ferritin molecules form an interesting comparison with the structures of Radiolaria and of the viruses. The polyhedral shapes of the siliceous exo-skeletons of the Radiolaria appear to be governed by their internal organic structure (D'Arcy Thompson 1952). In tobacco mosaic virus (Caspar and Klug 1962) or turnip yellow mosaic virus (Klug, Longley, and Leberman 1966) the form taken up by the nucleic acid within the protein results from its interaction with the protein subunits so that its symmetry mimics that of the quaternary structure.

The particulate, microcrystalline nature of ferritin mineral cores poses the question of their formation *in vivo* and, indeed, of how whole ferritin molecules are assembled. Two alternatives may be considered, a cumulative model or a once-for-all model. At about the time we were examining diffraction from ferritin iron-cores, Saltman and co-workers were investigating the properties of polynuclear iron complexes which formed when alkali was gradually added to solutions of ferric nitrate. Mossbauer spectra and X-ray scattering properties of Saltman's 'iron-balls', as they became known, resembled those of ferritin and their molecular weights approached those of ferritin iron-cores. Saltman proposed that preformed 'iron-balls' could form a template around which ferritin subunits could assemble (Spiro and Saltman 1969). Against this, however, there were several observations. It had been shown that iron stimulates the biosynthesis of ferritin and when ^{14}C-leucine was injected after a dose of iron the radioactive label was incorporated first into apoferritin or iron-poor molecules and later was found in iron-rich molecules (Fineberg and Greenberg 1955; Drysdale and Munro 1966). This suggested a cumulative process. Our structural studies, which showed that the mineral and protein structures were not specifically related, also made it seem unlikely that pre-formed iron-cores would act as templates for subunit assembly, especially if the mineral had a variety of particle sizes and shapes. Moreover, not only could iron be removed from intact molecules by the action of reducing agents, but the reverse process had been shown to occur. A red–brown solution resembling ferritin was obtained when ferrous ammonium sulphate was added to apoferritin under atmospheric oxygen (Bielig and Bayer 1955). We began to study this process in more detail. First we found that the iron was indeed entering the apoferritin shell to form a ferritin-like molecule. Crystalline reconstituted ferritin gave both low- and high-angle patterns like ferritin and 'micelles' were also observed by electron microscopy (Harrison *et al.* 1967; Macara, Hoy, and Harrison 1972). Examining different reconstitution conditions we found that not only were the amount of iron taken up and the crystallinity of the product dependent on the buffer and oxidant used, but the crystalline structure was always ferritin-like, even though, in the absence of apoferritin, α-FeOOH or γ-FeOOH might be precipitated. Hence apoferritin seemed to influence the atomic arrangement of the mineral, even though it did not align it.

The next development in the story was the finding, originally by Niederer, that apoferritin acted as a catalyst in iron-oxidation (Niederer 1970; Macara *et al.* 1972, 1973; Bryce and Crichton 1973). The question arose, 'How does apoferritin act and how do the crystalline particles form inside the shell?' Our X-ray evidence had suggested that the mineral is laid down layer by layer, so that molecules of higher iron content have larger average crystallite size (sometimes but not always containing only one crystal). Examining reconstituted ferritins in the analytical ultra-centrifuge, I remember being struck by the observation that under conditions favouring rapid iron accumulation a marked 'all-or-none' tendency was evident. 'To those that hath shall be given.' Molecules which already contain some iron must compete successfully for iron with those that have none. Thus ferritin formation from apoferritin had crystal growth characteristics: molecules which have successfully nucleated have surfaces on to which more mineral could be rapidly deposited (Macara *et al.* 1972; Harrison and Hoy 1973). A paper by Miller and Perkins (1969) seemed to clinch this idea. In measuring the net transfer of iron from transferrin to ferritins of different iron contents they found that the largest net transfer was to roughly half-full ferritin molecules. Suddenly everything fell into place: iron accumulation took place within the confines of the apoferritin shell and so the crystal surface available for iron deposition first increased and then decreased as it made contact with the protein.

A great deal of detailed work followed. Ian Macara found 'sigmoidal' or 'hyperbolic' iron uptake progress curves, depending on whether more or less iron per molecule had been added (Macara *et al.* 1972). Iron added in small steps was taken into ferritin at rates which first increased and then decreased, and similar behaviour has been found for iron uptake or release (with a variety of reagents) by isolated ferritin fractions of different iron content (Harrison, Hoy, Macara, and Hoare 1974; Harrison *et al.* 1978). We also found this dependence on iron content and confirmed the accumulation model *in vivo* when we isolated rat-liver ferritin after injecting the animals with small amounts of ^{59}Fe (Hoy and Harrison 1975, 1976). We interpreted the first slow stage of ferritin formation as heteronucleation on apoferritin. Preferential chelation of Fe(III) by carboxyl groups on apoferritin could provide a thermodynamic driving force for Fe(II)-oxidation and the same may apply to the oxy sites on the mineral surface which increase in number as the crystallite grows. Not all 24 subunits or all molecules nucleate simultaneously, and both intra- and inter-molecular competition for iron develops so that several molecules contain only one crystal and others contain none. The protein shell prevents the crystallites from coalescing and precipitating, thus ensuring that free surface is available (on complete ferritin molecules) at which iron can be deposited or released. The presence of neighbouring crystal sites, which first bind Fe(II), could be a factor in promoting greater efficiency of oxidation and explain how conditions which favour rapid iron uptake into ferritin also give maximum ratios of Fe(II) oxidized/O_2 molecule ($=4$) (Treffry, Sowerby, and Harrison 1978).

Although more complex models have been considered (Macara *et al.* 1972; Crichton and Roman 1978), this simple picture (see Fig. 28.6) seems to explain most of the data very satisfactorily. The lack of structural relationship between protein and mineral may simply result from a limited contact region, which allows the 'FeOOH' to grow in different directions, to move or even to become detached from its original site. A prediction of the model is that iron uptake and release should obey a 'last-in-first-out' principle and this we have observed by means of a radioactive label (Hoy, Harrison, and Shabbit 1974*c*; Treffry and Harrison, 1978), see Fig. 28.6(e).

Experiments on ferritin iron-uptake and release helped to keep us going during the initial slow stage of structure determination. In the last five years, this too has entered a growth phase. Horse-spleen apoferritin, although relatively easy to isolate and crystallize, is not a protein which readily forms heavy-atom derivatives suitable for phase determination by isomorphous replacement methods. Possibly this has something to do with the presence of Cd(II) of crystallization, which may compete with other cations for binding sites. Out of a large number tested, Richard Hoare found only two suitable derivatives. A covalent Hg derivative obtained with PCMB (*p*-chloromer-curibenzoate) was used for the calculation of the first electron-density map at 6 Å resolution (which included anomalous scattering data), see Fig. 28.7(b). With the aid of difference Fouriers based on this derivative UO_2^{2+} was located in a $UO_2(NO_3)_2$ derivative and this was included in the next 6 Å map, Fig. 28.7(c) and (d). This map (Hoare, Harrison, and Hoy 1975*a*, *b*) confirmed apoferritin as a hollow shell which was so compact that the subunit boundaries were not clearly delineated (perhaps explaining its remarkable stability) and showed it to contain several rods of high electron density, which, if α-helix, would account nicely for the ca. 60 per cent helix content estimated from ORD measurements (Listowsky, Betheil, and England 1967). Most excitingly it showed the presence of cation-binding sites on the inner surface of the shell, which might indicate locations for Fe-binding and mineral nucleation, and six channels through the shell about 10 Å diameter (around the fourfold axes) which could provide the means of access for these cations and for iron itself. The presence of these channels had been predicted as a means of access to the cavity (Harrison 1959), but now we were able to see them for the first time (see Fig. 28.7(c) and (d)) and to get some idea of their selectivity for ions and small molecules. The electron-density map

---

FIG. 28.6. Diagrammatic representation of the formation of ferritin from apoferritin (based on Macara *et al.* 1972; Harrison *et al.* 1974). Ferritin can be reconstituted from apoferritin *in vitro* by addition of Fe(II) in the presence of an oxidant (e.g. atmospheric O_2). Our model of this process is as follows:

(a) Apoferritin is a multisubunit protein with inter-subunit channels which allow Fe(II) and oxidant to enter the molecule (arrows). The inner surface of apoferritin contains sites at which Fe(II) may be bound and oxidized to Fe(III).

(b) A site with Fe(III) bound becomes a nucleation centre for the further addition of Fe and a crystallite of hydrous ferric oxide starts to grow inside the shell. Fe(II) may then be added directly to the crystallite surface and oxidized without involvement of an 'active site' on the protein.

(c) As the first crystallite grows its surface available for Fe deposition grows and the rate of Fe uptake into ferritin increases. A second crystal nucleus may form at an equivalent site within the shell, but established crystallites will tend to win the competition for further Fe addition.

(d) As the molecule fills there are now fewer available surface sites for Fe deposition and Fe uptake rates drop.

It is observed experimentally that when Fe is added in successive increments, initial rates of Fe uptake first increase and then decrease. X-ray analysis (see Fig. 28.5(b)) indicates that the mineral and protein structures are not specifically related at the atomic level.

(e) Because ferritin Fe is present in micro-crystals or 'micelles' of hydrous ferric oxide its atoms are not all equivalent. Fe recently added at surface sites (full arrow) is more available for release (dashed arrow) than Fe at internal sites. 'Last-in-first-out' behaviour is found experimentally. We used this schematic shell structure with channels to depict apoferritin, even before electron-density maps (Fig. 28.7(b) to (f)) were calculated.

also showed peaks in the intermolecular contacts which we attribute to cadmium ions of crystallization.

Intensity data for the 6 Å structure were obtained by measuring precession photographs with a Joyce Loebl microdensitometer. For high-resolution work these methods become both inefficient and time-consuming. Fortunately we have been able to purchase an Arndt–Wonacott rotation camera and to use the Bristol Optronics scanner and data reduction programs. In the summer of 1977 the high-resolution structure determination went ahead rapidly and a 2.8 Å resolution electron-density map was calculated by Stephen Banyard and David Stammers, Fig. 28.7(e) and (f). It was now clear that the rods seen at low resolution were indeed helical. Each subunit contains

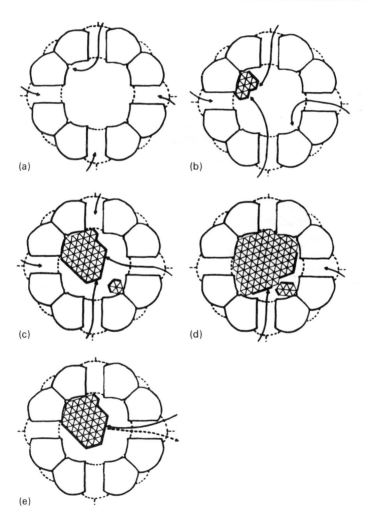

(a)

(b)

(c)

(d)

(e)

four long α-helices and a short one (3–4 turns) accounting for about 100 out of the 165 amino acids present in the subunit (see Fig. 28.8). The long helices are nearly parallel (or anti-parallel) and lie perpendicular to the radius vector. The inner helices are somewhat shorter (18–21 residues) than the outer pair (27–29 residues), so that the subunit is a truncated segment. The short helices lie near the fourfold axes and roughly parallel to the radius. These define the channels to the central cavity, Fig. 28.7(c) and (f). Across the subunit surface, near to and roughly perpendicular to the twofold axes, lies a long extended loop. It approaches its symmetry-related neighbour with which it seems to form a short region of anti-parallel pleated sheet around the dyads. A system of eight nearly parallel helices is also formed by this symmetry operation. The connectivities of the loop and between helices are uncertain because they lie in weak regions of electron density. Model building in the Richard's Box has suggested alternative ways of forming a complete subunit, the two most favoured being shown in Fig. 28.8(a) and (b). One of these, Fig. 28.8(a), was originally found by inspection of the small-scale map. Its C-terminal end is on the inside of the shell, when the subunit is placed in its quaternary structure. This must make the C-terminus inaccessible to digestion by carboxypeptidase. As it has been reported (Chrichton, private communication) that carboxypeptidase B releases arginine without prior denaturation of the molecule it now seems as if this alternative (Fig. 28.8(a)) is unlikely. This objection would not apply to Fig. 28.8(b). The apoferritin subunit conformation is rather similar to that in at least three other proteins met-haemerythrin (Stenkamp, Sieker, and Jensen 1976) and myohaemerythrin (Hendrickson, Klippenstein, and Ward 1975), Cytochrome b562 (Scott Mathews,

---

FIG. 28.7. Electron-density maps of horse-spleen apoferritin at successive stages of structural analysis of cubic crystals.

(a) 26 Å resolution. Section through molecular centre ($z=0$). Phases calculated for a uniform spherical shell, $r_1 = 62$ Å, $r_2/r_1 = 0.625$ (Harrison 1963).

(b) 6 Å resolution. Section through molecular centre ($z=0$). Phases based on a single isomorphous replacement by PCMB (R. J. Hoare, unpublished). Channels through the shell around fourfold axes were seen for the first time.

(c) and (d) 6 Å resolution. Spherical shell structure based on low resolution electron-density map (two derivatives, Hoare et al. 1975a, b). The top of the shell has been cut away (near a threefold axis) to reveal rods of electron density (identified as α-helix at 2.5 Å resolution, see (e) and (f)). Note rods around channels. The top slice, (d), viewed from the inside, shows a number of cation binding sites, some of which, on the inside surface, may represent potential Fe binding sites involved in ferritin formation (see Fig. 28.6(a)).

(e) and (f) 2.8 Å resolution. Approx. 10 Å thick slices of electron density (eight sections) perpendicular to a fourfold axis (Banyard et al. 1978). In (e), a slice near the molecular centre, parts of α-helices can be seen within the shell lying nearly perpendicular to a radius vector. In (f), a slice near the top of the molecule, four short helices, seen end-on, surround one of the six fourfold channels. Long helices lying within the shell can also be seen.

Bethge, and Czerwinski 1979) and the tobacco mosaic virus protein subunit (Champness, Bloomer, Bricogne, Butler, and Klug 1976; Bloomer, Champness, Bricogne, Staden, and Klug 1978) although the polypeptide chain length in haemerythrin and cytochrome b562 is only about two-thirds that of apoferritin, Fig. 28.8. Although some details of the subunit conformation remain to be resolved we are now able to define its shape (very roughly a cylinder of diameter 27 Å and length 54 Å) and packing. An idealized drawing of the quaternary structure, Fig. 28.9(b), shows it to be quite similar to the original snub-cube, Fig. 28.9(a).

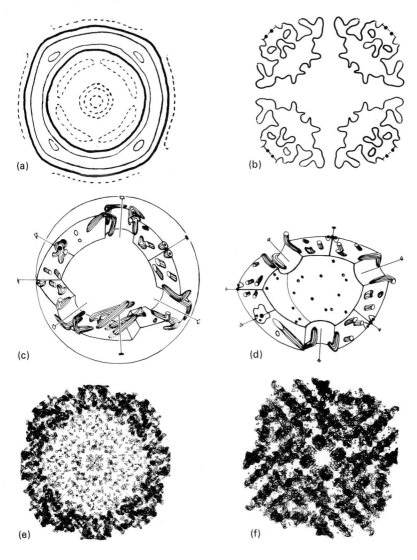

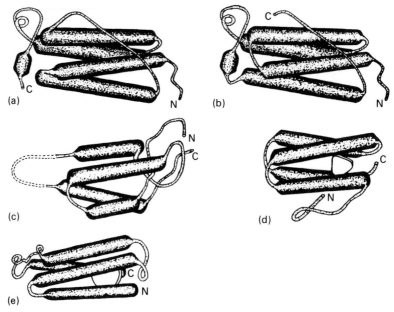

FIG. 28.8. Alternative conformations of horse-spleen apoferritin subunits compared with those of other proteins. Rods represent helices. (a) and (b) show alternative connectivities in apoferritin (subunit mol. wt. 18 500). (a) is our first interpretation of the 2.8 Å resolution electron-density map (Banyard *et al.* 1978) and (b) to an alternative derived by model building in a Richards Comparator (G. A. Clegg and P. M. Harrison, unpublished). (c), (d), and (e) represent respectively the tobacco mosaic virus disc subunit (mol. wt. 17 500), myohaemerythrin (mol. wt. 13 000) and cytochrome b562 (mol. wt. 12 000). Redrawn from Bloomer *et al.* (1979), Hendrickson *et al.* (1975), and Scott-Mathews *et al.* (1979). Note the similarity in folding in apoferritin (particularly conformation (b)) to that in the other proteins. In (b) both N- and C-termini lie on the outside of the apoferritin shell, whereas in (a) the C-terminus is inside.

This is by no means the end of the story, but the end of a chapter. The sequence determination of horse-spleen apoferritin by Bob Crichton and colleagues in Belgium is now nearing completion, and although we have already carried out preliminary fitting of partial sequences to our electron density, it would be premature to discuss this here. Sequence and structure determination for human apoferritins are in hand. Recently, molecules resembling ferritin have been isolated from *Azotobacter vinelandii* (Stiefel and Watt 1979) and from *E. coli* (Yariv *et al.* 1979). Haem has been found in the former in addition to iron-micelles and both molecules have been crystallized. Work on ferritin seems now to be just beginning, and we hope, by a detailed examination of horse-spleen apoferritin and related structures, within the next few years to be able to specify what residues bind iron inside the protein shell, what reductants or chelators may pass through the shell to release iron

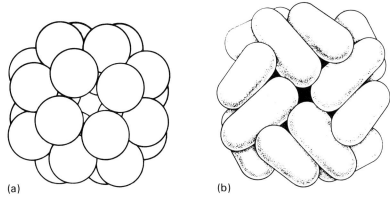

(a)   (b)

FIG. 28.9. Diagrammatic representations of horse-spleen apoferritin quaternary structure.

(a) Twenty-four spherical subunits at apices of a snub-cube originally proposed in Harrison (1959) and Hoy *et al* (1974*b*). (b) Quaternary structure based on the 2.8 Å resolution electron-density map of Banyard *et al.* (1978) (see Fig. 28.7(e) and (f)).

An alternative arrangement of twenty subunits at the apices of a pentagonal dodecahedron is now known to be incorrect (see Harrison 1963; Hoy *et al.* 1974*a, b*).

from the cavity, what interactions are important in assembly, and to what extent the structure is conserved in evolution. If we can relate structure and function at the atomic level our goal will have been achieved.

Acknowledgements

This short account has inevitably been selective and has concentrated on the work of my own research group. I should like to thank first of all Dorothy herself for introducing me to ferritin, and the following who have carried out much of the work I have described (in chronological order): T. G. Hoy, I. G. Macara, R. J. Hoare, S. H. Banyard, A. Treffry, D. K. Stammers, and G. A. Clegg. I should also like to thank the Medical Research Council and Science Research Council for many years' support.

References

ACKERVALL, K., STRANDBERG, B., ROSSMAN, M., BENGTSSON, V., FRIDBERG, K., JOHANNISEN, H., KANNAN, K. K., LÖVGREN, S., PETEF, G., ÖDBERG, B., EAKER, D., HJERTEN, S., RYDEN, L., and MORING, I. (1971). *Cold Spring Harb. Symp. quant Biol.* **36**, 469–83, 487–8.

BANYARD, S. H., STAMMERS, D. K., and HARRISON, P. N. (1978). *Nature (Lond.)* **271**, 282–4.

BIELIG, H. J. and BAYER, E. (1955). *Naturwissenschaften* **42**, 125–6.

BJORK, I. and FISH, W. (1971). *Biochemistry* **10**, 2844–8.

BLOOMER, A., CHAMPNESS, J. N., BRICOGNE, G., STADEN, R., and KLUG, A. (1978). *Nature, Lond.* **276**, 362–8.

BRYCE, C. F. A. and CRICHTON, R. R. (1971). *J. biol. Chem.* **246**, 4198–205.

——(1973). *Biochem. J.* **133**, 301–9.

CASPAR, D. L. D. and KLUG, A. (1962). *Cold Spring Harb. Symp. quant. Biol.* **27**, 1–24.

CHAMPNESS, J. N., BLOOMER, A. C., BRICOGNE, G., BUTLER, P. J. G., and KLUG, A. (1976). *Nature, Lond.* **259**, 20–4.

CRICHTON, R. R. and ROMAN, F. (1978). *J. molec. Catalysis* **4**, 75–82.

DEAL, W. C., RUTTER, W. J. and VAN HOLDE, K. E. (1963). *Biochemistry* **2**, 246–51.

DRYSDALE, J. W. and MUNRO, H. N. (1966). *J. biol. Chem.* **241**, 3630–7.

FANKUCHEN, I. (1943). *J. biol. Chem.* **150**, 57–9.

FARRANT, J. L. (1954). *Biochim. biophys. Acta* **13**, 569–76.

FINEBERG, R. A. and GREENBERG, D. M. (1955). *J. biol. Chem.* **214**, 97–106.

FISCHBACH, F. A., HARRISON, P. M. and HOY, T. G. (1969). *J. molec. Biol.* **39**, 235–8.

GRANICK, S. (1946). *Chem. Rev.* **38**, 379–403.

——(1951). *Physiol. Rev.* **31**, 489–511.

—— and MICHAELIS, L. (1943). *J. biol. Chem.* **147**, 91–7.

HAGGIS, G. H. (1965). *J. molec. Biol.* **14**, 598–602.

——(1966). *Proc. Sixth International Congress for Electron Microscopy*, Maruzen, Kyoto, Japan, p. 127.

HARRISON, P. M. (1959). *J. molec. Biol.* **1**, 69–80.

——(1963). *J. molec. Biol.* **6**, 404–22.

—— and HOFMANN, T. (1962). *J. molec. Biol.* **4**, 239–50.

—— and HOY, T. G. (1973). In *Inorganic biochemistry* (ed. G. L. Eichhorn) Chapter 8. Elsevier, Amsterdam.

—— HOFMANN, T., and MAINWARING, W. I. P. (1962). *J. molec. Biol.* **4**, 251–6.

—— FISCHBACH, F. A., HOY, T. G., and HAGGIS, G. H. (1967). *Nature, Lond.* **216**, 1188–90.

—— HOY, T. G., MACARA, I. G., and HOARE, R. J. (1974). *Biochem. J.* **143**, 445–51.

—— BANYARD, S. H., CLEGG, G. A., STAMMERS, D. K., and TREFFRY, A. (1978). In *Transport by proteins* (ed. G. Blauer and H. Sund) pp. 259–72. de Gruyter, Berlin.

HENDRICKSON, W. A., KLIPPENSTEIN, G. L. and WARD, K. B. (1975). *Proc. natn. Acad. Sci. U.S.A.* **72**, 2160–4.

HOARE, R. J., HARRISON, P. M., and HOY, T. G. (1975a). *Nature, Lond.* **255**, 653–4.

——————(1975b). In *Proteins of iron storage and transport in biochemistry and medicine* (ed. R. R. Crichton) pp. 231–6, North-Holland, Amsterdam.

HODGKIN, D. C. (1950). *Cold Spring Harb. Symp. quant. Biol.* **14**, 65–78.

HOFMANN, T. and HARRISON, P. M. (1963). *J. molec. Biol.* **6**, 256–67.

HOY, T. G. and HARRISON, P. M. (1975). In *Proteins of iron storage and transport in biochemistry and medicine* (ed. R. R. Crichton) pp. 279–86. North Holland, Amsterdam.

——————(1976). *Br. J. Haematol*, **33**, 497–504.

—————— and HOARE, R. J. (1974a). *J. molec. Biol.* **84**, 515–22.

——————(1974b). *J. molec. Biol.* **86**, 301–8.

—————— and SHABBIR, M. (1974c). *Biochem. J.* **139**, 603–7.

KLUG, A. (1971). *Cold Spring Harb. Symp. quant. Biol.* **36**, 483–7.

—— LONGLEY, W., and LEBERMAN, R. (1966). *J. molec. Biol.* **15**, 315–43.

LAUFBERGER, V. (1937). *Bull. Soc. Chim. biol.* **19**, 1575–82.

LISTOWSKY, I., BETHEIL, J. J., and ENGLAND, S. (1967). *Biochemistry* **6**, 1341–8.

MACARA, I. G., HOY, T. G., and HARRISON, P. M. (1972). *Biochem. J.* **126**, 151–62.

——————(1973). *Biochem. J.* **135**, 343–8.

MASSOVER, W. H. and COWLEY, J. M. (1973). *Proc. natn. Acad. Sci.* **70**, 3847–51.

MAZUR, A and SHORR, E. (1950). *J. biol. Chem.* **182**, 607–27.

MILLER, J. P. G. and PERKINS, D. J. (1969). *Eur. J. Biochem.* **10**, 146–51.
NIEDERER, W. (1970). *Experientia* **26**, 218–20.
ROSSMANN, M. G. and BLOW, D. M. (1962) *Acta crystallogr.* **15**, 24–31.
ROTHEN, A. (1944). *J. biol. Chem.* **152**, 679–93.
SCOTT-MATHEWS, F., BETHGE, P. H., and CZERWINSKI, E. W. (1979). *J. biol. Chem.* **254**, 1699–706.
SPIRO, T. G. and SALTMAN, P. (1969). *Struct. Bond.* **6**, 116–56.
STELLWAGEN, E. and SCHACHMAN, H. K. (1962). *Biochemistry* **1**, 1056.
STENKAMP, R. E., SIEKER, L. C., and JENSEN, L. H. (1976). *J. molec. Biol.* **100**, 23–34.
STIEFEL, E. I. and WATT, G. D. (1979). *Nature, Lond.* **279**, 81–3.
THOMPSON, D'ARCY W. (1952). *Growth and form* (2nd edn). Cambridge University Press.
TREFFRY, A. and HARRISON, P. M. (1978). *Biochem. J.* **171**, 313–20.
——SOWERBY, J. M. and HARRISON, P. M. (1978). *FEBS Lett.* **95**, 221–4.
WELLS, A. F. (1945). *Structural inorganic chemistry*, p. 342. Clarendon Press, Oxford.
YARIV, J., SPERLING, R., BAUMINGER, E. R., COHEN, S. G., LEVY, A., and OFER, S. (1979). *J. Phys. Colloque* C2, suppl. 3, **40**, C2–523.

29. The structure of leghaemoglobin

B. K. VAINSHTEIN

ALONGSIDE the appearance of the oxygen atmosphere on earth, approximately 2 billion years ago, the proteins performing the function of storage and transport of oxygen were arising in living organisms. Inherent in many animals as well as in man are proteins of the myoglobin–haemoglobin type. It is an iron–porphyrin system of haem that performs in them the oxygen-binding function.

It has long been considered that such proteins exist only in animals. However, in 1939 the Japanese scientist Kubo discovered in the root nodules of nitrogen-fixing plants a pigment with spectroscopic properties similar to those of haemoglobin.

The reason for the particular interest in legume plants (soya, kidney beans, lupins, broad beans) was prompted by their unique capacity for binding atmospheric nitrogen. This process takes place in nodules which coexist symbiotically with some strains of bacteria *Rhizobium*. Inside the nodules the bacteria undergo morphological changes and are transformed into the so-called 'bacteroid tissue'. At first, hypotheses were put forward that leghaemoglobin (Lg) performs the nitrogen-binding function. But these theories soon proved to be wrong: it is nitrogenase that fulfils the function of nitrogen-binding. As for leghaemoglobin, it carries on the function of all proteins of this type, i.e. that of oxygen-binding. Leghaemoglobin provides bacteroids with oxygen which participates in oxidative phosphorylation. The ATP generated is believed to be necessary for nitrogenase functioning.

Leghaemoglobins have been intensively investigated biochemically in many laboratories (see reviews by Virtanen 1947; Appleby 1974; Ellfolk 1972; Zhiznevskaya 1972). The molecular weight of approximately 17 000–18 000 daltons and other data showed that this protein appears to be homologous in structure to myoglobin and monomeric haemoglobins; but no direct evidence was available, and an X-ray study to solve the problem was still awaited. The results of X-ray studies of leghaemoglobin from *Lupinus luteus* that have been carried out by the author and his co-workers since 1974 are summarized in the present work.

Primary structure

The primary structure is known for a number of leghaemoglobins (Ellfolk and Sievers 1972; Lehtovaara and Ellfolk 1975; Richardson, Dilworth, and Scawen 1975).

The homogeneous preparation of leghaemoglobin from *Lupinus* nodule extracts is obtained by chromatography on Al_2O_3- and DEAE-cellulose at pH 7.0; the protein is eluted by ammonium acetate buffer (Kuranova, Grebenko, Konareva, Syromyatnikova, and Barynin 1976). Two main fractions I and II are observed in order of elution. The primary structure for both of them has been recently determined (Egorov *et al.* 1976; Egorov, Kazakov, Shakhparonov, Feigina, and Kostetsky 1978).

TABLE 29.1

Divergence in primary structure (%)

	Various globins					Leghaemoglobins		
Mb sperm whale	88					soya	55	
Hb man	84	78				kidney beans	51	20
Hb lamprey	87	75	75			broad beans	57	38	35
Hb Glycera	90	76	82	80		lupin (fraction I)	soya	kidney beans
Hb Chyronomus	83	79	88	79	85			
Lg soya	55	85	83	86	90	86		
	Lg lupin	Mg sperm whale	Hb man	Hb lamprey	Hb Glycera	Hb Chyronomus		

The divergences in primary structure for some globins of animals and plants, as well as between leghaemoglobins themselves, are given in Table 29.1.

As is seen from Table 29.1 the primary structure of plant globins differs from that of animal globins by 83–90 per cent. Such a large divergence in primary structure is not surprising since the common ancestor of these proteins existed about 1.3 billion years ago when plant and animal kingdoms came to develop differently. However, differences up to 85–88 per cent are also observed among animal globins (Holmquist, Jukes, and Moise 1976). At the same time, leghaemoglobin sets us another evolutionary riddle—it is present only in the symbiotic system of legume plants that appeared about 200 million years ago. The question arises in what organisms has the leghaemoglobin gene survived during this period 1.3–0.2 billion years ago—and up to the present.

Naturally, within the group of leghaemoglobins the divergence in primary structure is considerably smaller, being of the order of 20–57 per cent. Fractions I and II of Lg from *Lupinus luteus* differ in primary structure by 20 residues out of the total 153, i.e. by 13 per cent.

X-ray analysis

We studied the structure of monoclinic crystals of acetate-ferri-leghaemoglobin II *Lupinus luteus* at 5 Å (Vainshtein *et al.* 1974 *a, b*; 1975), 2.8 Å (Vainshtein *et al.* 1978), 2.0 Å resolution (Harutyunyan, Kuranova, Vainshtein, and Steigemann 1979) as well as the complex ferri-leghaemoglobin with nicotinic acid. We also found the molecular packing in the orthorhombic crystals of acetate-ferri-leghaemoglobin I (Harutyunyan, Kuranova, Grebenko, and Voronova 1977).

The brown–red crystals of Lg elongated along the twofold axis appeared from 1.7 per cent buffer solution, pH 7.0, oversaturated for the protein with ammonium sulphate. In three to five weeks the crystals were observed to achieve the dimensions of $0.5 \times 0.6 \times 1$ mm (Fig. 29.1). They proved suitable for X-ray diffraction studies and for search of heavy-atom derivatives. The

1 mm

FIG. 29.1. Leghaemoglobin crystals.

TABLE 29.2

X-ray data on leghaemoglobin

	Space group	Unit cell	Resolution (Å)	Number of reflections	Heavy-atom derivatives, or method	m, R
Acetate-ferri Lg II	B2	$a=93.33$ $b=38.40$ $c=52.40$	5	840	HgI_3^-, UO_2^{2+}, I_2, $3HgCl$-pyridine	$m=0.90$
		$\gamma=99.0°$	2.8	4600	UO_2^{2+} (1), UO_2^{2+} (2),	
		$Z=4$ $V=180\,000$ Å³	2	12 000	I_2, mersalyl refinement	$m=0.76$ $R=19\%$
Nicotinate-ferri Lg II	B2	$a=92.92$ $b=38.64$ $c=52.36$ $\gamma=99.6°$ $Z=4$ $V=182\,000$ Å³	2.8	4600	I_2, 3-ClHg-pyridine, mersalyl	$m=0.88$
Acetate-ferri Lg I	$B22_12$	$a=91.69$ $b=75.03$ $c=55.04$ $Z=8$ $V=382\,000$ Å³	5 2.8	800 4500	non-local search molecular replacement	$R=29\%$

data on the unit cell and heavy-atom sites are given in Table 29.2. The asymmetric unit contains one molecule.

The electron-density map at 5 Å resolution showed the obvious and unambiguous course of the polypeptide chain of the protein as well as the position and orientation of the haem group. The close similarity in tertiary structure with myoglobins and haemoglobins studied before has been

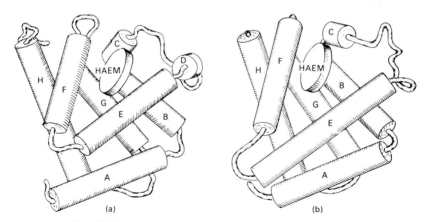

FIG. 29.2. Homology of structures of Mb (a) and Lg (b).

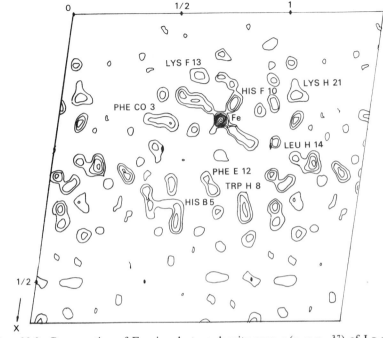

FIG. 29.3. Cross-section of Fourier electron-density map $\rho(x, y, z = \frac{37}{60})$ of Lg at 2 Å resolution, passing through the Fe atom.

FIG. 29.4. Model of the leghaemoglobin structure.

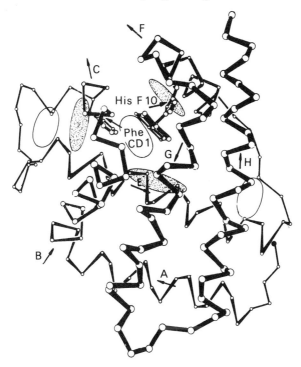

FIG. 29.5. Chain of α-carbon atoms in projection on to the *xy*-plane. The haem, residues CD1 Phe, F10 His, hydrophobic clusters (shaded), and cavities of the molecule are shown. Reading from left to right, the cavities are II, I, III. The left-most cluster is I, the cluster upper right is II, and the cluster lower right is III.

revealed. We observe the same 'myoglobin' type of chain folding that was first established by Kendrew with co-workers (1960) during the investigation of sperm-whale myoglobins (Mb). The similarity of both structures is shown in Fig. 29.2. The electron-density map was then obtained at 2.8 Å resolution and a preliminary model built up. After that we constructed a Fourier synthesis at 2.0 Å resolution (Fig. 29.3) and carried out the structure refinement. The model is shown in Fig. 29.4, the α-carbon chain in Fig. 29.5.

Tertiary structure of Lg and the comparison with Mb

Despite the fact that the general motif of myoglobin folding in structures of Lg and Mb is similar, nevertheless they show some essential differences; the majority of these may be explained in terms of divergence in primary structure. The comparison within the primary and secondary structures (Vainshtein, Kuranova, Harutyunyan, and Egorov 1980) is given in Table 29.3

TABLE 29.3

Comparison of primary and secondary structures of Mb and Lg

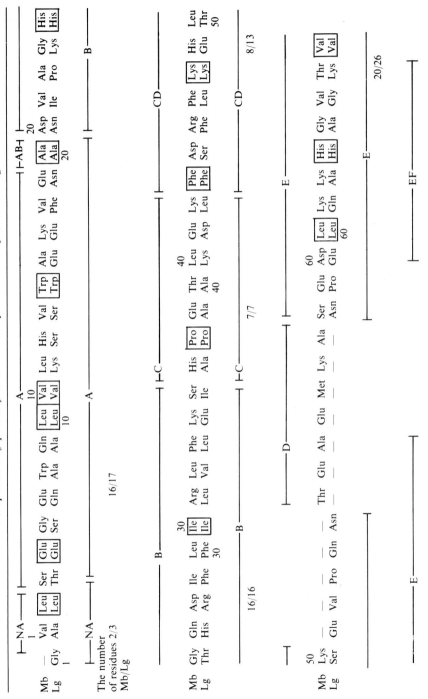

Mb	Leu	Thr	Ala	Leu	Gly	Ala	Ile	Leu	Lys	—	—	—	—	—	—	Lys	Lys	Gly	His	His	Glu	Ala	Glu	—
Lg	Phe	Lys	Leu	Val	Tyr	Glu	Ala	Ala	Ile	Gln	Leu	Gly	Val	Thr	Gly	Val	Val	Val	Ser	Asp	—	—	—	Ala
			70									80												

E — EF —

F — FG — G —

					90													100						
Mb	—	Leu	Lys	Pro	Leu	Ala	Gln	Ser	His	Ala	Thr	—	—	Lys	His	Lys	Ile	Pro	Ile	Lys	Tyr	Leu	Glu	Phe	Ile
Lg	Thr	Leu	Lys	Asn	Leu	Gly	Ser	Val	His	Val	Ser	Lys	Gly	Val	Ala	—	—	—	Asp	Ala	His	Phe	Pro	Val	Val
		90										100													110

F — FG — G —

10/14

G — GH — H —

			110												120									130
Mb	Ser	Glu	Ala	Ile	Ile	His	Val	Leu	His	Ser	Arg	—	—	His	Pro	Ala	Asp	Phe	Gly	Ala	Asp	Ala	Gln	Gly	Ala
Lg	Lys	Glu	Ala	Ile	Leu	Lys	Thr	Ile	Lys	Glu	Val	Val	Gly	—	—	Ala	Lys	Trp	Ser	Glu	Glu	Leu	Asn	Ser	Ala
										120												130			

G — GH —

19/20

H — HC —

										140											150			
Mb	Met	Asn	Lys	Ala	Leu	Glu	Leu	Phe	Arg	Lys	Asp	Ile	Ala	Ala	Lys	Tyr	Lys	Glu	—	—	Leu	Gly	Tyr	Gln	Gly
Lg	Trp	Thr	Ile	Ala	Tyr	Asp	Glu	Leu	Ala	Ile	Val	Ile	Lys	Lys	Glu	Met	Asp	Asp	Ala	Ala					
							140										150								

H —

25/27

where the helical and non-helical regions are denoted according to the Kendrew nomenclature. The E- and F-helices being substantially longer in Lg than in Mb and the absence of the D-helix may be regarded as peculiarities in the structure of Lg. The non-helical regions EF and FG are considerably shorter in Lg, but the CD region is longer. The E-helix elongation is due to the insertion of five additional amino-acid residues into the primary structure at its C-terminal.

The majority of the helical segments represent the classical α-helix $s = \frac{18}{5}$ (A, B, D, E, H). However, on most of the segments of α-helices the ϕ- and ψ-angles somewhat deviate from the standard values $\phi = -47°$ and $\psi = -57°$; ψ increases, being within the range $-47°$ to $-27°$; ϕ decreases: $-57°$ to $-77°$. (The same deviations were observed in Mb and erythrocruorin.) Slight deviations from the α-configuration also occur in the A- and E-helices (the second turn in A is closer to the π-helix). The C-helix belongs to type 3_{10}; however, the C3 Ala carbonyl group does not form an interhelical hydrogen bond. A segment close to 3_{10} may also be found in the F-helix. In the G-helix the first turn is 3_{10}, then Pro follows and, finally, α-helix.

The mutual orientation of A, B, C, E, F, and H helices in Lg is rather similar to that in Mb; owing to this, the pairs of residues with the same Kendrew index in both polypeptide chains participate in most of the contacts between the helices. The total number of residues in the G-helix of Lg that are in contact with the haem is the same as in Mb. But due to the azimuthal shift of the G-helix about its axis, the fifth residue, which moves towards the haem in Mb, comes near to the H-helix in Lg, and it is the G4 residue that approaches the haem. It should be noted that the NA1–2 and CD5–9 segments of the chain are poorly seen on the map, the fact reflecting, apparently, their considerable conformational flexibility. In general, the thermal and conformational movements of the atoms on the periphery of the molecule are considerably more extensive than at its centre, near the haem and the hydrophobic core.

Out of 153 residues of Lg and Mb, 25 residues are invariant (boxed in Table 29.3), 13 of them being important for supporting the structure or for the function: these are the proximal F10 and distal E7 histidine, CD1 Phe, its aromatic ring being parallel to the haem plane, and C2 Pro, which is responsible for a sharp bend of the chain in the transition from the B- to C-helix. NA3 Leu is also such a residue, fixing the N-terminal on the surface; A8 Val, E4 Leu, E11 Val, G11 Ile, F8 Ser, H10 Ala entering into hydrophobic clusters, and A12 Trp located in the slit between the A- and E-helices.

It is quite probable that the majority of invariant residues that are most conservative in evolution are responsible for preservation of properties that are common throughout the haemoglobin family. Nevertheless, as Table 29.3 shows, most of the residues are different, but contacts are observed between residues which occupy equivalent positions in both proteins (this fact is well known for globins). In many cases the hydrophobic and hydrophilic character of a residue is retained. As a result, about 75 per cent of the substitutions considered have no significant effect on chain configuration. In some cases there are changes in polarity of both residues which are in contact with each other. These 'compensated' substitutions also contribute to the preservation of the α-helical pattern.

Quite important for supporting the structure are the hydrophobic sides of helices that 'stick together', and hydrophobic clusters combining to form the hydrophobic core of the molecule (Kendrew *et al.* 1960). Takano (1977) described in Mb three main clusters that we also observed in Lg (Fig. 29.5). Cluster I (near the distal histidine) includes hydrophobic residues between distal histidine, B-helix and CD-segment and encloses the cavity occupied by the sixth Fe ligand; this latter bears a strong resemblance to that in Mb with the exception of B14 Leu (B14 Phe in Mb) which is displaced towards the C-helix and does not enter into the cluster.

Cluster II forms the proximal histidine environment. In Lg it has a number of specific features. Although it consists, as in Mb, of eight amino-acid residues, only five of them are positionally equivalent in both molecules, i.e. they have the same index in the secondary structure. The G5 Leu and FG4 Ile, residues in Mb fulfilled the same function as G4 Phe and FG2 Ala in Lg. Owing to the shortening of the FG segment, G4 Phe in Lg penetrates the cluster at a greater depth than G5 Leu in Mb. As for H23 Tyr, it is located in Mb near the cluster, whereas in Lg H23 Met enters into it directly. Here the side-chain of methionine fills up the space between the F-, H-, and G-helices and the FG segment.

The outlines of cluster III (part of the haem pocket cavity) are very similar in both molecules. The most striking differences are as follows: in Lg it comprises A15 Phe instead of B5 His in Mb, and instead of B10 Phe (Lg) it is B11 Ile (Mb) which separates the cavity inside the CD region from the haem pocket cavity.

Thus comparison of the clusters shows that they are formed, in the main, by amino-acid residues which occupy spatially the same or similar positions in the two molecules. Owing to this, the clusters retain their general outlines.

The Lg structure has three cavities. Later on we shall consider the structure of the main cavity on the distal side of the haem. The second cavity is located in the CD-region, while the third one is between the EF-region and H-helix (Fig. 29.5). The cavities are characterized as ellipsoids having dimensions of 5–8 Å. Such empty space in the molecule may seem to provide conformation freedom for side groups.

It may be noted that structure refinement revealed positions of a number of water molecules, mainly on the protein surface. All of them form hydrogen bonds to polar groups of the protein. In some cases, especially in the region of intermolecular contacts, a chain or network of water molecules is formed.

Structure of the haem pocket and functional features specific for Lg

The main specific features of Lg is its high affinity for oxygen (Wittenberg, Appleby, and Wittenberg 1972). The oxygen-binding curve is hyperbolic as is characteristic for monomers, but the binding constant is high: for Lg it is $K = 1.2 \times 10^8 M^{-1} s^{-1}$, while for Mb $K = 1 \times 10^7 M^{-1} s^{-1}$.

The functional peculiarities in proteins of this type, as elucidated in investigations of M. Perutz (see his review, 1976), are associated with the haem environment, the valency and spin states of the iron atom, the distances from the iron atom to proximal and distal histidine, the dimensions of the haem pocket on the distal side and the arrangement of neighbouring amino-acid residues in it. Apoprotein may also influence the haem through its side groups and, as a result, alter the electron state of the four pyrrole nitrogen atoms surrounding the iron atom, thereby exerting additional influence on the binding of iron to the sixth ligand.

Analysis of the structure of Lg at 2.8 Å resolution and its 2 Å refinement carried out according to Diamond's (1971) method made it possible to answer some of these questions.

The structural peculiarities in the haem environment are as follows: on the distal side of the haem the cavity is significantly broader than in other haemoglobins. The distances between the centres of the nearest atoms of pairs of residues facing one another: CD1 Phe and E15 Val (12 Å), E10 Val and G7 Val (7 Å and the distance from the cavity of the haem to B10 Phe located above it (7Å) give an idea of the volume of this cavity which may accommodate molecules larger than oxygen. This was evidenced by some physico-chemical tests. Thus in 1972 Ellfolk observed the inclusion of aliphatic acid molecules into Lg. Nicotinic acid can also be introduced in the haem pocket (see below). The time measurements of magnetic relaxation of proteins carried out by Vuk-Pavlovic *et al.* (1976) also give evidence for such a conclusion. Nevertheless, the haem pocket in Lg has no channel sufficiently large to allow the oxygen molecule to penetrate from outside without disturbance of the protein side chains. The entrance to the haem pocket is covered, on the side of the C- and G-helices, by the side group of G3 His, but on rotation of this group about the C_β—C_γ bond the haem pocket may become easy of access.

The difference map at 2.0 Å resolution showed that an acetic acid molecule (Figs. 29.6 and 29.7) is the sixth ligand in crystals of Lg which were investigated by us, and not a water molecule as we had thought at first. The presence of the acetic acid molecule is readily explained since crystallization proceeds from ammonium acetate buffer. The acetate group lies in a plane almost

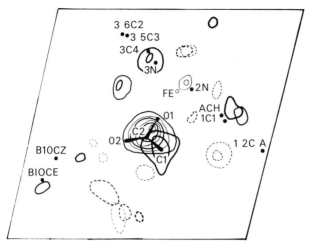

F IG. 29.6. Difference Fourier map showing position of the acetate group.

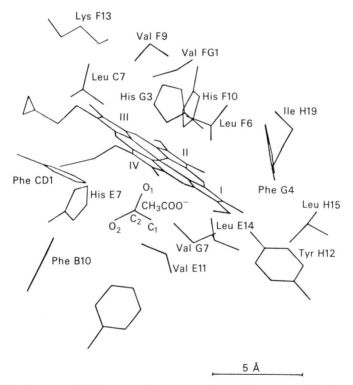

F IG. 29.7. Structure of the haem pocket, surroundings of the haem with the nearest amino-acid residues, and position of the acetate group.

perpendicular to the haem plane. The $Fe-O_1-O_2$ angle is equal to $158°$. The oxygen molecule is hydrogen bonded to N_ε of the distal histidine E7 (O—N $= 2.60$ Å). As a result, E7 His is found to be displaced towards the edge of the porphyrin ring (Fig. 29.7). The acetate group is stabilized additionally by interaction with the side groups of residues CD1 Phe, E11 Val, G7 Val, B9 Phe, and B10 Phe.

A very interesting observation has been made that in acetate-ferri-Lg there is a small displacement of the iron atom from the porphyrin-ring plane towards the distal His (but not proximal His, as in the majority of proteins of this type). This displacement has the value 0.05–0.10 Å.† The emergence of the Fe atom on the other side of the porphyrin ring is, apparently, defined by a relatively weak bond to the proximal histidine, and strong bond to the O_1 atom of the acetate group (*trans*-influence of the axial ligand). In fact, the $Fe-N_{(prox)}$ distance in other globins is within the 2.0–2.2 Å range (Perutz 1976), whereas in our case it is equal to 2.44 Å. As for the (acetate) $Fe-O_1$ distance, it appears to be 2.11 Å.

The displacement of Fe towards the distal side due to strong interaction with the sixth ligand which we found in Lg was observed in some other cases, e.g. a shift of Fe towards distal tyrosine in abnormal Nb-Boston (Pulsinelli, Perutz, and Nagel 1973).

As is known, M. Perutz explained the position of iron in or out of the porphyrin-ring plane taking into account the spin state of this atom. According to Ehrenberg and Ellfolk's data (1963) ferri-acetate of Lg from soya is a high-spin complex. But here the iron atom is trivalent, Fe^{3+}, and not bivalent, Fe^{2+}. Therefore in the high-spin (HS) state, on the one hand, the d-shell is larger as compared with the low-spin (LS) state: in other words, the radius of the Fe atom (LS) is smaller than that of Fe (HS). This fact explains the Perutz trigger mechanism. On the other hand, however, in the d-shell of Fe^{3+} there are five electrons, but not six as in Fe^{2+}; therefore the radius of iron decreases. Such a dependence on spin and charge state is known for the atoms of iron (and other transition metals) in inorganic compounds. Here the radii are as follows: Fe^{2+} (HS) 0.77 Å, Fe^{2+} (LS) 0.61 Å, Fe^{3+} (HS) 0.65 Å, Fe^{3+} (LS) 0.55 Å (Shannon and Prewitt 1969). It goes without saying that these radii cannot be applied mechanically to the estimation of Fe—N distances of pyrrole atoms of porphyrin-ring nitrogen, but they do suggest the qualitative conclusion that the size of Fe^{3+} (HS) almost equals (and only slightly exceeds) the size of Fe^{2+} (LS). Thus in our case the presence of Fe^{3+} (HS) in or near the porphyrin-ring plane is quite possible. It should be noted that the ring is slightly strained. Actually, the refinement shows that the haem is not absolutely flat. The deviations of atoms on the periphery of the porphyrin core from the plane do not exceed 0.1–0.2 Å and may be described as the formation of a surface of double curvature.

† In constrained refinement the value is 0.10 Å, in restrained refinement 0.05 Å.

Thus a small displacement of iron towards the distal side, the large size of the haem pocket on this side, as well as the weakening of the Fe—$N_{\varepsilon(prox)}$ bond are the stereochemical factors that make it possible to understand the high affinity of Lg for oxygen. The final conclusion, however, may be made only after the investigation of the structure of oxy- and deoxy- forms of this protein.

Lg–nicotinic acid complex

Appleby, Wittenberg, and Wittenberg (1973) showed that nicotinic acid can be bonded to Lg. They pointed out that nicotinic acid

may be considered as a natural ligand of Lg with especially high affinity for its met-form. Nicotinic acid lowers the affinity of oxygen for Lg. This observation was one of the first indications of the large size of the haem pocket.

As may be seen from Table 29.2, the crystals of the Lg + NA complex are almost isomorphous with those of Lg + AA. The presence of a heavy-atom derivative with 3Cl Hg-pyridine, which is close stereochemically to nicotinic acid, suggested a way for nicotinic acid to enter into Lg. Owing to limited isomorphism we could calculate the difference map (Lg + NA) − (Lg + AA) only at 5 Å resolution. This map showed that there occurred conformational

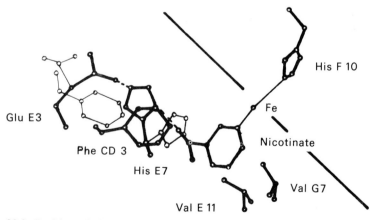

FIG. 29.8. Position of nicotinate and some other groups near the haem (solid lines). Position of E7 His, CD3 Phe in acetate-form (in the absence of nicotinic acid) is shown by thin lines.

changes in the Lg molecule. To carry out more accurate studies on these changes, we determined the phases for the Lg + NA complex at 2.8 Å resolution, using four heavy-atom derivatives (Table 29.2). We found that nicotinic acid actually enters into the haem pocket on the distal side of the haem and that this pocket is capable of accommodating such a bulky ligand. The nicotinic acid contacts through its nitrogen atom directly with the iron atom of the haem and occupies the site of distal histidine E7 His (Fig. 29.8). Some side groups are found to be displaced. E7 His itself shifts aside due to rotation about C_α—C_β bond and gives room for the nicotinic acid. N_δ of E7 His forms the hydrogen bond to one of the carboxyl oxygen atoms of nicotinic acid. The second hydrogen bond is formed between the second oxygen atom of nicotinate and the oxygen atom of E11 Val of the main chain. The other contacts of nicotinic acid are: B9 Phe, B10 Phe, CD1 Phe, CD3 Phe, E11 Val, A7 Val. The entrance of nicotinate into the haem pocket gives rise to some shift of the haem itself and the helices adjoining it (Harutyunyan, Tovlis, Czelenko, Vozonova, Nekzasov, and Vainshtein 1980).

Structure of Lg I crystals

As may be seen from Table 29.2, the orthorhombic crystals of Lg I have a unit cell whose a and c parameters are close to the parameters of the Lg II monoclinic cell. The b parameter (and the whole volume) is doubled (Harutyunyan, Kuranova, Grebenko, and Voronova 1977). Therefore the structure of Lg I may be regarded as the result of twinning of the unit cell of Lg II (Fig. 29.9). The symmetry increases owing to the appearance of twofold axes between the layers of Lg II molecules, i.e. between 'ex'-monoclinic cells.

The positions of molecules in the structure of Lg I were determined by two methods. One of them is the non-local search (Vainshtein, Gelfand, Kayushina, and Fedorov 1963) earlier applied only to small organic molecules. This method involves description of the molecule in terms of a small number of parameters. At 5 Å resolution this can be done by approximating the molecule as a system of continuous cylindrical rods of electron density similar to Fig. 29.2 (Borovikov, Vainshtein, Gelfand, and Kalinin 1979). Then such a molecule (with some freedom of rod mobility) is placed into the orthorhombic cell; the structure factors F are calculated for it and the R-type (reliability factor) function is minimized. This gives the position and orientation of the molecule. The R-value obtained for such a rough model at 5 Å resolution is 29 per cent.

In addition, utilization of data at 2.8 Å resolution made it possible to find the position of the molecule by the method of molecular replacement. The results of the two methods turned out to be almost the same. The centre of gravity of the Lg molecule in the orthorhombic cell has the following co-ordinates: $x = 25.9$, $y = 11.95$, $z = 19.0$ Å.

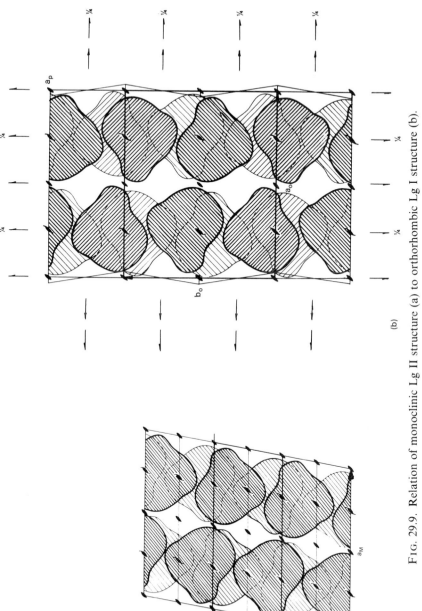

Fig. 29.9. Relation of monoclinic Lg II structure (a) to orthorhombic Lg I structure (b).

Conclusion

X-ray structure analysis of the structure of the plant globin lupin leghaemo-globin has revealed its homology to other proteins of the myoglobin–haemo-globin family. However, a number of peculiarities in the structure of the Lg molecule have been found, especially in the surroundings of the haem, that allow us to suggest an explanation of some properties of this protein: its high affinity for oxygen, its ability to bind large ligands.

Acknowledgements

This work has been carried out in the Laboratory of Protein Structure, Institute of Crystallography of the Academy of Sciences of the USSR by the author with co-workers: E. H. Harutyunyan, I. P. Kuranova, V. V. Borisov, N. I. Sosfenov, Yu. V. Nekrasov, A. I. Grebenko, and N. V. Konareva. The structure refinement of Lg II and the localization of the molecule in Lg I by the molecular replacement method were performed in collaboration with W. Steigemann of the Laboratory of Protein Structure, Max-Planck Institut für Biochemie, Munich.

References

APPLEBY, C. A. (1974). In *The biology of nitrogen fixation*, Ch. 11, 5. North Holland, Amsterdam.
——WITTENBERG, B. A., and WITTENBERG, J. B. (1973). *Proc. natn. Acad. Sci. U.S.A.* **70**, 564.
BOROVIKOV, V. A., VAINSHTEIN, B. K., GELFAND, I. M., and KALININ, D. I. (1979). *Kristallografiya* **24**, 227.
DEISENHOFER, J. and STEIGEMANN, W. (1975). *Acta crystallogr.* **B31**, 238–50.
DIAMOND, R. (1971). *Acta crystallogr.* **A27**, 436.
EGOROV, TS. A., FEIGINA, M. YA., KAZAKOV, V. K., SHAKHPARONOV, M. I., MITALEVA, S. I., OVCHINNIKOV, YU. A. (1976). *Bioorgan. Chem.* **2**, 125.
——KAZAKOV, V. K., SHAKHPARONOV, M. I., FEIGINA, M. YU., and KOSTETSKY, P. V. (1978). *Bioorgan. Chem.* **4**, 476.
EHRENBERG, A. and ELLFOLK, N. (1963). *Acta chem. scand.* **17**, S343.
ELLFOLK, N. (1972). *Endeavour* **31**, 139.
——and SIEVERS, G. (1972). *Acta chem. scand.* **26**, 1155.
HARUTYUNYAN, E. H., KURANOVA, I. P., GREBENKO, A. I., and VORONOVA, A. A. (1977). *Kristallografiya* **22**, 364; *Soviet Phys. Crystallogr.* **22**, 362.
————VAINSHTEIN, B. K., and STEIGEMANN, W. (1980). *Kristallografiya* **25**, 800.
————TOVLIS, A. B., CRELENKO, A. I., VORONOVA, A. A., NEKRASOV, YU. V., and VAINSHTEIN, B. K. (1980). *Kristallografiya* **25**, 526.
HOLMQUIST, R., JUKES, T. H., and MOISE, H. (1976). *J. molec. biol.* **105**, 39.
KENDREW, J. C., DICKERSON, R. E., STRANDBERG, R. E., HART, R. A., DAVIES, D. R., PHILLIPS, D. C., and SHORE, V. C. (1960). *Nature, Lond.* **185**, 442.
KURANOVA, I. P., GREBENKO, A. I., KONAREVA, N. V., SYROMYATNIKOVA, I. F., and BARYNIN, V. V. (1976). *Biokhimiya* **41**, 1603.
LEHTOVAARA, F. and ELLFOLK, N. (1975). *Eur. J., Biochem.* **54**, 577.
PERUTZ, M. F. (1976). *Br. med. Bull.* **32**, 195.

PULSINELLI, P. D., PERUTZ, M. F., and NAGEL, R. L. (1973). *Proc. natn. Acad. Sci. U.S.A.* **70,** 3870.

RICHARDSON, M., DILWORTH, M. J., and SCAWEN, M. D. (1975). *FEBS Lett.* **51,** 33.

SHANNON, R. D. and PREWITT, C. T. (1969). *Acta crystallogr.* **B25,** 1925.

TAKANO, T. (1977). *J. mol. Biol.* **110,** 537.

VAINSHTEIN, B. K., GELFAND, I. M., KAYUSHINA, R. L., and FEDOROV, YU. R. (1963). *Dokl. Akademii nauk SSSR* **153,** 93.

—— KURANOVA, I. P., HARUTYUNYAN, E. H., and EGOROV, TS. A. (1980). *Bioorgan. Chem.* **6,** 684.

—— HARUTYUNYAN, E. H., KURANOVA, I. P., BORISOV, V. V., SOSFENOV, N. I., and PAVLOVSKY, A. G. (1978). *Kristallografiya* **23,** 517; *Soviet Phys. Crystallogr.* **23,** 287.

—— —— —— —— —— —— GREBENKO, A. I., and KONAREVA, N. V. (1974a). *Kristallografiya* **19,** 963; *Soviet Phys. Crystallogr.* **19,** 598.

—— —— —— —— —— —— —— —— (1974b). *Kristallografiya* **19,** 971; *Soviet Phys. Crystallogr.* **19,** 602 (1975).

—— —— —— —— —— —— —— (1975). *Nature, Lond.* **254,** 163–4.

VIRTANEN, A. I. (1947). *Biol. Rev.* **23,** 239.

VUK-PAVLOVIC, S., BENKO, B., MARICIC, S., LAHAJNAR, G., KURANOVA, I. P., and VAINSHTEIN, B. K. (1976). *Int. J. Peptide Protein Res.* **8,** 427.

WITTENBERG, J. B., APPLEBY, C. A., and WITTENBERG, B. A. (1972). *J. biol. Chem.* **247,** 527.

ZHIZNEVSKAYA, G. I. (1972). *Copper, molybdenum and iron in the nitrogen metabolism of legume plants.* Science, Moscow.

30. Relationships between dehydrogenase structures

MARGARET J. ADAMS, IAN G. ARCHIBALD,
JOHN R. HELLIWELL, SUSAN E. JENKINS,
AND STEPHEN W. WHITE

A LARGE number of redox reactions of biological importance use nicotina-mide adenine nucleotide (NAD) or nicotinamide adenine nucleotide phos-phate (NADP) [I] as coenzyme.

$$R = H \quad NAD^+$$
$$R = PO_3^{2-} \quad NADP^+$$

[I]

Oxidation or reduction occurs by hydride transfer at position 4(*) of the nicotinamide ring.

The three-dimensional structures of four NAD-dependent dehydrogenases have been known for several years now, while the molecular geometry of

NADP-dependent enzymes has only been elucidated in the last two years. This paper first describes the probable chain conformation of the NADP-dependent enzyme 6-phosphogluconate dehydrogenase, and then compares it with the NAD- and the other NADP-dependent enzymes of known structure.

6-Phosphogluconate dehydrogenase

The NADP-dependent enzyme 6-phosphogluconate dehydrogenase (6-PGDH) (EC 1.1.1.44) from sheep liver was crystallized by Silverberg and Dalziel (1973) and its structure has been studied in Oxford since that time. The enzyme is an oxidative decarboxylase catalysing the conversion of 6-phosphogluconate to ribulose 5-phosphate.

$$
\begin{array}{ccc}
COO^- & NADP^+ \quad NADPH & CO_2 \\
OH & & CH_2OH \\
HO & \xrightarrow{\quad 6-PGDH \quad} & O \\
OH & & OH \\
OH & & OH \\
CH_2OPO_3^{2-} & & CH_2OPO_3^{2-}
\end{array}
$$

It is dimeric with a subunit molecular weight 50 000. The enzyme crystallizes from ammonium sulphate in space group $C222_1$, $a=72.7$ Å, $b=148.2$ Å, $c=102.9$ Å with a single subunit in the asymmetric unit. The dimer contains two subunits related by the crystallographic twofold axis parallel to b.

A study of the enzyme at 6 Å resolution using the method of multiple isomorphous replacement has been made by Adams, Helliwell, and Bugg (1977). We have recently calculated an electron-density map containing data to 2.7 Å resolution. The data were collected, using an Arndt–Wonacott rotation camera, from the native enzyme and two heavy-atom derivatives, potassium dicyanoaurate (I), $KAu(CN)_2$, and potassium tetracyanoplatinate (II), $K_2Pt(CN)_4$. Data for a further derivative, diamminedichloroplatinum (II), were collected to 6 Å resolution on the four-circle diffractometer. Phases obtained from the two-derivative set at 2.7 Å were combined with 6 Å resolution three-derivative phases by the method of Hendrickson and Lattman (1970). Parameters for the heavy-atom compounds are given in Table 30.1. The overall mean figure of merit was 42 per cent.

From the resulting best-electron-density map, the course of the α-carbon chain could be followed. There are two or three places where the density is very weak and connectivity was inferred, but the secondary structural units are unambiguous. A representation of the chain trace is shown in Fig. 30.1;

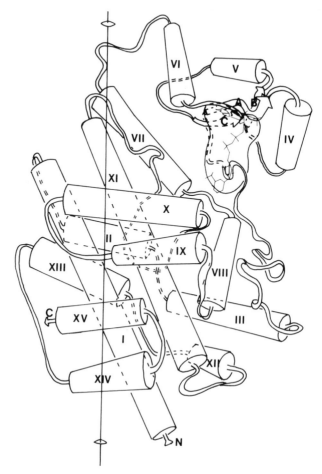

Fig. 30.1. Chain trace of 6PGDH viewed (a) down *a* and (b) (*opposite*) up *c*. The coenzyme is indicated in (a).

the normal convention of arrows for β-sheet strands and barrels for α-helices has been followed.

It is immediately apparent that the main elements of secondary structure are α-helices. The 15 helices account for 60 per cent of the α-carbon residues and two of them are particularly long, helix I—36 residues and helix XI—32 residues; they run antiparallel with axes 17° from one another. There is very little β-sheet. A small, three-stranded parallel sheet is connected by helices IV and V. The subunit has two domains. A small one, containing residues 90–220, has helices IV, V, VI, and VII and the small parallel sheet. The large domain, which includes residues 1–90 and 220–437, contains the

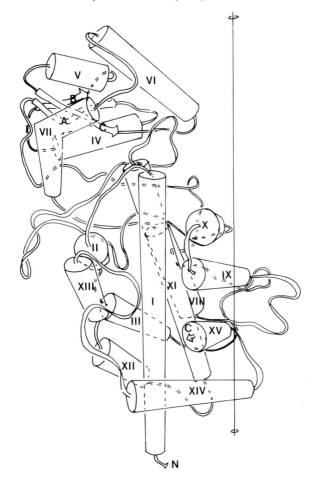

(b)

eleven remaining helices, eight of which partly wrap around the two long helices. Subunit contacts are made by one face of the helical domain.

The position of the coenzyme, the binding of which has been studied at 6 Å resolution (Adams, Archibald, and Helliwell 1977; Abdallah *et al.* 1979), is shown in Fig. 30.1(a). The adenine is at the C-terminus of two sheet strands, the ribose 2′-phosphate is fairly close to the N-terminus of helix IV, and the nicotinamide is at the junction of the large and small domains. Possible conformations of the coenzyme have been explored and the one compatible with preferred dihedral angles, giving a reasonable fit to the electron density, is slightly less open than the NAD^+ conformation in the lactate dehydrogenase ternary complex (Chandrasekhar, McPherson, Adams, and Rossmann 1973). The enzyme is being sequenced by Dr A. Carne in Cambridge, but this work is not complete and no attempt has yet been made to fit peptides

TABLE 30.1

Heavy atom parameters for 6PGDH structure determination

Resolution (Å)	Compound	Site	Occupancy	x	y	z	B	RMS F_H†	RMS E
2.7	$K_2Pt(CN)_4$	A	0.54	0.3266	0.0882	0.4495	1.5	198	123
		B	0.22	0.0225	0.1168	0.2353	0		
		C	0.33	0.1196	0.2026	0.4994	−7.0		
2.7	$KAu(CN)_2$	1	0.30	0.2273	0.0759	0.2321	−7.6	167	128
		2	0.39	0.4927	0.2264	0.1479	−15.0		
6	$K_2Pt(CN)_4$	A	0.54	0.333	0.088	0.445	(15)‡	247	138
		B	0.41	0.030	0.115	0.237	(15)		
		C	0.48	0.276	0.206	0.503	(15)		
6	$KAu(CN)_2$	1	0.76	0.229	0.079	0.229	(15)	238	159
		2	0.38	0.497	0.223	0.143	(15)		
6	$Pt(NH_3)_2Cl_2$	1	0.67	0.264	0.165	0.423	(15)	240	192
		2	0.47	0.337	0.104	0.210	(15)		

Overall mean figure of merit = 0.42.

† RMS F_H = root mean square value of calculated heavy-atom structure factor $|F_H|$; RMS E = root mean square value of lack of closure error ($|F_{HP}| - |F_H + F_P|$) where F_{HP} = structure factor of heavy atom derivative and F_P = structure factor of native protein.

‡ Values in parentheses are not refined.

to the electron density. It is unfortunate that the interpretation of the electron density was most difficult in the coenzyme binding region.

A comparison of the NAD-dependent dehydrogenases

The structures of the enzymes lactate dehydrogenase (LDH) (White *et al.* 1976), soluble malate dehydrogenase (MDH) (Hill, Tsernoglou, Webb, and Banaszak 1972), alcohol dehydrogenase from horse liver (LADH) (Eklund *et al.* 1976), and glyceraldehyde phosphate dehydrogenase (GAPDH) from both lobster and *Bacillus stearothermophilus* (Beuhner, Ford, Moras, Olsen, and Rossman 1974; Biesecker, Harris, Thierry, Walker, and Wonacott 1977) have been described at high resolution. The structures of these enzymes have been compared and the binding of NAD⁺ (ADP ribose for LADH) has been studied (Rossmann, Liljas, Brändèn, and Banaszak 1975; Chandrasekhar *et al.* 1973; Webb, Hill, and Banaszak 1973; Biesecker *et al.* 1977; Eklund, Nordström, Zeppezauer, Söderlund, Boiwe, and Brändèn 1974). The enzymes all transfer hydride directly between substrate and coenzyme. The molecular weights and subunit structures are shown in Table 30.2 and the fold of the subunits in Fig. 30.2. The enzyme subunits can each be described as having two domains, a coenzyme binding domain and a catalytic domain, indicated in Fig. 30.2. The similar structures of the coenzyme binding domains has been described previously; all show the same right-handed $\beta\alpha\beta$ connections described by Sternberg and Thornton (1976) and the same

TABLE 30.2

Molecular weights, subunit structures, and coenzymes for some dehydrogenases of known structure

Enzyme	EC no.	Subunit MW	Subunit structure	Coenzyme
Lactate dehydrogenase (LDH)	1.1.1.27	35 000	Tetramer	NAD
s-Malate dehydrogenase (s-MDH)	1.1.1.37	36 000	Dimer	NAD
Glyceraldehyde phosphate dehydrogenase (GAPDH)	1.2.1.1	36 000	Tetramer	NAD
Liver alcohol dehydrogenase (LADH)	1.1.1.1	40 000	Dimer	NAD
Dihydrofolate reductase (DHFR)	1.5.1.3	20 000	Monomer	NADP
Glutathione reductase (GR)	1.6.4.2	50 000	Dimer	NADP, FAD
p-Hydroxybenzoate hydroxylate (p-HBH)	1.14.13.2	43 000	(Dimer)	NADP, FAD
6-Phosphogluconate dehydrogenase (6PGDH)	1.1.1.44	47 000	Dimer	NADP

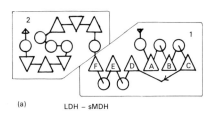

(a)   LDH – sMDH

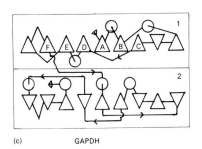

(c)   GAPDH

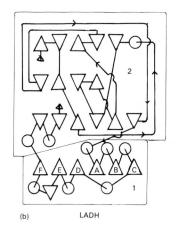

(b)   LADH

FIG. 30.2. Fold of (a) lactate dehydrogenase and *s*-malate dehydrogenase, (b) liver alcohol dehydrogenase, and (c) glyceraldehyde phosphate dehydrogenase. 1 = coenzyme binding domain, 2 = catalytic domain. These diagrams taken from Sternberg and Thornton (1977). ⊖ helix end on, ⊙ helix viewed along its length. $\Delta = \beta$ strand from N-terminus, $\nabla = \beta$ strand from C-terminus.

sequence of sheet strands. The hypothesis of divergence from a single nucleo-tide binding protein has been suggested (Rossmann *et al.* 1975; Rossmann, Moras, and Olsen 1974).

It is interesting to note that, while only a few residues are identical in the nucleotide binding sites of the three dehydrogenases for which complete sequences are known (LDH, GAPDH, and LADH), some residues that are important in the binding of coenzyme are invariant. Two glycines, one at the carboxyl end of sheet strand A (28 in LDH) and one just after strand D (99 in LDH), make close contacts with the coenzyme and are conserved, and an aspartate at the C-terminus of strand B (53 in LDH) which hydrogen bonds with the adenine ribose 2′-hydroxyl group is invariant.

The NADP-dependent enzymes

The NADP-dependent enzymes studied to a resolution of 3 Å or better are also shown in Table 30.2.

Dihydrofolate reductase (DHFR) has been solved for both *Lactobacillus casei* and *Escherichia coli* (Matthews *et al.* 1977, 1978). This is the smallest of the dehydrogenases studied so far and is monomeric. Sequences are known for several species—the protein fold is shown in Fig. 30.3. The enzyme con-sists of a single sheet with eight strands, the six N-terminal ones being parallel. There is a single domain to which both coenzyme and substrate bind. NADP binds at the C-terminus of the sheet in a manner similar to, but not the same as, NAD binding in LDH. The 2′-phosphate interacts with an arginine residue (43) at the carboxyl end of strand B, a position comparable with the invariant aspartate of the NAD-dependent dehydrogenases. The whole coen-zyme is, however, translated by one strand so that a glycine (96) at the end of strand E, which might be compared with glycine (99) of LDH, is close to the pyrophosphate and not the nicotinamide ribose. The whole coenzyme is slightly more extended than is bound NAD.

Glutathione reductase (GR) from human erythrocytes has been solved at 3 Å resolution (Schulz, Schirmer, Sachsenheimer, and Pai 1978). The enzyme is dimeric, of subunit molecular weight 50 000 (Table 30.2), and requires fla-vin adenine dinucleotide (FAD) and NADP to reduce glutathione. The fold is shown in Fig. 30.3; there are three domains, outlined in the figure; each is based on a β-sheet. Domains 1 and 2 with mainly parallel sheets bind FAD and NADP respectively at their C-termini; the N-terminal region of helix II contains a redox active disulphide bridge; the third domain forms the subunit contact region. The folds for strands A–G (domain 1) and H–N (domain 2) are similar enough to suggest gene duplication. The conformation of NADP is essentially the same as that of NAD in the NAD-dependent enzymes (Zappe, Krohne-Ehrich, and Schulz 1977), but the nicotinamide

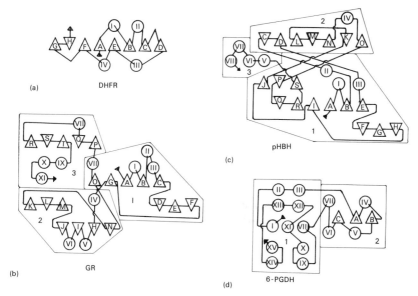

FIG. 30.3. Fold of (a) dihydrofolate reductase, (b) glutathione reductase, (c) *p*-hydroxy-benzoate hydroxylase, (d) 6-phosphogluconate dehydrogenase. Conventions are as in Fig. 30.2 and domains are indicated.

must be moved in the active enzyme in order to transfer electrons to the iso-alloxazine ring of FAD.

A second FAD, NADP-dependent enzyme, *p*-hydroxybenzoate hydroxy-lase from *Pseudomonas fluorescens* has recently been solved (Wierenga, deJong, Kalk, Hol, and Drenth 1979). The subunit molecular weight is 43 000 and it is suggested that it is dimeric in solution. There are again three domains, two based on sheet–helix interactions and a third only on helices. The fold is shown in Fig. 30.3 It is complex; domains 1 and 2 are each formed from two discontinuous parts of the polypeptide chain. Domain 3 is the inter-monomer contact region while FAD binds with the adenine and adenine ribose portions interacting with βA, αI, βB of domain 1 and the flavin in the regions between domains 1 and 2. Binding of NADP (or NADPH) has not yet been accomplished without change of conformation of the enzyme.

The FAD-binding regions of GR and *p*-HBH are similar, and it is of inter-est to note the excursion αII, αIII in GR, which contains the active cysteines, while the *p*-HBH excursion to domain 2 αII, βC, βD contains residues which bind the flavin. This fold is not repeated, however, in *p*-HBH and so the binding of NADP cannot be the same in both enzymes. The pyrophosphate of FAD interacts with the N-terminus of a helix in each enzyme in a way suggested by Hol, van Duijnen, and Berendsen (1978).

The fold of 6PGDH is also shown in Fig. 30.3 and the two domains indicated. The contrast between this enzyme and the other NADP- or NAD-dependent enzymes is immediately obvious in the lack of β-sheet. The absence of β-sheet in the large domain is unusual not only for dehydrogenases but for any enzyme this large. The only comparable structure determined so far is that for the coenzyme A-dependent enzyme citrate synthase (Wiegand, Kukla, Scholze, Jones, and Huber 1979), the 370 carboxyl terminal residues of which consist of 15 α-helices. The detailed fold of this enzyme is different from that of 6PGDH.

The dehydrogenases and divergent evolution

The similarity of the coenzyme binding regions of the NAD-dependent dehydrogenases has been used as evidence that these enzymes have diverged from a common precursor nucleotide-binding protein. It has been suggested that this might itself have been the result of gene duplication of a mononucleotide-binding protein. LDH, LADH, and GAPDH would then have arisen by gene fusion with different catalytic domains while MDH and LDH would have diverged after fusion. This explanation of the relationship of these four enzymes is still tenable.

The much wider variation of the structure of the NADP-dependent enzymes studied so far precludes extension of this idea to these enzymes. The divergence of a dinucleotide binding protein into NAD- and NADP-binding proteins and further divergence for substrate specificity of the NADP enzymes is not supported by the four enzymes studied so far.

While three of these enzymes have a predominantly parallel β-sheet of five or more strands, the adenine portion of FAD is bound in both GR and *p*-HBH. NADP in GR binds to a four-stranded sheet with strand sequence the same as the first four strands of the NAD enzymes and helices in comparable positions. There are, however, no strands five and six. While the four strands BCDE of the sheet in DHFR have the same sequence and NADP binds in this region, the helices are differently placed. The sheet of 6PGDH is vestigial and, although it is connected by anti-parallel helices, its resemblance to the other enzymes is small.

It is interesting to notice the wider range of function of the NADP enzymes as well as the more varied structures. It might perhaps be expected that different classes of enzymes will emerge, the FAD, NADP ones on the one hand and those where hydride transfer is direct to substrate on the other. However, the two enzymes known so far of the second group, 6PGDH and DHFR, are very different from each other in size, subunit composition, and predominant secondary structure, and might themselves be examples of different classes. It does not appear at present that these enzymes have even converged to bind NADP in the same structure.

The similarities of the NAD enzymes and the variety of the NADP ones should be a great stimulus to further activity. The structural requirements for NADP binding and the specificity exhibited by a large number of enzymes for NAD or NADP need elucidation. It is especially important to investigate different NAD enzymes to explore the possibility of structural variability, and more NADP enzymes to discover further examples of some of the structural classes found so far.

Acknowledgements

This work was begun in Dorothy Hodgkin's laboratory. We thank her and Professor Sir David Phillips for encouragement. We acknowledge MRC support and MRC and SRC research scholarships. M. J. A. is a fellow of Somerville College and a member of the Oxford Enzyme Group.

References

ABDALLAH, M. A., ADAMS, M. J., ARCHIBALD, I. G., BIELLMANN, J.-F., HELLIWELL, J. R., and JENKINS, S. E. (1979). *Eur. J. Biochem.* **13,** 121–30.

ADAMS, M. J., ARCHIBALD, I. G., and HELLIWELL, J. R. (1977). *Pyridine nucleotide dependent dehydrogenases* (ed. H. Sund) pp. 72–83. Walter de Gruyter, Berlin.

——HELLIWELL, J. R., and BUGG, C. E. (1977). *J. molec. Biol.* **112,** 183–97.

BIESECKER, G., HARRIS, J. I., THIERRY, J.-C., WALKER, J. E., and WONACOTT, A. J. (1977). *Nature, Lond.* **266,** 328–33.

BUEHNER, M., FORD, G. C., MORAS, D., OLSEN, K. W., and ROSSMANN, M. G. (1974). *J. molec. Biol.* **90,** 25–49.

CHANDRASEKHAR, K., MCPHERSON, A. J., ADAMS, M. J., and ROSSMANN, M. G. (1973). *J. molec. Biol.* **76,** 503–18.

EKLUND, H., NORDSTRÖM, B., ZEPPEZAUER, E., SÖDERLUND, G., BOIWE, T., and BRÄNDEN, C. I. (1974). *FEBS Lett.* **44,** 200–4.

——————OHLSONN, I., BOIWE, T., SÖDERBERG, B. O., TAPIA, O., BRÄNDEN, C. I., and ÅKESON, Å. (1976). *J. molec. Biol.* **102,** 27–59.

HENDRICKSON, W. A. and LATTMANN, E. E. (1970). *Acta crystallogr.* **B26,** 136–43.

HILL, E., TSERNOGLOU, D., WEBB, L., and BANASZAK, L. J. (1972). *J. molec. Biol.* **72,** 577–91.

HOL, W. D. J., VAN DUIJNEN, P. T., and BERENDSEN, H. J. C. (1978). *Nature, Lond.* **273,** 443–6.

MATTHEWS, D. A., ALDEN, R. A., BOLIN, J. T., FREER, S. T., XUONG, N.-H., KRAUT, J., POE, M., WILLIAMS, M., and HOOGSTEEN, K. (1977). *Science, N.Y.* **197,** 452–4.

——————FILMAN, D. J., FREER, S. T., HAMLIN, R., HOL, W. G. J., KISLIUK, R. L., PASTORE, E. J., PLANTE, L. T., XUONG, N.-H., and KRAUT, J. J. (1978). *J. biol. Chem.* **253,** 6946–54.

ROSSMANN, M. G., MORAS, D., and OLSEN, K. W. (1974). *Nature, Lond.* **250,** 194–9.

——LILJAS, A., BRÄNDEN, C. I., and BANASZAK, L. J. (1975). *The enzymes,* 3rd edn (ed. P. D. Boyer) Vol XI, pp. 61–102. Academic Press, New York.

SCHULZ, G. E., SCHIRMER, R. H., SACHSENHEIMER, W., and PAI, E. F. (1978). *Nature, Lond.* **273,** 120–4.

SILVERBERG, M. and DALZIEL, K. (1973). *Eur. J. Biochem.* **38,** 229–38.

STERNBERG, M. J. E. and THORNTON, J. M. (1976). *J. molec. Biol.* **105**, 367–82.
———(1977). *J. molec. Biol.* **110**, 269–83.
WEBB, L. E., HILL, E. J., and BANASZAK, L. J. (1973). *Biochemistry* **12**, 5101–9.
WHITE, J. L., HACKERT, M. L., BUEHNER, M., ADAMS, M. J., FORD, G. C., LENTZ, P. J., JR, SMILEY, I. E., STEINDEL, S. J., and ROSSMANN, M. G. (1976). *J. molec. Biol.* **102**, 759–79.
WIEGAND, G., KUKLA, D., SCHOLZE, H., JONES, A. T., and HUBER, R. (1979). *Eur. J. Biochem.* **93**, 41–50.
WIERENGA, R. K., DEJONG, R. J., KALK, K. H., HOL, W. G. J., and DRENTH, J. (1979). *J. molec. Biol.* **131**, 55–73.
ZAPPE, H., KROHNE-EHRICH, G., and SCHULZ, G. E. (1977). *J. molec. Biol.* **113**, 141–52.

31. Comparison of the refined structures of two sulphydryl proteases: actinidin and papain

E. N. BAKER

Introduction

COMPARATIVE studies of related proteins can be of great value in pointing out common structural features which may be of critical importance to the biological function, or differences which may explain functional differences. Such comparisons also provide a valuable insight into the features which are most important in determining the structure and stability of these molecules.

Actinidin and papain are representatives of a class of proteolytic enzymes which depend on a free sulphydryl group for their catalytic activity. Such enzymes are found in plants (e.g. papain, actinidin, ficin, bromelain), bacteria (e.g. streptococcal proteinase), and animals (e.g. cathepsins B1 and B2), and appear to share many common features (for reviews see Glazer and Smith 1971; Lowe 1976). Extensive chemical and physical studies have been carried out on papain, and its full amino-acid sequence (Mitchell, Chaiken, and Smith 1970) and three-dimensional structure (Drenth, Jansonius, Koekoek, and Wolthers 1971) are known. Neither the sequences nor the three-dimensional structures of any of the other sulphydryl proteases were known, however, until the recent concurrent amino-acid sequence (Carne and Moore 1978) and X-ray crystallographic studies (Baker 1977) on actinidin were undertaken.

The crystallographic studies on actinidin showed that the polypeptide chain conformation was almost identical to that of papain. The amino-acid sequences could then be aligned, and on the assumption that, in actinidin, two extra residues were inserted between residues 59 and 60 (papain numbering), one extra between 78 and 79, four extra between 168 and 169, and two extra at the C-terminal end and one residue (194) deleted, a total of 110 out of 212 residues in papain (52 per cent) were found to be different in actinidin.

The purpose of this article is to comment on some effects of these changes on structure and function. Moreover, although both of the original X-ray structure analyses were at reasonably high resolution (2.8 Å), models at this resolution are still likely to contain errors or uncertainties, e.g. in the conformations of individual amino-acid residues, and in description of solvent structure, in and around the protein molecules. Both structures have therefore been refined, in preparation for a detailed comparison, and some results are presented here.

Crystallographic refinement of the structures

The same method was used to refine both structures, viz. the reciprocal-space fast-Fourier least-squares method developed by Agarwal (1978), and first used in the refinement of insulin (Isaacs and Agarwal 1978). Atoms were refined individually, with the protein geometry being regularized at intervals using the model-fitting routines of Dodson, Isaacs, and Rollett (1976). For actinidin, data to a resolution of first 2.0 Å, and later 1.7 Å, were available for the refinement, while for papain data were limited to 2.5 Å resolution (the original set collected by Drenth, Jansonius, Koekoek, Swen, and Wolthers 1968, on a linear diffractometer). Collection of higher-resolution data for papain is in progress so that the structure can be refined further. It is of interest, however, to compare the refinement done so far on papain, at 2.5 Å resolution, with that of actinidin.

The refinement of actinidin, from an initial R value of 0.429, for 2.0 Å data, to a final R value of 0.171, for 1.7 Å data, has been described fully elsewhere (Baker and Dodson 1980). The final model contained, in addition to 1657 protein atoms, a total of 163 solvent molecules (regarded as water), and had an estimated error in atomic positions of 0.1 Å. During the refinement manual intervention, via difference electron-density maps, was required at intervals, to make major corrections to the structure (two peptide units and 16 side-chains received radically revised conformations in this way, and one change to the amino-acid sequence was necessary; Asp 86→Glx 86). Many large shifts were, however, made quite automatically by the least-squares procedures, giving a mean shift in main-chain atom positions of 0.50 Å, and for side-chain atoms 0.83 Å. Individual B values for the atoms were also refined; with 1.7 Å data these assumed generally sensible values, consistent within groups of atoms, and within different parts of the structure. They also proved invaluable indicators of errors in the structure.

The initial co-ordinate set for papain, kindly provided by Professor J. Drenth, was that obtained from the original 2.8 Å electron-density map, and subsequently regularized by the model-building procedures of Diamond (1966). The progress of the refinement, from an initial R value of 0.452, through 16 cycles of xyz refinement and nine regularizations, to a final R value of 0.259 is depicted graphically in Fig. 31.1. The mean shift in main-chain atom positions was 0.59 Å, and in side-chain atom positions 0.92 Å. Throughout the refinement, 2.5 Å data (10 088 reflections) were used, and the final uncertainty in atomic positions, estimated from the size of shifts in least-squares cycles and in regularizations, is about 0.2 Å.

Compared, with actinidin, the main differences in the refinement of papain arose from the significantly lower resolution of the data, viz. differences in the rate of convergence, the size of shifts calculated in least-squares cycles, and in the treatment of B values. As expected, it also proved more difficult to rebuild suspect portions of the structure from difference maps, because of the poorer definition.

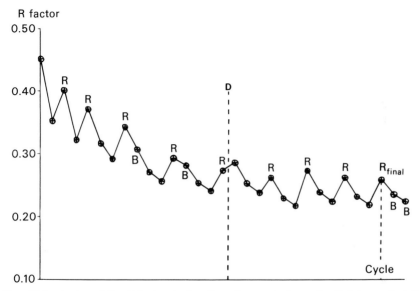

FIG. 31.1. A plot of the R factor during the refinement of papain. Points marked R represent regularizations of the structure; those marked B represent B refinement cycles; unmarked points represent *xyz* refinement cycles. A difference map, to rebuild suspect structure, was calculated at D. After the termination of refinement (R final) two B refinement cycles were calculated just to see what their effect would be.

Throughout refinement, least-squares shifts were considerably higher than those seen for actinidin at comparable R values. In the initial cycles, shifts averaged 0.5 Å (compared with 0.3 Å for actinidin) and it was necessary to proceed more cautiously, allowing no shift to exceed two times the average, and regularizing the geometry of the structure frequently (after each least-squares cycle at first). Later, shifts averaged 0.2–0.3 Å per cycle, and 0.15–0.2 Å in regularizations. This led to considerable oscillation (see Fig. 31.1). Nevertheless, the computer time involved was small (each least-squares cycle requiring only 7–8 minutes CPU time on the University of Groningen CDC Cyber 74–16 computer), there was very little manual intervention (only one difference map was calculated) and the reliability of the structure has clearly been greatly improved.

An interesting aspect of the refinement was the treatment of B values. Initially, B values of $15.0\,\text{Å}^2$ (slightly less than the overall value of $18.0\,\text{Å}^2$ obtained from a Wilson plot) were assumed for all atoms, and were kept fixed. Later (see Fig. 31.1), individual B values were refined, mainly to identify any badly misplaced portions of the structure. All groups containing any atom with $B > 25\,\text{Å}^2$ (42 side-chains and 7 peptide units) were then omitted from the model and scrutinized in a difference map. Revised conformations were given to 17 side-chains and 2 peptide units (one of the latter rotated through 180°); other groups simply had weak density (so that the high B values were understandable); for others, however, the B values appeared to be high because the groups lay on the edge of, or outside, good density, and needed to be bodily shifted into it, i.e. they had not yet refined to their correct positions. In retrospect, therefore, although the refinement of B values at a resolution of $\sim 2.5\,\text{Å}$ can usefully identify wrongly placed structure, such refinement should not be begun too early. If the xyz refinement has not converged, when refinement of B values is begun, the latter may be misleading, and further refinement of the atomic positions may be inhibited.

It was also clear that, at $2.5\,\text{Å}$ resolution, B values were somewhat inconsistent (e.g. varying considerably within aromatic rings). They were therefore returned to overall values of $12.5\,\text{Å}^2$ for protein atoms and $20\,\text{Å}^2$ for solvent molecules (this increased the value of R by about 0.03), and will be refined when higher-resolution data are available. The final model, with a regularized protein structure (1655 atoms) and 57 solvent molecules (taken as water), gave an R value of 0.259.

Comparison of actinidin and papain

In the following discussion residue numbers refer to the numbering for papain, unless otherwise indicated.

Polypeptide chain conformations

Both structures are organized in the same way (see Fig. 31.2, where the polypeptide chain conformations of the two are superimposed). Both are binuclear molecules, with roughly the first 100 residues (18–112) folded up to form Domain 1, and the second 100 residues (113-207) forming Domain 2. The amino-terminal end (residues 1–17) is closely associated with Domain 2, and the carboxy-terminal end (residues 208–212) with Domain 1, i.e. each crosses over into the other domain, to act as 'straps' in helping to hold the two halves of the molecule together. Domain 1 is largely helical, as far as secondary structure is concerned, with three significant lengths of α-helix, residues 25–42, 50–56, and 67–78, while Domain 2 is based on a substantial, if irregular, piece of twisted β-structure. (See the original articles on papain (Drenth *et al.* 1971) and actinidin (Baker 1977) for more detail.)

To obtain a quantitative comparison of the folding in the two molecules, the co-ordinates for papain were transformed so that the main-chain atoms of papain could be superimposed on those of actinidin. The positions of three equivalent pairs of α-carbon atoms (chosen by reference to stereo diagrams) were used to obtain a preliminary transformation matrix. A more accurate transformation matrix was then calculated using a non-linear least-squares method based on that of Rao and Rossmann (1973). The structures proved to be so similar that the transformation was the same whether all main-chain

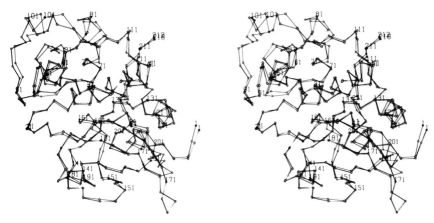

FIG. 31.2. Stereo plot of the polypeptide chain conformations (Cα positions), of the two proteins superimposed. Actinidin is represented by double lines, papain by single lines.

atoms (Cα, C, O, N) or just Cα atoms were used. (Positions determined in the two ways differed on average by less than 0.02 Å.) The following figures have been obtained using a transformation matrix obtained from the positions of all main-chain atoms (Cα, C, O, N), excluding those of residues which are insertions in one molecule or the other, and excluding residues 97–101, whose conformation is radically different in the two proteins, i.e. a total of 824 out of 848 main-chain atoms were used.

The root mean square difference between atomic positions in the two proteins (for 824 out of 848 main-chain atoms) is 0.65 Å, indicating a very striking similarity. Moreover, the median difference is even less, 0.48 Å. Considering that the standard deviations in atomic positions are of the order of 0.1 Å (actinidin) and 0.2 Å (papain), the two polypeptide chain conformations can be regarded as identical along most of their length. This is in spite of the fact that 52 per cent of the amino-acid side-chains are different. A distribution of differences (in Cα positions) between actinidin and papain is shown in

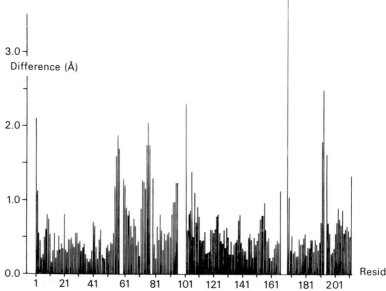

F IG. 31.3. Distribution of differences between actinidin and papain main-chain atom positions, after superimposing one on the other. Residue numbers refer to papain. Gaps represent positions at which extra residues are inserted in actinidin (59–60, 78–79, 168–169) or in papain (194). There is also a gap at the one position (97–101) where the conformation is grossly different.

Fig. 31.3. Apart from the difference at residues 97–101, noted earlier, the main differences are clearly at the two chain termini (residues 1–2 and 212) and around each of the positions where insertions or deletions occur. Generally, however, the insertions and deletions only affect the conformations of residues two or three on either side.

Since the refinement of papain is not yet complete, a more detailed analysis of the similarities and differences is not warranted at this stage. However, it is of note that when the refined structure of actinidin was compared with the unrefined structure of papain the r.m.s. difference was 0.95 Å. After eight cycles of refinement on papain this was reduced to 0.70 Å, and on convergence (16 cycles) 0.65 Å. This tends to confirm the validity of the refinement done on papain, even with limited resolution data, and it also suggests that for other pairs of related proteins, in which the differences have appeared to be greater, refinement may result in the reduction or even disappearance of apparent differences.

Distribution of amino-acid sequence changes

Although overall 48 per cent of the amino-acid residues are the same in actinidin and papain, this figure disguises the way in which changes are distributed.

In some parts of the structure there is a very high degree of homology. For example, in the central α-helix, where residues 25–40 contribute side-chains to the hydrophobic core of Domain 1, and to the interface between Domains 1 and 2, 12 out of 16 of the side-chains (75 per cent) are the same in the two proteins. Likewise, in the N-terminal 'strap', residues 1–17, 10 are the same (60 per cent), and amongst the residues which make up the twisted β-structure of Domain 2, and at the same time contribute side-chains to its hydrophobic core, 65 per cent are the same. Thus regions which provide specific interactions important to the stability of the structure are highly conserved.

On the other hand, residues 79–127, which form a long piece of extended chain, stretching over the surface of Domain 1, across to Domain 2, and culminating in an α-helix on the outside of Domain 2, include only 11 which are the same (23 per cent). This part of the chain makes very few specific internal interactions, and a large amount of variation is apparently possible. It contains many of the substitutions responsible for the low isoelectric point of actinidin (3.1), compared with that of papain (8.75)—in actinidin there are one basic and nine acidic side-chains in this section, and in papain seven basic and two acidic side-chains. This very variable region also includes the one substantial difference in conformation (apart from insertions or deletions), viz. residues 97–101. In papain, $O\gamma$ of Ser 97 makes a hydrogen bond to an internal carboxyl group, that of Glu 52. The corresponding residue in actinidin (100) is Val, so that such an interaction is not possible. Rearrangement of the polypeptide chain occurs and Glu 52 hydrogen-bonds instead to a buried water molecule. Other interactions of a less specific nature probably contribute to this change, but this seems the most obvious explanation, and the rearrangement certainly has no effect on the rest of the molecule.

At the interface between the two domains most of the side-chains are the same. In fact, out of 24 side-chains which contribute to this interface, only six are different in actinidin and papain. Furthermore, two of these, Val 32 and Ala 162, which in papain are adjacent across the interface, become Ala 32 and Val 165 in actinidin—an example of a co-ordinated, or compensating, change. A striking feature of this interface, seen in both proteins, is a group of four charged side-chains, two from Domain 1 (Glu 35 and Glu 50) and two from Domain 2 (Lys 17 and Lys 174), all of them partly or completely buried in the molecular interior. In actinidin, refinement of the structure showed that these hydrophilic groups are associated with a network of eight internal water molecules extending along the interface. Although the refinement of papain is not yet sufficiently advanced to identify solvent molecules with certainty, at least five of them appear to be present in papain in very similar positions.

Somewhat surprisingly, there does not seem to be such a high degree of conservatism in the hydrophobic cores of the two domains. As can be seen

Comparison of actinidin and papain

from Table 31.1, only five out of 14 of the side-chains which make up the core of Domain 1 are the same, and seven out of 15 in Domain 2. Many of the changes are to very similar side-chains (e.g. Ile→Val), and there is at least one more example of a co-ordinated change involving two side-chains in contact (Ile 34 and Val 75 in papain become Val 34 and Ile 77 in actinidin). Although there are many van der Waals contacts of 3.5–4.5 Å between these internal, non-polar side-chains, 'holes' also exist and the overall impression is that the packing is not uniformly efficient. For example, in actinidin there is a 'hole' ~ 5 Å in diameter at the centre of the non-polar core of Domain 2 (flanked by the side-chains of Leu 137, Val 180, Met 194, Ile 196, Ile 208),

TABLE 31.1

Residues in the non-polar cores of papain and actinidin

Domain 1				Domain 2			
Papain		Actinidin		Papain		Actinidin	
Trp 26	†	Trp	26	Leu 121	†	Leu	124
Ala 30	†	Ala	30	Ile 125		Val	128
Val 31		Ile	31	Val 130	†	Val	133
Ile 34		Val	34	Val 132	†	Val	135
Tyr 48		Leu	48	Leu 134	†	Leu	137
Leu 53	†	Leu	53	Phe 141	†	Phe	144
Ala 71		Gly	73	Val 161		Ile	164
Leu 72		Phe	74	Ala 163		Ile	166
Leu 74		Phe	76	Ile 171		Trp	178
Val 75		Ile	77	Ile 173		Val	180
Ile 80	†	Ile	83	Trp 181	†	Trp	188
Ala 105		Val	108	Ile 187		Met	194
Thr 107		Ile	110	Ile 189	†	Ile	196
Pro 209	†	Pro	215	Leu 202		Ile	208
				Tyr 203		Ala	209

† Residues identical in both proteins.

and a similar one at the centre of Domain 1. These probably have the result that amino-acid substitutions can be accommodated without affecting the main-chain conformation.

The active site

The broad similarity between the catalytic sites of actinidin and papain has been remarked on before (Baker 1977). After refinement of the structures it can be seen (Fig. 31.4) that not only are the same residues present, but their relative orientations are almost identical. This is in spite of the fact that the two proteins were crystallized from very different media (papain from a methanol–water solution containing 60–70 per cent w/v methanol, and actinidin from 20 per cent saturated ammonium sulphate solution) and at different pH values (papain 9.3, actinidin 5.9). Thus the observed geometry of the active site does not seem to be affected at all by the surrounding medium or the crystal environment.

The two residues believed to be directly involved in catalysis, Cys 25 and His 159, lie adjacent, with the sulphur atom ($S\gamma$) of Cys 25 close to one nitrogen atom ($N\delta1$) of His 159 (3.6 Å away in papain, 3.3 Å away in actinidin). The other nitrogen atom ($N\varepsilon2$) of His 159 forms a hydrogen bond (2.8 Å in papain, 2.8 Å in actinidin) with the side-chain amide oxygen atom of the buried asparagine, Asn 175. The main role of this hydrogen bond is probably to give the appropriate orientation to the imidazole ring of His 159. Similar orientating interactions appear to be commonly found for active site histidine residues in other enzymes. For example, Argos, Garavito, Eventoff, Rossmann, and Branden (1978) have pointed out that where His is a ligand for zinc in zinc-containing enzymes, it is always hydrogen-bonded to an 'orienter' (e.g. Asp, Glu, Asn, Gln, Ser). Another example is seen in the serine proteases, where the catalytic triad Ser----His----Asp (internal) is very similar to that of Cys----His----Asn (internal) found here in the sulphydryl proteases. While the Asp----His interaction in the former is usually seen in terms of its role in the proposed charge-relay system (Blow, Birktoft, and Hartley 1969), the analogy with other enzymes suggests its orientational role may be at least as important.

Another point of interest concerns the orientation of $S\gamma$ of Cys 25. For maximum interaction with $N\delta1$ of His 159 it would be expected to lie in the plane of the imidazole ring, yet in both actinidin and papain it lies some distance out of this plane (2.7 Å in actinidin, 2.3 Å in papain). Although this is complicated by the fact that in both proteins the sulphydryl group is partially oxidized, it is interesting to note (Matthews, Alden, Birktoft, Freer, and Kraut 1977; Brayer, Delbaere, and James 1978) that in the serine proteases the normal position for O_γ of the active serine residue also appears to be out of the imidazole plane, some 1.5–2.5 Å away from an ideal hydrogen-bonding position.

While the involvement of Cys and His residues in catalysis by the sulphydryl proteases is generally agreed upon, some kinetic studies continue to implicate a carboxyl group as well (Zannis and Kirsch 1978). The only acidic residue near the active site in both proteins is Asp 158, whose carboxyl group is ∼ 6 Å away from the imidazole ring of His 159. As can be seen in Fig. 31.4, however, it has exactly the same orientation in both proteins, turned away from the active site to hydrogen-bond to the main-chain NH group of residue 136. Thus it is probably too far from the catalytic site, and with the wrong orientation, for any direct role in catalysis. Suggestions that Asp 158 may be involved in catalysis via a bound water molecule linking it to His 159 (Allen, Stewart, Johnson, and Wettlaufer 1978), or that His 159 undergoes a pH-dependent conformational change, forming an ion pair with

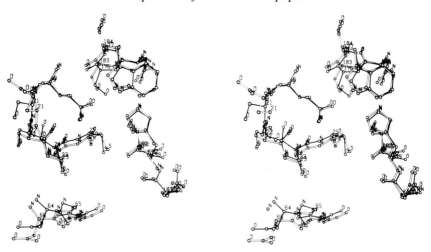

FIG. 31.4. Residues in the active sites of papain (single line) and actinidin (double line) superimposed. The catalytically important residues are Cys 25 and His 159 (162 in actinidin); Gln 19 is implicated in substrate binding; the ring of Trp 178 (184) covers the hydrogen bond between His 159 (162) and Asn 175 (182); and Asp 158 (161), sometimes implicated in proposed mechanisms, is at bottom right.

Asp 158 at neutral pH (Angelides and Fink 1978) are also not supported by the structures. No bound water molecule is seen in this area in actinidin (though many are found in other parts of the structure) nor is there any obvious water molecule hereabouts in papain at the present stage of refinement. Likewise the active site structure of actinidin at pH 5.9 is the same as that of papain at pH 9.3.

Other similarities in the active sites of the two proteins make it clear that they share the same mode of substrate binding as well as catalysis. The 'oxanion hole', formed by the $-NH_2$ group of the side-chain of Gln 19 and the main-chain NH of residue 25, is the same in both proteins, and main-chain C=O and NH groups on either side of the active site have the same orientations. The rings of Trp 177, which covers the His----Asn hydrogen bond, and His 159, also have the same orientation in both proteins, differing by only 5–6°. In both the imidazole ring is in the position shown by Drenth, Kalk, and Swen 1976 to be optimal for transfer of a proton to the leaving group of a substrate.

Conclusions

It is clear that there is a remarkable similarity between the two protein structures, which probably extends to other enzymes of this type. Further refinement of papain should allow a proper comparison of the solvent structure in and around the molecules, and it is hoped that the refined structures can be used to interpret further binding studies and gain a more detailed understanding of the catalytic process.

Acknowledgements

I am very grateful to the many people who helped with this work; especially to Professor Jan Drenth for his constant interest and support, and to Drs Wim Hol and Jan-Harm Ploegman, and other colleagues in the Laboratory for Structural Chemistry, University of Groningen, for their help during a very pleasant three-month stay in Groningen, during which most of this work was done; to Mrs Eleanor Dodson and Dr Guy Dodson of the University of York for their assistance in the refinement; to Professors Dick Batt and Geoff Malcolm and others at Massey University, for supporting, and encouraging the establishment of, protein crystallography at Massey University, to Drs Chris Moore and Alan Carne for enjoyable collaboration during the sequence and X-ray analysis of actinidin; to the Massey University Council for granting study leave and supporting my visit to York and Groningen; and last but not least to Professor Dorothy Hodgkin who first imbued me with the determination to undertake protein crystallography in New Zealand.

References

AGARWAL, R. C. (1978). *Acta crystallogr.* **A34**, 791–809.

ALLEN, K. G. D., STEWART, J. A., JOHNSON, P. E., and WETTLAUFER, D. G. (1978). *Eur. J. Biochem.* **87**, 575–82.

ANGELIDES, K. J. and FINK, A. L. (1978). *Biochemistry* **17**, 2659–68.

ARGOS, P., GARAVITO, R. M., EVENTOFF, W., ROSSMANN, M. G., and BRÄNDEN, C. I. (1978). *J. molec. Biol.* **126**, 141–58.

BAKER, E. N. (1977). *J. molec. Biol.* **115**, 263–77.

—— and DODSON, E. J. (1980). *Acta crystallogr.* (in press).

BLOW, D. M., BIRKTOFT, J. J., and HARTLEY, B. S. (1969). *Nature, Lond.* **221**, 337–40.

BRAYER, G. D., DELBAERE, L. T. J., and JAMES, M. N. G. (1978). *J. molec. Biol.* **124**, 261–83.

CARNE, A. and MOORE, C. H. (1978). *Biochem. J.* **173**, 73–83.

DIAMOND, R. (1966). *Acta crystallogr.* **21**, 253–66.

DODSON, E. J., ISAACS, N. W., and ROLLETT, J. S. (1976). *Acta crystallogr.* **A32**, 311–15.

DRENTH, J., KALK, K. H., and SWEN, H. M. (1976). *Biochemistry* **15**, 3731–8.

—— JANSONIUS, J. N., KOEKOEK, R., and WOLTHERS, B. G. (1971). *Adv. Protein Chem.* **25**, 79–115.

———— SWEN, H. M., and WOLTHERS, B. G. (1968). *Nature, Lond.* **218**, 929–32.

GLAZER, A. N. and SMITH, E. L. (1971). *The enzymes*, 3rd edn (ed. P. D. Boyer) Vol. 3, pp. 501–46. Academic Press, New York.

ISAACS, N. W. and AGARWAL, R. C. (1978). *Acta crystallogr.* **A34**, 782–91.

LOWE, G. (1976). *Tetrahedron* **32**, 291–302.

MATTHEWS, D. A., ALDEN, R. A., BIRKTOFT, J. J., FREER, S. T., and KRAUT, J. (1977). *J. biol. Chem.* **252**, 8875–83.

MITCHELL, E. J., CHAIKEN, I. M., and SMITH, E. L. (1970). *J. biol. Chem.* **245**, 3485–92.

RAO, S. T. and ROSSMANN, M. G. (1973). *J. molec. Biol.* **76**, 241–56.

ZANNIS, V. I. and KIRSCH, J. F. (1978). *Biochemistry* **17**, 2669–74.

32. The tertiary structure of penicillopepsin: towards a catalytic mechanism for acid proteases

MICHAEL N. G. JAMES, I-NAN HSU,
THEO HOFMANN, AND ANITA R. SIELECKI

Introduction

ACID proteases figure prominently in the early development of protein crystallography. Pepsin, although not the first protein to be crystallized, was the
first protein whose crystals were examined by X-ray crystallographic techniques (Bernal and Crowfoot 1934). It was possible to deduce from these
diffraction patterns that the pepsin molecules were globular, and, perhaps
more importantly, they gave promise of examining the structures of all crystalline proteins in great detail. The early accounts of this exciting research
have been documented for us by Dorothy Hodgkin (Hodgkin and Riley
1968). J. D. Bernal's own account of this historic moment is found in the
volume compiled by P. P. Ewald (1962) to commemorate fifty years of X-ray
diffraction. There, Bernal modestly describes the monumental contribution of Dorothy Crowfoot Hodgkin and himself in a single sentence in the
chapter describing the early history of British crystallography: 'In 1934 the
first successful photographs were taken of protein single-crystals due to a
tactical breakthrough of examining them in the wet state.' It was these two
pioneers who thus inspired the desire to determine the structures of molecules
like pepsin, haemoglobin, and insulin after those early days in Cambridge.

In the present chapter we wish to explore the three-dimensional structure
of penicillopepsin, a fungal acid protease closely related to pepsin. We
attempt to relate the structural results which are becoming known at 2.0Å
resolution to its possible mechanism of catalytic action, and by extension,
to those of the other acid proteases.

Acid proteases are distributed widely throughout the living kingdom. The
best-known representatives are the mammalian gastric enzymes, such as pepsin, gastricsin, and chymosin, whose function is the initial breakdown of

ingested protein. However, recent studies have shown that the intracellular enzymes cathepsins D and E (Mycek 1971) and the angiotensin-releasing enzyme renin (Inagami *et al.* 1977) also belong to this class of enzymes. Acid proteases have also been isolated from many different fungi and from plants and have been shown to be present in a protozoon, *Tetrahymena* (see review by Hofmann 1974). In view of the intracellular occurrence of enzymes such as cathepsin which presumably function at or near neutral pH, the term 'acid proteases' may well be a misnomer. Since a common feature of this enzyme is the presence of two aspartic acid residues at the active site, the term 'aspartyl proteases' suggested by many others is undoubtedly more appropriate, but this nomenclature has gained little general acceptance so far.

A brief look at the distribution of the aspartyl proteases shows that they are special among proteolytic enzymes because they are required to function physiologically over a very wide range of pH, from pepsin, which normally functions at $pH \sim 1.0–1.5$, to renin, which functions in the blood at pH 7.4. The pH optima for these enzymes range from 1.5 (pepsin with ovalbumin as substrate) to 8.3 (mouse submaxillary gland renin with sheep plasma angiotensinogen as substrate (Inagami *et al* 1977)). This places severe demands on the design of the catalytic apparatus. Even for a single enzyme (pepsin) pH optima have been observed between 1.5 and 4.5, depending on the nature of the substrate (Fruton 1976).

The detailed knowledge of the tertiary structure of penicillopepsin, combined with knowledge on specificities and kinetics of pepsin and other enzymes, gives us an important basis for the full understanding of the action of the aspartyl proteases.

Molecular structure of aspartyl proteases

With today's technology, X-ray crystallography is the only approach capable of providing a definitive solution to a structural study. In the case of protein structures the definition may be blurred depending upon the resolution limit; nevertheless, one is able to 'see' the molecule. Developments in the techniques for determining protein structures since the pioneering work of Green, Ingram, and Perutz (1954) have reached such a state of refinement that the rate-limiting step in solving a structure is becoming in many cases the growth of suitable crystals of the native protein. Much of the forty-three years that separate the first recording of the diffraction pattern from pepsin crystals to the visualization of the acid protease structure is due to the development of these crystallographic methods. The successful solution of the crystal structure of a protein or enzyme brings with it a view of the molecular architecture rich in its detail, and a promise of deducing the relationship of that structure to function.

The crystal structures of three microbial aspartyl proteases have been determined at medium resolution (Subramanian *et al.* 1977*b*; Hsu, Delbaere, James, and Hofmann 1977*a*). Also, from a recent publication from the USSR (Andreeva *et al.* 1978) we learn that the pepsin molecule has a polypeptide chain conformation similar to those of the three fungal enzymes. Most of the present discussion centres around the penicillopepsin molecule, the structure of which has been described at 2.8 Å resolution (Hsu *et al.* 1977*a, b*).

Penicillopepsin is closely related to pepsin. Not only are many chemical and enzymatic properties very similar (Sodek and Hofmann 1970; Mains, Takahashi, Sodek, and Hofmann 1971), but also the amino-acid sequence shows extensive homology (Cunningham *et al.* 1976; Hsu *et al.* 1977*a*). In fact, the sequence is 30 per cent identical with those of bovine chymosin and porcine pepsin. These enzymes, thus, belong clearly to the pepsin super-family (Dayhoff 1978).

Completion of the entire Watson–Kendrew molecular model of penicillo-pepsin has indicated minor revisions to the published tentative sequence (Hsu *et al.* 1977*a*). Firstly, the molecule consists of 323 amino-acid residues. An additional amino acid has been included in the region 249–258 and this addition removes the proposed deletion relative to porcine pepsin at position 264. Secondly, the whole region from residue 249 to residue 264 is tentatively revised to: Asx—Cys—Ser—Ser—Ser—Val—Pro—Asx—Phe—Ser—Val—Ser—Ile—Ser—Gly—Tyr—. Thirdly, residues 299 and 302 are both interpreted as Phe. Further revisions to this amino-acid sequence will only be made subject to the results from higher-resolution least-squares refinement currently in progress.

Penicillopepsin: the general folding

The molecular structure of penicillopepsin has been described in terms of an extensive β structure consisting primarily of antiparallel stretches of polypeptide chains (Hsu *et al.* 1977*a*). In general, the acid proteases are made up of two prominent hydrophobic domains with a well-defined active site cleft (see also Subramanian *et al.* 1977*b*). Comparison of the tertiary structures of the three microbial acid proteases determined by X-ray diffraction shows extensive structural homology. The root mean square deviations between topologically equivalent α-carbon atoms (about 200 in each enzyme) are of the order of 1.35 Å. The structural comparison of penicillopepsin and endothiapepsin is shown in the α-carbon drawing of Fig. 32.1. (We thank Professor T. L. Blundell for providing the preliminary α-carbon atom co-ordinates of endothiapepsin from which this comparison was made.) It can be seen that the topological equivalence is really remarkable and similar in extent to that displayed by serine proteases from differing species (see the comparison of the bacterial serine protease structure SGPA with that of α-chymotrypsin by Brayer, Delbaere, and James 1978).

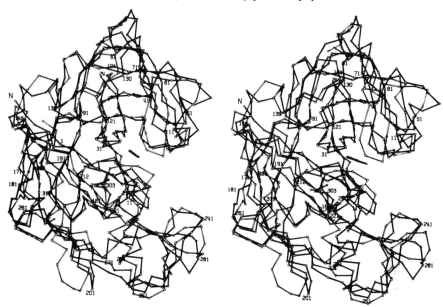

FIG. 32.1. A stereo-drawing of the α-carbon atoms of penicillopepsin (solid virtual bonds) and endothiapepsin (open virtual bonds) after maximizing the topological equivalence between these two molecules. The program used to compare the structures was that written by W. Bennett of Yale University, and it was based on the algorithm of Rossmann and Argos (1977). The view of these molecules is along the extended binding cleft (marked by the solid black line) with the active site located centrally in the drawing. The N-terminal domain is at the top of the drawing; the C-terminal domain at the bottom. The extensive hydrophobic core connects the N- and C-terminal domains and is surrounded by the antiparallel β-sheet conformation of the polypeptide backbone. The numbering scheme is that of penicillopepsin (Hsu *et al.* 1977a).

Further comparison indicates a marked topological equivalence of the two domains (Tang, James, Hsu, Jenkins, and Blundell 1978). There is an approximate intramolecular twofold symmetry axis which relates these two topologically similar domains. In penicillopepsin 61 residues of each lobe have their α-carbon atom positions in topologically equivalent positions after the operation of this pseudosymmetry element (Fig. 32.2). The rotation angle is 174.6° and there is a 1.0 Å translation parallel to the rotation axis in order to maximize the topological relatedness (see Rossmann and Argos (1976, 1977) for definitions of the rules for deriving the structural equivalences). In spite of considerable structural reorganization within each domain of the acid proteases (Fig. 32.2) the topological equivalence is highly suggestive of gene duplication, divergence and gene fusion in the evolutionary past of the acid proteases (Tang *et al.* 1978; see also Chapter 33).

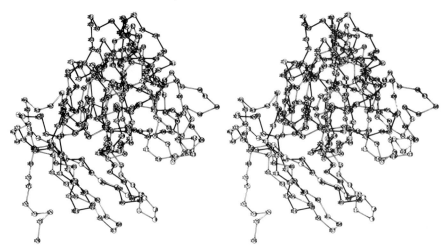

FIG. 32.2. A stereo-drawing of the α-carbon atoms in the N-terminal domain of penicil-lopepsin (open virtual bonds) superimposed on the α-carbon atoms of the C-terminal domain (filled virtual bonds between α-carbons). The superposition of the two domains was achieved by the comparison program of W. Bennett. The N-terminal domain is orientated to correspond to the view of the whole molecule in Fig. 32.1. Sixty-one residues of each lobe are topologically equivalent (for details see Tang *et al.* 1978).

Amino-acid residues in the penicillopepsin active site

Table 32.1 contains the amino-acid sequences for eight stretches of polypep-tide chain from the three acid proteases that have been completely sequenced, fungal penicillopepsin (Hsu *et al.* 1977*a*), porcine pepsin (Tang *et al.* 1973), and bovine prochymosin (Foltmann, Pedersen, Jacobsen, Kauffman, and Wybrandt 1977). These stretches of polypeptide chain were chosen from the complete sequences by the fact that the residues have the greatest degree of primary structural homology in all three enzymes. In addition, it can be seen from Figs. 32.1 and 32.3 that the residues in Table 32.1 are those residues which comprise the active site of penicillopepsin. Of the 64 residues from these three enzymes listed in Table 32.1, 39 (61 per cent) are chemically identi-cal and topologically equivalent. Certainly, it is clear from Fig. 32.1 that there is extensive tertiary structural homology for these residues in the active site regions of penicillopepsin and endothiapepsin.

The catalytic apparatus of the aspartyl proteases consists, in part, of the carboxyl groups of two aspartate residues (see reviews by Knowles 1970 and Fruton 1976). These aspartyl residues were identified in the linear sequence of porcine pepsin as Asp 32 and Asp 215 (Tang *et al.* 1973). It can be seen in Table 32.1, that these two residues are located in two highly homologous stretches of the polypeptide chain (residues 29–40 and residues 213–221). Not only do these two stretches show a prominent sequence homology to one

The tertiary structure of penicillopepsin

TABLE 32.1

Residues in the active site of the aspartyl proteases

	11	12	13	14	15
pen	D	E	E	Y	I
pep	D	T	E	Y	F
chy	D	S	Q	Y	F

	29	30	31	32	33	34	35	36	37	38	39	40
pen	L	N	F	D	T	G	S	A	D	L	W	V
pep	V	I	F	D	T	G	S	S	N	L	W	V
chy	V	L	F	D	T	G	S	S	D	F	W	V

	71	72	73	74	75	76	77	78	79
pen	W	S	I	S	Y	G	D	G	S
pep	L	S	I	T	Y	G	T	G	S
chy	L	S	I	H	Y	G	T	G	S

	118	119	120	121	122	123	124	125
pen	D	G	L	L	G	L	A	F
pep	D	G	I	L	G	L	A	Y
chy	D	G	I	L	G	M	A	Y

	150	151	152	153	154	155
pen	L	F	A	V	A	L
pep	L	F	S	V	Y	L
chy	L	F	S	V	Y	M

	213	214	215	216	217	218	219	220	221
pen	I	A	D	T	G	T	T	L	L
pep	I	V	D	T	G	T	S	L	L
chy	I	L	D	T	G	T	S	K	L

	300	301	302	303	304	305	306	307	308
pen	(L	I	F)	G	D	I	F	L	K
pep	W	I	L	G	D	V	F	I	R
chy	W	I	L	G	D	V	F	I	R

	311	312	313	314	315	316
pen	Y	V	V	F	D	S
pep	Y	T	V	F	D	R
chy	Y	S	V	F	D	R

The amino-acid sequences of penicillopepsin (pen), pepsin (pep), and chymosin (chy) are given in the one-letter code (*Biochem. J.* **113**, 1–4 (1969). The numbering scheme adopted is that for pepsin (Tang *et al.* 1973), so that the important residues, Asp 32 and Asp 215, retain their familiar sequence numbers. Those residues which are enclosed in boxes are chemically identical in all three enzymes (61 per cent of the residues in this active site region).

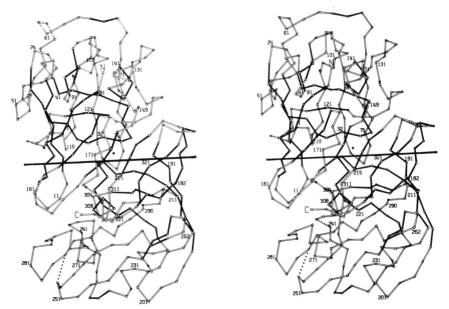

F IG. 32.3. The α-carbon drawing of the penicillopepsin molecule when viewed parallel
to the approximate twofold axis relating the two domains (see text). The 61 residues
which have the closest topological equivalence are connected by solid virtual bonds.
The majority of these topologically equivalent residues are intimately associated with
the residues comprising the active site and are listed in Table 32.1. The horizontal line
serves to define the approximate location of the extended binding cleft of penicillopepsin
(as in Fig. 32.1). This view is approximately perpendicular to that in Fig. 32.1.

another, but also they display a marked degree of topological equivalence
(see Figs. 32.2 and 32.3). Even though the two segments of the polypeptide
chain which contain these two aspartyl residues have very similar conformations, the two carboxyl groups have quite different environments. Asp 32
is not readily accessible to the solvent milieu, whereas the carboxyl of Asp
215 is. Residues Leu 29, Phe 31, Leu 38, Val 39, Leu 220, and Leu 221 form
part of the extensive hydrophobic core which lies behind the active site cleft
at the junction of the two domains.

Another segment of polypeptide chain which displays significant sequence
homology is that from Trp 71 to Ser 79. Within this portion of chain is Tyr
75, a residue which is also implicated in the catalytic mechanism as discussed
later. The stretch of chain from residue 71 to residue 79 is part of a large
hairpin loop in the acid protease structure (Fig. 32.1). This loop is part of
a three-stranded antiparallel β-sheet on the surface of the molecule. This sheet
(residues 70–77, 78–83, and 105–108) constitutes the 'flap' (James, Hsu, and
Delbaere 1977) which has been implicated in a conformational change
induced on activator binding to pencillopepsin (Wang, Dorrington, and

Hofmann 1974). The conformation of the flap seems to differ slightly when comparing the penicillopepsin structure to that of endothiapepsin (Fig. 32.1). Conformational lability of the flap in penicillopepsin is limited by the crystal packing as this region is close to a crystallographic twofold axis in the space group C2. Conformational flexibility of this region of the acid protease from *Rhizopus chinensis* has also been suggested, as the associated electron-density distribution is low (Subramanian, Liu, Swan, and Davies 1977).

There are two stretches of polypeptide chain of penicillopepsin which exhibit parallel β-sheet structures. The segment from Asp 118 to Leu 123 runs parallel to the chain from Leu 29 to Thr 33 and the other has the region from Leu 300 to Asp 304 which runs parallel to the strand, Gly 212 to Thr 216. The two strands (118–125 and 300–308 in Table 32.1) are internal strands of polypeptide chain which transverse the hydrophobic core. The completed structures which are made up of the six strands exhibit the internal approximate twofold symmetry remarkably well (Figs. 32.2 and 32.3). They have been termed ψ structures because of their close resemblance to that Greek character (Tang *et al.* 1978).

Two further strands in Table 32.1 (Leu 150–Leu 155 and Tyr 311–Ser 316) form part of a six-stranded antiparallel β-sheet which comprises the wall of the hydrophobic interior opposite to the active site cleft. This sheet is made up of polypeptide chain segments, a, j, i, q, r, and k in the nomenclature of Hsu *et al.* (1977*a*) and the two strands with highest sequence homology among the three proteases are those listed in Table 32.1 (150–155 and 311–316). Hydrophobic residues protruding into the interior of the molecule from these strands are Phe 151, Val 153, Leu 155, Val 312, and Phe 315. There are two views of the residues constituting this six-strand β-pleated sheet structure in Fig. 32.4. The molecular twofold axis runs approximately perpendicular to it and is between strands i (residues 150–156) and q (residues 311–316).

The above discussion and reference to the Figures serve to illustrate two main points. Firstly, the polypeptide chain of penicillopepsin has a β-sheet conformation which surrounds an extensive hydrophobic core (Hsu *et al.* 1977*a*, *b*). As this core is continuous from the N-terminal domain to the C-terminal domain, we feel that it is unlikely that substrate binding would disrupt the relative disposition of one domain to the other. In addition, there is evidence in support of this premise as deduced from inhibitor binding studies (Knowles, Sharp, and Greenwell 1969). The pH-independence of the K_i of a series of competitive inhibitors indicates that pepsin undergoes no major conformational changes (at least, none that affects inhibitor binding at the active site) over the range from 0.2 to 5.8. The marked sequence and tertiary structural homology between pepsin and penicillopepsin suggests that this result for the former is applicable to the latter. Secondly, the aspartate residues 32 and 215 are located very nearly at the geometrical centre of

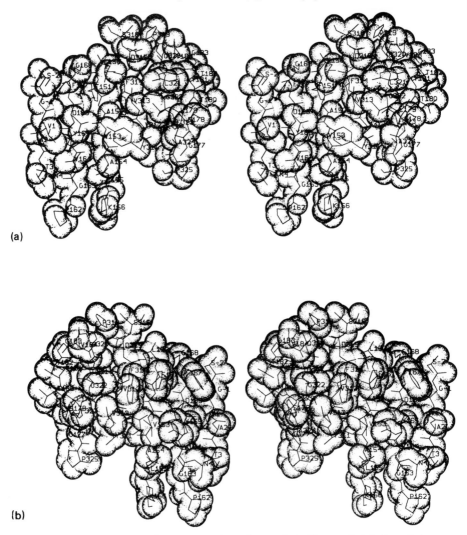

FIG. 32.4. (a) A stereoscopic representation of part of the 18-stranded β-pleated sheet structure of penicillopepsin. In the nomenclature of Hsu *et al.* (1977*a*) these are strands a, j, i, q, r, k. Strands j and k are connected by a segment of polypeptide chain (10 residues) not shown in this diagram. The van der Waals radii of C, N, and O atoms have been used to produce the atom representation. The residues are labelled with the one-letter code for the amino acids and the sequence residue number (Hsu *et al.* 1977*a*). The vantage-point of this view is from above the external predominantly hydrophilic surface of the sheet. The molecular pseudo-twofold axis runs between strands i (residues 150–156) and q (residues 311–316). (b) The same β-sheet as shown in Fig. 32.4(a), but viewed from the internal hydrophobic surface of the sheet.

the molecule and are accessible only via the extended binding cleft (see Fig. 32.1). The proximity of the surrounding polypeptide chains leaves the carboxylate of Asp 32 even less accessible than that of 215. Major conformational changes would be required to alter significantly the environment of the side-chain of Asp 32.

Finally, the very marked degree of primary structural homology among the three acid protease molecules in Table 32.1 convincingly confirms the premise that the tertiary structures are also similar and that deductions regarding the catalytic mechanism of one should be applicable to the other acid proteases.

The catalytic residues and binding cleft of penicillopepsin

The electron-density map with the corresponding model of penicillopepsin in the region of the active site is shown in Fig. 32.5. This map and the superposed protein model were generated from the partially refined penicillopepsin structure at 2.0 Å resolution. The present agreement factor, R, for the 14141 data from 5.0–2.0 Å resolution is 0.247; the model does not yet include solvent and further refinement is in progress. It can be seen in Fig. 32.5 that the carboxyl groups of Asp 32 and Asp 215 are in intimate contact and most probably share a bound proton. There are several bound water molecules in the vicinity of both Asp 32 and Asp 215; these sites were detected on the 2.8 Å multiple isomorphous replacement (m.i.r.) map (James *et al.* 1977).

The overall view of the binding cleft in penicillopepsin is illustrated in Fig. 32.6. The two stereoviews in this figure show the whole penicillopepsin molecule in 32.6(a) and then the polypeptide backbone alone in 32.6(b). The binding cleft is remarkably extended. This feature of the pepsin molecule was deduced by Fruton, who has carried out extensive analyses of the kinetic parameters of pepsin with a number of synthetic substrates (Fruton 1970, 1976). The cleft is sufficiently long to accommodate comfortably 7–8 amino-acid residues of a protein substrate. The crystal structure thus confirms the estimate of the length of the extended binding site made on the basis of substrate-modifier studies (Wang and Hofmann 1977). It was suggested there that the part of the binding site consisting of the secondary site A (as defined by Takahashi and Hofmann 1975) and the primary sites S_1† and S_1', but excluding the secondary site B, should be equivalent to a penta- or hexapeptide chain. A similar estimate was obtained for pig pepsin (Wang and Hofmann 1976). We discuss in a subsequent section the binding of a pentapeptide inhibitor to penicillopepsin. In light of that study, there does not seem to be a well-defined mode of substrate binding as one observes for the serine

† Binding sites on the enzyme are denoted by the system proposed by Schechter and Berger (1967). The substrate, P_n–P_n', binds to sites on the enzyme denoted as S_n–S_n' in this system and the scissile bond P_1–P_1' is bound by sites S_1, S_1'.

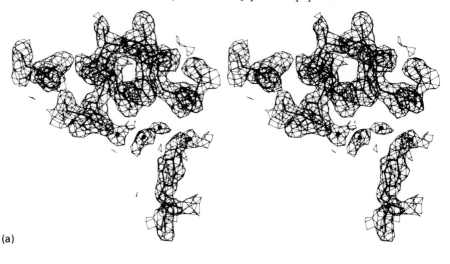

(a)

(b)

F IG. 32.5. The electron-density distribution associated with groups in the active site
of penicillopepsin. This map, depicted in (a), has Fourier coefficients of $2|F_o|-|F_c|$ and
phases, α_c, which have resulted from restrained parameter least-squares refinement
(Hendrickson and Konnert 1980). The carboxyl groups of Asp 32 (on the right) and
Asp 215 (on the left) are clearly evident. The close interactions of these groups to the
peptide bond from Thr 216 to Gly 217 can also be discerned. The residue labelling of
the model is shown below in (b) for clarity. The positions of four water molecules and
the tentative hydrogen-bonding network are detailed in part (b). This figure and
Figs. 32.6, 32.10, 32.11, and 32.12 were produced by the MMS-X graphics system and
drawn on the Calcomp plotter.

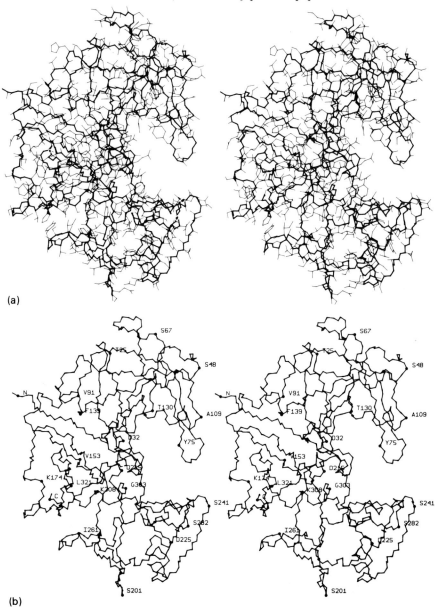

(a)

(b)

FIG. 32.6. (a) All 2364 atoms of the penicillopepsin molecule displayed in a stereoscopic view. The view direction of this figure is approximately that of the α-carbon drawing in Fig. 32.1. The extended binding cleft is the apparent 'empty' region between the two domains which runs almost parallel to the direction of view. (b) The same view of penicillopepsin as in (a) but with the polypeptide backbone atoms only. Some of the α-carbon atom positions of important residues are labelled.

proteases. These latter enzymes have a segment of polypeptide chain which makes an antiparallel β-sheet interaction with the peptide backbone of the putative substrate. There is no doubt that the acid proteases have defined substrate binding subsites, but the absolute specificity for various amino acids binding in these sites is broad (Powers, Harley, and Myers 1977).

An extended charge relay system

Catalysis of the cleavage of peptide bonds by enzymes usually involves the presence of an electrophilic grouping of atoms and a nucleophilic centre. In the case of the serine proteases, the electrophilic grouping comprises two N—H bonds of the polypeptide chain (from Gly 193 and from Ser 195) and the potential nucleophile is the Oγ of Serine 195 (but only after it has given up its proton). The electrophilic grouping of the metalloproteases (carboxypeptidase A and thermolysin) is the enzyme-bound Zn^{2+} ion. The potential 'nucleophile' seems to be a bound water molecule which presumably deposits one of its protons with the negatively charged carboxylate of a glutamic acid residue (Glu 270 in carboxypeptidase A and Glu 143 in thermolysin). These mechanisms have been discussed at length by Lipscomb *et al.* (1968), Lipscomb, Hartsuck, Quiocho, and Reeke (1969), Lipscomb, Reeke, Hartsuck, Quiocho, and Bethge (1970), Quiocho and Lipscomb (1971), and Kester and Matthews (1977). A second electrophile is also involved in these latter mechanisms via the donation of a proton to the amide nitrogen of the scissile bond. This second electrophilic group is Tyr 248 in carboxypeptidase A and His 231 in thermolysin.

Acid proteases are catalytically active at low pH values, generally with pH optima < 5.0. Therefore, it is reasonable to assume that the electrophilic grouping of the catalytic machinery will involve a proton associated with a carboxyl group as has been discussed by Knowles (1970) and Knowles *et al.* (1970). Two carboxyl groups are at the active site of the acid proteases and these carboxyl groups have different pK_as. The carboxylate with the lower pK_a is Asp 32; this is not unreasonable in view of its surrounding environment. It is the recipient of four hydrogen bonds (Hsu *et al.* 1977*a*, *b*). The other carboxyl group is Asp 215 and this has a pK_a of 4.7 in porcine pepsin (Cornish-Bowden and Knowles 1969). Such an association of two carboxyl groups (see Fig. 32.5) produces one pK_a of the system which is much lower than that of an isolated carboxyl group. The remaining proton is shared and therefore should have a relatively high affinity for these two carboxylates (the two pK_as of the similarly disposed carboxylates of maleic acid are 1.85 and 6.08, respectively). It is therefore necessary to examine the molecular environment of the two carboxylates in the aspartyl proteases in order to deduce a mechanism for an enhanced electrophilicity of the shared proton.

The enhanced electrophilicity of the proton bound between the two carboxylates can be achieved by a charge relay mechanism (James *et al.* 1977).

This charge relay is shown schematically in Fig. 32.7. The motive force is a proposed conformational change associated with substrate binding. It is well established that acid proteases have an extended substrate binding site (see above). Also, kinetic work (Wang and Hofmann 1976) on the activator Leu—Gly—Leu shows that this molecule binds in sites S_4–S_2 or S_3–S_1, and both binding modes bring the N-terminal leucyl residue into close contact with the hairpin loop containing Asp 11. A relatively minor conformational change in the enzyme structure could therefore disrupt the existing ion pair between Asp 11 and Lys 308 (this residue is an Arg residue in porcine pepsin and bovine prochymosin).

The importance of residue 308, which has its lysyl side-chain buried in the hydrophobic interior of penicillopepsin, is established by additional data. There are very few basic residues (Arg and Lys) in the acid proteases and the only position in the three proteases completely sequenced which is absolutely conserved as a basic one is position 308. Chemical evidence for the importance of this basic group comes from an experiment by Kitson and Knowles (1971), who modified one of the two arginines of pepsin with phenylglyoxal and observed a loss of proteolytic activity. Experimental difficulties have so far prevented identification of the arginine involved. Huang and Tang (1972) treated pepsin with butanedione and observed a loss of Arg 316 with concomitant loss of activity. However, subsequent extended studies of this reaction showed that the inactivation was due to a photosensitized oxidation of tryptophan and possibly tyrosine (Gripon 1978; Gripon and Hofmann, unpublished). This latter conclusion was strengthened by the observation that penicillopepsin is also inactivated by butanedione, although it is devoid of arginine (Hofmann and Gripon, unpublished). Ethoxyformic anhydride has been found to inactivate penicillopepsin by reacting with the ε-amino group of Lys 308 (Mains, Fenje, and Hofmann, unpublished).

In order to overcome a situation which would leave a buried positive charge in the hydrophobic interior (the side-chain of Asp 11 is sufficiently close to the surface that it could become solvated or, in penicillopepsin, interact with the imidazolium ring of His 157) we considered the immediate environment of this side-chain. The only other acidic group in the vicinity, other than Asp 11, is the carboxyl of Asp 304, a residue which is also conserved in primary sequence studies of the acid proteases (Table 32.1) and buried in the hydrophobic interior. A sequence of events as depicted in Fig. 32.7 which involves protonation of the peptide bond between Thr 216 and Gly 217, thereby creating a very strong acid, could subsequently occur and render ultimately an enhanced electrophilic character to the proton between Asp 32 and Asp 215. It is unlikely that such a charge relay system actually results in the complete transfer of the imino proton from the NH of Gly 217 to the carboxylate of Asp 32. Such a situation would alter the pK_as of these two acidic groups too drastically.

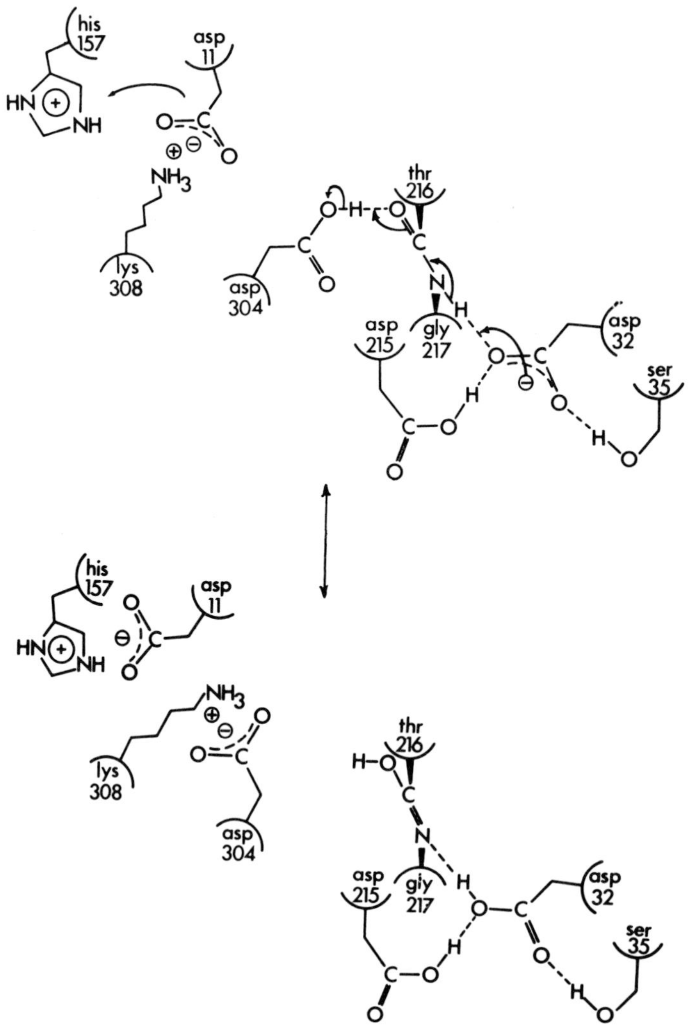

FIG. 32.7. Proposed charge relay in the acid proteases. The net effect of the disruption of the ion pair between Asp 11 and Lys 308 is the relay of the positive charge resident on Nε of Lys 308 out towards the solvent to enhance the electrophilicity of the proton shared by Asp 32 and Asp 215. It is energetically expensive to maintain an unpaired electrostatic charge in the hydrophobic interior of a protein molecule.

Recent results from the refinement of penicillopepsin at 2.0 Å resolution show that the N—H group of Gly 217 is approximately equidistant to two of the oxygen atoms of Asp 32 and Asp 215. The implication of this is a more direct effect on the electrophilicity of the proton in question than is shown in Fig. 32.7. Also of interest is the electrophilic environment of the two aspartyl side-chains of residues 32 and 215. If one examines closely the stereo-drawing presented in Fig. 32.8, it is evident that there are five peptide bonds in the immediate vicinity of the two active site carboxyl groups which have the positive ends of their dipole moments† directed approximately at the shared proton. The peptide bonds involved are: Asp 32—Thr 33, Thr 33—Gly 34, Asp 215—Thr 216, Thr 216—Gly 217 and Gly 217—Thr 218.

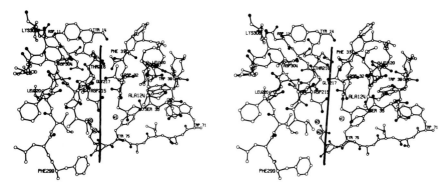

FIG. 32.8 An ORTEP plot of the amino acids in the vicinity of the active site. The substrate binding cleft runs parallel to the superposed straight line in this drawing. The anionic binding site, Asp 37, lies to the right of the line in the N-terminal domain; the hydrophobic site, S'_1, lies to the left and is seen only in part (Ile 213, Ile 301, Phe 299).

The orientation of all five of these dipoles certainly would have a profound effect on the pK_a of the two carboxyl groups; the directions of the dipole moments depress the pK_as more than they already are by their intimate contact. It is difficult to assess the quantitative effect that this orientation of dipole moments has on the catalytic mechanism, but by analogy with the three other proteolytic enzyme types it should be considerable. Clearly, this is an area where careful and detailed quantum mechanical calculations could be of enormous benefit.

† We have recently been reminded of the considerable permanent dipole moment associated with the peptide bond (Hol, van Duijnen, and Berendsen 1978). This dipole is equivalent to a full unit of charge separated by 0.72 Å and has a directionality from the partially positively charged proton on the nitrogen to the partially negatively charged oxygen atom of the carbonyl. The orientation of the dipole moments of peptide bonds is an important factor in protein structure and could play a prominent role in catalysis as suggested here.

Chemical and kinetic studies

Enzyme specificity

Traditionally, the specificity of proteases has been defined in terms of the amino-acid side-chains that surround the scissile bond. For many enzymes such as chymotrypsin, trypsin, and thermolysin the dominant feature that renders a substrate a 'good' substrate (high values of k_{cat}) is the nature of the amino acids on the carboxyl or amino side of the bond that is cleaved, although secondary interactions of amino-acyl residues not involved in the scissile bond are often important. Increasing the chain length of a typical substrate of elastase, Ac—Ala—OMe, by additional alanine residues has a large effect on binding but not on k_{cat} (Gertler and Hofmann 1970). Conversely, for the bacterial serine proteases SGPA and SGPB, additional residues toward the N-terminus from the scissile bond have a large effect upon both K_M and k_{cat} (Bauer 1978).

The dominant feature that determines the catalytic efficiency of acid proteases, on the other hand, is the necessity of extensive secondary binding on *both* the amino and the carboxyl sides of the cleavage site (Fruton 1970). Secondary binding has large effects on the k_{cat} values, but only small effects on K_M. The extensive studies on pepsin with numerous synthetic substrates, carried out especially by Fruton, and which led to this conclusion, have been reviewed (Fruton 1970, 1976; Clement 1973). The primary specificity of pepsin is determined by a high preference for Phe, Met, Leu, and Trp, in that order, for the position P_1 and a less well-defined preference for hydrophobic residues for position P_1'. This is evident both from studies with synthetic substrates and from the summary by Powers *et al.* (1977) of the action of pepsin on proteins and polypeptides, information that was extracted from sequence studies.

Much less information is currently available on the specificity of penicillopepsin. A comparison of the action of penicillopepsin on the S-sulpho B-chain of insulin with that of pepsin and other acid proteases shows a significant similarity (Mains *et al.* 1971). In spite of this, penicillopepsin cleaved only poorly, or not at all, such pepsin substrates as Z—Glu—Tyr (Z = benzyloxy-carbonyl) and Z—His—Phe—Trp—OEt (no hydrolysis), and Z—Gly—Gly—Phe—Phe—OP4P (OP4P = 3-(4-pyridyl)propyl ester) ($k_{cat} = 0.02\,\mathrm{s}^{-1}$ compared to $71.8\,\mathrm{s}^{-1}$ for pepsin (Sachdev and Fruton 1969)). One large difference between penicillopepsin and other fungal acid proteases on the one hand, and the mammalian acid proteases on the other, is the ability of the microbial proteases to activate trypsinogen (Sodek and Hofmann 1971). The activation involves the cleavage of a single Lys—Ile bond, which occurs optimally at pH 3.4 and shows a high catalytic efficiency ($k_{cat}/K_M = 1450\,\mathrm{s}^{-1}$ mM^{-1} at 1 °C (Hofmann and Shaw 1964)). This compares favourably with the value for the 'best' pepsin substrate so far studied, Z—Ala—Ala—Phe—

Phe—OP4P, $k_{cat}/K_M = 7050\,s^{-1}mM^{-1}$ at 37 °C (Sachdev and Fruton 1970). Morihara and Oka (1973) have studied a variety of acid proteases (not including penicillopepsin) and showed that there was a close relationship between the ability to activate trypsinogen and the ability to cleave bonds on the carboxyl side of lysine in synthetic peptides. They also found that the highest cleavage rates were obtained with the long substrate Z—(Ala)$_2$—Lys—(Ala)$_3$. When we found that the latter was also a good substrate for penicillopepsin, we synthesized a similar substrate that would be hydrolysed at high rates, would be amenable to spectrophotometric assay, would be readily soluble in aqueous solvents over a wide range of pH and would not undergo protonation over the pH range 1–7. (As Clement (1973) has pointed out, almost all the substrates used for pepsin suffer disadvantages because of ionization in the pH region of pepsin activity, poor solubility over the whole or part of the pH range or the need for organic solvents which are known to inhibit pepsin.)

The peptide Ac—Ala—Ala—Lys—Nph—Ala—Ala—NH$_2$ (T–II) (Nph = 3-nitrophenylalanine) was synthesized accordingly and initial studies show that it meets all the criteria set out above (Hofmann, James, and Hodges, unpublished). In the pH range 3–6 the peptide is rapidly cleaved at the Lys—Nph bond (see Fig. 32.9), with a specificity constant $k_{cat}/K_M \simeq 1000s^{-1}mM^{-1}$, at 20°C, pH 5.25; this compares quite well with the best pepsin substrate mentioned above. The peptide T-II is readily soluble in water; the cleavage of the Lys—Nph bond is accompanied by a decrease in absorbance with a maximum at 296 nm and $\varepsilon_M = 1800$, thus enabling assays to be carried out spectrophotometrically. (3-Nitrophenylalanine was introduced by Inouye and Fruton (1967) into the P$_1$ position of pepsin substrates, but it has not been used so far in the P$_1'$ position.) The study of peptide T-II shows one significant difference between penicillopepsin and pepsin, namely the presence of an anionic binding site for the lysine side-chain which is clearly shown in the pH dependence of K_M (Fig. 32.9). At pH > 5, K_M drops to low values (< 0.1 mM), but as the pH decreases there is a dramatic increase in K_M. Experimental difficulties have so far prevented measurements below pH 3. (Pepsin has a very low tolerance for lysine in position P$_1$ (Powers *et al.* 1977), but it does hydrolyse peptide T-II slowly by cleaving the —Nph—Ala—bond.) The pH dependence of K_M suggests the presence of a carboxyl-group that binds the ε-ammonium group of lysine and has an apparent pK$_a$ of 3.5 (or below).

We propose that the anionic site is the side-chain of Asp 37. If one fits a model of the substrate into the extended binding cleft, such that the substrate side-chains protrude from the extended main-chain alternating to either side, and with the Lys—Nph bond in close proximity to the active site aspartates, then Asp 37 is in the appropriate position to interact with an extended Lys side-chain. In order that an ideal ionic bond forms from

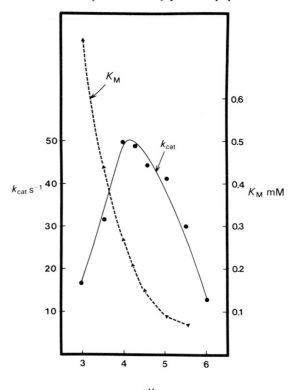

FIG. 32.9. pH Dependence of k_{cat} and K_M for hydrolysis of Ac—Ala—Ala—Lys—Nph—Ala—Ala—NH$_2$ by penicillopepsin. Spectrophotometric assays at 296 or 306 nm, 25 °C, 0.02 M sodium formate (pH 3–4.5), 0.02 M sodium acetate (pH 4.25–5.5), or 0.01 M sodium phosphate (pH 5.5 and pH 6.0).

the ammonium group to the carboxylate, a rotation of $\sim 120°$ for χ^1 of Asp 37 is required and is possible. There are no other carboxyl groups in sterically acceptable positions to act as an anionic binding site.

A further point which is in support of Asp 37 acting as the S$_1$ determinant is the presence of the hydrophobic grouping which has been equated with the S$_1'$ binding site (James *et al.* 1977). The residues involved in this binding site for a hydrophobic P$_1'$ residue are Phe 189, Ile 213, Phe 299, and Ile 301. Their position, located on the C-terminal domain (the S$_1$ site is in the N-terminal domain), would represent the binding site of the nitrophenylalanine residue of the peptide. Thus for the extended conformation of the substrate main-chain, consecutive side-chain binding sites alternate from the left to the right side of the cleft. In agreement with a previous prediction (James *et al.* 1977) the N-terminus of the substrate would lie close to the loop containing Asp 11 and Glu 12. The failure of pepsin to hydrolyse the Lys—Nph bond can be due to the fact that the residue at position 37 is an asparagine (Tang *et al.* 1973).

The pH dependence of K_M suggests that the main determinant for binding is the lysine side-chain. It is therefore surprising to find that of the two products of the hydrolysis of T-II, Ac—Ala—Ala—Lys, and Nph—Ala—Ala—NH_2, only the latter is a significant inhibitor. At pH 5.25 Nph—Ala—Ala—NH_2 inhibits the reaction competitively with $K_I = 0.07$ mM, while Ac—Ala—Ala—Lys inhibits only weakly ($K_I = 10$–20 mM). This would indicate that the apparent strong binding of the lysine in T-II occurs only co-operatively with the hydrophobic nitrophenylalanine, although it is possible that the free carboxylate ion of Ac—Ala—Ala—Lys is repelled by the net negative charge on the catalytic carboxyl groups. The strong binding of Nph—Ala—Ala—NH_2 indicates that the hydrophobic binding requirement at the S_1' site deduced from studies on the insulin B-chain (Mains *et al.* 1971; Hofmann 1974) appears to be satisfied by peptide T-II, and that hydrophobic binding in S_1' may also be an important feature of this substrate.

pH dependence of the catalytic step

Discussions of the pH dependence of the hydrolytic reaction catalysed by pepsin, and by implication by other acid proteases, stress the fact that these reactions show low pH optima and that k_{cat} depends on two ionizing groups with pK_as near 1.5 and 4.7. The information comes from many experiments with neutral and acidic dipeptides (reviewed by Clement 1973) and from studies of the pH optima with protein substrates. However, any mechanism of action of pepsin and other acid proteases has to take into account the fact that many of them act physiologically at pHs as high as 7.4 (e.g. renin). It is, therefore, not surprising to find that with good substrates of pepsin the pH-dependence differs markedly. Thus Hollands and Fruton (1968) showed that the pH optimum of k_{cat} for Z—His—Phe—Phe—OEt is 4.5 with pK_as for the groups involved in catalysis of $\simeq 3.5$ and $\simeq 5.2$, while between pH 1 and 3 the values of k_{cat} and K_M are invariant. For the substrate Z—Gly—Ala—Phe—Phe—OP4P both k_{cat} and K_M remain unchanged within experimental error over the whole pH range (pH ~ 1.7 to ~ 4.4) over which measurements were made (Sachdev and Fruton 1970). Other substrates, such as Gly—Gly—Phe—Phe—OEt, also show deviations from the bell-shaped curves obtained with the dipeptides, and pH optima that are above 4.

The pH dependence of k_{cat} for the penicillopepsin-catalysed hydrolysis of substrate T-II is shown in Fig. 32.9. The optimum of k_{cat} is around pH 4.5, and the pK_a values for the ionizable catalytic groups are around 3.5 and 5.5. (At present insufficient data are available for a more accurate determination of these values.)

The question now arises, as to the significance of the low pK_a values of pepsin obtained with dipeptide substrates (low k_{cat}s, < 0.1 s^{-1} at 25–37 °C) on the one hand and the higher values obtained with 'good' substrates ($k_{cat} > 10\,s^{-1}$ at 25–37 °C). A strong possibility is that the low pK_a values represent, or are close to, the pK_a values of the two catalytic carboxyl groups in the free enzyme, whereas the higher values are those of the enzyme–substrate complex. This is a reasonable assumption if one considers that the 'good' substrates differ from the 'poor' substrates structurally only in chain length and not in primary specificity, and kinetically only in k_{cat} and not in affinity. It can then be inferred that the secondary binding induces a shift of the pK_a values of the catalytic groups to higher values.

Transpeptidation reactions catalysed by acid proteases

Amino transfer

Neumann, Levine, Berger, and Katchalski (1959) first showed that when pepsin hydrolysed Z—Glu—Tyr the dipeptide Tyr—Tyr was formed as a minor product. Fruton *et al.* (1961) postulated that the tyrosyl-residue of the substrate formed a covalent amino intermediate with the enzyme since radioactive free tyrosine was not incorporated into the transpeptidation product.

Wang and Hofmann (1976, 1977) studied the transpeptidation by aminotransfer of a variety of compounds with both pepsin and penicillopepsin and found that substrates with leucine in the P_1' position gave high yields of the transpeptidation reactions. Thus the reaction products of Z—Phe—Leu with pepsin consisted of 70 per cent Leu—Leu—Leu, 20 per cent Leu—Leu, and only about 10 per cent free leucine (Wang and Hofmann 1976). Incubation of pepsin with 2 mM Z—Phe—Leu in the presence of 10 mM [^{14}C]-leucine failed to incorporate radioactive leucine into the transpeptidation products. The available evidence suggested a mechanism for the transpeptidation reactions in which the C-terminal amino acid of the substrate (e.g. Z—Phe—Leu) forms a covalent intermediate with one of the carboxyl groups at the active site. Another substrate molecule subsequently acts as acceptor to give the putative product Z—Phe—Leu—Leu which is either cleaved at the Phe—Leu bond or acts as an acceptor for another leucyl group to give Z—Phe—Leu—Leu—Leu, which is then cleaved to give Z—Phe + Leu—Leu—Leu. The proposed presence of a covalent amino intermediate in these transpeptidation reactions has been used to postulate the involvement of such a covalent intermediate in the mechanism of *all* pepsin catalysed reactions (Fruton 1970; Clement, Snyder, Price, and Cartmell 1968).

Acyl transfer

Takahashi, Wang, and Hofmann (1974) observed that incubation of Leu—Tyr—NH$_2$ with pepsin or penicillopepsin led to the formation of Leu—Leu via the detectable intermediate Leu—Leu—Tyr—NH$_2$. In the case of pepsin, the tripeptide Leu—Leu—Leu was also formed in significant amounts. As in the case of amino transfer an excess of [^{14}C]-leucine failed to exchange with enzyme-bound leucine as judged from the failure of leucine to be incorporated into the transpeptidation products (Wang and Hofmann, unpublished). This transpeptidation therefore constitutes an 'antipolar' image of the amino transfer and appears to involve a covalent intermediate with the enzyme (anhydride intermediate). By analogy with the amino transfer, the acyl transfer is observed only with a single amino acid that carries a free α-amino group.

Amino and acyl transfer

Takahashi and Hofmann (1975) showed that Leu—Leu was formed from the substrate Leu—Tyr—Leu. The available evidence suggested that the main reaction occurred via an acyl transfer of the N-terminal leucine, but did not exclude a transpeptidation involving an amino transfer of the C-terminal leucine.

Newmark and Knowles (1975) used [14C]-Leu—Tyr—[3H]-Leu as substrate and thus were able to show that both acyl and amino transfers occurred with the same substrate and at rates that, although different, were of the same order of magnitude. They also obtained evidence that Leu—Leu was not formed by condensation of two substrate molecules and subsequent cleavage, because no [3H]-Leu—[14C]-Leu was found. This was subsequently confirmed by Wang and Hofmann (1976), who used Leu—Trp—Met, another substrate in which both transfer reactions occur. The peptides Leu—Leu, Leu—Leu—Leu, Met—Met, and Met—Met—Met were detected in the reaction, but there was no evidence for Met—Leu.

Enzyme activation

As has been pointed out above, Fruton (1970) showed that secondary interactions of long substrates are important in inducing high catalytic efficiency. The secondary interactions are important not only for pepsin, but also for its homologues, such as chymosin and cathepsin-D (Sampath-Kumar and Fruton 1974), penicillopepsin (Wang *et al.* 1974), and other microbial proteases (Oka and Morihara 1973). Wang *et al.* (1974) and Wang and Hofmann (1976, 1977) showed that small peptides could bring about many-fold activations of the cleavage of small substrates. The small peptides affect only k_{cat} and not K_M of the substrate. However, the same activator peptides act as inhibitors of the hydrolysis of large substrates, including proteins. This shows that the activator peptides bind in the substrate binding groove. They presumably simulate the effects of elongating poor substrates. The study of the interaction of one of the activators, Leu—Gly—Leu, with penicillopepsin in the absence of substrate showed changes in the circular dichroism spectrum of the enzyme, i.e. a decrease in ellipticity of two bands assigned to tryptophan and one band assigned to tyrosine. These changes could reflect conformational shifts of the enzyme, or solvent displacement from the tryptophan by the activator, since the tryptophan and tyrosine residues affected are most likely those that are located in the secondary site A (subsites S_4–S_2) in which Leu—Gly—Leu most probably binds.

Inhibitor binding to penicillopepsin in the crystalline form

The binding of inhibitor molecules or virtual substrates to a variety of proteolytic enzymes has provided great insight into not only the details of enzyme–substrate (or inhibitor) interactions, but also the possible ways in which the enzymes might catalyse the hydrolytic reaction. There are a number of specific inhibitors of aspartyl proteases, some of them synthetic and some naturally occurring. Pepstatin is a naturally occurring hexapeptide inhibitor of the acid proteases which has remarkably high association constants (pepsin, 1×10^{10} M^{-1} (Kunimoto *et al.* 1974); penicillopepsin $5(\pm 2) \times 10^9$ M^{-1} (Hofmann, unpublished)). The molecular mechanism for inhibition of the aspartyl proteases by pepstatin, although unknown, is thought to be associated with the presence of an unusual amino acid, statine (4-amino-3-hydroxy-6-methyl-heptanoic acid). It has been suggested that the statyl residue is a transition state analogue for the acid protease catalytic pathway (Marciniszyn, Hartsuck, and Tang 1977).

Another inhibitor of the acid proteases is the synthetic epoxide, 1,2-epoxy-3-(*p*-nitrophenoxy)propane(EPNP). Studies on the inhibition of porcine pepsin by EPNP have shown that two molecules of this reagent react covalently with both Asp 32 and Asp 215 (Chen and Tang 1972). Crystallographic evidence for the reaction of EPNP with penicillopepsin confirms the fact that both aspartyl residues have reacted covalently with EPNP molecules (James *et al.* 1977).

A third inhibitor is a dipeptide analogue 3-mercapto-2-D-methyl-propanoyl-L-proline (SQ-14, 225). This inhibitor was obtained through the generosity of the Squibb Institute for Medical Research, which produces it as a potential antihypertensive agent. It is a very potent inhibitor of angiotensin-converting enzyme in plasma (Lanzillo and Fanburg 1977).

Finally we have considered the binding of the pentapeptide Ac—Pro—Ala—Pro—Ala—Phe—COOH to penicillopepsin. This peptide was very kindly provided by Dr C.-A. Bauer of the University of Lund, Sweden, for a study with the bacterial serine proteases SGPA and SGPB. It has proved informative with the aspartyl proteases as well.

We consider below the binding of the latter three inhibitors to penicillopepsin and point out the specific enzymic inhibitor interactions deduced from the difference electron-density maps. The three difference electron-density maps were computed with the amplitude differences $(|\,F_I\,| - |\,F_P\,|)$† with the

---

† $|\,F_I\,|$ = structure factor amplitudes measured from inhibited crystals of penicillopepsin. $|\,F_P\,|$ = structure factor amplitudes from native penicillopepsin crystals.

2.8 Å resolution m.i.r. phases. Where possible, we have attempted to fit a molecular model to the positive difference density present on these maps.

EPNP binding to penicillopepsin

The difference electron-density map contoured in the region of the active site of penicillopepsin is shown in Fig. 32.10. Two main points are evident on this map. Firstly, there are two molecules of EPNP bound in the active site and both of these molecules have reacted covalently with the carboxyl groups of Asp 32 and Asp 215. The electron density associated with each EPNP molecule is continuous with that of the two aspartyl residues in the native map (the native electron density is not shown in Fig. 32.10). This result provides definitive confirmation of the findings of Chen and Tang (1972) and of Hofmann (1974) with a number of other acid proteases.

The second point of importance concerns the conformational change of Tyr 75 associated with the binding of EPNP. It is seen in Fig. 32.10 that there is a negative trough (dashed lines) coincident with the native position of the phenoxy ring of Tyr 75 and a neighbouring positive peak (solid contour

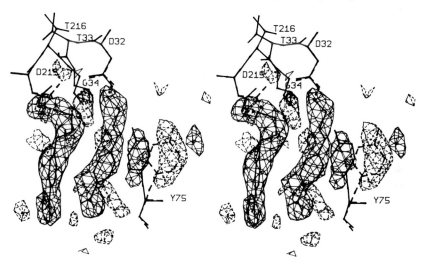

FIG. 32.10. Difference electron-density map showing the EPNP binding to penicillo-pepsin. Only the residues in the immediate vicinity of the two aspartyl residues, Asp 32 and Asp 215, have been included in the drawing (Tyr 75 is shown as well). The native enzyme structure is displayed with solid black lines except for those atoms which undergo a conformational change (Asp 32, Asp 215, and the side-chain of Tyr 75). The original position is shown in dashed lines and the new position in solid lines. The positive electron-density contour surfaces are displayed with a solid network whereas the negative regions are shown by dashed lines. The models of the EPNP molecules are shown fitted into the positive electron-density contours. The covalent bonds from the carboxyl groups of Asp 32 and Asp 215 to the Cl atoms of the propane moieties are denoted by dotted lines.

surface) which represents the new position of this side-chain. This conformational movement (predominantly a rotation about the C_α—C_β bond) suggested a possible electrophilic role for the —OH of Tyr 75 in the acid protease mechanism (James *et al.* 1977). It also suggested that a more extensive conformational change might occur on substrate binding. The larger conformational movement would involve the entire 'flap' region (the antiparallel β bend from residues 72–81). However, this movement has not been observed in the present crystalline modification of penicillopepsin. Conformational lability of this region is consistent with the slightly different orientation of the polypeptide chain of the flap in the endothiapepsin molecule (Fig. 32.1).

The EPNP molecules react with the two active-site carboxyl groups of penicillopepsin in a manner which was not expected. The mechanism of ring opening of 1,2-epoxides in acid solution favours the products which have the nucleophile covalently bonded to the 2-position of the propane moiety (Parker and Isaacs 1959). It can be seen in Fig. 32.10 that our present interpretation involves a covalent attachment of an oxygen atom from each of the carboxyl groups of Asp 32 and Asp 215 to the 1-carbon atom of the propane portion of EPNP. In addition, the negative trough between the two carboxyl groups suggests a conformational change of these groups and this has been incorporated in fitting the EPNP difference electron-density map. In an earlier publication we had indicated that Asp 215 reacted with C2 of the propane moiety (James *et al.* 1977). This model could not be accommodated when fitting the electron density. However, the relevant conclusion, that the COO⁻ group of Asp 32 is inaccessible, is still valid.

The conformational change of Tyr 75 brings the plane of the phenol ring parallel to the *p*-nitro-phenoxy ring of the EPNP bound to Asp 32. The planes of the *p*-nitro-phenoxy rings in the present model make an angle of $\sim 60°$ to one another.

SQ-14,225 binding to penicillopepsin

The difference electron-density map at 2.8 Å resolution of this inhibitor bound in the active-site region is shown in Fig. 32.11. There is no covalent attachment of the inhibitor to the penicillopepsin molecule. We have interpreted the chief positive density in terms of a molecular model of the inhibitor, 3-mercapto-2-D-methylpropanoyl-L-proline. The most intense peak in this map was chosen as the sulphur atom and there is good interpretable electron density for all of the D-cysteine analogue. However, the density which should correspond to the proline group is poorly defined and an extended peak for which there is no possible model is also present. The latter feature could be enhanced binding of solvent. There is another peak, again probably a solvent molecule which is between the two carboxylate groups of penicillopepsin.

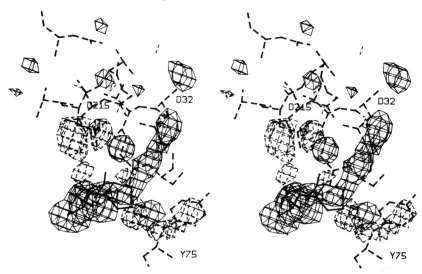

FIG. 32.11. The difference electron-density map resulting from binding SQ-14,225 to penicillopepsin. The native enzyme positions are depicted by dashed lines and the model of the inhibitor is in solid lines. Electron-density contour surfaces are denoted as for Fig. 32.10. It can be seen that the 3-mercapto-2-methylpropanoyl portion of the inhibitor fits well into the positive density whereas there is little if any density corresponding to the L-proline portion. There is no conformational change associated with Tyr 75 although Asp 77 moves away from the inhibitor (shown as the negative/positive pair of peaks close to the proline ring). The side-chain of Thr 218 is in a large negative trough in this map. The positive peak between the two carboxyl groups at the active site is not explained, but could represent a bound solvent molecule.

There is an apparent conformational change of Thr 218 (negative trough in Fig. 32.11) which occurs on inhibitor binding. The resultant position (i.e. a corresponding positive peak) is not evident on this map. Another minor conformational change can be seen just below the proline ring of the inhibitor. This movement is associated with the side-chain of Asp 77 which moves away from the proline moiety on binding.

The sulphydryl group of the inhibitor interacts with Ile 301. Such an interaction is presumably hydrophobic in nature.

If the present interpretation of the difference electron-density distribution is correct, then there seems to be another reason for this molecule acting as an inhibitor, over and above the fact that the cysteinyl analogue is *D*. The solution binding and kinetic data mentioned earlier for the acid proteases indicate that the binding sites on the enzyme, S_4–S_1, lie in the cleft from the active aspartyl groups towards the large hairpin loop containing Asp 11— Tyr 14 (see following section). The N-terminal portion of a substrate binds in this region (Figs. 32.1, 32.3, and 32.6). This being the case, it is evident that this inhibitor is binding back to front in the active site cleft. Such a binding mode is clearly non-productive, yet the enzyme–inhibitor interactions must be sufficiently strong for this compound to act as an inhibitor. It is interesting to note that the possibility of non-productive reverse binding has been suggested for the pepsin substrate Z—His—Trp—Phe—OMe by Clement (1973). The steric arguments regarding the inhibitory action of SQ-14,225 deduced from this relatively poor 2.8 Å electron-density map are rather soft, but hopefully will be clarified as the result of refinement at 2.0 Å resolution.

Acetyl-prolylalanylprolylalanylphenylalanine binding

The difference electron-density map resulting from the soaking of a penicillo-pepsin crystal in a solution of this pentapeptide at pH 4.4 and 2.5M Li_2SO_4 is shown in Fig. 32.12. The positions of the N and C termini of the pentapeptide

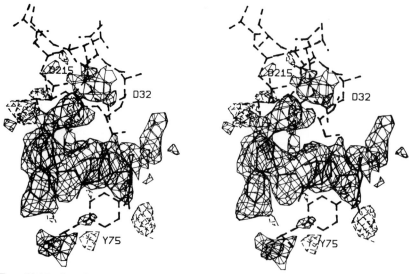

FIG. 32.12. Ac—Pro—Ala—Pro—Ala—Phe—OH binds in an extended conformation near the active site of penicillopepsin (the enzyme is depicted with dashed lines, the model of the pentapeptide inhibitor with solid lines). The elongated positive difference electron-density peak which corresponds to the bound inhibitor lies parallel to the extended substrate binding cleft (see Figs. 32.1 and 32.6) with the C-terminal Phe close to the active site aspartyl residues. The N-terminal Ac—Pro residue is close to the β-bend near residues 11–13 in penicillopepsin. Thr 218 lies in a negative trough as it did in the SQ-14,225 difference map, and there does not seem to be a corresponding positive peak for a possible movement of that side chain. The fitting of the model of Ac—Pro—Ala—Pro—Ala—Phe to the electron-density contour surface leaves parts of the two proline rings out of density. The density which lies between the two active-site aspartyl groups is too large for a carboxyl group, but the orientation corresponds to that expected for the pathway shown in Fig. 32.16.

are unambiguous owing to the prominent density present for the C-terminal phenylalanine residue. This peptide is bound so that the C-terminal
carboxyl group is directed towards the two active-centre aspartyl residues
as predicted previously for the acid proteases (James *et al.* 1977). The
N-terminus of the pentapeptide lies close to the hairpin loop containing
Asp 11, Glu 12, and Glu 13.

The conformational change of Tyr 75 evident from the EPNP binding experiment (Fig. 32.10) does not occur with this inhibitor. Nevertheless, such
a movement of the tyrosyl side-chain is not only sterically possible, but also
the resultant position of the —OH group of Tyr 75 is remarkably well
positioned for this group to function as the second electrophilic species, in
supplying a proton to the —NH of the scissile bond. This movement could
also provide a binding interaction for a P_1 phenylalanine residue as the planes
of the phenyl groups (Tyr 75 and the substrate) would lie approximately
coplanar.

There seems to be relatively little conformational change of the enzyme
associated with the binding of Ac—Pro—Ala—Pro—Ala—Phe—OH,
although the side-chain of Thr 218 lies in a negative trough as with SQ-
14,225. It is evident that access to the extended binding cleft of penicillopepsin
is possible without the conformational movement of the flap alluded to previously. However, the binding mode of this peptapeptide is such that only the
N-terminal sites S_1–S_4 are occupied. With an intact substrate, comprising
a length to fill seven or eight subsites, it may be necessary to have the conformational change of the flap region in order to bind the substrate properly
(see Wang *et al.* 1974 and James *et al.* 1977).

A catalytic mechanism for acid proteases

Historically, the details of the catalytic mechanism for the acid proteases have
been elusive. Nevertheless, several proposals which were based primarily on
chemical and kinetic data from pepsin have been made (Knowles 1970;
Knowles *et al.* 1970; Fruton 1970, 1971, 1976; Hofmann 1974). The major
problem, until 1977, had been the lack of knowledge of the disposition of
groups at the active sites of these proteases. A possible mechanism was then
proposed (James *et al.* 1977) and it was based on the results of the 2.8 Å
resolution structure of penicillopepsin. Implicit in making that proposal was
the close homology of pepsin on which the majority of the solution studies
were made, and penicillopepsin for which we had determined the detailed
tertiary structure. The present discussion will consider the several possible
alternatives briefly, and then present the arguments for preferring a mechanism which does not involve covalent intermediates.

Proteolytic enzymes, whether they are members of the serine, sulphydryl-,
metallo-, or aspartyl-protease families, *all* share a common feature; they

have all been designed, through evolutionary pressures, to cleave peptide bonds (C_α—CO—NH—C_α). There are only a limited number of ways whereby this chemical cleavage can be accomplished and it appears that nature, given the limited number of potential catalytic groups among the 20 amino acids, has evolved all possible ways, depending in part upon the milieu in which the protease is required to function. The mammalian serine proteases function in the alkaline conditions of the small intestine and have evolved a mechanism resembling that of general base catalysis. The metallo-proteases have a mechanism predominantly electrophilic in character. The aspartyl proteases, typified by pepsin, are capable of functioning in the strongly acid medium of the stomach (pH ~ 2). The mechanism of the aspartyl proteases resembles, understandably, a general acid catalysed hydro-lysis (electrophilic). It has been pointed out that, in spite of the absence of sequence homology and the lack of similarity of catalytic groups at the active site, the kinetic mechanism of carboxypeptidase A resembles in many respects that of pepsin (Fruton 1974). We should keep in mind these possibilities, then, when considering the following structurally based mechanisms.

There are a number of experimental observations made for the aspartyl proteases, with which any catalytic mechanism must be consistent. A number of these have already been alluded to, but we wish to restate them here.

The catalytic hydrolysis is mediated by the two carboxyl groups of Asp 32 (low pK_a) and Asp 215 (high pK_a). It is clear from the preceding discus-sions on the structure and mode of binding of inhibitors to penicillopepsin that these are indeed the catalytic aspartyl residues. The very low pK_a of Asp 32 seems to be a result of the tight hydrogen-bonded environment in which it is located (Hsu *et al.* 1977*a*, *b*).

The rate-limiting step of the hydrolytic pathway occurs subsequent to the formation of the Michaelis complex and is predominantly electrophilic in character (Knowles 1970). We have shown in a previous discussion and in Fig. 32.7 how the proton which is shared by Asp 32 and Asp 215 may provide this electrophilic part of the pathway. The hydrogen-bonding is such that the NH of Gly 217 seems to interact with both active-site carboxyl groups. Increasing the electrophilic character is accomplished by the necessity of driv-ing the uncomfortably buried positive charge of Lys 308 out, towards the solution, via Asp 304, the peptide bond joining Thr 216 and Gly 217 and the carboxyl groups of Asp 32 and Asp 215.

Secondary binding sites play an important role in catalysis (Fruton 1976). The interesting point in this regard, especially in connection with the trigger of the extended charge relay system we have proposed, is that increasing the length of the substrate towards the N-terminal binding sites (S_3 and S_4, secon-dary site A) has little effect on K_M, but a pronounced effect on k_{cat}.

Finally, not only must the catalytic mechanism of the acid proteases account for hydrolysis, but it should also explain the phenomenon of both

acyl and amino transpeptidations. Whereas transpeptidation may be an arte-
factual phenomenon, the catalytic apparatus must have within its framework
the necessary latitude to accommodate these reactions.

It must also be kept in mind that many attempts have been made to trap
any covalent intermediate which might occur on the catalytic pathway of the
acid proteases. None of these attempts have succeeded (Fruton 1976).

Figure 32.13 shows, in summary form, three alternative pathways for acid
protease catalysis. There is no direct experimental evidence which allows for
a clear-cut choice amongst these possibilities.

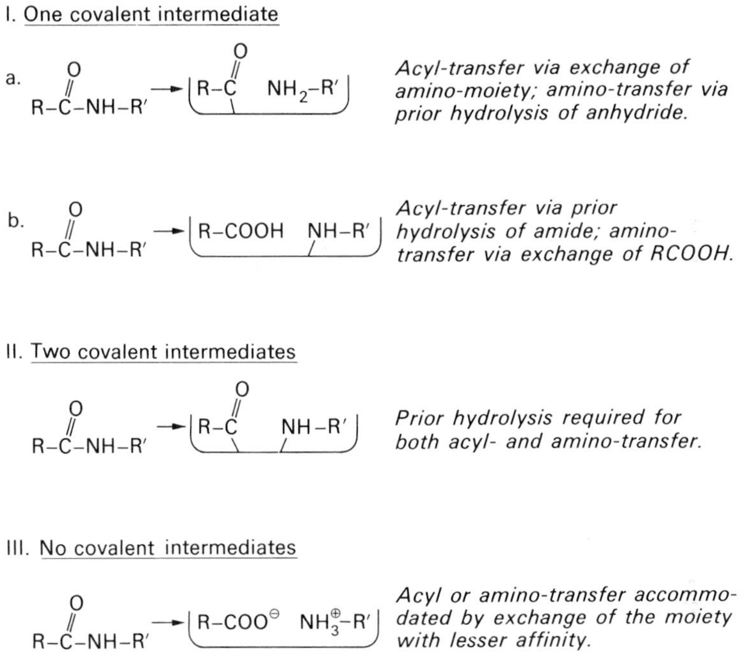

I. One covalent intermediate

a. $R-\overset{O}{\overset{\|}{C}}-NH-R' \longrightarrow \boxed{R-\overset{O}{\overset{\|}{C}} \quad NH_2-R'}$

*Acyl-transfer via exchange of
amino-moiety; amino-transfer via
prior hydrolysis of anhydride.*

b. $R-\overset{O}{\overset{\|}{C}}-NH-R' \longrightarrow \boxed{R-COOH \quad NH-R'}$

*Acyl-transfer via prior
hydrolysis of amide; amino-
transfer via exchange of RCOOH.*

II. Two covalent intermediates

$R-\overset{O}{\overset{\|}{C}}-NH-R' \longrightarrow \boxed{R-\overset{O}{\overset{\|}{C}} \quad NH-R'}$

*Prior hydrolysis required for
both acyl- and amino-transfer.*

III. No covalent intermediates

$R-\overset{O}{\overset{\|}{C}}-NH-R' \longrightarrow \boxed{R-COO^\ominus \quad NH_3^\oplus-R'}$

*Acyl or amino-transfer accommo-
dated by exchange of the moiety
with lesser affinity.*

FIG. 32.13. Alternative catalytic mechanisms for the carboxyl proteases. Only the
intermediates which result from substrate–enzyme interaction are shown.

The first pathway I (Fig. 32.13) involves the possibility of one covalent
attachment. From our knowledge of the behaviour of porcine pepsin and
penicillopepsin with appropriate substrates (e.g. Leu—Tyr—Leu as discussed by Takahashi and Hofmann (1975) or by Newmark and Knowles
(1975) and Leu—Trp—Met by Wang and Hofmann (1976, 1977)) one can
observe in the same reaction vessel transfer of the acyl moiety (which may
or may not involve an explicit acyl-enzyme, Ia) and transfer of the amino-
moiety (which may or may not involve an explicit amino-enzyme, Ib). Thus
in considering the possibility of one covalent intermediate, we must consider
both a covalent acyl-enzyme Ia (anhydride bond) and a covalent amino-

enzyme Ib (amide bond). For each of these pathways, acyl and amino transfer can occur.

The second pathway, II, involves two covalent intermediates. It is unlikely that such a pathway exists for the aspartyl proteases because of the relatively restricted access to the reactive carboxyl groups. In order to accommodate acyl or amino transfer one must have prior hydrolysis occurring. Pathway II is an unlikely possibility, and we shall not consider it further.

Finally the third mechanism has no covalent intermediates. As will become evident this pathway has simplicity, symmetry, and can accommodate either *acyl* or *amino* group transfer in transpeptidation reactions simply by which moiety departs first.

There are three points pertinent to all three possibilities which we note. We have been consistent in the choice of the potential nucleophile, Asp 32, primarily because of the lower apparent pK_a of this group. The primary electrophilic component (that responsible for the rate-limiting step) is the proton shared by Asp 32 and Asp 215. Its electrophilicity is enhanced by the proposed extended charge relay system. Thirdly, the second electrophilic component is the OH of Tyr 75 which can approach closely to the amino group of a scissile bond.

Covalent acyl-enzyme intermediate

The chemical steps for pathway Ia are outlined in Fig. 32.14. The first step involves protonation of the carbonyl oxygen atom of the substrate, R—CO—NH—R′, by Asp 215 and concomitant direct nucleophilic attack on the carbonyl carbon atom by Asp 32. The tetrahedral intermediate so formed would then collapse following the electrophilic involvement of Tyr 75 in donating its phenolic proton to the leaving group to give an acyl-enzyme intermediate and a non-covalently bound free amino moiety (NH_2—R′). Figure 32.14 shows two possible positions for the attack by water on the anhydride, either (i) the carbonyl carbon of the substrate or (ii) the carbonyl carbon of Asp 32. The two ensuing tetrahedral intermediates of deacylation are shown. Collapse of these tetrahedral intermediates could then occur, as shown at the bottom of Fig. 32.14, with the regeneration of the native enzyme E and release of the acyl portion of the substrate, RCOOH.

Transpeptidation involving acyl transfer could be accommodated by exchange of the amino moiety NH_2—R′ with NH_2—R″. Transpeptidation involving amino transfer could occur by R′—NH_2 having a strong affinity to the enzyme and remaining non-covalently bound to the active site, whilst R—COO^- is hydrolysed via reaction (i) or (ii) and replaced by R‴—COO^- (not shown). This latter possibility is not plausible.

The major argument against this pathway is the inaccessibility of the carboxylate of Asp 32. We have pointed out that there is not sufficient room for a covalently bound tetrahedral intermediate at Asp 32 without major

PATHWAY Ia

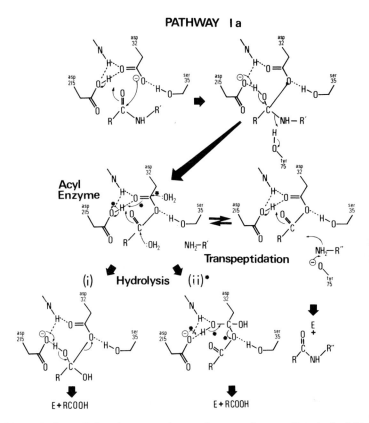

FIG. 32.14. Pathway Ia involves a covalent acyl-enzyme intermediate (anhydride) with the carboxyl of Asp 32 and the carbonyl-carbon of the substrate R—CO—NH—R′. Hydrolysis of the anhydride is shown following one of two possible pathways, depending upon (i) whether attack of water is on the substrate carbonyl-carbon, or (ii) attack of water is at the carboxyl of Asp 32. This latter pathway is denoted by an asterisk and is the least likely hydrolytic pathway on steric grounds. Transpeptidation involving acyl transfer is shown occurring via an exchange of the amino moiety R′—NH$_2$ replaced by R″—NH$_2$. The transpeptidation reaction is thus the microscopic reverse of the initial formation of the acyl-enzyme through the tetrahedral intermediate shown with Asp 32 as the nucleophile. The electrophilic involvement of the OH of Tyr 75 in donating its proton to the amide nitrogen is shown following the breakdown of the tetrahedral intermediate. The resulting Tyr—O⁻ is the base on the protein which presumably accepts a proton from the water following its attack on the anhydride.

reorientation of the two domains of penicillopepsin. Such a conformational change would be energetically expensive. The hydrolysis pathway Ia (ii) is also most unlikely as this would give rise to a tetrahedral carbon atom in a severely sterically hindered position.

Covalent amino-enzyme intermediate

The sequence of possible chemical events in pathway Ib are shown in Fig. 32.15. Following an identical step as in pathway Ia, we arrive to the same initial tetrahedral intermediate. However, conversion of this intermediate into the covalent amino-enzyme could be accomplished by a four-centre re-action (Delpierre and Fruton 1965; also advanced by Knowles 1970). In our opinion, however, there is no chemical advantage to hydrolyse one peptide link only to form another with the enzyme.

The hydrolysis of the amide link from enzyme to substrate could be accomplished by attack of water at the carbonyl carbon of Asp 32, as this is now the only centre which can be hydrolysed. Release of the amino group $R'-NH_2$ could then be facilitated by the electrophilic OH of Tyr 75.

Transpeptidation via amino transfer is handled by the exchange of the acyl moiety RCOOH with R″COOH and reversing the steps required to cleave the first peptide bond. A transpeptidation reaction in which one has acyl transfer involves the retention of the acyl moiety RCOOH while the amino moiety is hydrolysed to give $R'-NH_2$ and exchange with $R'''-NH_2$, etc. (not shown).

Similar objections to this pathway may be raised as were raised to pathway Ia. The steric requirements at Asp 32 are severe and the only centre for attack by water (not usually a strong nucleophile at pH 2!) is the carbonyl carbon of this amino acid.

Neither Ia nor Ib has made use of the very well-ordered water molecules bound near the carboxyl groups of the two active-site aspartyl residues.

Pathway III: no covalent intermediates

In the description of this pathway, we will make use of the extended charge relay system to enhance the electrophilicity of the shared proton, and we will involve a water molecule which is observed within a hydrogen-bond dis-tance of Asp 32 in the native enzyme structure. The close similarity of this mechanism to that of the Zn-carbonyl mechanism proposed by Lipscomb *et al.* (1968) for carboxypeptidase A should be evident. It is of interest that a mechanism also resembling that of carboxypeptidase A for the metalloen-zyme thermolysin was proposed by Kester and Matthews (1977).

The schematic representation of the sequence of events for pathway III is illustrated in Fig. 32.16. The formation of the Michaelis complex is shown as the first step and this has initiated the transfer of positive charge (Lys 308) out toward the active site. In addition, Asp 32 is not assumed to activate the bound water, because its pK_a of ~ 1 is considerably lower than

PATHWAY Ib

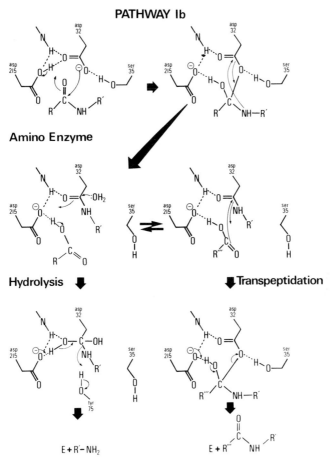

Amino Enzyme

Hydrolysis ⬇

Transpeptidation ⬇

$E + R'-NH_2$

$E + R'''$NH$

F IG. 32.15. Pathway Ib provides an alternative pathway which has been advanced (Knowles 1970) in order to account for the possibility of an amino-enzyme intermediate. Formation of this species occurs via a four-centre exchange reaction of the first tetrahedral intermediate. The enzyme–substrate amide bond is cleaved in the hydrolysis reaction following an initial attack of water on the carbonyl-carbon of Asp 32 to give the tetrahedral intermediate at that carbon atom. The electrophilic component of Tyr 75 operates at the level of collapse of this tetrahedral species. Steric limitations at Asp 32 also detract from this catalytic pathway.

that of water. Rather, Asp 32 is the recipient of a proton which water has to discard before it can form the tetrahedral covalent intermediate with the strongly positively charged carbonyl-carbon atom of the substrate. Collapse of this tetrahedral intermediate into a free carboxylate and new amino group is facilitated by the second electrophilic component, Tyr 75, shown protonating the amino nitrogen. This step can only take place after the

PATHWAY III

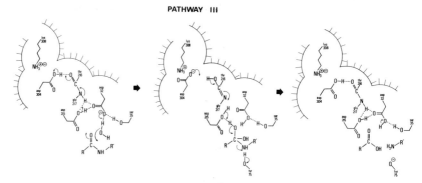

FIG. 32.16. In this diagram depicting the main elements of Pathway III, we have attempted to include the details of the extended charge-relay system which we propose to enhance the electrophilicity of the proton shared by Asp 32 and Asp 215. Pairing of the positive charge on Lys 308 by the carboxylate of Asp 304 induces the alternate tautomer of the peptide bond, Thr 216 to Gly 217, shown in the central panel. We show a concerted reaction of the positively charged carbonyl-carbon of the substrate with a water molecule which in turn deposits its proton on the carboxylate of Asp 32. Thus, the formation of the non-covalent tetrahedral intermediate of the central panel. Collapse of this intermediate is enhanced by the electrophilic component of Tyr 75 to yield the non-covalently bound free carboxyl (R—COOH) and free amino (H₂N—R') groups. Acyl- or amino-transfer transpeptidations can be accommodated and are predominantly directed by the affinity of each moiety to the enzyme. This phenomenon is discussed more fully in the text. The negative counter-ion to Lys 308 in the native enzyme is Asp 11. At present, there is no experimental evidence for the conformational change of Asp 304 and the effect could be achieved by inductive transfer through the intricate hydrogen-bonding network.

nitrogen of the scissile bond has rehybridized to an sp^3 state as the partial double-bond character of the peptide bond is lost.

Although transpeptidation reactions could be accommodated in this pathway by the relative affinity of either the acyl or the amino moiety to the enzyme and simply by which moiety leaves first, the failure of [14C]-leucine to exchange with the enzyme–leucine complex formed from amino-transfer substrates (Z—Phe—Leu) or from acyl-transfer substrates (Leu—Tyr—NH₂) is not readily explained. This suggests that the leucine, in either case, is held onto the enzyme very tightly but not necessarily covalently. Thus, in experiments carried out with Z—Phe—Leu in the presence of a fivefold excess of [14C]-leucine of high specific activity, no radioactive leucine was incorporated into the major product Leu—Leu—Leu. Leu—Leu—Leu is presumably formed by the following reaction:

$$Z—Phe—Leu—Leu + Enz—Leu \rightarrow Z—Phe—Leu—Leu—Leu + Enz$$

$$Z—Phe + Leu—Leu—Leu,$$

where Z—Phe—Leu—Leu, the acceptor, is the product of a transfer of an enzyme–bound leucine to the substrate acting as acceptor. The steady-state concentration of Z—Phe—Leu—Leu is presumably quite low. The failure of the enzyme-bound leucine to be released (the total formation of free leucine in the reaction is only about 10 per cent of the total leucine cleaved from the substrate), or to exchange with [^{14}C]-leucine, shows that the rate of complex formation between Enz—Leu and Z—Phe—Leu—Leu must be very much faster than the rate of release of leucine. Leucine itself has only low affinity for the active site regions of the enzyme since it inhibits the enzyme only weakly. It can therefore be concluded that the enzyme–leucine complex formed by the reaction with substrate is different from an equilibrium complex between enzyme and free leucine. This 'nascent' complex is either covalent, as has been assumed hitherto, or else the 'nascent' leucine is trapped in the active site by conformational constraints that were induced either by substrate binding or during the substrate cleavage. This 'nascent' complex can then react with suitable acceptors even at low concentrations without significant release of the leucine moiety.

Exactly equivalent arguments apply to the acyl-transfer reactions. Evidence indicates that the nature of the enzyme–leucine complexes is different for the two reactions, amino transfer and acyl transfer. This is shown by the absence of formation of [^{3}H]-Leu—[^{14}C]-Leu from [^{14}C]-Leu—Tyr—[^{3}H]-Leu (Newmark and Knowles 1975) which means that the amino-terminal leucine can only act as acceptor for enzyme-bound leucine that arises from the N-terminal position, and the C-terminal leucine can only act as acceptor for leucine from the C-terminal position. This experiment, too, argues for tight and specific association between 'nascent' leucine and the enzyme. Some preliminary transpeptidation experiments carried out in urea suggest that trapping of the 'nascent' leucine to the enzyme by conformational constraints is a likely explanation for the stability of the complex. While the rate of formation of leucine from Leu—Trp—Met—Arg is only slightly inhibited by up to 4 M urea, transpeptidation is not observed at urea concentration > 2.5 M (Chow and Hofmann, unpublished).

There is a further phenomenon associated with the transpeptidation reactions. Amino-transfer reactions have only been observed with substrates where the amino acid in position P$_1'$ carries a free α-carboxylate group. Thus, Silver and Stoddard (1972), using Ac—[^{3}H]-Phe as acceptor, observed transpeptidation only with Ac—Phe—Phe—COO$^-$, but not with the closely related substrates Ac—Phe—Phe—OEt or Ac—Phe—Phe—Tyr—NH$_2$ although all three peptides are hydrolysed. If the hydrolysis of these substrates proceeded via a covalent amino intermediate then it is difficult to account for the fact that an enzyme —Phe—COO$^-$ intermediate readily transfers the amino–acyl moiety to an acceptor while the intermediates enzyme —TyrCO<u>NH</u>$_2$ and enzyme—PheCOOEt do not. Similarly, acyl-transfer reactions have only been observed with substrates in which the N-terminal amino acid that is transferred has a free α-ammonium group.

These observations force us to consider seriously the possibility that the transpeptidation reactions do not offer any clues that are relevant to our understanding of the pepsin mechanism.

The binding of the pentapeptide inhibitor as observed in the crystals of penicillopepsin is most consistent with the events envisaged for pathway III.

Conclusions

The crystal structure determination of penicillopepsin has provided the details of the molecular geometry at the active site of acid proteases. In conjunction with this structure and in light of the many important chemical studies on the pepsin mechanism, we are able to formulate a plausible reaction pathway which is in accord with all of this factual detail.

It is worth while to reiterate the analogy of the acid protease mechanistic proposal with that of the metalloproteases. Nature has provided an environment for the production of an electrophilic centre required to polarize the carbonyl bond of a substrate by the acid proteases when they are in a strongly acid medium. The proton between the two aspartate groups has this electrophilic character and is analogous to the strong Lewis acid, Zn^{2+}, of the metalloenzymes. We propose, primarily on steric arguments, that Asp 32 does not act as a nucleophile directly, but it serves to accept the proton from a water molecule as the resulting hydroxide ion reacts to form a covalent bond with the positively charged carbonyl carbon of the substrate. The analogous residue in carboxypeptidase A is Glu 270, in thermolysin, Glu 143. Finally, the second electrophilic component is Tyr 75, which protonates the iminonitrogen of the scissile bond in a manner analogous to Tyr 248 which was suggested by Lipscomb and co-workers.

Proteolytic enzymes, because they perform the identical chemical reaction, have many features of their reaction mechanisms similar to one another. Only the actors have changed.

Acknowledgements

One of us (M. N. G. J.) would like to express his sincere gratitude to Dorothy Hodgkin for introducing him to the wonderful world of biological macromolecules, and the techniques whereby one can 'see' them. We thank Colin Broughton for providing the ease and facility to make images of these molecules on the MMS-X interactive graphics system. This research has been supported by grants from the Medical Research Council of Canada to the Group in Protein Structure and Function (M. N. G. J.) at the University of Alberta, and to T. H. at the University of Toronto.

References

ANDREEVA, N. S., FEDOROV, A. A., GUSHCHINA, A. E., RISKULOV, R. R., SHUTSKEVER, N. E., and SAFRO, M. G. (1978). *Molekulyarnaya Biologiya* **12**, 922–36.

BAUER, C. A. (1978). *Biochemistry* **17**, 375–80.

BERNAL, J. D. and CROWFOOT, D. (1934). *Nature, Lond.* **133**, 794–5.

BRAYER, G. D., DELBAERE, L. T. J., and JAMES, M. N. G. (1978). *J. molec. Biol.* **124**, 261–82.

CHEN, K. C. S. and TANG, J. (1972). *J. biol. Chem.* **247**, 2566–74.

CLEMENT, G. E. (1973). *Prog. Bioorg. Mechan.* **2**, 177–238.

——SNYDER, S. L., PRICE, H., and CARTMELL, R. (1968). *J. Am. chem. Soc.* **90**, 5603–10.

CORNISH-BOWDEN, A. J. and KNOWLES, J. R. (1969). *Biochem. J.* **113**, 353–62.

CUNNINGHAM, A., WANG, H.-M., JONES, S. R., KUROSKY, A., RAO, L., HARRIS, C. I., RHEE, S. H., and HOFMANN, T. (1976). *Can. J. Biochem.* **54**, 902–14.

DAYHOFF, M. O. (1978). *Atlas of protein sequence and structure*, Vol. 5, Suppl. 3, pp. 104–7. National Biomedical Research Foundation, Washington.

DELPIERRE, C. R. and FRUTON J. S. (1965). *Proc. natn. Acad. Sci. U.S.A.* **54**, 1161–7.

EWALD, P. P. (1962). *Fifty years of X-ray diffraction*, p. 380. N.V.A. Oosthoek's Uitgeversmaatschappij, Utrecht.

FOLTMANN, B., PEDERSEN, V. B., JACOBSEN, H., KAUFFMAN, D., and WYBRANDT, G. (1977). *Proc. natn. Acad. Sci. U.S.A.* **74**, 2321–4.

FRUTON, J. S. (1970). *Adv. Enzymol.* **33**, 401–43.

——(1971). In *The enzymes*, 3rd edn (ed. P. D. Boyer) Vol. 3, pp. 119–64. Academic Press, New York.

——(1974). *Acc. Chem. Res.* **7**, 241–6.

——(1976). *Adv. Enzymol.* **44**, 1–36.

——FUJI, S., and KNAPPENBERGER, M. H. (1961). *Proc. natn. Acad. Sci. U.S.A.* **47**, 759–61.

GERTLER, A. and HOFMANN, T. (1970). *Can. J. Biochem.* **48**, 384–6.

GREEN, D. W., INGRAM, V. M., and PERUTZ, M. F. (1954). *Proc. R. Soc.* **A225**, 287–307.

GRIPON, J.-C. (1978). Ph.D. thesis, Université de Caen, France.

HENDRICKSON, W. A. and KONNERT, J. H. (1980). In *Biomolecular structure, conformation, function and evolution* (ed. R. Srinivasan) Vol. 1, pp. 43–57. Pergamon Press, Oxford.

HODGKIN, D. C. and RILEY, D. P. (1968). In *Structural chemistry and molecular biology* (ed. A. Rich) pp. 15–28. Freeman, San Francisco.

HOFMANN, T. (1974). *Adv. Chem. Ser.* **136**, 146–85.

——and SHAW, R. (1964). *Biochim. biophys. Acta* **92**, 543–57.

HOL, W. G. J., VAN DUIJNEN, P. T., and BERENDSEN, H. J. C. (1978). *Nature, Lond.* **273**, 443–6.

HOLLANDS, R. T. and FRUTON, J. S. (1968). *Biochemistry* **7**, 2045–53.

HSU, I.-N., DELBAERE, L. T. J., JAMES, M. N. G., and HOFMANN, T. (1977a). *Nature, Lond.* **266**, 140–5.

————————(1977b). In *Acid proteases, structure, function and biology* (ed. J. Tang) pp. 61–81. Plenum Press, New York.

HUANG, W.-Y. and TANG, J. (1972). *J. biol. Chem.* **247**, 2704–10.

INAGAMI, T., MURAKAMI, K., MISONO, K., WORKMAN, R. J., COHEN, S., and SUKETA, Y. (1977). In *Acid proteases, structure, function and biology* (ed. J. Tang) pp. 225–48. Plenum Press, New York.

INOUYE, K. and FRUTON, J. S. (1967). *Biochemistry* **6**, 1765–77.

JAMES, M. N. G., HSU, I.-N., and DELBAERE, L. T. J. (1977). *Nature, Lond.* **267**, 808–13.

KESTER, W. R. and MATTHEWS, B. W. (1977). *Biochemistry* **16**, 2506–16.

KITSON, T. M. and KNOWLES, J. R. (1971). *Biochem. J.* **122**, 249–56.

KNOWLES, J. R. (1970). *Phil. Trans. R. Soc.* **B257**, 135–46.

——SHARP, H. C., and GREENWELL, P. (1969). *Biochem. J.* **133**, 343–51.

——BAYLISS, R. S., CORNISH-BOWDEN, A. J., GREENWELL, P., KITSON, T. M., SHARP, H. C., and WYBRANDT, G. B. (1970). In *Structure–function relationships of proteolytic enzymes* (ed. P. Desnuelle, H. Neurath, and M. Otteson) pp. 237–50. Munksgaard, Copenhagen.

KUNIMOTO, S., AOYAGI, T., NISHIZAWA, R., KOMAI, T., TAKEUCHI, T., and UMEZAWA, H. (1974). *J. Antibiot.* **27**, 413–18.

LANZILLO, J. J. and FANBURG, B. L. (1977). *Biochemistry* **16**, 5491–5.

LIPSCOMB, W. N., HARTSUCK, J. A., QUIOCHO, F. A., and REEKE, G. N., JR (1969). *Proc. natn. Acad. Sci. U.S.A.* **64**, 28–35.

——REEKE, G. N., JR, HARTSUCK, J. A., QUIOCHO, F. A., and BETHGE, P. H. (1970). *Phil. Trans. R. Soc.* **B257**, 177–214.

——HARTSUCK, J. A., REEKE, G. N., QUIOCHO, F. A., BETHGE, P. J., LUDWIG, M. L., STEITZ, T. A., MUIRHEAD, H., and COPPOLA, J. C. (1968). *Brookhaven Symp. Biol.* **21**, 24–90.

MAINS, G., TAKAHASHI, M., SODEK, J., and HOFMANN, T. (1971). *Can. J. Biochem.* **49**, 1134–49.

MARCINISZYN, J., JR, HARTSUCK, J. A., and TANG, J. (1977). In *Acid proteases, structure, function, and biology* (ed. J. Tang) pp. 199–210. Plenum Press, New York.

MORIHARA, K. and OKA, T. (1973). *Archs biochem. Biophys.* **157**, 561–72.

MYCEK, M. J. (1971). *Meth. Enzymol.* **19**, 285–314.

NEUMANN, H., LEVINE, J., BERGER, A., and KATCHALSKI, E. (1959). *Biochem. J.* **73**, 33–41.

NEWMARK, A. and KNOWLES, J. R. (1975). *J. Am. chem. Soc.* **97**, 3557–9.

OKA, T. and MORIHARA, T. (1973). *Archs biochem. Biophys.* **156**, 552–6.

PARKER, R. E. and ISAACS, N. S. (1959). *Chem. Rev.* **59**, 737–99.

POWERS, J. C., HARLEY, A. D., and MYERS, D. V. (1977). In *Acid proteases, structure, function and biology* (ed. J. Tang) pp. 141–57. Plenum Press, New York.

QUIOCHO, F. A. and LIPSCOMB, W. N. (1971). In *Advances protein chemistry* (ed. C. B. Afinsen, J. T. Edsall, and F. M. Richards) pp. 1–78. Academic Press, New York.

ROSSMANN, M. G. and ARGOS, P. (1976). *J. molec. Biol.* **105**, 75–95.

——(1977). *J. molec. Biol.* **109**, 99–129.

SACHDEV, G. P. and FRUTON, J. S. (1969). *Biochemistry* **8**, 4231–8.

————(1970). *Biochemistry* **9**, 4465–70.

SAMPATH-KUMAR, P. S. and FRUTON, J. S. (1974). *Proc. natn. Acad. Sci. U.S.A.* **71**, 1070–2.

SCHECHTER, I. and BERGER, A. (1967). *Biochem. Biophys. Res. Commun.* **27**, 157–62.

SILVER, M. S. and STODDARD, M. (1972). *Biochemistry* **11**, 191–200.

SODEK, J. and HOFMANN, T. (1970). *Can. J. Biochem.* **48**, 425–31.

————(1971). *Meth. Enzymol.* **19**, 372–96.

SUBRAMANIAN, E., LIU, M., SWAN, I. D. A., and DAVIES, D. R. (1977a). In *Acid proteases, structure, function and biology* (ed. J. Tang) pp. 33–60. Plenum Press, New York.

——SWAN, I. D. A., LIU, M., DAVIES, D. R., JENKINS, J. A. TICKLE, I. J., and BLUNDELL, T. L. (1977b). *Proc. natn. Acad. Sci. U.S.A.* **74**, 556–9.

TANG, J., JAMES, M. N. G., HSU, I.-N., JENKINS, J. A., and BLUNDELL, T. L. (1978). *Nature, Lond.* **271,** 618–21.

—— SEPULVEDA, P., MARCINISZYN, J., CHEN, K. C. S., HUANG, W.-Y., TOO, N., LIU, D., and LANIER, J. P. (1973). *Proc. natn. Acad. Sci. U.S.A.* **70,** 3437–9.

TAKAHASHI, M. and HOFMANN, T. (1975). *Biochem. J.* **147,** 549–63.

—— WANG, T.-T., and HOFMANN, T. (1974). *Biochem. biophys. Res. Commun.* **57,** 39–46.

WANG, T. T. and HOFMANN, T. (1976). *Biochem. J.* **153,** 691–9.

———— (1977). *Can. J. Biochem.* **55,** 286–94.

—— DORRINGTON, K. J., and HOFMANN, T. (1974). *Biochem. biophys. Res. Commun.* **57,** 865–9.

33. Symmetry in the structure and organization of proteins

TOM BLUNDELL, TREVOR SEWELL,
AND BILL TURNELL

ALTHOUGH biological organisms have evolved by natural selection, they
have at all times been constrained by the laws of physics and chemistry. This
was recognized in 1917 by D'Arcy Thompson who in his book *Growth and
form* beautifully described the symmetrical arrangement of the cells and skeletons of many organisms—helical, dodecahedral, hexagonal close-packed,
and so on. In the 1930s these ideas influenced crystallographers such as
J. D. Bernal and his student Dorothy Crowfoot who realized that their work
could lead to an understanding of 'growth and form' at an even more fundamental level—that of large biological molecules. Thanks to their confidence
and vision, we are now able to describe these molecules in detail, and it
appears that biomolecular structure and organization involve an even more
fascinating use of symmetry than was at first believed. In this paper we discuss
recent discoveries made by X-ray crystallographic techniques, concerning not
only the arrangement of subunits but also the folding of polypeptide chains
of globular proteins.

X-ray crystallographic investigations have shown that proteins have a hier-
archy of structures, which is indicated for one protein in Fig. 33.1. The *primary structure* or *sequence* defines the order of the amino acids in the polypeptide chain, and this leads to a preferred conformation and hydrogen-bonding
pattern in the main chain—α-helix (α), β-sheet (β), or β-turn—known as the
secondary structure. Consecutive segments of secondary structure fold
together to give *folding units* such as αα, αβ, or ββ. Quite often a number
of different folding units are associated to give *supersecondary structures*; this
term is often used where extended β-sheets are involved. In general, however,
the folding units are self-associated into well-defined globular *domains* which
have close-packed hydrophobic cores and have an organization and three-
dimensional structure which appear independent of other parts of the polypeptide chain. Indeed, in many cases—the immunoglobulins and acid proteinases for example—the polypeptide may be proteolytically cleaved to leave

Sequence

↓

Secondary   structure

β-sheet

α -helix

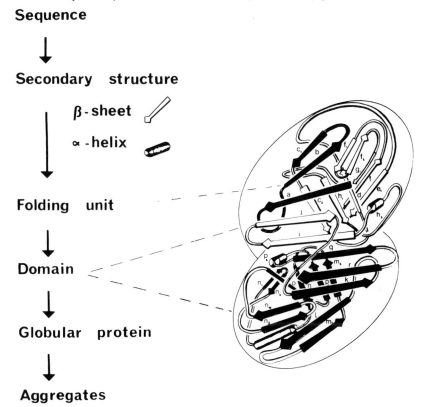

Folding   unit

↓

Domain

↓

Globular   protein

↓

Aggregates

FIG. 33.1. The hierarchical organization of protein structures using the acid proteinase
from *Endothia parasitica* as an example.

a stable domain. Some small proteins comprise only one domain, for example
insulin; for large proteins two or three domains are usually observed in the
tertiary structure. Frequently the active site or allosteric effector binding sites
lie between two adjacent domains, as in the acid proteinase shown in Fig.
33.1 where the deep active-site cleft lies between the two well-defined
domains. Finally, several identical proteins or protomers can be assembled
together to give an oligomer; the arrangement of the protomers in the oligomer is known as the *quaternary structure*. We now discuss examples of
the symmetrical arrangements of protein structures, starting at the level of
folding units, proceeding to consider *domains*, and finally describing protomer
organization in quaternary structure.

Symmetry between folding units

The folding of two adjacent helical segments—the αα folding unit—is found
quite frequently, and most especially in α-helical domains without β-sheet.

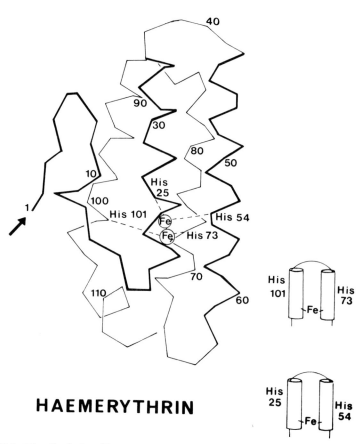

HAEMERYTHRIN

FIG. 33.2. The C_α chain of haemerythrin, showing the two pairs of helices, drawn so that the pseudo-diad between the pairs is approximately vertical. (Reproduced with permission from Hendrickson and Ward 1977.)

A good example is haemerythrin, an invertebrate oxygen carrier, which is illustrated in Fig. 33.2. Hendrickson and Ward (1977) showed that the structure can be considered as two α–α folding units, each binding an iron atom through a histidine side-chain on each helix. The folding units are related by a pseudo-twofold axis as shown.

Similar relationships exist between proteins with $\alpha\beta$ folding units. Rossmann, Moras, and Olsen (1974) were the first to draw attention to the existence of a pseudo-twofold symmetry between the $\alpha\beta$ structures which bind the adenine and nicotinamide moieties of the coenzyme NAD in dehydrogenases, where the supersecondary structure so formed is an extended parallel twisted sheet, flanked by α-helices. In both triosephosphate isomerase (Banner, Bloomer, Petsko, Phillips, and Wilson 1976) and pyruvate kinase (Levine, Muirhead, Stammers, and Stuart 1978) a closely related $\alpha\beta$ secondary structure exists comprising a barrel of eight strands of parallel β-sheet surrounded by α-helices as shown in Fig. 33.3. The barrels have eight α–β units related by a very approximate eightfold rotation symmetry; the barrel has a more exact diad.

In our laboratory we have found examples of symmetry between folding units involving β-sheet. The γ-crystallins (Slingsby *et al.*, unpublished results) from mammalian eye lenses have mainly antiparallel β-sheet structure. In γ-crystallin II, each domain comprises two folding units of

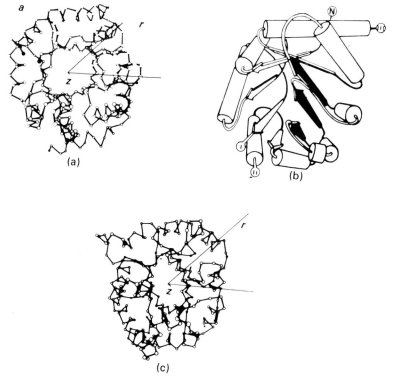

(a)

(b)

(c)

FIG. 33.3. The C_α chains of pyruvate kinase (a) and (b) (Levine *et al.* 1978) and triose phosphate isomerase (c) (Banner *et al.* 1976) viewed from equivalent positions and showing the $\alpha\beta$ barrels. (Reproduced with permission from Levine *et al.* 1978.)

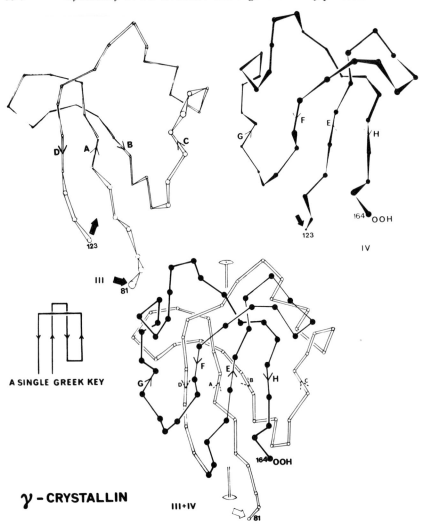

FIG. 33.4. The C$_\alpha$ chain of the C-terminal domain of bovine γ-crystallin II, showing the pseudo-diad (vertical) between the third and fourth Greek-key motifs. (From Slingsby *et al.* unpublished data.)

the *Greek-key* motif (Richardson 1977) as in Fig. 33.4, which shows the C-terminal domain. The folding units are individually very similar, each with approximately 40 amino acids, and are related together by pseudo-twofold symmetry.

Each domain of the acid proteinases, shown in Figs. 33.1 and 33.6, also has a pseudo-diad, but here the symmetry is less extensive. It has been quantified in pepsin (Andreeva and Gustchina 1979) and in the acid proteinase from

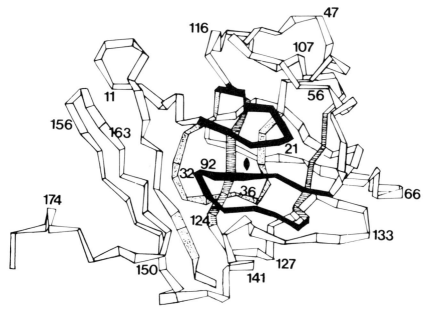

FIG. 33.5. The C_α ribbon of the acid proteinase from the first domain of *Endothia parasitica*, looking down the pseudo-diad. (Reproduced with permission from Tang *et al.* 1978.)

Endothia parasitica (Blundell, Sewell, and McLachlan 1979) which is illustrated in Figs. 33.5 and 33.6. The pseudo-diad relates residues in four strands of antiparallel β-sheet (a, b, c, d) to residues in four topologically equivalent strands (e, f, g, h_1). In the second domain four strands (k, l, m, n) are symmetry-related to strands (n_1, n_2, o, p). The residues in each of the four strands

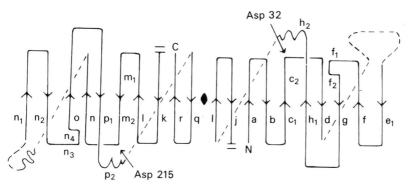

FIG. 33.6. A schematic representation of the polypeptide backbone of the acid proteinase from *Endothia parasitica* laid flat about the pseudo-diad between the two domains. (From Blundell *et al.* 1978.)

are topologically equivalent and the α-carbons of about 20 amino acids overlap with those of equivalent residues with an r.m.s. deviation of ~ 2 Å when
the twofold symmetry operation is applied.

In the examples described so far, the three-dimensional structures show
striking similarities, but, with the exception of haemerythrin, there is little
evidence for this in the primary structures. Indeed, the sequences alone did
not suggest internal homology, and what little exists was identified retrospec-
tively once the X-ray crystal structures of these proteins were determined.

Symmetry between domains

In contrast, the duplication of sequences within one polypeptide chain did
suggest topologically equivalent domains before the three-dimensional
structures were elucidated. For example, the heavy chain of IgG immunoglobulin contains four homologous sequences while the light chains contain two.
Edelman (1970) suggested that these sequences probably represent equivalent
three-dimensional domains and the later X-ray structural studies confirmed
this (Poljak 1975). However, the topologically equivalent domains are not
related by a simple rotation axis of symmetry and the regions of polypeptide
between the domains appear to be flexible and act as hinge regions.

On the other hand, the two domains in rhodanese (Ploegman, Drent, Kalk,
and Hol 1978), the acid proteinases (Tang, James, Hsu, Jenkins, and Blundell
1978), and the γ-crystallins (Slingsby *et al.*, unpublished results) are topologically quite closely equivalent and are related by a good twofold symmetry.
In the acid proteinases, Tang *et al.* (1973) had noticed that the residues
around the active-site aspartate residues (Asp 32 and Asp 215 in the pepsin
sequence of Table 33.1) are homologous but they could not find further evidence from the sequence in support of their contention that the two halves
of the enzymes evolved by gene duplication and fusion. The three-
dimensional structure shows convincingly that topological equivalence does

TABLE 33.1

*The primary sequences of symmetrically related sides of the active site of three species
of acid proteinase. (For review, see Tang* et al. *1978)*

	30	31	32	33	34	35	36	37	38	39	40	41
Pig pepsin:	Ile	Phe	*Asp*	Thr	Gly	Ser	Ser	Asn	Leu	Trp	Val	Pro
Calf chymosin:	Leu	Phe	*Asp*	Thr	Gly	Ser	Ser	Asp	Phe	Trp	Val	Pro
Penic.pepsin:	Asp	Phe	*Asp*	Thr	Gly	Ser	Ala	Asp	Leu	Trp	Val	Pro
Pig pepsin:	Ile	Val	*Asp*	Thr	Gly	Thr	Ser	Leu	Leu	Thr	Gly	Pro
Calf chymosin:	Ile	Leu	*Asp*	Thr	Gly	Thr	Ser	Lys	Leu	Val	Gly	Pro
Penic.pepsin:	Ile	Ala	*Asp*	Thr	Gly	Thr	Thr	Leu	Leu	Leu	Leu	Asx
	213	214	215	216	217	218	219	220	221	222	223	224

indeed exist. Figure 33.6 shows schematically the secondary structure of the acid proteinases comprising 20 strands; ten strands in each lobe are topologically equivalent. The two lobes are hydrogen-bonded together through antiparallel β-sheet. In the acid proteinase from *Endothia parasitica* the best fit between domain 1 and domain 2 by a diad axis gives 63 structurally equivalent atom pairs with an r.m.s. deviation of 2.06 Å (Blundell *et al.* 1979) and the match corresponds closely with the one reported for penicillopepsin (Tang *et al.* 1978; see also Chapter 32). The pseudo-diad runs through the active site cleft so that the two topologically equivalent active-site aspartates are brought into close juxtaposition and appear to be hydrogen-bonded. The diad axes relating the folding units described earlier make angles of about 45° to the inter-domain axis and the three axes lie almost in a plane.

A rather different and, to date, exceptional situation occurs in γ-crystallin II. Analysis of the primary squence of the monomer did not, and does not, predict good local symmetry either between the folding units, the Greek-key motifs, or between the domains, the β-barrels. The two barrels are each formed by 82 residues, leaving six residues for the single connecting peptide. Each barrel is made up of a pair of right-hand twisted, four-stranded antiparallel sheets, with each sheet laid out as a Greek-key motif as described above. Thus the whole molecule is a stack of four such sheets.

Beginning at the second residue from the N-terminus, the α-carbon atoms of the 38 residues of motif I may be rotated about 180° into the 38 C_α atoms of motif II, with an r.m.s. error of fit of 3.54 Å. Similarly, after the connecting peptide, the 41 C_α atoms of motif III are related about a local diad to those of motif IV, with an r.m.s. error of 3.62 Å. These r.m.s. errors are close to the resolution of the map from which the C_α co-ordinates have been taken. Moreover, the central 76 C_α atoms of the N-terminal barrel (motifs I and II) are topologically equivalent to those of the C-terminal barrel (motifs III and IV) about a third non-crystallographic diad, with an r.m.s. error of 2.58 Å. This third diad bisects the angle of 40° between the other two, all three diads being approximately coplanar.

In rhodanese there are no diads within the two domains, but a diad relates 117 of the approximately 140 C_α atoms in each domain with an r.m.s. deviation of 1.95 Å. The angle of rotation deviates by $\sim$ 1° from 180° and the translation parallel to the axis is less than 1 Å. Each of the domains comprises a five-stranded parallel β-sheet flanked by five helices; the structural similarity between the two domains in rhodanese is much greater than between rhodanese and the nucleotide-binding domain of dehydrogenase (Ploegman *et al.* 1978).

Quaternary structure

For soluble oligomeric proteins, point-group symmetries of the types shown in Fig. 33.7 are most probable (for a review see Matthews and Bernhard

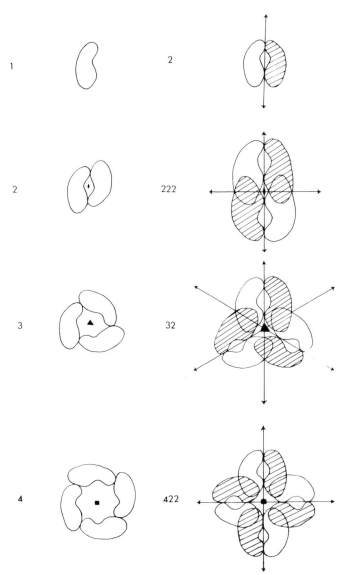

FIG. 33.7. Point group symmetries of some common protein assemblies.

1973). Of course, symmetries of higher order than four are found, but are comparatively rare. The most interesting questions concern (i) whether oligomers have simple rotational symmetries or n2 symmetry (diads perpendicular to a further rotation axis), and (ii) whether the symmetry will be perfect or only approximate.

The large number of oligomeric protein structures determined by X-ray analysis shows that symmetries of the n2 type are most common. This may relate to the fact that if one favourable interaction of the type A_1–B_2 is formed, a further interaction of the type B_1–A_2 may be possible without disturbing the first, so leading to a diad. This is the isologous interaction of Monod, Wyman, and Changeux (1965). The self-assembly of oligomers in stages through dimers leading to a dimer of dimers (222), a trimer of dimers (32), a tetramer of dimers (422), etc., may be an efficient mechanism (i.e. with a low activation energy). Furthermore, if there is a relatively wide channel around the 3, 4, or higher-order axis, all interactions are around diads—interactions of this kind can be all isologous.

Dimers often form through antiparallel β-pleated sheet (diad perpendicular to the sheet) as was first observed by Hodgkin and her co-workers for insulin (Adams *et al.* 1969; Blundell *et al.* 1971). It has been more recently discovered in alcohol dehydrogenase dimers and concanavalin A and prealbumin tetramers. Other common interactions arise from the packing of α-helices together at an angle as was also found in the B-chain helices of insulin, and the packing of two β-sheets on top of each other (diad parallel to the plane of the sheet) as in prealbumin.

Many oligomeric proteins crystallize with more than one protomer in the crystallographic asymmetric unit, indicating that the symmetry may not be exact. In porcine 2Zn insulin hexamers the symmetry is only approximately a diad, but in dimeric hagfish insulin (Cutfield *et al.* 1974) and cubic dimeric insulin there is a crystallographic diad. In many crystallographic studies the electron densities of the protomers are averaged to improve the phasing and this may obscure subtle differences in conformation which may be important to structure and function. However, it remains an often impossible task to distinguish whether the lack of perfect symmetry is due to interactions within the dimer and so represents the structure found in solution, or whether the lack of symmetry is the result of the differing crystal packing of the protomers within the asymmetric unit. Conversely an apparent crystallographic diad may result from an averaging along the pseudo-diad of an asymmetric dimer as may be the case in the crystal structure of the dimeric tRNA synthetase studied by Blow and his colleagues (Irwin, Nyborg, Reid, and Blow 1976).

Although pure rotational symmetries of three and higher orders are less common, they do exist. Crystalline glucagon exists as a cubic array of molecules related by threefold axes (see Fig. 33.8) and it appears to exist as a trimer at certain concentrations in solution (Sasaki *et al.* 1975). However, glucagon is an unusual polypeptide comprising an α-helical secondary structure without a proper tertiary structure. The existence of an organized secondary structure is dependent upon quaternary interactions. The cylindrical shape of the glucagon molecule may make the trimer more stable than a dimer where less extensive interactions between helices would be possible.

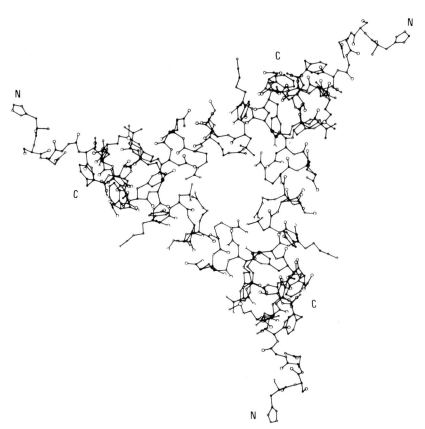

FIG. 33.8. The helices and side-chains of the porcine glucagon trimer, looking down the threefold rotation axis. (Reproduced with permission from Sasaki *et al.* 1975.)

Two other trimers of more conventional globular proteins have been identified by crystallography in a bacterial aldolase (Mavrides and Tulinsky 1976) and the bacterio-chlorophyll protein (Matthews, Fenna, and Remington 1977). In the latter case a diad may be functionally less useful than a threefold axis where the electronic transition dipoles in the symmetry equivalent chlorophyll molecules would always be orientated differently and should be more efficient in absorbing light.

Functional advantages of symmetry and its implications for evolution

The widespread existence of symmetrical oligomeric proteins indicates that such arrangements are thermodynamically stable. Similar criteria may apply to the symmetrical arrangements of folding units and domains. For instance, the high symmetry of γ-crystallins may relate to the need for stable proteins which last the lifetime of the animal in the denucleated cells of the lens. Although symmetry may be functionally advantageous in oligomers by allowing positive co-operativity as pointed out by Monod *et al.* (1965), unsymmetrical structures are required to explain negative co-operativity (Kirtley and Koshland 1967). In a similar way, symmetrical arrangements of folding units may have advantages in binding dinucleotide coenzymes (NAD to dehydrogenases), but substrates in an active site need to be in an asymmetric environment if a push–pull mechanism is involved. Thus in the acid proteinases the two aspartates are topologically equivalent and hydrogen-bonded together. However, the amino-acid side-chains close to the two aspartates in the three-dimensional structure are quite different and are not related by twofold symmetry. This allows Asp 32 to be closely hydrogen-bonded, giving it a low pK while Asp 215 is more accessible to solvent and has a more normal pK. The requirement for an asymmetric active site in rhodanese is achieved by having it between the two topologically equivalent domains but not on the diad axis.

The topological equivalence of structures within the polypeptide chain may indicate convergent evolution towards thermodynamically stable structures, but it may also be the result of divergent evolution after gene duplication followed by fusion. In most symmetrically arranged equivalent structures discussed here the sequence homology is very low and is generally not indicative of the three-dimensional homology. Clearly topologically equivalent structures can be assembled from quite different sequences.

With the exception of haemerythrin, the arguments in favour of evolution by gene duplication appear to be stronger for domains. In the acid proteinases an ancestral protomer of molecular weight $\sim 17\,000$ with no protease activity giving dimers hydrogen-bonded through antiparallel pleated sheet may have given rise after gene duplication and divergence to an asymmetric dimer with proteolytic activity. Gene fusion to give an acid proteinase of molecular weight $\sim 35\,000$ would then be selectively advantageous (Tang *et al.* 1978). However, in rhodanese the active site cleft for small substrates is almost entirely in one domain and it is possible that some rhodaneses (for example, that from rat liver) may be smaller proteins corresponding to one domain. The bilobal rhodanese may then have arisen as suggested by Ploegman *et al.* (1978) by gene duplication, divergence, and gene fusion. The bilobal enzyme then would be selectively advantageous in recognizing and binding larger substrates. In fact, differences between the two domains of rhodanese are largest between the active site and the topologically equivalent region.

If the γ-crystallin II molecule evolved via gene duplication and fusion then the topological equivalence-fit errors suggest that the first duplication/fusion resulted in the ancestor of one barrel, and that a subsequent duplication/fusion of that whole sequence produced the younger, two-barrelled structure. The connecting peptide lying off the inter-barrel diad is consistent with this scheme. Unfortunately, the whole process requires a single Greek-key β-sheet to provide the initial ancestral structure, and single, four-stranded motifs have not as yet been found on their own. However, once duplicated, the local diads of the γ-crystallin domains enable the structure to be close-packed as pseudo-isologous pairs of units, thus exceptionally stable, giving the γ-crystallin molecule a lifetime of years in the mammalian eye lens.

References

ADAMS, M. G., BLUNDELL, T. L., DODSON, E. J., DODSON, G. G., VIJAYAN, M., BAKER, E. N., HARDING, M. M., HODGKIN, D. C., RIMMER, B., and SHEET, S. (1969). *Nature, Lond.* **224**, 491–2.

ANDREEVA, N. S. and GUSTCHINA, A. E. (1979). *Biochem. biophys. Res. Commun.* **87**, 32–42.

BANNER, D. W., BLOOMER, A. C., PETSKO, G. A., PHILLIPS, D. C., and WILSON, I. A. (1976). *Biochem. biophys. Res. Commun.* **72**, 146–55.

BLUNDELL, T. L., SEWELL, B. T., and MCLACHLAN, A. D., (1979). *Biochim. biophys. Acta* (in press).

——CUTFIELD, J. F., DODSON, G. G., DODSON, E. J., HODGKIN, D. C., MERCOLA, D., and VIJAYAN, M. (1971). *Nature, Lond.* **231**, 506–9.

CUTFIELD, J. F., CUTFIELD, S. M., DODSON, E. J., DODSON, G. G., HODGKIN, D. C., and SABESAN, M. N. (1974). *J. molec. Biol.* **87**, 23–30.

EDELMAN, G. M. (1970). *Scient. Am.* **223**, 34–42.

HENDRICKSON, W. A. and WARD, K. B. (1977). *J. biol. Chem.* **252**, 3012–18.

IRWIN, M. J., NYBORG, J., REID, B. R., and BLOW, D. M. (1976). *J. molec. Biol.* **105**, 577–86.

KIRTLEY, M. E. and KOSHLAND, D. E. (1967). *J. biol. Chem.* **242**, 4192.

LEVINE, M., MUIRHEAD, H., STAMMERS, D. K., and STUART, D. I. (1978). *Nature, Lond.* **271**, 626–30.

MATTHEWS, B. W. and BERNHARD, S. A. (1973). *A. Rev. Biophys. Bioeng.* **2**, 257–317.

——FENNA, R. E., and REMINGTON, S. J. (1977). *J. ultrastruct. Res.* **58**, 316–30.

MAVRIDES, I. M. and TULINSKY, A. (1976). *Biochemistry* **15**, 4410–17.

MONOD, J., WYMAN, J., and CHANGEUX, J. P. (1965).. *J. molec. Biol.* **12**, 88–118.

PLOEGMAN, J. H., DRENT, G., KALK, K. H., and HOL, W. G. T. (1978). *J. molec. Biol.* **123**, 557–94.

POLJAK, R. J. (1975). *Nature, Lond.* **256**, 373–6.

RICHARDSON, J. S. (1977). *Nature, Lond.* **268**, 495–500.

ROSSMANN, M. G., MORAS, D., and OLSEN, K. W. (1974). *Nature, Lond.* **250**, 194–9.

SASAKI, K., DOCKERILL, S., ADAMIAK, D. A., TICKLE, I. J., and BLUNDELL, T. L. (1975). *Nature, Lond.* **257**, 751–7.

TANG, J., JAMES, M. N. G., HSU, I.-N., JENKINS, J. A., and BLUNDELL, T. L. (1978). *Nature, Lond.* **271,** 618–21.
—— SEPULVEDA, P., MARCINISZYN, J., JR, CHEN, K. C. S., HUANG, W.-Y., TAO, N., LIU, D., and LANIER, J. P. (1973). *Proc. natn. Acad. Sci. USA* **70,** 3437–9.
THOMPSON, D. (1952). *On growth and form.* Cambridge University Press.

V Insulin: biology, chemistry, and structure

34. Insulin precursors

DONALD F. STEINER

Perfection is finally attained not when there is no longer anything to add, but when there is no longer anything to take away.

Antoine De Saint-Exupéry

WITH the elucidation by Sanger and his associates (Ryle, Sanger, Smith, and Kitai 1955; Sanger 1959) of the unique double-chain structure and the amino-acid sequence of the insulin molecule one of the cornerstones of molecular biology was put in place, and the way was cleared for the elucidation of the genetic code and for further studies on the three-dimensional structure of protein molecules through X-ray crystallographic analysis. The elucidation of the structure of insulin to 1.9 Å by Dorothy Hodgkin and her team of brilliant young associates (Adams *et al.* 1969; Hodgkin 1974) represented an important advance in the field of protein hormone chemistry and provided the key with which the chemical, and hopefully the biological, secrets of this fascinating molecule might at last be unlocked. This work provided an intense stimulus to a field which had begun to wither for lack of progress. Indeed, the X-ray structure immediately shed light on much of the chemical and immunological data on insulin that had been amassed for many years and it prompted much further work on insulin derivatives of all kinds. It also provided a clear basis for understanding the conservation of certain primary structural features in the insulin molecule throughout the whole of vertebrate evolution; some of these features are the positions of the 3 disulphide bonds, the N-terminal and C-terminal regions of the A chain, and the hydrophobic residues clustered in several regions in the B chain. As had been surmised, conservation of certain amino acids turned out to be essential for the maintenance of the secondary and the tertiary structure of the hormone, and even, to some extent, of its quaternary structure (Blundell, Dodson, Hodgkin, and Mercola 1972). However, it has been difficult to clearly identify an active centre in the insulin molecule because of its parsimonious construction. The topography of its binding region seems to be so closely dependent on the hormone's overall structure that it is difficult to perturb the structure without at the same time perturbing function. No residues thus stand out as uniquely contributing only to a binding site. This aesthetically rather pleasing and satisfying aspect of insulin's chemistry has thus proved to be an obstacle to the facile solution of its biological mechanism of binding and action. Perhaps in consequence the gainful employment of a whole generation of bioscientists has been spared, and this problem remains to challenge imaginative and undaunted young insulin researchers.

A somewhat less intractable problem has diverted my colleagues and me from the study of insulin action over the past decade, and in this work the knowledge provided by the elucidation of the structure of insulin by the combined efforts of Professors Sanger and Hodgkin and their co-workers has also proved to be invaluable in the design of experiments and the interpretation of results. This is the problem of how the two-chain insulin structure is assembled in nature and how it is then stored within the granules of islet cells so as to be available in suitable quantities to meet the constantly fluctuating demands of metabolism. Studies on these problems have helped to expand our horizons with regard to several aspects of insulin, one of the most interesting of these being the gradual identification of other biologically active protein molecules that are related to insulin but derived through the evolutionary permutation of its single-chain precursor form, proinsulin (Steiner and Oyer 1967; Chance, Ellis, and Bromer 1968). Among these are nerve growth factor (Bradshaw, Hogue-Angeletti, and Frazier 1974), insulin-like growth factor or IGF (Rinderknecht and Humbel 1978), relaxin (Isaacs *et al.* 1978), and probably others as yet unknown.

The identification of proinsulin and, more recently, of preproinsulin has also altered our concepts of the nature and structure of the genes for insulin as we now recognize that the 51 residue insulin molecule is derived from a single gene which codes for a polypeptide containing 110 amino acids. Within the additional 59 amino acids of preproinsulin is the information required to direct its sequestration and ultimate storage within secretory granules (Steiner *et al.* 1979), as well as the means to direct its efficient folding and the correct oxidation of its sulphydryls (Steiner and Clark 1968). The elucidation of the structure of preproinsulin has utilized both the classical methods of protein chemistry derived from the work of Fred Sanger and of Per Edman as well as the more recently developed methods for microsequencing tiny amounts of biosynthetically labelled peptide material (Tager, Emdin, Clark, and Steiner 1973; Kemper, Habener, Ernst, Potts, and Rich 1976) and the very new indirect sequencing methods based on the elucidation of the nucleotide coding sequence of the messenger RNA molecules (Ullrich *et al.* 1977; Villa-Komaroff *et al.* 1978; Nakanishi *et al.* 1979).

Proinsulin (Steiner 1967; Chance *et al.* 1968; Grant and Reid 1968) provided the first glimpse of a steadily enlarging group of precursors which share some structural features at cleavage sites and appear to be processed by specialized intracellular proteinases localized in or near the Golgi apparatus and

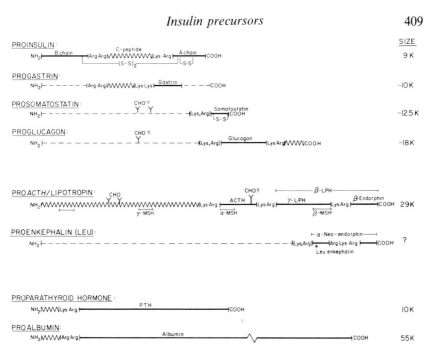

FIG. 34.1. Structures of known proproteins showing paired basic residues at cleavage sites. Heavy lines indicate regions which appear as biologically active or useful products. Unsequenced regions are indicated by the dashed lines. Data sources are as follows: proinsulin (Steiner *et al.* 1972; Chance *et al.* 1968); progastrin (Noyes, Mevarech, Stein, and Agarwal 1979); prosomatostatin (Patzelt *et al.* 1980); proglucagon (Tager and Steiner 1973; Patzelt *et al.* 1979); proACTH/lipotropin (Mains *et al.* 1977; Nakanishi *et al.* 1979); proenkephalin (Kanagawa and Matsuo 1979); proparathyroid hormone (Hamilton *et al.* 1974); proalbumin (Russell and Geller 1975).

the developing secretion granules (Steiner 1978). The complete or partial structures of eight such proproteins are represented diagrammatically in Fig. 34.1. Proinsulin contains the A and B chains of insulin linked together as a larger single chain by a highly species-variable C-segment containing at most 35 residues (Steiner, Kemmler, Clark, Oyer, and Rubenstein 1972; Steiner, Kemmler, Tager, and Peterson 1974; Steiner *et al.* 1976). The connecting segment or C-peptide clearly facilitates the formation of the correct disulphide-linkages (Steiner and Clark 1968). However, this role cannot explain the great variability and the rather excessive length of the connecting segment which exceeds by many times the 8–10 Å gap between the COOH-terminus of the B chain and the NH$_2$-terminus of the A chain in insulin (Blundell *et al.* 1972). Moreover, its structure-inducing function can be reproduced by simple non-peptide bifunctional cross-linking reagents (Brandenburg and Wollmer 1973; Busse, Hansen, and Carpenter 1974). It is therefore more likely that the connecting segment also functions as a highly variable spacer

(Steiner 1978), serving to enlarge the peptide chain to a length of 65–70 residues so that it can span the distance from the site of peptide chain synthesis between the large and small ribosomal subunits to the luminal side of the RER (Patzelt *et al.* 1978*a*). This 'minimum length hypothesis' allows us to predict that many other small secreted peptides and hormones containing less than about 50 residues will require spacer regions for biosynthesis and that their enlarged forms will be similar in size to proinsulin. The constantly enlarging body of information on prosecretory forms tends to support this notion (see Fig. 34.1).

Most of the proproteins contain pairs of basic residues at their sites of cleavage, as shown in Fig. 34.1, and arginine rather than lysine seems to be preferred on the carboxyl side of the pair. In the case of proinsulin a combined mechanism involving both an endoprotease and an exopeptidase, e.g. a mixture of ordinary cationic trypsin and pancreatic carboxypeptidase B, can readily reproduce the cleavage patterns seen *in vivo* to yield the known correct products, i.e. insulin, C-peptide, 3 arginine, 1 lysine (Kemmler, Peterson, and Steiner 1971). Similar considerations apply as well to proparathyroid hormone conversion (Habener, Chang, and Potts 1977). The secretion and metabolic fate of the C-peptide has been studied extensively by Rubenstein and his co-workers and has proved useful for many diagnostic purposes as well as for studies of the natural history of diabetes (Rubenstein *et al.* 1977). Several families have been identified with high levels of circulating proinsulin (Gabbay, Deluka, Fisher, Mako, and Rubenstein 1976) or proalbumin (Brennan and Carrell 1978). The affected individuals evidently carry point mutations which result in the replacement of one of the basic residues by a neutral residue in sites of cleavage, thereby lowering their susceptibility to converting enzymes (Brennan and Carrell 1978). The proinsulin defect appears to involve only the cleavage site between the B chain and the C-peptide (Gabbay *et al.* 1977).

Recent studies by Dr Christoph Patzelt have provided additional information on the biosynthesis of two other important hormones of the pancreatic islets (Patzelt *et al.* 1979, 1980). Analysis of tryptic maps of their putative precursors has shown that a trypsin-like enzyme would be sufficient to cleave the COOH-terminal somatostatin sequence from its precursor, but that both trypsin-like and carboxypeptidase B-like enzymes would be necessary to cleave the internally located glucagon sequence from its prohormone. The structures we have proposed for these precursors are shown in Fig. 34.1, and further studies on their isolation and characterization are now under way.

Preproinsulin was first identified through translation of insulin mRNA in a cell-free system prepared from wheatgerm (Chan, Keim, and Steiner 1976). Subsequent studies have shown that preproinsulin can be detected at low levels in intact islets of Langerhans where it turns over with a half-life of a few minutes (Patzelt *et al.* 1978*b*). As in the case of the precursors of immunoglobulin light chains first identified by Milstein and his co-workers (Milstein, Brownlee, Harrison, and Matthews 1972), and preproparathyroid hormone (Kemper *et al.* 1976), preproinsulin contains an additional peptide segment located at its N-terminus (Chan *et al.* 1976). The presecretory proteins, as these forms with extended N-termini have been designated, are rapidly cleaved by microsomal peptidases (Jackson and Blobel 1977), and they clearly function to guide secretory proteins across the membrane of the rough endoplasmic reticulum in preparation for their intracellular transport to the Golgi area where they are (usually) packaged into storage granules. The mechanism of transmembrane transfer of the presecretory proteins is

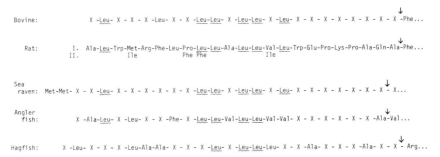

FIG. 34.2. Partial amino-acid sequence of the N-terminal extension of preproinsulin of several vertebrate species. Data sources are as follows: Bovine (Lomedico, Chan, Steiner, and Saunders, 1977); Rat (Chan *et al.* 1976, and unpublished data; Ullrich *et al.* 1977; Villa-Komaroff *et al.* 1978); Sea Raven and Angler fish (Shields and Blobel 1977); Hagfish (S. J. Chan, unpublished data). The positions of tryptophan, arginine, lysine, glutamic acid, and glutamine residues in the rat I prepeptide are based on nucleotide sequencing data alone. All other assignments were made on the basis of radiosequencing results.

not clearly understood but is believed to depend on the strongly lipophilic nature of the prepeptide. Several mechanisms have been proposed and discussed in detail elsewhere (Blobel and Dobberstein 1975; Steiner *et al.* 1980; Inouye and Halegouya 1979).

Dr Shu Jin Chan's initial studies on the structure of the rat prepeptide were carried out by a sensitive radiosequencing procedure in which the preproinsulin was first labelled during its biosynthesis in the wheatgerm system with amino acids of high specific activity. Only one, or at most two, radioactive amino acids were incorporated at a time. After selective immunoprecipitation of the preproinsulin by insulin antisera the radioactive material was placed in the spinning cup of the Beckmann Sequencer and subjected to 30–40 cycles of Edman degradation. The PTH derivatives from each cycle were then counted and further analysed, if necessary, to identify the positions along the chain occupied by labelled amino acids such as leucine, phenyl-alanine, proline, or others that had been incorporated biosynthetically into the preproinsulin (Chan *et al.* 1976). These experiments, in our laboratory and others, have resulted in the partial elucidation of the amino-acid sequences of the prepeptides of preproinsulins from a broad selection of vertebrates (Fig. 34.2). The results, although incomplete, show that preproinsulin occurs in the most primitive vertebrates as well as in the teleost fishes. They also indicate that the structure of the prepeptide is reasonably well conserved, perhaps especially the hydrophobic central region extending from residues -8 to -17, which may be the functionally most critical part of the prepeptide in its interaction with biological membranes (Steiner *et al.* 1980).

Since it is well established that eukaryotic proteins are all initiated with methionine, we were worried when our initial sequencing results did not identify an N-terminal methionine residue. We knew that initiating methionine residues followed by neutral amino acids such as alanine would most likely be removed by an amino peptidase in the wheatgerm system, so that this seemed a reasonable explanation for the absence of this residue. However, to be certain that some of the initial translation product had not been lost via proteolytic degradation during incubation in the wheatgerm system, we designed experiments, in collaboration with Drs Eric Ackerman and Paul B. Sigler of the Department of Biophysics, to determine unequivocally the location of the initiator methionine residue of preproinsulin. To accomplish this, it would be necessary to block the amino group of the initiating methionine residue so that it no longer could be removed by the wheatgerm amino-peptidase. This was done by charging yeast initiator Met–tRNA with ^{35}S-methionine and blocking the amino group with [10]N-formyl tetrahydrofolate and *E. coli* transformylase. A large excess of formylated initator Met-tRNA was then added to the wheatgerm system during the translation of rat insulin mRNA. The positions of phenylalanine in the cell-free product were labelled by adding ^{3}H-phenylalanine. After synthesis the purified cell-free product was dansylated to inactivate peptides initiated by competing non-formylated endogenous initiator tRNA, and the formylated peptide was deblocked selectively at the N-terminus by mild acid hydrolysis. The product was then radio-sequenced and yielded the results shown in Fig. 34.3. As expected, a peak of ^{35}S-Met was obtained on the first cycle of degradation and the phenyl-alanine residues at position 6 as well as at the beginning of the insulin B chain (position 24) were shifted to the right by one cycle. These and similar experiments which will be reported in greater detail elsewhere (Chan, Ackerman, Quinn, Sigler, and Steiner 1979*a*) have established that methionine indeed initiates preproinsulin synthesis and that this residue is located just before the N-terminal alanine previously identified; thus the overall length of the rat prepeptide is 24 residues.

We were faced with another, less readily resolved, difficulty in studying the structure of rat preproinsulin, due to the presence of two non-allelic genes coding for insulin in rats, as well as in mice (Clark and Steiner 1969; Markussen 1971). Thus, in our experiments we were preparing and sequencing mixtures of two closely related proteins whose messenger RNAs were almost identical in size and not easily resolvable. We had recognized and discussed this problem in our initial radiosequencing studies on the preproinsulin and had, in fact, identified several positions where more than one amino acid appeared to occupy the same position (Chan *et al.* 1976). With the advent

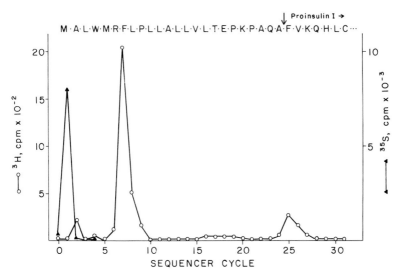

FIG. 34.3. Identification of the initiator methionine residue in rat preproinsulin. Preproinsulin was prepared in the wheatgerm system and immunoprecipitated as described in the text. Note that while ^{35}S-methionine was released on the first degradation cycle, subsequent fractions were not labelled, including position 5, the location of an internal residue of methionine. The positions of phenylalanine are shifted one position to the right, as compared with their positions shown in Fig. 34.2.

of recombinant DNA techniques a new and powerful method for attacking the problem of resolving admixtures of closely related mRNAs became available. We accordingly set out to make complementary DNA strands to partially purified insulin mRNA, in preparation for the incorporation of this material into a suitable plasmid for transfection into *E. coli*. While our experiments were in progress Ullrich *et al.* (1977) and Villa-Komaroff *et al.* (1978) succeeded in cloning and sequencing nearly all of the coding sequence of rat preproinsulin I. Their elegant results confirmed much of our published (as well as unpublished) radiosequencing data on rat preproinsulin, added new information, and allowed us to begin to sort out those positions where dual amino-acid assignments had been made.

However, their work left unconfirmed the first three residues of rat preproinsulin I, as well as the entire prepeptide of rat preproinsulin II. We therefore proceeded with our cloning studies using special procedures to ensure
the cloning of nearly complete transcripts of the rat insulin mRNA and the
unequivocal identification of colonies containing as much of the 5′ region
of the mRNA as possible (Chan, Noyes, Agarwal, and Steiner 1979*b*). In
order to accomplish this goal we first prepared double-stranded cDNA from
insulin mRNA using an oligo dT primer and reverse transcriptase for the
first strand synthesis, followed by *E. coli* polymerase I to complete the complementary DNA strand. The double-stranded DNA was treated with S-1
nuclease and then electrophoresed on a polyacrylamide gel. Material ranging
in estimated size from 400 to 600 nucleotides was eluted from the gel and

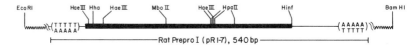

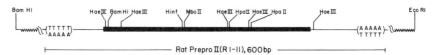

FIG. 34.4. Partial restriction maps of plasmids pRI-7 and pRI-11 containing preproinsulin I and II cDNAs. Construction, selection, and characterization of the plasmids
from rat insulinoma poly(A)-RNA are described elsewhere (Chan *et al.* 1979*b*). Orientation shown is that of the mRNA sense strand, 5′ to 3′ from left to right. The thick
line indicates the coding region for preproinsulin.

tailed with poly dA/T for insertion into the cloning vehicle pBR 322.
Approximately 400 transformants were obtained when the recombinant plasmids were transfected into the host organism X1776. To select insulin-
related sequences a cDNA probe was generated from insulin mRNA using
a synthetic deoxydecanucleotide primer prepared by Dr Kan Agarwal in
this department and having a nucleotide sequence corresponding to the
region coding for amino acids 11–13 plus the first nucleotide of the codon
for residue 14 of the B chain of rat insulin I as reported by Ullrich *et al.*
(1977). After purification by gel electrophoresis, a 170 nucleotide insulin
mRNA derived component was used in colony hybridization assays to
select 16 positive colonies. Restriction enzyme analysis of plasmid DNA
from each of the 16 clones indicated that seven contained nearly complete
rat insulin I DNA sequences while the remaining nine contained the DNA
sequences of rat insulin II (Fig. 34.4).

Plasmid pRI-11 corresponding to rat insulin II was partially sequenced

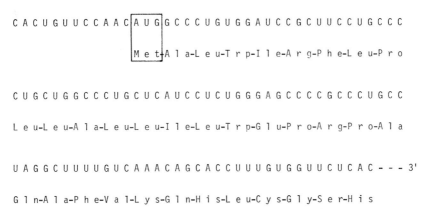

5' - - - - G A C C U G C U U G C U G A U G G U U U C C G A C U A U A G C U G G U

C A C U G U U C C A A C A U G G C C C U G U G G A U C C G C U U C C U G C C C

Met-Ala-Leu-Trp-Ile-Arg-Phe-Leu-Pro

C U G C U G G C C C U G C U C A U C C U C U G G G A G C C C C G C C C U G C C

Leu-Leu-Ala-Leu-Leu-Ile-Leu-Trp-Glu-Pro-Arg-Pro-Ala

U A G G C U U U U G U C A A A C A G C A C C U U U G U G G U U C U C A C - - - 3'

Gln-Ala-Phe-Val-Lys-Gln-His-Leu-Cys-Gly-Ser-His

FIG. 34.5. Nucleotide sequence from the 5′ region of the mRNA for rat preproinsulin II. The amino-acid sequence of the prepeptide and a portion of the B chain are shown below the corresponding codons. The initiator methionine codon is enclosed in the box. The presence of serine at position 9 of the B chain confirms that this sequence is from preproinsulin II.

by Dr Barbara Noyes to confirm its identification and obtain information on the amino-acid sequence of the prepeptide of rat preproinsulin II as well as on the nucleotide sequence in the 5′ untranslated region of its mRNA. The results summarized in Fig. 34.5 confirmed that this clone contained the complete coding sequence for preproinsulin II as well as much of the 5′ untranslated region of the mRNA (Chan *et al.* 1979*b*). The prepeptide is 24

FIG. 34.6. Complete primary structures of the two rat preproinsulins. Residues enclosed in boxes differ between the two proteins. Arrows show points of cleavage in the conversion to insulin. Data sources are as follows: residues −24 to +1 (Chan *et al.* 1976; Chan *et al.* 1979*b*; Ullrich *et al.* 1977; Villa-Komaroff 1978); residues 1–30 (Smith 1966; Clark and Steiner 1969); residues 31–65 (Markussen and Sundby 1972; Sundby and Markussen 1972; Tager and Steiner 1972); residues 66–86 (Smith 1966).

residues long, including the initiator methionine residue, and this analysis thus confirms our previous assignment of the initiator methionine residue just before the N-terminal alanine residue identified in radiosequencing studies of the cell-free translation product of insulin mRNA (Fig. 34.3). A comparison of the amino-acid sequences of the two rat prepeptides (Fig. 34.6) shows that these are well conserved, being identical in length and differing only at three positions where conservative substitutions have occurred. Further DNA sequence analysis will be required to elucidate the complete nucleotide sequence of the two rat insulin mRNAs, but the results already at hand allow us to complete the amino-acid sequences of both rat preproinsulins and have verified the validity of much of our previous radiosequencing results for the presequences. The two sequences shown in Fig. 34.6 differ at only seven positions out of a total of 110, indicating that the gene duplication that led to their creation was probably a relatively recent evolutionary event. There are indications, however, of further amino-acid differences in the prepeptide sequences among various rat strains that we have studied, suggesting the possible existence of additional gene copies or of allelic forms of one or both insulin genes in rat populations.† Further studies now in progress in several laboratories on the chromosomal genes for insulin using cloned cDNAs corresponding to insulin mRNA sequences as probes for their identification should shed further light on this and related questions regarding the structure and evolution of the genes for insulin. Thus it is most likely that the year 1979 will mark not only the culmination of our efforts of many years to determine the initial form in which insulin is synthesized, but also the isolation and structural elucidation of the chromosomal genes which code for this hormone. The availability of this information should open the way to a better understanding of insulin production in the organism and may in the future provide new insights into the nature and cause of human diabetes as well as new approaches to its treatment.

† Such differences could explain our earlier finding of phenylalanine residues at positions 8 and 9 of the rat prepeptides (Fig. 34.2), as well as of lysine residues at positions 3 and 11 (Chan *et al.* 1976). We have recently confirmed that the two phenylalanines appear only in the cell-free translation product of insulin mRNA from Sprague–Dawley rats. The studies described here were on NEDH strain rats.

Acknowledgement

It is a great pleasure to dedicate this paper to Professor Dorothy Hodgkin who has inspired and stimulated several generations of students and scientists throughout the world with her brilliant research work, her remarkable and genuine humanitarianism, and her great personal integrity, warmth, and humility. I also wish to salute the many fine students and associates who have contributed in so many important ways to the elucidation of the insulin biosynthetic pathway. These include Philip Oyer, Dennis Cunningham, Jeffrey Clark, Bradley Aten, Lilian Spigelman, Charles Birdwell, James

McKenzie, Franco Menani, Wolfgang Kemmler, Susan Terris, Hiroyuki Sando, Simon Pilkis, Jo Borg, Sooja Cho Nehrlich, James D. Peterson, John Duguid, Raymond Carroll, Allen Labrecque, Chris Nolan, Emmanuel Margoliash, Peter Lomechico, Glen Hortin, and Arthur H. Rubenstein. For the work discussed in this paper I am especially indebted to Shu Jin Chan, Barbara E. Noyes, Paul Quinn, Christoph Patzelt, Pamela Keim, Robert Heinrikson, Eric Ackerman, Paul B. Sigler, Howard S. Tager, and Kan Agarwal. It is a pleasure also to acknowledge the fruitful collaboration of Stefan Emdin and Sture Falkmer in studies on the structure and biosynthesis of hagfish insulin (cr. Chapter 35). I also thank Lise McKean for assisting in the preparation of this manuscript.

Work in this laboratory has been supported by grants AM 13914, AM 20595, and AM 17046 of the USPHS and by grants-in-aid from the Lolly Couston Memorial Fund, the Kroc Foundation, the Novo Research Laboratories, the Lilly Research Laboratories, and Cetus Corporation.

References

ADAMS, M. J., BLUNDELL, T. R., DODSON, E. J., DODSON, G. G., VIJAYAN, M., BAKER, E. N., HARDING, M. M., HODGKIN, D. C., RIMMER, B., and SHEAT, S. (1969). *Nature Lond.* **224**, 491–5.

BLOBEL, G. and DOBBERSTEIN, B. (1975). *J. cell. Biol.* **67**, 835–51.

BLUNDELL, T., DODSON, G., HODGKIN, D., and MERCOLA, D. (1972). *Adv. Protein Chem.* **26**, 279–402.

BRADSHAW, R. A., HOGUE-ANGELETTI, R. A., and FRAZIER, W. A. (1974). *Recent Prog. Horm. Res.* **30**, 575–96.

BRANDENBURG, D. and WOLLMER, A. (1973). *Hoppe-Seyler's Z. Physiol. Chem.* **354**, 613–27.

BRENNAN, S. O. and CARRELL, R. W. (1978). *Nature, Lond.* **274**, 908–9.

BUSSE, W. D., HANSEN, S. R., and CARPENTER, F. H. (1974). *J. Am. chem. Soc.* **96**, 5949–50.

CHAN, S. J., KEIM, P., and STEINER, D. F. (1976). *Proc. natn. Acad. Sci. U.S.A.* **73**, 1964–68.

—— ACKERMAN, E., QUINN, P. SIGLER, P., and STEINER, D. F. (1979*a*). Manuscript in preparation.

—— NOYES, B. E., AGARWAL, K. L., and STEINER, D. F. (1979*b*). *Proc. natn. Acad. Sci. U.S.A.* **76**, 5036–40.

CHANCE, R. E., ELLIS, R. M., and BROMER, W. W. (1968). *Science N.Y.* **161**, 165–7.

CLARK, J. L. and STEINER, D. F. (1969). *Proc. natn. Acad. Sci. U.S.A.* **62**, 278–85.

GABBAY, K. H., WOLFF, J., BERGENSTAL, R., MAKO, M., and RUBENSTEIN, A. H. (1977). *Diabetes* **26** (Suppl. 1), 376.

—— DELUCA, K., FISHER, J. N., MAKO, M. E., and RUBENSTEIN, A. H. (1976). *New Engl. J. Med.* **294**, 911–15.

GRANT, P. T., and REID, K. B. M. (1968). *Biochem. J.* **110**, 281–8.

HABENER, J., CHANG, H. T., and POTTS, J. T., JR (1977). *Biochemistry* **16**, 3910–17.

HAMILTON, J. W., NIALL, H. D., JACOBS, J. W., KEUTMANN, H. J., POTTS, J. T., JR, and COHN, D. V. (1974). *Proc. natn. Acad. Sci. U.S.A.* **71**, 653.

HODGKIN, D. C. (1974). *Proc. R. Soc.* **A338**, 251–75.

INOUYE, M. and HALEGOUA, S. (1979). *Chemical Rubber Co. Reviews.* (in press).

ISAACS, N., JAMES, R., NIALL, H., BRYANT-GREENWOOD, G., DODSON, G., EVANS, A., and NORTH, A. C. T. (1978). *Nature, Lond.* **271**, 278–81.

JACKSON, R. C. and BLOBEL, G. (1977). *Proc. natn. Acad. Sci. U.S.A.* **74,** 5598–602.

KANGAWA, K. and MATSUO, H. (1979). *Biochem. biophys. Res. Commun.* **86,** 153–60.

KEMMLER, W., PETERSON, J. D., and STEINER, D. F. (1971). *J. biol. Chem.* **246,** 6786–91.

KEMPER, B., HABENER, J., ERNST, M. D., POTTS, J., JR, and RICH, A. (1976). *Biochemistry* **15,** 15–19.

LOMEDICO, P. T., CHAN, S. J., STEINER, D. F., and SAUNDERS, G. F. (1977). *J. biol. Chem.* **252,** 7971–8.

MAINS, R. E., EIPPER, B. A., and LING, N. (1977). *Proc. natn. Acad. Sci. U.S.A.* **74,** 3014–18.

MARKUSSEN, J. (1971). *Intl. J. Protein Research* **III,** 149–55.

—— and SUNDBY, F. (1972). *Eur. J. Biochem.* **25,** 153–62.

MILSTEIN, C., BROWNLEE, G. G., HARRISON, T. M., and MATHEWS, M. B. (1972). *Nature New Biol.* **239,** 117–20.

NAKANISHI, S., INOUE, A., KITA, T., NAKAMURA, M., CHANG, A. C. Y., COHEN, S. N., and NUMA, S. (1979). *Nature, Lond.* **278,** 423–7.

NOYES, B. E., MEVARECH, M., STEIN, R., and AGARWAL, K. L. (1979). *Proc. natn. Acad. Sci. U.S.A.* **76,** 1770–4.

PATZELT, C., TAGER, H. S., CARROLL, R. J., and STEINER, D. F. (1979). *Nature, Lond.* **282,** 260–6.

———————(1980). *Proc. natn. Acad. Sci. U.S.A.* (in press).

——CHAN, S. J., DUGUID, J., HORTIN, G., KEIM, P., HEINRIKSON, R. L., and STEINER, D. F. (1978a). In *Regulatory proteolytic enzymes and their inhibitors. Proc.* Vol. 47, Symp. A6 (ed. S. Magnusson) pp. 69–78, Pergamon Press, New York.

——LABRACQUE, A., DUGUID, J., CARROLL, R., KEIM, P., HEINRIKSON, R. L., and STEINER, D. F. (1978b). *Proc. natn. Acad. Sci. U.S.A.* **75,** 1260–4.

RINDERKNECHT, E. and HUMBEL, R. E. (1978). *J. biol. Chem.* **253,** 2769–76.

RUBENSTEIN, A. H., STEINER, D. F., HORWITZ, D. L., MAKO, M. E., BLOCK, M. B., STARR, J. I., KUZUYA, A. H., and MELANI, F. (1977). *Recent Prog. Horm. Res.* **33,** 435–68.

RYLE, A. P., SANGER, F., SMITH, L. F., and KITAI, R. (1955). *Biochem. J.* **60,** 541–56.

RUSSELL, J. H. and GELLER, D. M. (1975). *J. biol. chem.* **250,** 3409–13.

SANGER, F. (1959). *Science, N.Y.* **129,** 1340–4.

SHIELDS, D. and BLOBEL, G. (1977). *Proc. natn. Acad. Sci. U.S.A.* **74,** 2059–63.

SMITH, L. F. (1966). *Am. J. Med.* **40,** 662–6.

STEINER, D. F. (1967). *Trans. N.Y. Acad. Sci.* Ser. II, **30,** 60–8.

——(1978). *Diabetes* **27** (Suppl. 1), 145–8.

——and CLARK, J. L. (1968). *Proc. natn. Acad. Sci. U.S.A.* **60,** 622–9.

——and OYER, P. E. (1967). *Proc. natn. Acad. Sci. U.S.A.* **57,** 473–80.

—— KEMMLER, W., TAGER, H. S., and PETERSON, J. D. (1974). *Fedn Proc. Fedn Am. Socs exp. Biol.* **33,** 2105–15.

—— TERRIS, S., CHAN, S. J., and RUBENSTEIN, A. H. (1976). In *Insulin.* (ed. R. Luft) pp. 53–107. A. Lindgren & Søner AB, Stockholm.

—— KEMMLER, W., CLARK, J. L., OYER, P. E., and RUBENSTEIN, A. H. (1972). In *Handbook of physiology—endocrinology I.* (ed. D. F. Steiner and N. Freinkel) pp. 175–98. Williams and Wilkins, Baltimore.

—— QUINN, P. S., PATZELT, C., CHAN, S. J., MARSH, J., and TAGER, H. S. (1980). In *Cell biology: a comprehensive treatise* (ed. L. Goldstein and D. M. Prescott), Vol. IV, pp. 175–202. Academic Press, New York.

SUNDBY, F. and MARKUSSEN, J. (1972). *Eur. J. Biochem.* **25,** 147–52.

TAGER, H. S. and STEINER, D. F. (1972). *J. biol. Chem.* **247,** 7936–40.
———— (1973). *Proc. natn. Acad. Sci. U.S.A.* **70,** 2321–5.
——EMDIN, S. O., CLARK, J. L., and STEINER, D. F. (1973). *J. biol. Chem.* **248,** 3476–82.
ULLRICH, A., SHINE, J., CHIRGIVIN, J., PICTET, R., TISCHER, E., RUTTER, W. J., and GOODMAN, H. M. (1977). *Science, N.Y.* **196,** 1313–19.
VILLA-KOMAROFF, L., EFSTRATIADIS, A., BROOME, S., LOMEDICO, P., TIZARD, R., NABER, S. P., CHICK, W. L., and GILBERT, W. (1978). *Proc. natn. Acad. Sci. U.S.A.* **75,** 3727.

35. Insulin evolution

STURE FALKMER AND
STEFAN O. EMDIN

Introduction

ABOUT three years ago we made a brief review of insulin evolution (Emdin
and Falkmer 1977), written against the background of a concomitant survey
of the comparative endocrinology of the whole gastro-entero-pancreatic
(GEP) endocrine system (Falkmer and Östberg 1977). In these two accounts
it could be stated that no fundamental differences seemed to exist between
insulin and most other GEP hormones with regard to their evolution. Thus,
insulin-producing cells obviously exist, not only in the endocrine pancreas
or in a separate islet organ, but also as cells of open type in the mucosa of
the digestive tract and even in the central nervous system, as maintained by
Pearse in his so-called APUD concept (Pearse and Takor Takor 1976; Pearse
1978). Moreover, it could be shown that insulin is not only a polypeptide
hormone of considerable age but also that the insulin molecule seems to have
been kept surprisingly stable during evolution. Lastly, it was stated that the
most original vertebrate islet parenchyma was that of the jawless fish (the
cyclostomes), notably the islet organ of the hagfish, *Myxine glutinosa*. Consequently, particular attention was paid to the molecular biology of *Myxine*
insulin, which in these respects is now the best known of the non-mammalian
insulins.

The present report is an updated version of our preceding review (Emdin
and Falkmer 1977), giving also a survey of some current work on insulin
evolution in our laboratories and in those of our collaborators. It is written
against the background of a concomitant review of the comparative endocrinology of the whole GEP neuroendocrine system (VanNoorden and
Falkmer 1980).

Invertebrates

So far no insulin from any invertebrate species has been isolated and purified
to homogeneity in sufficient quantities to permit determination of the hor-

mone's amino-acid sequence, tertiary structure, receptor-binding affinity, or biological activity. For such analyses rather large quantities (at least about 10–20 mg of purified substance) are usually required, and the main reason for this lack of progress is obviously that an invertebrate species has still not been found in which it is possible to obtain sufficiently large amounts of tissue where the insulin cells are not so widely scattered as in the mucosa of the digestive tract (cf. Falkmer and Östberg 1977). Nevertheless some major steps forward have been taken during these last three years.

It is now well established that insects possess a hormone with insulin-like activity, claimed to be present in the alimentary tract, haemolymph, and—in particular—in the neuroendocrine system (Tager, Markese, Kramer, Speirs, and Childs 1976; Duve 1978; Duve, Thorpe, and Lazarus 1979). By experimental (Duve 1978) and immunocytochemical (Duve and Thorpe 1979) studies it could be shown that the hormone is produced by some cells in the median neurosecretory system in the brain of an insect, the blowfly, *Calliphora vomitoria*. About 120 000 blowflies were deep-frozen in bags. The bags were shaken vigorously, the heads of the flies detached, and collected by appropriate sieves. By this procedure about 460 g of fly heads could be used for partial purification of the hormone by means of acid–ethanol extraction, gel filtration, and ion-exchange chromatography. It cross-reacted with ox-insulin antibody, exhibited an R_f value identical with ox insulin on polyacrylamide gel electrophoresis, displaced specifically bound ^{125}I-insulin from its liver plasma membrane receptor, and displayed insulin-like biological activity on isolated rat fat-cells (Duve *et al.* 1979). After surgical removal of the median neurosecretory system, the blowflies became hypertrehalosaemic and hyperglycaemic (Duve 1978). The haemolymph levels of the two sugars were normalized by injection of the partially purified insulin-like substance, suggesting a physiological role of the hormone on carbohydrate metabolism (Duve *et al.* 1979). In the membrane-binding assays the blowfly insulin-like substance produced a displacement curve being steeper than that of ox insulin, indicating that the two hormones are conformationally different (Duve *et al.* 1979).

At the production site, it was observed that the neurosecretory cells, being immunoreactive with antisera to ox insulin, constituted only 6–8 out of the total of 24–26 cells in the median neurosecretory system (Duve and Thorpe 1979). The immunofluorescent cells were only faintly aldehydefuchsinophil, whereas strongly stained cells were non-immunoreactive. This observation indicates a heterogeneity of the median neurosecretory cells. It was hypothesized that the insulin-like substance, previously isolated from the corpora cardiaca/corpora allata complex of another insect, the tobacco hookworm, *Manduca sexta* (Tager *et al.* 1976), represents stored hormone, transported from the median neurosecretory system (Duve and Thorpe 1979). Whether the same holds true also for the insulin-like material observed in the alimentary tract (Ishay, Gitter, Galun, Doron, and Laron 1976) of insects and in honey-bee royal jelly (Kramer *et al.* 1977) is still unknown.

In any case, the localization of an insulin-like material in the neuroendocrine system of insects has several interesting evolutionary implications. It gives additional support to the theory that alimentary polypeptide hormones have originated from the neural ectoderm as neural transmitters or modulators, later to accept a new role as the vertebrate gastrointestinal tract developed (Pearse 1978; Barrington 1978; VanNoorden and Falkmer 1980).

The ability to detect insulin by immunological procedures, even in phylogenetically vastly separated phyla, has recently also been employed in studies on insulin-like substances in the alimentary tract of invertebrates. Thus, the cellular sites of the insulin production have been investigated in species from both the deuterostomian (Fritsch and Sprang 1977) and protostomian (Plisetskaya, Kazakov, Soltitskaya, and Leibson 1978) evolutionary lines. It was clearly shown that the ox- and pig-insulin immunoreactive invertebrate cells were tall, columnar epithelial cells of so-called open type in the digestive tract mucosa, equipped with secretion granules similar to those of many vertebrate insulin cells. They were light-microscopically fine-granular aldehydefuchsinophil cells, but, again, represented only a fraction of all these cells. The gut insulin cells in molluscs responded readily to a glucose load by degranulation, indicating that insulin was released from these cells, a hypothesis that was supported by the results of actual insulin assays in the haemolymph and in the gut mucosa (Plisetskaya *et al.* 1978). By experiments with administration of mammalian or teleostean insulins, as well as with excess anti-insulin sera in combination with blood glucose and tissue glycogen assays, it could also be concluded that the isolated hormone took part in the regulation of carbohydrate metabolism (Plisetskaya *et al.* 1978). Another interesting observation was that the immunoreactive insulin contents of the crystalline style and the hepatopancreas of two bivalve molluscs did not correspond to any actual presence of cells, showing immunoreactivity with ox–pig-insulin guinea-pig antisera. This finding was hypothesized to indicate that an insulin precursor, produced by the aldehydefuchsinophil cells in the gut mucosa, was secreted to the blood and/or the gut lumen, taken up by the style, transferred from the style to the stomach and then into the hepatopancreas, where it became partly split by proteolysis and transported in active form into the haemolymph (Plisetskaya *et al.* 1978). Cumbersome as it may seem, and also quite different from the biosynthetic route in mammalian B-cell, this hypothesis is in many respects similar to the idea put forward by Steiner *et al.* (1969) that insulin has evolved as a digestive proto-pro-insulin which was split into insulin in the gut and then entered circulation. This has later been supported by De Haën, Swanson, and Teller (1976), who found significant sequence homology between trypsin-related serine proteases and proinsulin. In addition, the disclosed sequences of several growth factors—nerve growth factor (Bradshaw, Hogue–Angeletti, and Frazier, 1974), insulin-like growth factors (Blundell, Dodson, Hodgkin, and Mercola 1972), and relaxin (James and Niall 1977)—indicate structural similarities between these and insulin.

It is therefore tempting to speculate in line with Steiner *et al.* (1969); Steiner, Chan, Terris, and Hoffman (1978); and De Haën *et al.* (1976), that at some point in time in a primitive invertebrate, before the occurrence of the vertebrate phyla, an anabolic protein evolved perhaps from a digestive serine protease. During the course of evolution the anabolic protein duplicated and the proteins' functional characteristics diversified into a family of growth-promoting substances and insulin.

Vertebrates

Whereas in the invertebrates the GEP endocrine system is restricted to the digestive-tract mucosa only, there are step-wise evolutions of the endocrine pancreas in the vertebrates. Thus, the first islet organ—still completely devoid of exocrine acinar pancreatic parenchymal cells—is formed already at the level of the most original vertebrates, viz. the jawless fish (Agnatha and Cyclostomata) (Falkmer and Östberg 1977). It contains only insulin cells (around 99 per cent) and somatostatin (around 1 per cent) cells (Falkmer, Elde, Hellerström, and Peterson 1978). The glucagon cells still remain in the gut mucosa as cells of open type. There seem to be no cells producing the newly discovered fourth-islet hormone, the pancreatic polypeptide (PP). The next major step from a two-hormone islet organ towards the ultimate four-hormone islets of Langerhans of higher vertebrates is taken at the level of the earliest cartilaginous fish, the holocephalan fish, the Atlantic and Pacific ratfish (*Chimaera monstrosa* and *Hydrolagus colliei*, respectively). Here, an actual pancreatic gland is present, closely attached to the spleen, but widely separated from the gut, emptying its secretory products via a long slender duct. In the large islets of Langerhans that occur in this pancreas, three kinds of islet cells are found: insulin, somatostatin, and glucagon cells. It is not until at the evolutionary stage of the most highly developed cartilaginous fish, such as the elasmobrachian sharks, rays, and skates, that also the PP-cells are present in the endocrine pancreas (Falkmer and Stefan 1978; Stefan and Falkmer 1978). After these three initial major steps to a four-hormone islet parenchyma, no radical changes occur during the evolution of the vertebrate endocrine pancreas (Falkmer and Östberg 1977; Van-Noorden and Falkmer 1980).

All these facts imply that in a functional evolution study of insulin and the islet parenchyma, key positions are occupied by the cyclostomes (hagfishes and lampreys), the holocephalan ratfish, and the elasmobranchs. As far as the insulins of the latter two are concerned, there is no major new information to offer in addition to that given in our preceding review (Emdin and Falkmer 1977). We shall then focus on the earliest vertebrate extant, the hagfish, and see how its insulin, which diverged from all other vertebrates some 500–600 million years ago (cf. Falkmer, Thomas, and Boquist 1974; Ostberg 1976), is processed, secreted, and handled. Attention will also be paid to hagfish insulin's structure and activity in certain mammalian test systems.

Insulin in the Atlantic hagfish

Biosynthesis

The way in which the Atlantic hagfish manufactures its insulin is in many respects similar to that of all other vertebrate species investigated (Emdin and Falkmer 1977). Hagfish islet mRNA translates, in a cell-free system, a preproinsulin with a hydrophobic N-terminal extension (Chan, Steiner, Emdin, unpublished) as does also rat islet mRNA (Steiner 1976). Like all other vertebrates studied, the hagfish has a proinsulin with a molecular weight of approximately 9000 daltons, and a preliminary amino-acid composition of this proinsulin has been given (Steiner *et al.* 1975*a*). It may be of interest to note that nature has exploited precursors in protein processing for numerous peptide hormones, a variety of enzymes, and also collagen and albumin. Such precursors can be assumed to be widely distributed throughout the animal kingdom, way below the level of the hagfish, since they are found even in such primitive instances as viral capsule proteins (cf. Steiner *et al.* 1975*b*).

As a whole, insulin biosynthesis is sluggish in the hagfish at its low ambient temperature (Emdin and Falkmer 1977). At 11 °C the half time of conversion is about ten times slower in the hagfish than in the rat. It takes a minimum of 20 h at 11 °C before any newly converted insulin has passed through the cell and is released. Since another 20 h are needed to synthesize proinsulin and convert it to insulin, a total of 40 h is needed before the newly processed insulin reaches the circulation (Fig. 35.1).

The hagfish B-cell differs from that of higher vertebrates inasmuch as its rate of proinsulin biosynthesis is unaffected by the glucose concentration of the medium (Emdin and Falkmer 1977). Neither did arginine, leucine, cAMP and phosphodiesterase inhibitors affect hagfish insulin's biosynthesis. It was, however, inhibited by more than 50 per cent by 2-4-dinitrophenol, an uncoupler of oxidative phosphorylation (Emdin, unpublished).

We have suggested that this original vertebrate islet parenchyma may lack a positive modulator of proinsulin synthesis (Emdin and Falkmer 1977). This is understandable against the background that any insulinotropic agent will need about 40 h at 11 °C before it can contribute to the peripheral insulin needs. It may be then that rate of synthesis is constant for a given temperature and that the mature granules that are not secreted, disintegrate, and their

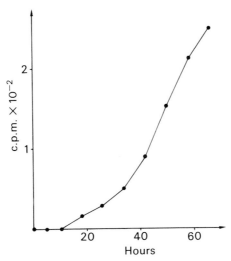

F IG. 35.1. Time-course of secretion of *de novo* synthesized proinsulin–insulin. Twenty-five islet organs were pulsed for 20 hours at 11 °C. This is a time-interval corresponding to the time period needed by the islet organs to synthesize sufficient amounts of labelled proinsulin, but not to convert it. The islet organs were then chased for the time periods indicated, and the chasing medium replaced at given intervals. The labelled insulin was then extracted, purified, and measured, and the accumulated amount of *de novo* synthesized insulin was plotted against time. No significant amounts of labelled proinsulin could be detected. The '0-time' indicates the beginning of the chase period. (Reproduced from Emdin and Falkmer (1977) with permission.)

insulin contents are degraded. Some morphological support for such a speculation may be given by the occurrence of cystic cavities of degenerative nature within the islet lobules (Winbladh and Hörstedt 1975; Thomas, Östberg, and Falkmer 1978), and the presence of lysosomal dense bodies (cf. Östberg 1976).

Conversion of hagfish proinsulin takes place in the Golgi apparatus and in the secretion granules (Emdin and Falkmer 1977). In none of these cellular compartments is hagfish insulin apparently crystalline. Moreover, its B_{21} residue does not possess a free carboxylic acid group as in most insulins. Such charged groups reduce the rate of tryptic hydrolysis of a juxtaposed basic residue. The B_{22}–B_{23} bond in hagfish insulin is split with trypsin about two orders of magnitude faster than the corresponding bond in pig insulin, both present in similar aggregation states (Emdin, unpublished). It is, therefore, not surprising that under mildest tryptic condition ^3H-labelled hagfish proinsulin is converted to a roughly equimolar mixture of insulin and the degradation product—desnonapeptide hagfish insulin (Emdin and Falkmer 1977). This argues strongly against a strictly tryptic enzyme operating in the conversion of hagfish proinsulin. In addition, trypsin will also presumably readily attack the basic residues A_9–A_{10} which are not buried in hagfish insulin's aggregation. The fact that many hormone precursor peptides are separated from the nascent protein by pairs of basic residues makes it tempting to speculate that an enzyme system with specificity for such basic pairs may be operating during the limited intracellular proteolysis (Steiner *et al.* 1975*b*; Emdin and Falkmer 1977).

Secretion

Hagfish insulin secretion has recently been studied *in vitro*, using a homologous radioimmunoassay (RIA) system (Emdin and Steiner 1980; Emdin, in preparation). It was observed that glucose stimulated insulin release with a half maximal response around 3 mM glucose. A stimulatory response was seen already at 1 mM, and the maximal response was at 5 mM glucose, where a threefold stimulation was seen. These glucose levels correspond to the levels seen in fasted and fed hagfish, respectively (Emdin, in preparation). The secretory response was potentiated by a phosphodiesterase inhibitor, isobutyl-methyl-xanthine. Sulphydryl reagents, such as *p*-chloro-mercuri-benzene-sulphonic acid, stimulated the release, and in all of the above cases the stimulatory response was markedly inhibited by omission of calcium. An inverse relationship between stimulated insulin release and temperature was also found. This phenomenon has also been observed in higher vertebrates (Dahl and Henquin 1978).

The onset of insulin release was found to be around 20 minutes. Such a slow onset may very well be explained by the low temperature. In addition, the presence of the peculiar connective-tissue secretion pattern in the endocrine organs of cyclostomes (cf. Falkmer *et al.* 1974), as well as the fact that a thick connective-tissue capsule occurs between the individual avascular islet lobules and the portal circulation, may contribute to a sluggish release pattern in the hagfish. We have in these studies also found that amino acids do not stimulate insulin release in the hagfish. This means either that the hagfish B-cell has lost its responsiveness to amino acids during the development of its suggested neotenous features (cf. Falkmer and Östberg 1977), or that such an ability has evolved after the gnathostomes diverged from the cyclostomes, since it has been shown that insulin secretion is stimulated by amino acids in teleost bony fish (cf. Ince and Thorpe 1977). An investigation of insulin secretion in species occupying other key positions in vertebrate evolution would perhaps clarify this point.

Effects of insulin in the hagfish

The role of insulin in the hagfish has earlier been studied in our laboratories (Falkmer and Winbladh 1964; Falkmer and Matty 1966). It was found to show considerable similarities with higher vertebrates. We have recently extended these investigations by a study of the effects of hagfish insulin on the metabolic rates of ^{14}C-leucine and ^{14}C-glucose in fed and fasting normal hagfish (Emdin, in preparation). The incorporation of label into glycogen, protein, and lipids was studied in liver and skeletal musculature. Animals were injected with tracers and either saline or hagfish insulin at 0.1 μg/g body weight: There was no obvious insulin effect on liver metabolism, apart from a slight stimulation of protein synthesis, using ^{14}C-leucine. On the other hand, both muscle glycogen and protein synthesis were enhanced two- to tenfold in the presence of insulin using both tracers, and the same was found to hold true for neutral lipids in muscle when ^{14}C-glucose was used as precursor. However, the significant enhancement of these synthetic pathways induced by hagfish insulin did not appear to be of great quantitative importance, because during the time of the experiment (33 h) only 10 per cent of the given radioactive dose was utilized by liver and musculature, and insulin's metabolic effects contributed to only about half of these 10 per cent. By means of the homologous RIA (Emdin and Steiner 1980) we have measured the insulin levels in hagfish fasted for a month. It was found to be 1.10 ± 0.1 nM (SEM, $n = 38$), whereas the insulin levels of newly caught and fed hagfish were 2.16 ± 0.2 nM (SEM, $n = 40$). If starved hagfish were loaded with glucose, the insulin values rose to a similar value within 3 h, viz. 1.90 ± 0.2 nM (SEM, $n = 18$), demonstrating that glucose stimulated insulin release not only *in vitro* but also *in vivo*, and to a level corresponding to the one seen in fish fed *ad libitum* (Emdin, in preparation).

In another experiment with ^{14}C-glucose and ^{14}C-leucine, animals were pretreated with glucose and amino acids for three days before the isotopes were injected (Emdin, in preparation). The given glucose did, as expected, increase the serum insulin levels to about 2 nM. Moreover, these endogenously elevated insulin levels were high enough to evoke stimulatory effects on the incorporation of the label into muscle. Therefore, it was concluded that, in the intact hagfish, glucose is able to stimulate insulin secretion about twofold, and this elevated insulin level elicits an enhancement of glycogen, protein, and lipid synthesis in skeletal muscle, whereas the liver remains relatively unaffected by the hormone. Insulin's quantitative effect appears to be smaller in the hagfish than in higher vertebrates, and it gives a general picture of a slow-reacting animal with a narrow spectrum of metabolic adaptations. Similar results have recently been presented also in the Pacific hagfish, *Eptatretus stouti* (Inui, Yu, and Gorbman 1978; Inui and Gorbman 1978). Nevertheless, the physiological role of insulin in the hagfish is still poorly known and a large number of important points have not yet been studied at all in any myxinoid species.

Structure of hagfish insulin

A study of the primary structure of hagfish insulin and a number of insulins from mammals, birds, and fish shows that insulin is a polypeptide hormone that has undergone surprisingly few alterations during the vertebrate evolution (Peterson, Steiner, Emdin, and Falkmer 1975). Nevertheless, hagfish insulin is one of the most highly substituted, naturally occurring insulins. There are in hagfish insulin altogether 19 amino-acid residues out of 51 different from those of pig insulin, and 16 of these, mostly located in the B-chain, have previously not been observed in these positions. Surveying these differences, it is noteworthy that hagfish insulin does not appear to be readily related to insulins in either mammals, birds, or fish. This is in conformity with the phylogenetic fact (Falkmer *et al.* 1974) that the cyclostomes diverged from the gnathostomian vertebrate evolution lines before the ancestors of the other groups diverged from each other. On the other hand, all the so-called invariant amino-acid residues, i.e. those that have been present in all

FIG. 35.2. Hagfish insulin crystals (slightly less than 1 mm), grown in acetone–citrate buffer at pH 6.0 in the absence of zinc.

the mammalian, avian, and teleostean insulins studied up to now, have been demonstrated also in hagfish insulin, with one minor exception. From the sequence it can be deduced that the residues, important in stabilizing the insulin monomer and dimer, were present already in hagfish insulin, and have been entirely conserved in most species throughout the evolution. This has been supported by the observation that the dimerization constant at neutral pH of hagfish and pig insulins are indistinguishable within one order of magnitude (Bruce H. Frank, Eli Lilly Research Laboratories, Indianapolis, USA). However, the amino-acid sequence of hagfish insulin implies that a zinc-coordinated hexamer is unlikely.

The evident similarity in structure and preservation of sequence in the region of the hormone considered to be involved in biological activity has

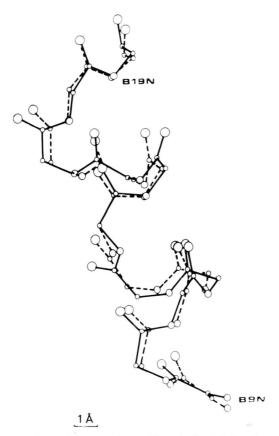

B19N

B9N

1 Å

FIG. 35.3. A comparison of the atomic positions in hagfish insulin (solid lines) and 2Zn pig insulin (dashed lines), molecule 2 for the B-chain residues in the α-helix (B_9–B_{19}). (Reproduced from Cutfield *et al.* (1979) with permission.)

prompted the study of the zinc-free tetragonal hagfish insulin crystal (Fig. 35.2). Some results have been reported earlier (Cutfield, Cutfield, Dodson, Dodson, and Sabesan 1974), and the data herein refer to a study at 3.1 Å (Cutfield *et al.* 1979). A refinement of the structure at 1.9 Å is at present under way (Dodson and Reynolds, in preparation).

Hagfish insulin crystallized in the absence of zinc with one molecule per asymmetric unit and was organized as a symmetric dimer, lying on a twofold crystallographic axis. The space group was $P4_12_12$, $a=b=38.4$ Å and $c=85.3$ Å. Despite the differences in amino-acid sequence and the difference in aggregation, pig and hagfish insulins had similar structures. The similarities in structure extended beyond the general folding of the molecule, and many of the side-chains proved to be in the same positions in the two insulins. The most striking match was seen in the B-chain alpha-helix B_9B_{19} (Fig.

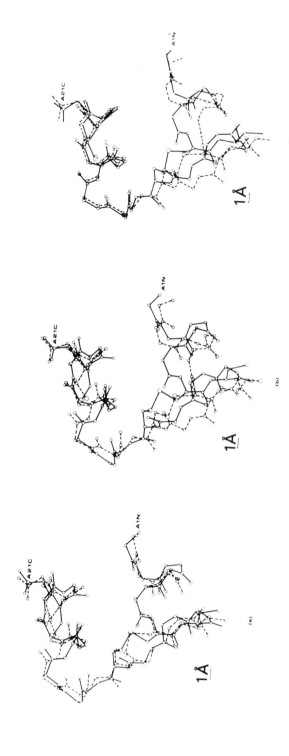

Fig. 35.4. A comparison of the atomic positions for the A-chain residues (A_1–A_{21}) in (a) hagfish insulin (solid lines) and 2Zn pig insulin (dashed lines), molecule 2, (b) hagfish insulin (solid lines) and 2Zn pig insulin (dashed lines), and (c) 2Zn pig insulin, molecule 1, and molecule 2 (solid lines) and molecule 1 (dashed lines). (Reproduced from Cutfield et al. (1979) with permission.)

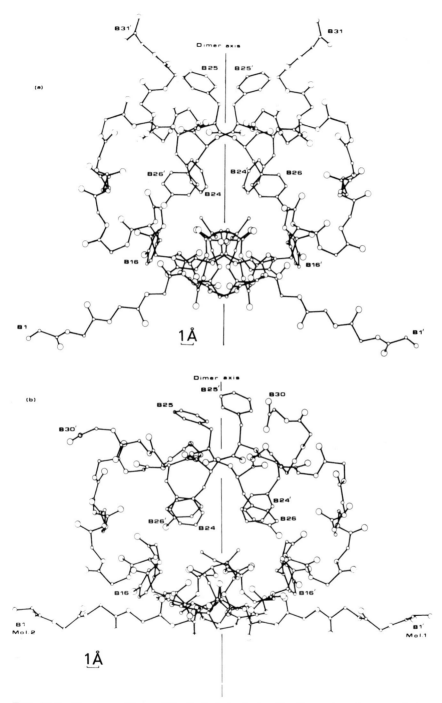

FIG. 35.5. Atomic positions of the B-chain in the insulin dimer viewed perpendicular to the twofold axis for (a) hagfish insulin and (b) 2Zn pig insulin. Side-chains for the residues involved in dimer formation (B_{12}, B_{16}, B_{24}, B_{25}, and B_{26}) are shown. (Reproduced from Cutfield *et al.* (1979) with permission.)

35.3) and the extended C-terminal residues B_{23}–B_{26}. These residues are invariant, and include residues buried by dimer formation. The A-chain in hagfish insulin displayed generally similar folding of the backbone to that in pig insulin (Fig. 35.4). Some of the side-chains were displaced up to 3 Å. There was evidence of minor structural movement of about 2 Å at the disulphide bond A_6–A_{11}. This shift extended further down the A-chain and was most obvious at the A_{13} leucine residue and the A_{14} tyrosine residue. These changes could reflect hagfish insulin's lower aggregation state, since these residues in 2Zn pig insulin make close contacts between dimers.

In 2Zn pig insulin the two molecules in the dimer exhibit differences in conformation. In all these features hagfish insulin was closer to pig insulin molecule 2 than pig insulin molecule 1. The hagfish insulin dimer appeared to be constructed from two molecules almost identical to pig insulin molecule 2. In fact the tertiary structure of hagfish insulin was more similar to pig insulin molecule 2 than were pig insulin molecules 1 and 2 (Fig. 35.5).

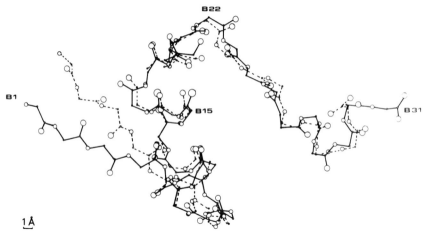

FIG. 35.6. Atomic positions of the B-chain for hagfish insulin (solid lines) and 2Zn pig insulin (dashed lines). The differences in structure of the B-chain termini are shown.

Two regions existed where pig and hagfish insulin differed considerably, and there were in these regions also pronounced differences between the primary structures. One was the region B_1–B_4 which in pig insulin is involved in hexamer formation (Fig. 35.6). Its structural alteration in hagfish insulin is, therefore, understandable. This region in mammalian insulins does not seem to be of crucial importance for high biological activity (cf. Blundell *et al.* 1972). At the other end of the B-chain, residues B_{28}–B_{31} in the hagfish turn toward A_1-glycine (Fig. 35.6). There are no corresponding intermolecular contacts within the pig insulin hexamer, and this altered structure was interpreted as an inherent feature of the monomer, reflecting the changes

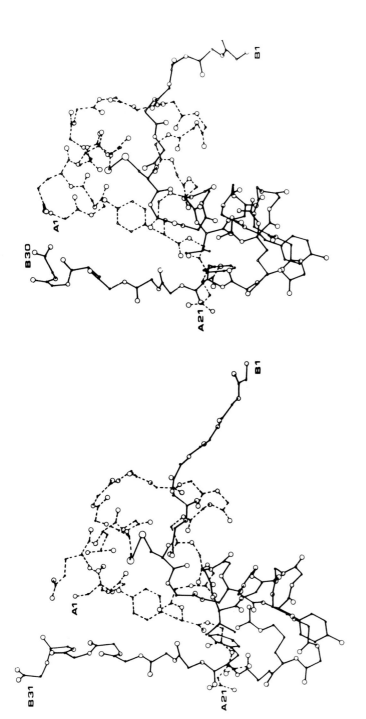

FIG. 35.7. A comparison of the atomic positions of the A- and B-chains in (a) hagfish insulin and (b) 2Zn pig insulin, molecule 2 viewed parallel to the twofold axis. Side-chains for the residues implicated in activity are shown. (A_1, A_4, A_5, A_7, A_{19}, A_{21}, B_7, B_{12}, B_{16}, B_{22}, and B_{24}). (Reproduced from Cutfield et al. (1979) with permission.)

in sequence in this region. A comparison of the assembled pig and hagfish insulin monomers is shown in Fig. 35.7.

Hagfish insulin's behaviour in mammalian test systems

Despite the great structural similarities between hagfish and pig insulins in the crystalline state, there are substantial differences in the biological activities in mammalian test systems. More than a decade ago, Weitzel, Strätling, Hahn, and Martini (1967) reported that hagfish insulin had a biological activity of 7 per cent of that of ox or pig insulins in the rat epididymal fat pad assay. More recently, a figure of 5 per cent was arrived at in the rat free-fat-cell assay by us and in collaboration with others (Emdin, Gammeltoft, and Gliemann 1977; Muggeo *et al.* 1979*b*). In our first study (Emdin *et al.* 1977) the binding of hagfish insulin was about 25 per cent. Hagfish insulin was the first insulin to display a discrepancy between receptor binding and biological activity. Hagfish insulin could therefore be defined as a partial antagonist on the receptor but, on the whole cell, as a full antagonist owing to the presence of a large number of spare receptors on the rat fat-cell. This discrepancy has since been questioned (cf. Muggeo *et al.* 1979*a*). We have, however, tested the intrinsic binding properties of trace ^{125}I-hagfish and ^{125}I-pig insulins iodinated in parallel (Emdin, Sönne, Gliemann 1980). The binding of ^{125}I-hagfish insulin was still around 20–25 per cent of that of pig insulin in several different experiments. An analysis of the rates of ^{125}I-hagfish insulin association and dissociation from this receptor showed that hagfish insulin dissociated and—even more so—associated slower than pig insulin did. Therefore, we feel that the discrepancy between receptor binding affinity and biological activity in this system is real and shows that these two cellular events can be separated. We do not know the structural and molecular mechanisms of this discrepancy. The removal of B_{31}-methionine and B_{30}-lysine residue did not eliminate the discrepancy. Gliemann and Sonne (1978) have recently shown that the fat cell insulin receptors degrade insulin. Attractive as the idea may seem, as a possible explanation for the discrepancy between binding and action, we found no differences in the fractional degradation of ^{125}I-pig and hagfish insulins (Emdin *et al.* 1980). It is likely, however, that about five times fewer glucose carriers are activated per hagfish insulin receptor complex than with other insulins (Emdin *et al.* 1977). It is tempting to speculate that evolution beyond the hagfish has led to insulins being more efficient per unit binding. This would cause an increase in sensitivity but not necessarily an increase in the maximum response. In a recent report by Kahn, Baird, Jarret, and Flier (1978), it has been shown that monovalent fragments from an insuin receptor antibody are fully active in binding to the fat cell insulin receptor, but still have only little biological activity. This observation gives additional support to the idea that molecular

TABLE 35.1

Hagfish insulin: summary of biological behaviour relative to pig insulin

Rat fat cells:		*References*
Biological activity	4–7%	(Emdin *et al.* 1977)
		(Muggeo *et al.* 1979*b*)
		(Weitzel *et al.* 1967)
Receptor binding as:		
Inhibition of ^{125}I-pig insulin binding	23%	(Emdin *et al.* 1977)
	5–10%	(Muggeo *et al.* 1979*b*)
^{125}I-hagfish insulin binding	25%	(Emdin *et al.* 1980)
Degradation as:		
Inhibition of ^{125}I-pig insulin degradation	12 nM (hagfish)	(Emdin *et al.* 1977)
by the membrane associated protease	130 nM (pig)	
Receptor mediated degradation	80–100%	(Emdin *et al.* 1980)
Rat liver cells:		
Receptor binding as:		
Inhibition of ^{125}I-pig insulin binding	3–7%	(Gammeltoft *et al.* 1978)
		(Terris and Steiner, 1975)
Degradation as:		
Inhibition of ^{125}I-pig insulin degradation	15 nM (hagfish)	(Gammeltoft *et al.* 1978)
by the membrane associated protease	120 nM (pig)	
Degradation velocity	4%	(Terris and Steiner 1975)
Hagfish erythrocytes:		
Inhibition of ^{125}I-pig insulin binding	25%	(Muggeo *et al.* 1979*b*)
Human IM 9 lymphocytes:		
Inhibition of ^{125}I-pig insulin binding	5–10%	(Muggeo *et al.* 1979*b*)
^{125}I-hagfish insulin binding	23%	(Emdin *et al.* 1980)
Ability to induce negative co-operativity	5%	(Muggeo *et al.* 1979*b*)

events involved in insulin's binding and action can be separated (Piron, De Meyts, Horuk, and Emdin 1980).

Although most of the studies on the biological effects of hagfish insulin have been made with isolated rat fat cells, there is additional information on its binding in other test systems. This information is presented in Table 35.1.

Hagfish insulin deviates from other insulins in yet another respect. During recent years a phenomenon called negative co-operativity of insulin receptors has been introduced (De Meyts, Van Obberghen, Roth, Wollmer, and Brandenburg 1976). The interpretation of the phenomenon has been controversial (cf. Pollet, Standaert, and Haase 1977). Regardless of its physical interpretation, it has been found that insulin binding to receptor sites under certain conditions causes a loss of affinity of other receptor sites for insulin owing to an accelerated dissociation rate of the insulin receptor complex. This acceleration is dependent on insulin concentration, but the negative co-operativity diminishes at higher insulin concentrations. The latter phenomenon has been interpreted as being caused by insulin dimerization since during the transition of the hormone from monomer to dimer the co-operative site is supposedly covered. As a consequence, the idea has been put forward that insulin dimers bind and compete with insulin monomers for binding to the receptors (De Meyts *et al.* 1978). The concentration-dependent decrease was not observed with tetranitro- and guinea-pig insulins which both do not dimerize. Hagfish insulin had about 5 per cent relative potency to induce negative co-operativity. There was, however, no fall off of the negative co-operativity as the insulin concentration increased (De Meyts *et al.* 1978), despite the fact that hagfish insulin dimerizes as pig insulin does. A possible explanation (Muggeo *et al.* 1979*b*) was that the hagfish insulin dimer was structurally different and did not cover the co-operative site, or that hagfish insulin dimers do not bind to the receptor. Neither has it been proved that the fall off in negative co-operativity is due to dimerization despite the evidence supporting the view, nor that dimers bind to the receptor. The dimer-forming region has in fact been implicated as being a substantial part of the receptor-binding region of the insulin (Blundell *et al.* 1972; Pullen *et al.* 1976). We see hagfish insulin's behaviour within the scope of the phenomenon negative co-operativity is still unsettled and intriguing.

It has been shown that the insulin receptor evolves even slower than the homologous insulin, and thus the affinity of a given receptor for a given insulin is independent of the kind of insulin that is endogenous in that organism (Muggeo *et al.* 1979*a*). In recent studies of the hagfish erythrocyte receptor (Muggeo *et al.* 1979*b*) similar absolute affinity and rank order of preference of insulin and insulin analogues were observed as in other species studied (Muggeo *et al.* 1979*b*). The affinity of hagfish insulin was 25 per cent of that of pig insulin. The hagfish erythrocyte receptor also showed the same type of changes in affinity in response to changes in pH and temperature as other insulin receptors, and this receptor also displayed the negative co-operativity phenomenon.

Structure v. biological activity: a problem

Hagfish insulin has all the residues implemented for high biological activity (Blundell *et al.* 1972; Pullen *et al.* 1976), and a structure in the crystal which is remarkably similar to 2Zn pig insulin molecule 2 (see Chapter 42). Hagfish insulin even has a feature correlated with high activity, viz. an A_8 histidine residue (Pullen *et al.* 1976), and the changes in structure seen at the B-chain termini do not seem to be responsible for hagfish insulin's reduced activity. Yet hagfish insulin must possess structural changes that add up to the small drop in energy (1–2 kcal/mol, assuming the system is in equilibrium) needed to go from full binding to 5–25 per cent (Cutfield *et al.* 1979).

Clearly minute conformational changes could produce the reduction in binding affinity seen in hagfish insulin. A hitherto unexplored possibility is the presence of a non-polar valine residue in position B_{21}. This residue is in all other insulins a charged residue and in most cases a carboxylic acid, and the position of this residue is immediately outside the putative receptor-binding region.

It is also possible that molecular flexibility rather than a specific residue or conformation is of importance in expression of binding and action. Pig and ox insulins can undergo substantial rearrangements in going from 2Zn to the 4Zn structure (Bentley, Dodson, Dodson, Hodgkin, and Mercola 1976). The conformational change involves the B-chain N-terminus, separation of B-chains' C-termini, as well as those of the A-chain. In the hagfish, there are substantial sequence changes at the B-chain N-terminus and the nearby A-chain residues 8, 9, and 10. Residues in this region vary considerably between different insulins, but they may possess a certain amount of flexibility in highly potent insulins, necessary for full expression of biological activity which, in turn, may be absent in hagfish insulin.

The importance of molecular flexibility on the receptor is hard to study experimentally. In the case of hagfish insulin there is, however, indirect evidence that could possibly support the idea of flexibility in a simple-minded way. Firstly, hagfish insulin, in comparison with pig insulin, has difficulties in associating with the receptor. Once on the receptor, hagfish insulin is relatively more fixed in this position and less prone to dissociate. Whilst being on the receptor, hagfish insulin displays partial antagonism and is less able than all other insulins to activate a certain number of glucose carriers. Secondly, on the surface of the fat cell there is membrane-associated soluble enzyme that degrades insulin. Here, hagfish insulin behaves as a more 'rigid' structure, since the rate constant of the irreversible product formation is about ten times lower for hagfish insulin than for pig insulin (Emdin *et al.* 1977).

Concluding remarks on hagfish insulin's evolution

We can describe the situation from two different viewpoints. In terms of molecular structure there is a remarkable conservatism throughout the evolution of the insulin molecule. The fact that pig and hagfish insulins, that have evolved separately for several hundred millions of years, are still so similar in their chemical and structural features shows that the primary structure— and even more so the tertiary structure—of insulin must exist under extremely strict structural–functional constraints, even though evolution beyond the hagfish has led to more efficient insulins. On the other hand, the processing, secretion, and utilization of hagfish insulin in the hagfish displays differences at several points; proinsulin biosynthesis does not seem to be modulated, secretion is not stimulated by amino acids, the liver seems to be of relatively minor importance, and the overall quantitative importance of insulin's peripheral effects also appears less obvious than in higher mammals. Hence, the hagfish is relatively less well developed in these respects. Our position is, however, one of ambiguity, since biosynthesis, insulin release and action in the hagfish have many features in common with the corresponding processes in higher vertebrates and we cannot with certainty describe our observations as being neotenous, 'primitive', or adaptive.

Acknowledgements

This work was supported by grants from the Swedish Medical Research Council (Project No. 12X–718), the Swedish Diabetes Association, the Medical Faculty of Umeå, and the Kroc Foundation, USA.

References

BARRINGTON, E. J. W. (1978). In *Comparative endocrinology* (ed. P. J. Gaillard and H. H. Boer) pp. 381–96. Elsevier/North Holland Biomedical Press, Amsterdam.

BENTLEY, G., DODSON, E. J., DODSON, G. G., HODGKIN, D. C., and MERCOLA, D. A. (1976). *Nature, Lond.* **261**, 166–8.

BLUNDELL, T. L., BEDARKAR, S., RINDERKNECHT, E., and HUMBEL, R. E. (1978). *Proc. natn. Acad. Sci. U.S.A.* **75**, 180–4.

—— DODSON, G. G., HODGKIN, D. C., and MERCOLA, D. A. (1972). *Adv. Protein Chem.* **26**, 279–402.

BRADSHAW, R. A., HOGUE-ANGELETTI, R. A., and FRAZIER, W. A. (1974). *Recent Prog. Horm. Res.* **30, 575.**

CUTFIELD, J. F., CUTFIELD, S. M., DODSON, E. J., DODSON, G. G., and SABESAN, M. N. (1974). *J. molec. Biol.* **87**, 23–30.

———— DODSON, G. G., DODSON, E. J., EMDIN, S. O., and REYNOLDS, C. D. (1979). *J. molec. Biol.* **132**, 85–100.

DAHL, G. and HENQUIN, J.-C. (1978). *Cell Tiss. Res.* **194**, 387–98.

DE HAËN, C., SWANSON, E., and TELLER, D. C. (1976). *J. molec. Biol.* **196**, 639–61.

DE MEYTS, P., BIANCO, A. R., and ROTH, J. (1976). *J. biol. Chem.* **251**, 1877–88.

—— VAN OBBERGHEN, E., ROTH, J., WOLLMER, A., and BRANDENBURG, D. (1978). *Nature, Lond.* **273**, 504–9.

DUVE, H. (1978). *Gen. comp. Endocr.* **36**, 102–10.

—— and THORPE, A. (1979). *Cell Tiss. Res.* **200**, 189–91.

—— and LAZARUS, N. R. (1979). *Biochem. J.* **184**, 221–7.

EMDIN, S. O. and FALKMER, S. (1977). *Acta paediat. Stockh.* Suppl. **270**, 15–25.

—— and STEINER, D. F. (1980). *Gen. comp. Endocr.* (in press).

—— GAMMELTOFT, S., and GLIEMANN, J. (1977). *J. biol. Chem.* **252**, 602–8.

—— SÖNNE, O., and GLIEMANN, J. (1980). *Diabetes* (in press).

FALKMER, S. and MATTY, A. J. (1966). *Gen. comp. Endocr.* **6**, 334–46.

—— and ÖSTBERG, Y. (1977). In *The diabetic pancreas* (ed. B. W. Volk and K. F. Wellman) pp. 15–60. Plenum, New York.

—— and STEFAN, Y. (1978). *Scand. J. Gastroenterol.* **13**, Suppl. **49**, 59.

—— and WINBLADH, L. (1964). In *The structure and metabolism of the pancreatic islets* (ed. S. E. Brolin, B. Hellman, and H. Knutson) pp. 33–43. Pergamon Press, Oxford.

—— THOMAS, N. W., and BOQUIST, L. (1974) In *Chemical zoology* (ed. M. Florkin and B. T. Scheer) Vol. VIII, pp. 195–257. Academic Press, New York.

——Elde, R. P., Hellerström, C., and Peterson, B. (1978). *Metabolism* **27**, Suppl. **1**, 1193–6.

Fritsch, H. A. R. and Sprang, R. (1977). *Cell Tiss. Res.* **177**, 407–13.

Gammeltoft, S., Østergaard-Kristensen, L., and Sestoft, L. (1978). *J. biol. Chem.* **253**, 8406–13.

Gliemann, J. and Sonne, O. (1978). *J. biol. Chem.* **253**, 7857–63.

Ince, B. W. and Thorpe, A. (1977). *Gen. comp. Endocr.* **31**, 249–56.

Inui, Y. and Gorbman, A. (1978). *Comp. Biochem. Physiol.* **60A**, 181.

——Yu, J. Y.-L., and Gorbman, A. (1978). *Gen. comp. Endocr.* **36**, 133–41.

Ishay, J., Gitter, S., Galun, R., Doron, M., and Laron, Z. (1976). *Comp. Biochem. Physiol.* **54A**, 203–6.

James, R. and Niall, H. (1977). *Nature, Lond.* **267**, 544–6.

Kahn, C. R., Baird, K. L., Jarret, D. B., and Flier, J. S. (1978). *Proc. natn. Acad. Sci. U.S.A.* **75**, 4209–13.

Kramer, K. J., Tager, H. S., Childs, C. N., and Speirs, R. D. (1977). *J. Insect Physiol.* **23**, 293–5.

Muggeo, M., Ginsberg, B. H., Roth, J., Kahn, C. R., De Meyts, P., and Neville, D. M., Jr (1979a). *Endocrinology* **104**, 1393–402.

——Van Obberghen, E., Kahn, C. R., Roth, J., Ginsberg, B. H., De Meyts, P., Emdin, S. O., and Falkmer, S. (1979b). *Diabetes* **28**, 175–81.

Östberg, Y. (1976). *Umeå University Medical Dissertations* **15**, 1–41.

Pearse, A. G. E. (1978). In *Surgical endocrinology: clinical syndromes* (ed. F. R. Friesen) pp. 18–34. Lippincott, Philadelphia.

——and Takor Takor, T. (1976). *Clin. Endocr. Suppl.* **5**, 229–44.

Peterson, J. D., Steiner, D. F., Emdin, S. O., and Falkmer, S. (1975). *J. biol. Chem.* **250**, 5183–91.

Piron, M.-A., De Meyts, P. Horuk, R., and Emdin, S. O. (1980). In *Insulin. Chemistry, structure, and function of insulin and related hormones. Proc 2nd Int. Insulin Symp.*, Aachen, 4–7 September 1979. Walter de Gruyter, Berlin (in press).

Plisetskaya, E., Kazakov, V. K., Soltitskaya, L., and Leibson, L. B. (1978). *Gen. comp. Endocr.* **35**, 133–45.

Pollet, R. J., Standaert, M. L., and Haase, B. A. (1977). *J. biol. Chem.* **252**, 5828–34.

Pullen, R. A., Lindsay, D. G., Wood, S. P., Tickle, I. J., Blundell, T. L., Wollmer, A., Krail, G., Brandenburg, D., Zahn, H., Gliemann, J. and Gammeltoft, S. (1976). *Nature, Lond.* **259**, 369–73.

Stefan, Y. and Falkmer, S. (1978). *Diabetologia* **15**, 272.

Steiner, D. F. (1976). *Diabetes* **26**, 322–40.

——Chan, S. J., Terris, S., and Hoffman, C. (1978). In *Hepatotropic factors*. Ciba Foundation Symposium **55** (new series), pp. 217–46. Elsevier/Excerpta Medica, North-Holland, Amsterdam.

——Terris, S., Emdin, S. O., Peterson, J. D., and Falkmer, S. (1975a). In *Early diabetes in early life* (ed. R. A. Camerini-Davalos) pp. 41–8. Academic Press, New York.

——Kemmler, W., Tager, H. S., Rubinstein, A. H., Lernmark, Å., and Zühlke, H. (1975b). In *Proteases and biological control* (ed. E. Reich, D. B. Rifkin, and E. Shaw) Cold Spring Harbor Laboratory, pp. 531–49.

——Clark, J. L., Nolan, C., Rubinstein, A. H., Margoliash, E., Aten, B., and Oyer, P. E. (1969). *Recent Prog. Horm. Res.* **25**, 207–82.

Tager, H. S., Markese, J., Kramer, K. J., Speirs, R. D., and Childs, A. N. (1976). *Biochem. J.* **156**, 515–20.

Terris, S. and Steiner, D. F. (1975). *J. biol. Chem.* **250**, 8389–98.

THOMAS, N. W., ÖSTBERG, Y., and FALKMER, S. (1978). *Acta Zool. Stockh.* **59,** 119–23.

VANNOORDEN, S. and FALKMER, S. (1980). *Invest. Cell Path.* **3,** 21–36.

WEITZEL, U. G., STRÄTLING, W. H., HAHN, J., and MARTINI, O. (1967). *Hoppe-Seyler's Z. physiol. Chem.* **348,** 523–32.

WINBLADH, L. and HÖRSTEDT, P. (1975). *Acta Zool. Stockh.* **56,** 213–16.

36. Novel semisynthetic approaches to insulin and its precursors

V. K. NAITHANI, H.-G. GATTNER,
E. E. BÜLLESBACH, AND H. ZAHN

SINCE the first historical synthesis of insulin (Meienhofer *et al.* 1963; Katsoyannis 1964; Kung *et al.* 1966), improved chemical techniques of peptide synthesis and their innovative application have led to an elegant route to human insulin (Sieber *et al.* 1974). With the discovery of the insulin precursors proinsulin (Steiner and Oyer 1967; Chance, Ellis, and Bromer 1968) and preproinsulin (Chan, Keim, and Steiner 1976; Shields and Blobel 1977; Ullrich *et al.* 1977; Chan, Noyes, Agarwal, and Steiner 1979), we have new goals to achieve. At present the Aachen group is attempting the semisynthesis of human insulin (Gattner, Danho, and Naithani 1979), human proinsulin (Naithani, Gattner, Büllesbach, Föhles, and Zahn 1979; Naithani, Gattner, and Zahn 1979) and a hybrid preproinsulin (Naithani *et al.* 1979c, d; Naithani, Büllesbach, and Zahn 1979). To meet this challenge, we have applied new semisynthetic approaches using mixed anhydride and enzyme catalysed acylation procedures described below.

Previous semisynthetic modifications (for details see Blundell, Dodson, Hodgkin, and Mercola (1972); Geiger (1976)) at the amino terminus of the A- and B-chains of insulin have revealed that the stepwise active ester-acylation procedure was of limited value for the semisynthesis of proinsulin. Other aspects of semisynthesis such as thiol protection and solubility and purification of products did not appear to pose any significant problem. For example, protected amino-acid-active esters could be satisfactorily condensed to A-chain (as S-sulphonate (Weinert, Brandenburg, and Zahn 1969; Krail, Brandenburg, and Zahn 1975a, b) and cyclic bis-disulphide (Naithani and Gattner, unpublished results) and to insulin (partially protected and native (Blundell *et al.* 1972; Geiger 1976; Friesen, Naithani, Gattner, and Zahn 1976; Friesen 1976) in a variety of solvent systems. Nevertheless, in the stepwise addition of amino acids, even with quantitative acylation, it becomes tedious to assemble a total sequence of 86 (human proinsulin) and 105 residues (hybrid preproinsulin consisting of rat prepeptide-I (Chan *et al.* 1979) and bovine proinsulin). Fragment condensation is therefore mandatory for the successful semisynthesis of these molecules. The azide method, as applied by Shimonishi (1970) and later by Naithani and Weinert (unpublished results), for condensing protected fragments to the A-chain, was found to be unsatisfactory owing to slow and incomplete acylations and the side reaction occurring as a result of Curtius rearrangement (Schnabel 1962).

We have recently applied mixed anhydrides for acylations of polypeptides and proteins in aqueous–organic solutions, and shown that they are very powerful and efficient reagents for semisynthesis (Naithani *et al.* 1979*a–d*). Using this procedure synthetic protected amino acid and peptides were attached to native polypeptides and proteins. Another potential tool for semisynthesis is the highly specific trypsin-catalysed peptide bond formation as demonstrated by Inouye *et al.* (1979) and Morihara, Oka, and Tsuzuki (1979) in the semisynthetic conversion of porcine to human insulin. Like the authors mentioned above we have reacted insulin derivatives with amino acids and peptides. Boc—Arg was also coupled by enzyme-catalysed reaction with native A-chain. The application of both procedures in the semisynthesis of insulin, proinsulin, and preproinsulin sequences are briefly described below.

Semisynthesis using mixed anhydrides

Mixed anhydride acylations in organic solutions have been widely used for peptide synthesis (Boissonnas 1951; Vaughan 1951; Wieland and Bernhard 1951; Anderson, Zimmerman, and Callahan 1967), because of high speed, yield, purity, and lack of racemization. Repetitive excess mixed anhydride method (REMA-method) with quantitative coupling at each step has led to the synthesis of several pure peptides without purification of the intermediates (Tilak 1970; von Zon and Beyerman 1973, 1976). Although acylations in aqueous–organic mixtures had been shown to proceed to give a fairly good yield (Boissonnas 1951; Vaughan 1951; Wieland and Bernhard 1951; Naithani, unpublished results) this procedure, as opposed to anhydrous solutions, had not been explored in peptide synthesis.

Mixed anhydride can yield two acylation products (Fig. 36.1): the desired peptide (A) and a permanently blocked urethane derivative (B). In practice, reaction (A) dominates, although reaction (B) has been observed as a minor side-reaction (Bodanszky and Tolle 1977).

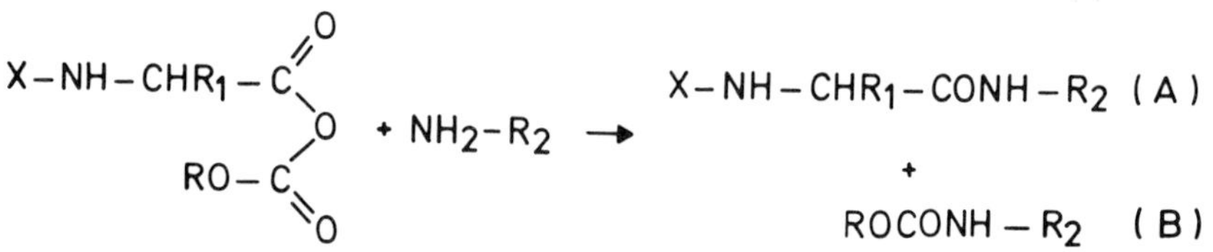

FIG. 36.1. Possible pathway of reaction with mixed anhydride.

Next we turned our attention to mixed anhydride acylation in aqueous–organic mixtures with the aim of developing the method for semisynthesis of proinsulin and preproinsulin. The advantages are that mixed anhydrides are highly reactive, easy to generate from protected amino acids or peptides, and readily used for acylations in aqueous–organic mixtures. Under these conditions the polypeptides and proteins as well as the mixed anhydrides can remain in homogeneous phase, thus facilitating the bimolecular acylation reaction. As A-chain and proinsulin do not irreversibly denature in organic or aqueous–organic mixtures, we have used various organic solvents at different concentrations without any difficulty.

The reported reactions, in which peptides with no side-chain functional groups (Boissonnas 1951; Vaughan 1951; Wieland and Bernhard 1951) or with protected side-chain groups (Naithani, unpublished results) were synthesized, were carried out under various conditions. The presence of water in the reaction mixture introduces nucleophilic OH^- ions although amino groups are stronger nucleophiles. The amount of water and the pH of the reaction mixture would therefore effect the acylation by causing the hydrolysis of mixed anhydrides and by affecting the reactivity of the amino groups. Thus the acylations should be best effected under conditions of maximal stability of anhydrides and minimal protonation of amino groups, which is dependent on the pK of the amino function. In order to optimize peptide-bond formation, it was essential to investigate the parameters of the reaction. As our primary aim was to carry out semisynthesis with polyfunctional polypeptides and proteins (A-chain, insulin, and proinsulin), the possibility of any side-reactions at side-chain functional groups required additional investigations.

The reaction parameters were investigated using as a model the reaction of the mixed anhydride of *t*-butyloxycarbonylmethionine and A-chain cyclic bis-disulphide. Methionine was chosen as it does not occur in the native molecule, and is therefore easy to estimate by amino-acid analysis. The reactions were carried out in aqueous–organic buffers. The presence of high concentrations of organic solvent made the absolute pH-determination difficult. The 'pH' of the freshly made buffers was adjusted using a glass electrode. Both 80 per cent DMF-water and 80 per cent DMSO-water were used for acylations; the latter was found to be a versatile solvent for dissolving proinsulin sequences and mixed anhydrides. Lower proportions of organic solvent gave lower extents of acylation. The pH-optimum for N-acylation in 80 per cent DMF-water was 6.5. Below pH 5.5 very little N-acylation was observed although considerable urethane formation took place. Little or no urethane formation was observed between pH 6.5–7.5. The acylation at $-10\,^{\circ}$C was rather slow. Maximum rates of acylation were found at room temperature. Studies at pH 7.5 showed that two equivalents of the reagent caused quantitative acylation in 80 per cent DMF or 80 per cent DMSO-water. The reaction was extremely rapid: with five equivalents of the reagent at pH 7.5, the reaction was complete in less than 30 s. The influence of pH, solvent, and temperature on the degree of stereospecificity of acylation will also be investigated. We have observed two side-reactions: (i) urethane formation, which is pronounced at pH 5.5, this reaction varies from amino acid to amino acid and peptide to peptide; and (ii) O-acylation, which is only significant with very high reagent excess (20-fold), longer reaction time (20 minutes), and lower water content. The presence of water in the reaction mixture is a deterrent to O-acylation and to possible anhydride transfer reaction as no cross-linked insulin or proinsulin products were observed. In view of these properties, the mixed anhydride acylations of proteins with hydroxyl and guanidine side-chains in aqueous–organic solution are of advantage.

Mixed anhydride application was exploited in two ways: (i) the partial acylations were carried out and the intermediates were purified at each step; and (ii) the use of REMA-method that caused total disappearance of the amine component at each step, with or without urethane formation. This side-product remains inert to deblocking as well as acylation reactions. Therefore, the next coupling could be carried out without the necessity of purifying the intermediates. At the end of semisynthesis the difference in molecular weight or net charge of the desired molecule from the urethane-blocked sequence would facilitate its purification.

The varying chain length, size, and solubility of the reagents which we planned to use during the semisyntheses were expected to influence the coupling yield. We realized that A-chain and proinsulin were dissimilar not only in their sizes and charges but also in their conformational characteristics and in the different pKs of their amino functions, and that each of these would cause differences in the acylation reaction. Such individual variations have been observed in practice.

Repeated excess mixed anhydride acylations of A-chain cyclic bis-disulphide without the purification of the intermediates were carried out as follows: 5 equivalents of the mixed anhydrides (protected amino acids and peptides) were reacted with the amine component for 1 minute in 80 per cent DMSO-water, pH 7.5 at room temperature. The reaction was stopped by adding hydroxylamine, the mixture desalted by gel filtration or by dialysis, and the products lyophylized. End-group determination, electrophoresis of protected and unprotected products, and amino-acid analysis were used to check the completion of the reaction and purity of the products. This reaction scheme was used in the syntheses of human proinsulin sequences.

Partial mixed anhydride acylation procedure was used for the preferential reaction at the α-amino function of proinsulin. The products were purified at each step and their characterization was done as mentioned above.

Semisynthesis of human proinsulin sequences

Human insulin differs from porcine insulin at position B30, the former having Thr and the latter having Ala. Thus native porcine A-chain served as the starting compound for the mixed anhydride coupling of synthetic human C-peptide fragments to obtain C-peptide–A-chain sequence of human proinsulin. Des—Ala-(B30)-porcine B-chain is available by the enzymatic digestion of isolated B-chain, if this could be coupled with the C-peptide–A-chain derivative, then the semisynthesis of human proinsulin is possible.

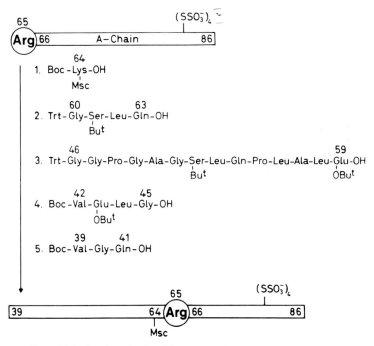

FIG. 36.2. Semisynthesis of human proinsulin sequence 39–86.

Arg–A-chain tetra S-sulphonate was obtained by the procedure of Weinert, Kircher, Brandenburg, and Zahn (1971). This product was then successively elongated with six C-peptide fragments, as shown in Fig. 36.2, by the mixed anhydride couplings to yield a 48-residue sequence (residues 39–86 of human proinsulin) (Naithani *et al.* 1979*b,c*).

The S-sulphonate groups of the A-chain were not stable over a long period of storage and under the conditions of peptide synthesis. The semisynthesis was therefore being repeated with Arg–A-chain cyclic bis-disulphide. Its preparation has been achieved by the trypsin-catalysed peptide synthesis described below.

In order to achieve the final coupling of the B-chain sequence to the C-peptide–A-chain sequence resulting in B-chain–C-peptide–A-chain (proinsulin), selective activation of the C-terminal carboxyl group of the B-chain derivative is essential. The following two approaches were applied: (i) Use of 1,29-di-Boc–B-chain cyclic disulphide which has three carboxyl groups, two side-chains (Glu-13 and Glu-21), and one C-terminal (Ala-30). The C-terminal carboxyl group is, in general, more reactive than the side-chain carboxyl groups. Owing to the 7–19 cyclic disulphide bond the middle region might form a loop and thereby hinder the reaction of the side-chain carboxyl

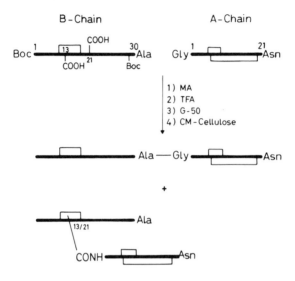

F I G . 36.3. Semisynthesis of B-chain–A-chain sequence.

groups. Glu-21 is adjacent to Arg-22 and thus may be hindered by ionic interaction from reacting. It was therefore hoped that the C-terminal carboxyl group was most likely to be activated by the mixed anhydride formation in a model reaction with 1,29-di-Boc–B-chain cyclic disulphide. The mixed anhydride activation of this derivative was followed by its reaction with A-chain and yielded B–A product in 5 per cent yield (Fig. 36.3) (Naithani *et al.* 1979*b*, *c*). Carboxy-peptidase-A digestion revealed it to contain 30 per cent of B–A products acylated at either Glu-13 or Glu-21 or both. (ii) Using 1, 29-di-Boc–B-chain cyclic disulphide—13,21 dimethyl ester in which the side-chain carboxyl groups are protected while the C-terminal group is free. Consequently the mixed anhydride activation of the free carboxyl group should be selective. Work is in progress to use this approach in the semisynthesis.

Semisynthesis of preproinsulin derivatives

In preproinsulin, a prepeptide of 24 residues extends from the amino terminus of proinsulin (Chan *et al.* 1976, 1979; Shields and Blobel 1977; Ullrich *et al.* 1977; see also Ch. 34). During biosynthesis the prepeptide is rapidly cleaved from the growing precursor chain eventually yielding proinsulin which is latter processed to insulin and C-peptide (Steiner and Oyer 1967). With the availability of bovine proinsulin and the knowledge of rat prepeptide sequences (prepeptide I and II) (Chan *et al.* 1979) we undertook the semisynthesis of a hybrid preproinsulin consisting of rat prepeptide-I and bovine proinsulin. Such a hybrid preproinsulin should have identical properties to

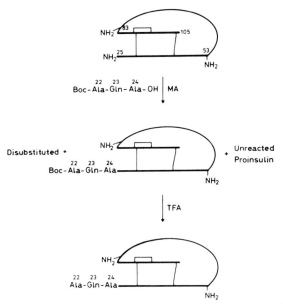

FIG. 36.4. Semisynthesis of (Ala—Gln—Ala)-proinsulin.

native preproinsulin as prepeptides of different species show a great deal of homology (Chan *et al.* 1976, 1979; Shields and Blobel 1977; Ullrich *et al.* 1977).

The significant differences in the pKs of the α-amino group (Phe-1) and the two ε-amino groups (Lys-29 and 59) of proinsulin enabled us to acylate the α-amino group preferentially (Naithani *et al.* 1979 *a, c, d*). The optimum pH for this reaction was 6; at this pH the ε-amino groups are essentially protonated. Acylation of proinsulin with 5 equivalents of Boc—Ala—Gln—Ala (the C-terminal tripeptide of the rat prepeptide sequence 22–24) yielded mainly a product substituted only at Phe-1 and some disubstituted product (Fig. 36.4). The purified and characterized (Ala—Gln—Ala)-proinsulin,

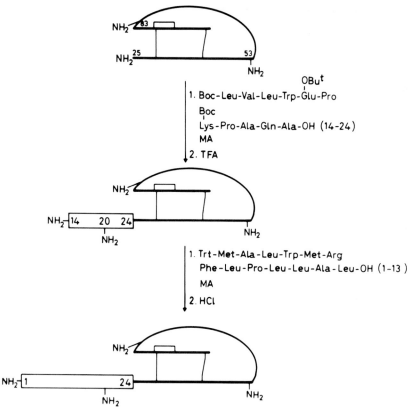

FIG. 36.5. Semisynthesis of preproinsulin (1–105) in the disulphide form.

represents a partial preproinsulin, des-(1–21)-preproinsulin. In the same way
Boc—Leu—Val—Leu—Trp—Glu (OBu^t)—Pro—Lys (Boc)—Pro—Ala—
Gln—Ala (residues 14–24 of preproinsulin) was preferentially attached to
proinsulin to yield the desired α-aminosubstituted derivative (Fig. 36.5). The
protecting groups were removed with trifluoroacetic acid and ethanedithiol.
The product contained an α-amino group (Leu-14) and three ε-amino groups
(20, 53, and 83 corresponding to hybrid preproinsulin). A second acylation
under the same conditions with Trp—Met—Ala—Leu—Trp—Met—
Arg—Phe—Leu—Pro—Leu—Leu—Ala—Leu (1–13) was also preferential
for the α-amino group.

   The purification of the derivative 1–105 turned out to be difficult. The solubility was low and the protein had a strong tendency to adsorb irreversibly
to ion-exchange columns. An attempt to purify it on DEAE-cellulose at
pH 8.1 in an isopropanol-containing buffer proved unsatisfactory. HPLC-
purification of 22–105 and 14–105 revealed these products to be essentially

pure.† The preproinsulin (1–105), however, was contaminated with impurities. Discelectrophoresis at pH 6.9 showed 22–105 to run identically to proinsulin while 1–105 remained at the start. The amino-acid analysis of 22–105 and 14–105 agreed with the theoretically expected values.

Trypsin-catalysed semisynthesis

Recently Inouye *et al.* (1979) and Morihara *et al.* (1979) described the semisynthesis of human insulin by trypsin-catalysed couplings Di—Boc—des—octapeptide (B23–30)-porcine insulin was condensed with synthetic human octapeptide (B23–30), and des—Ala (B30)-porcine insulin with Thr—OBu^t in 50 per cent DMF-water at pH 6.5. The advantages of the enzyme-catalysed semisynthesis are simple operation, high yields, and racemization-free acylations.

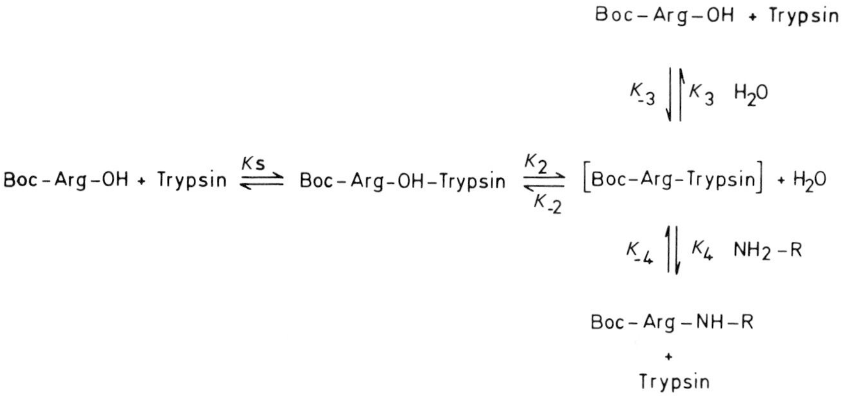

$$Boc-Arg-OH + Trypsin$$

$$K_{-3} \Big\Vert K_3 \quad H_2O$$

$$Boc-Arg-OH + Trypsin \xrightleftharpoons{Ks} Boc-Arg-OH-Trypsin \xrightleftharpoons[K_{-2}]{K_2} \Big[Boc-Arg-Trypsin\Big] + H_2O$$

$$K_{-4} \Big\Vert K_4 \quad NH_2-R$$

$$Boc-Arg-NH-R$$
$$+$$
$$Trypsin$$

SCHEME 36.1. Diagrammatic representation of the enzyme catalysed acylation.

The basis of trypsin-catalysed couplings is illustrated in Scheme 36.1. The acyl-enzyme intermediate (Boc–Arg–Trypsin) undergoes aminolysis with NH_2-R as well as hydrolysis. As $K_3 \gg K_{-3}$ and $K_4 \gg K_{-4}$, the ratio of aminolysis to hydrolysis clearly depends on K_4/K_3, and thus on the concentration of the nucleophile (NH_2-R) which is in competition with water. Addition of a suitable organic solvent like DMF or DMSO diminishes the concentration of water and favours formation of the peptide bond. Thus it has been demonstrated (Inouye *et al.* 1979) that at pH 6.5 in 50 per cent DMF or 50 per cent DMSO the equilibrium between peptide-bond hydrolysis and peptide-bond formation in the presence of trypsin was shifted towards

† HPLC-purifications of preproinsulin sequences were kindly performed by Dr J. Shield, Eli Lilly Corporation.

the latter. We had also been working on these lines and had used trypsin-catalysed couplings as follows:

Semisynthesis of Arg–A-chain cyclic bis-disulphide

An attempt to prepare this derivative by reaction of A-chain cyclic bis-disulphide with NCA–Arg, analogous to Weinert *et al.* (1971), yielded poorly soluble reaction products as a result of overreaction. Therefore, trypsin-catalysed semisynthesis using *t*-butyloxycarbonylarginine and A-chain cyclic bis-disulphide was attempted.

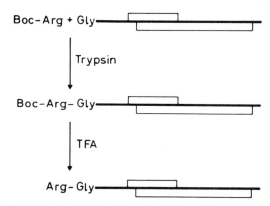

FIG. 36.6. Semisynthesis of Arg–A-chain cyclic bis-disulphide.

The semisynthesis (Gattner *et al.* 1979; Naithani *et al.* 1979c) was carried out with tenfold molar excess of *t*-butyloxycarbonylarginine over A-chain cyclic bis-disulphide in 50 per cent DMF-water buffer at pH 6.5 essentially according to the procedure of Inouye *et al.* (1979) (Fig. 36.6). Amino-acid analysis and electrophoresis indicated about 60 per cent coupling yield. Pure Arg–A-chain cyclic bis-disulphide was obtained after purification in 40–50 per cent yield.

Semisynthesis of human insulin from porcine insulin

Gattner *et al.* (1979) carried out a trypsin-catalysed condensation of des— Ala (B30)-insulin with a 50-fold excess of Thr—OMe using the system described above (Fig. 37.7). A yield of 67 per cent of human insulin (B30) mono-methyl ester was obtained as judged by HPLC. This derivative was purified by ion-exchange chromatography on DEAE-cellulose. The fraction containing monomethyl ester was chromatographed on Sephadex-G-25 in 0.05 M NH_4HCO_3 and immediately saponified in the same buffer at pH 9.5 for 72 h at room temperature. After purification a final yield of 40 per cent of human insulin was obtained. To our surprise Arg—Gly (B22–23) remained intact during the semisynthesis.

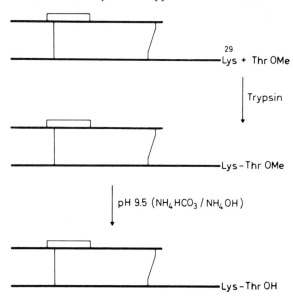

$$\text{—Lys} + \text{Thr OMe}$$

Trypsin

$$\text{—Lys–Thr OMe}$$

pH 9.5 ($NH_4 HCO_3$ / $NH_4 OH$)

$$\text{—Lys–Thr OH}$$

FIG. 36.7. Semisynthesis of human insulin.

Our method and that of Morihara *et al.* (1979) offer simple alternatives for carrying out modifications and elongations to obtain insulin analogues and proinsulin intermediates.

Conclusion

We have demonstrated the usefulness of two novel methods in the semisynthesis of polypeptides and proteins. The use of mixed anhydrides in aqueous–organic solvents allows a rapid and efficient extension of both A-chain derivatives and of proinsulin. In the latter case the preferential acylation of α-amino group was achieved at pH 6, at which ε-amino groups are essentially protonated. The method has been used in the semisynthesis of preproinsulin in the disulphide form. Trypsin can be used to catalyse the formation of peptide bonds under particular conditions. This has allowed the acylation of the A-chain with Boc—Arg, and also the synthesis of human insulin from porcine insulin via des—Ala (B30)-porcine insulin. Both methods are versatile and promise to have enormous potential in future semi-synthetic work.

Acknowledgements

This work is dedicated to Dorothy Hodgkin for the inspiration and motivation she has provided to·all those attempting to understand the insulin molecule. It was financially supported by the Ministerium für Wissenschaft und Forschung des Landes Nordrhein-Westfalen and the Deutsche Forschungsgemeinschaft. We are thankful to Dr R. E. Chance, Eli Lilly Corporation, for providing commercial b-component, to Dr J. Shield, Eli Lilly Corporation, for the HPLC-purification of preproinsulin sequences, and to Dr D. Saunders for his valuable suggestions in preparing this manuscript.

References

ANDERSON, G. W., ZIMMERMAN, J. E., and CALLAHAN, F. M. (1967). *J. Am. chem. Soc.* **89,** 5012–17.

BLUNDELL, T. L., DODSON, G. G., HODGKIN, D. C., and MERCOLA, D. A. (1972). *Adv. Protein Chem.* **26,** 279–402.

BODANSZKY, M. and TOLLE, J. C. (1977). *Int. J. Peptide Protein Res.* **10,** 380–4.

BOISSONNAS, R. A. (1951). *Helv. chim. Acta* **34,** 874–9.

CHAN, S. J., KEIM, P., and STEINER, D. F. (1976). *Proc. natn. Acad. Sci. U.S.A.* **73,** 1964–8.

—— NOYES, B. E., AGARWAL, K. L., and STEINER, D. F. (1979). *Proc. natn. Acad. Sci. U.S.A.* (in press).

CHANCE, R. E., ELLIS, R. M., and BROMER, W. W. (1968). *Science, N.Y.* **161,** 165–7.

FRIESEN, H.-J. (1976). Ph.D. thesis, RWTH, Aachen.

—— NAITHANI, V. K., GATTNER, H.-G., and ZAHN, H. (1976). *10th Int. Cong. Biochemistry*, Hamburg, 25–31 July.

GATTNER, H.-G., DANHO, W., and NAITHANI, V. K. (1979). *2nd Int. Insulin Symp.* Aachen, 4–7 September.

GEIGER, R. (1976). *Chemikerzeitung* **100,** 111–29.

INOUYE, K., WATANABE, K., MORIHARA, K., TOCHINO, Y., KANAYA, T., EMURA, J., and SAKAKIBARA, S. (1979). *J. Am. chem. Soc.* **101,** 751–2.

KATSOYANNIS, P. G. (1964). *Diabetes* **13,** 339–48.

KRAIL, G., BRANDENBURG, D., and ZAHN, H. (1975a). *Hoppe-Seyler's Z. physiol. Chem.* **356,** 981–96.

—— —— (1975b). *Makromol. Chem.* Suppl. **1,** 7–22.

KUNG, Y. T., DU, Y. C., HUANG, W. T., CHEN, C. C., KE, L. T., HU, S. C., JIANG, R. Q., CHU, S. Q., NIU, C. I., HSU, J. Z., CHANG, W. C., CHENG, L. L., LI, H. S., WANG, Y., LOH, T. P., CHI, A. H., LI, C. H., SHI, P. T., YIEH, Y. H., TANG, K. L. HSING, CHI, Y. (1966). *Scientia sin.* 544–61.

MEIENHOFER, J., SCHNABEL, E., BREMER, H., BRINKHOFF, O., ZABEL, R., SROKA, W., KLOSTERMEYER, H., BRANDENBURG, D., OKUDA, T., and ZAHN, H. (1963). *Z. Naturf.* **18B,** 1120–1.

MORIHARA, K., OKA, T., and TSUZUKI, H. (1979). *Nature, Lond.* **280,** 412–13.

NAITHANI, V. K., BÜLLESBACH, E. E. and ZAHN, H. (1979a). *Hoppe-Seyler's Z. physiol. Chem.* **360,** 1363–6.

—— GATTNER, H.-G., and ZAHN, H. (1979b). *2nd Int. Insulin Symp.*, Aachen, 4–7 September.

—— —— BÜLLESBACH, E. E., FÖHLES, J., and ZAHN, H. (1979c) *6th Am. Peptide Symp.* Washington, DC, 17–22 June.

—— BÜLLESBACH, E. E., SHIELD, J., CHANCE, R., ROOT, M. A., and ZAHN, H. (1979d). *2nd Int. Insulin Symp.*, Aachen, 4–7 September.

SCHNABEL, E. (1962). *Ann. Chem.* **659,** 168–84.

SHIELDS, D. and BLOBEL, G. (1977). *Proc. natn. Acad. Sci. U.S.A.* **74,** 2059–63.

SHIMONISHI, Y. (1970). *Bull. chem. Soc. Japan* **43,** 3251–6.

SIEBER, P., KAMBER, B., HARTMANN, A., JÖHL, A., RINIKER, B., and RITTEL, W. (1974). *Helv. chim. Acta* **57**, 2617–21.
STEINER, D. F. and OYER, P. E. (1967). *Proc. natn. Acad. Sci. U.S.A.* **57**, 473–80.
TILAK, M. A. (1970). *Tetrahedron Lett.* **11**, 849–54.
ULLRICH, A., SHINE, J., CHIRGWIN, J., PECTET, R., TISCHER, E., RUTTER, W. J., and GOODMAN, H. M. (1977). *Science, N.Y.* **196**, 1313–19.
VAUGHAN, J. R., JR (1951). *J. Am. chem. Soc.* **73**, 3547.
WEINERT, M., BRANDENBURG, D., and ZAHN, H. (1969). *Hoppe-Seyler's Z. physiol. Chem.* **350**, 1556–62.
—— KIRCHER, KL., BRANDENBURG, D., and ZAHN, H. (1971). *Hoppe-Seyler's Z. physiol. Chem.* **352**, 719–24.
WIELAND, T. and BERNHARD, H. (1951). *Ann. Chem.* **572**, 190–4.
VAN ZON, A. and BEYERMAN, H. C. (1973). *Helv. chim. Acta* **56**, 1729–40.
———— (1976). *Helv. chim. Acta* **59**, 1112–26.

37. The chemical synthesis of insulin and the pursuit of structure–activity relationships

PANAYOTIS G. KATSOYANNIS

Introduction

I RECALL that several years ago I considered the complicated structures that crystallographers have proposed to represent the three-dimensional form of proteins as exotic toys, beautiful to look at and interesting additions to the décor of departmental libraries. I must also confess that I considered their deductions from the study of such structures, regarding intramolecular interactions and their possible effect on the biological behaviour of proteins, as having at best only remote relevance to reality. The elucidation of the three-dimensional structure of insulin by Dorothy Hodgkin and her co-workers in 1969 has shattered all these misgivings. It was only a few years earlier that we had accomplished the chemical synthesis of insulin and had the methodologies in hand that permitted the synthesis of insulin analogues. We could now directly test the dependability of the predictions of the X-ray model. In my laboratory alone, more than forty insulin analogues have been synthesized to date. It is exciting and gratifying to state that the biological evaluation of these analogues verified to a remarkable degree most of the predictions of the three-dimensional structure of insulin with regard to intramolecular interactions and their possible effect on the biological behaviour of this protein. Without any doubt, the elucidation of the structure of insulin has now provided the basis for a more rational approach to the study of structure–activity relationships. I am most indebted to Dorothy Hodgkin for her great accomplishment.

It is a great pleasure and a privilege to contribute to this volume dedicated to Dorothy Hodgkin by her colleagues, students, and friends for her prodigious contributions to X-ray crystallography. In the following pages I will describe some of the work carried out in my laboratory on the chemical synthesis of insulin and insulin analogues. The selection of many of these analogues for synthesis was to a great extent predicated on the conformational relationships established by the crystallographic work of Dorothy Hodgkin.[†]

The chemical synthesis of insulin

Insulin occupies a unique position among proteins. It is the first protein to be recognized as a hormone and thus established the fact that a protein can be a hormone and vice versa; it is one of the first proteins to be crystallized; it is the first protein whose structure was elucidated and thus became the vanguard of protein structural analysis; it is the first protein-hormone whose three-dimensional structure was determined; it is the first protein-hormone whose precursor processing was recognized as playing an important role in its biosynthesis; and finally, it is the first protein to be chemically synthesized. This last aspect is of great importance in the clinical treatment of diabetes. It has been emphasized from different quarters that, in the not too distant future, the demand for insulin will most likely surpass the supply from natural sources and the need for synthetic insulin or related substances will become acute. Should that time arrive, I believe we are prepared to cope with this problem.

Insulin formation by recombination of A- and B-chains

The structure of insulin, as proposed by Sanger (1959), is shown in Fig. 37.1. We undertook the synthesis of insulin in 1959, making the assumption that if chemically synthesized A- and B-chains were available, it might be possible to obtain insulin by air oxidation of a mixture of the sulphydryl forms of

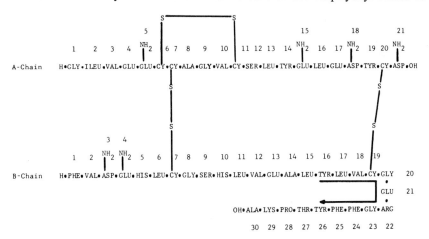

FIG. 37.1. Structure of sheep insulin.

[†] Synthetic studies on insulin carried out in two other laboratories are described elsewhere in this volume.

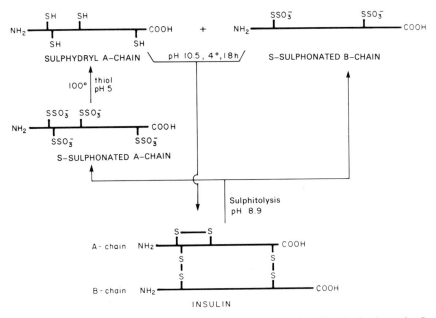

FIG. 37.2. Cleavage and resynthesis of insulin: sulphitolysis of insulin leads to the S-sulphonated A- and B-chain; the S-sulphonated A-chain is converted to its sulphydryl form; interaction of the sulphydryl form of the A-chain with the S-sulphonated B-chain affords insulin.

these chains. While the work on synthesis towards that goal was in progress, Dixon and Wardlaw (1960) and Du, Zhang, Lu, and Tsou (1961) confirmed our assumption. These investigators reported the cleavage of insulin to its two chains by oxidative sulphitolysis (Bailey and Cole 1959), the separation of the chains in the S-sulphonated form, and the subsequent regeneration of insulin activity by air oxidation of a mixture of the thiol forms of the chains. Originally the yield of insulin, following the above procedure, ranged from 1 to 10 per cent (Du *et al.* 1961) and eventually yields up to 50 per cent were reported (Du, Jiang, and Tsou 1965). In our laboratory we developed a different procedure for chain recombination (Katsoyannis and Tometsko 1966; Katsoyannis, Trakatellis, Johnson, Zalut, and Schwartz 1967; Katsoyannis, Trakatellis, Zalut, Johnson, Tometsko, Schwartz, and Ginos 1967) which is illustrated in Fig. 37.2. In our procedure the S-sulphonated A- and B-chains produced by sulphitolysis of insulin are separated, the S-sulphonated A-chain is converted to the sulphydryl form by treatment with a thiol, and this in turn is reacted with the S-sulphonated B-chain to produce insulin. The amount of insulin in the recombination mixture, as detected by the mouse convulsion assay method, corresponds to an approximate yield of 50 per cent, based on the amount of the B-chain used.

Sulphitolysis of insulin and separation of the A- and B-chains

During the course of our investigations on chain recombination (see above) we have developed highly efficient methods for the sulphitolysis of insulin, the complete separation of the S-sulphonated A- and B-chains produced (Katsoyannis, Tometsko, Zalut, Johnson, and Trakatellis 1967; Katsoyannis, Trakatellis, Johnson, Zalut, and Schwartz 1967), and the isolation of insulin from the recombination mixture of the A- and B-chains (Katsoyannis, Trakatellis, Johnson, Zalut, and Schwartz 1967; Katsoyannis, Trakatellis, Zalut, Johnson, Tometsko, Schwartz, and Ginos 1967). These methods were used subsequently in the synthesis of insulin from different species (Katsoyannis, Trakatellis, Zalut, Johnson, Tometsko, Schwartz, and Ginos 1967) and in the synthesis of insulin analogues, and will be briefly outlined. Insulin is sulphitolysed upon treatment with a mixture of sodium sulphite and sodium tetrathionate in 8 M guanidine hydrochloride solution at pH 8.9–9.2 for three hours.† Dialysis of the mixture and adjustment of the non-diffusible material to pH 5 causes the precipitation of the bulk of the S-sulphonated B-chain (B-SSO$_3^-$), whereas the bulk of the S-sulphonated A-chain (A-SSO$_3^-$) remains in the supernatant. Complete separation of the S-sulphonated A- and B-chains from these fractions (enriched A-SSO$_3^-$ and enriched B-SSO$_3^-$) is effectively accomplished by chromatography on CM-cellulose columns with urea–acetate buffer at pH 4.0. Figure 37.3 illustrates the chromatographic pattern obtained by chromatography of each of the above fractions. Dialysis and lyophilization of the column eluates affords the S-sulphonated chains in a highly purified form and in yields, based on the amount of insulin sulphitolysed, ranging from 60 to 70 per cent. The chemical and stereochemical homogeneity of the purified chains was ascertained by amino-acid analysis after acid and enzymatic hydrolysis, and by paper and thin-layer electrophoresis at different pH values. It is interesting to note that the insulin chains are devoid of activity, either as S-sulphonated derivatives or when reduced and air-oxidized individually (Katsoyannis, Tometsko, Zalut, Johnson, and Trakatellis 1967).

---

† Originally the sulphitolysis of insulin was carried out for 24 hours. We observed, however, that the length of the sulphitolysis reaction affected the structure of the A-chain produced and its potential ability to recombine with the B-chain to yield insulin. Infrared spectroscopy of the S-sulphonated A-chain produced after three hours sulphitolysis of insulin exhibited a single band at 1025 cm^{-1}. A 24-hour sulphitolysis resulted in the reduction of the intensity of the band at 1025 cm^{-1} and the concomitant appearance of another band at 1040 cm^{-1}. A further reduction of the intensity of the band at 1025 cm^{-1} and an increase of the intensity of the band at 1040 cm^{-1} resulted when the sulphitolysis was carried out for 48 hours. We were unable to detect any changes in the amino-acid composition of the S-sulphonated A-chains obtained under these conditions. We observed, however, that whereas the S-sulphonated A-chain obtained after three hours' sulphitolysis recombined with the B-chain to produce insulin in an approximate yield of 50 per cent (based on the B-chain used), the A-chain derivative obtained after 48 hours' sulphitolysis recombined with the B-chain in only 15–30 per cent yield to produce the hormone. This phenomenon is now under investigation in our laboratory.

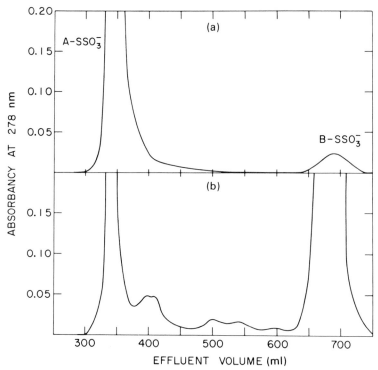

FIG. 37.3. Separation of ASSO$_3^-$ and BSSO$_3^-$ by chromatography on a 4×60 cm CM-cellulose column with urea-acetate buffer (pH 4.0). (a) Chromatography of the fraction enriched in ASSO$^-_3$. (b) Chromatography of the fraction enriched in BSSO$^-_3$.
  (Katsoyannis, Tometsko, Zalut, Johnson, and Trakatellis (1967), by permission.)

Isolation of insulin from combination mixture of A- and B-chains

In addition to our studies of insulin synthesis by recombination of the A- and B-chains, we have directed our efforts to the development of effective methods for the isolation of insulin from the recombination mixture and for the identification of the by-products of the recombination reaction. From these studies we have developed (Katsoyannis, Trakatellis, Johnson, Zalut, and Schwartz 1967; Katsoyannis, Trakatellis, Zalut, Johnson, Tometsko, Schwartz, and Ginos 1967) an isolation procedure which involves, as the initial step, the conversion of all components of the recombination mixture to the picrate and, eventually, to the hydrochloric acid salt. Separation of the insulin hydrochloride from the other products of the mixture was finally accomplished by chromatography on a CM-cellulose column with acetate buffer (pH 3.3) and an exponential sodium chloride gradient. The recovery of insulin by this isolation procedure, based on the activity present in the combination mixture, is approximately 50 per cent (Katsoyannis, Trakatellis,

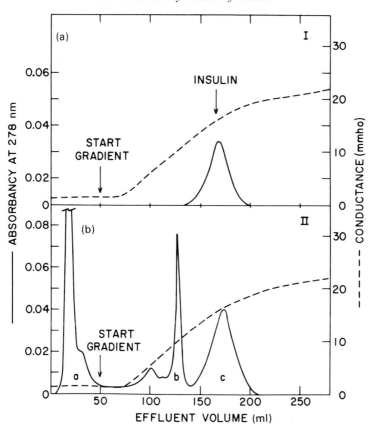

Fig. 37.4. Chromatography of (a) natural bovine insulin and (b) a combination mixture of the sulphydryl A-chain and S-sulphonated B-chain on a 0.9 × 23 cm CM-cellulose column with acetate buffer (pH 3.3) and an exponential NaCl gradient. (Katsoyannis, Trakatellis, Johnson, Zalut, and Schwartz (1967), by permission.)

Johnson, Zalut, and Schwartz 1967; Katsoyannis, Trakatellis, Zalut, Johnson, Tometsko, Schwartz, and Ginos 1967). Figure 37.4(a) shows the chromatogram of natural bovine insulin in this chromatographic system, and Fig. 37.4(b) illustrates the chromatographic pattern obtained from a combination mixture of natural bovine insulin chains. The regenerated insulin, isolated by this procedure was crystallized and found to be identical with the natural hormone with respect to amino-acid composition, biological activity, electrophoretic mobility on thin-layer plates, mobility on CM-cellulose columns, infrared pattern, and crystalline form. It is apparent, from Fig. 37.4(b), that in the combination mixture of the A- and B-chains, besides the regenerated insulin (component c), only two other major components (b and a) are present. Amino-acid analysis indicated that component a is unreacted derivative(s) of the A-chain, and component b, unreacted derivative(s) of the B-chain (Katsoyannis, Trakatellis, Johnson, Zalut, and Schwartz 1967). The unreacted derivative(s) of the A-chain (a) is readily isolated, resulphitolysed, and used in recombination experiments with B-chain to produce insulin. Thus, theoretically, all the available A-chain can be converted to insulin. The unreacted B-chain derivative(s), however, could not be converted to usable form for recombination with A-chain (unpublished data from this laboratory). It must be emphasized that our data demonstrated that insulin is the exclusive product formed, among the many possible isomers, by combination of A- and B-chains. Other products of the combination mixture are components derived from unreacted A-chain and components derived from unreacted B-chain. No other components consisting of A- and B-chain, except for insulin itself, were ever found. The implication therefore arises that, as far as insulin is concerned, the necessary information for folding and orientation of the chains in a manner that permits their spontaneous combination to form the protein is contained within the primary structure of the chains. A schematic presentation of the isolation of insulin from the combination mixture of A- and B-chains is shown in Fig. 37.5.

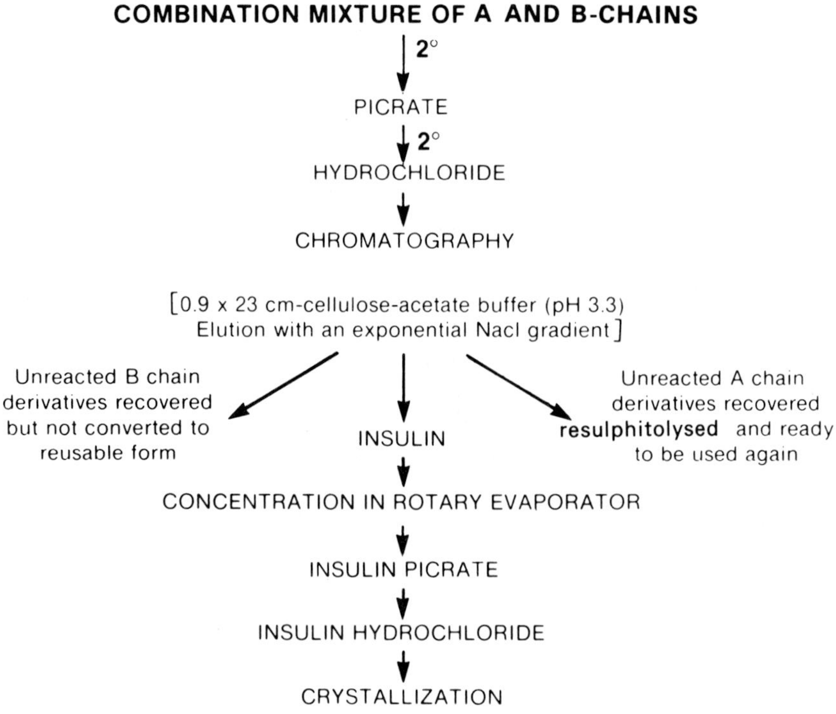

COMBINATION MIXTURE OF A AND B-CHAINS

$\downarrow$ 2°

PICRATE

$\downarrow$ 2°

HYDROCHLORIDE

$\downarrow$

CHROMATOGRAPHY

[0.9 x 23 cm-cellulose-acetate buffer (pH 3.3)
Elution with an exponential NaCl gradient]

Unreacted B chain derivatives recovered but not converted to reusable form ← INSULIN → Unreacted A chain derivatives recovered resulphitolysed and ready to be used again

$\downarrow$

CONCENTRATION IN ROTARY EVAPORATOR

$\downarrow$

INSULIN PICRATE

$\downarrow$

INSULIN HYDROCHLORIDE

$\downarrow$

CRYSTALLIZATION

FIG. 37.5. Schematic presentation of the isolation of insulin from the combination mixture of A- and B-chains.

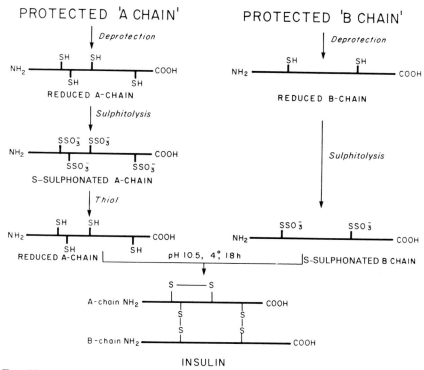

FIG. 37.6. Overall scheme for the chemical synthesis of insulin. The chemically synthesized protected A- and B-chains are deprotected and the resulting reduced chains are converted to the S-sulphonated derivatives by oxidative sulphitolysis. The S-sulphonated A-chain is converted to its sulphydryl form on treatment with a thiol (i.e. β-mercaptoethanol) and subsequently upon interaction with the S-sulphonated B-chain generates insulin.

Synthesis of the A- and B-chains of insulin

In view of the above considerations, it was apparent that the problem of insulin synthesis was, in essence, the problem of synthesis of the A- and B-chains in their S-sulphonated form. The chemical synthesis of polypeptides of the length and complexity of the insulin chains necessitates protection of the secondary functions of the amino-acid components during the various synthetic steps (for a review see Katsoyannis and Schwartz 1977). Hence the final product of synthesis will be A- and B-chains with their secondary functions, including the sulphydryl groups, protected. Removal of the protecting groups will lead to the reduced chains which, by sulphitolysis, will be converted to the S-sulphonated derivatives. Following the steps mentioned previously, synthetic insulin will be produced. Figure 37.6 summarizes the overall scheme for the chemical synthesis of insulin from chemically synthesized protected chains.

Employing classical methods of peptide chemistry, namely, step-wise elongation and fragment condensation (for a review see Katsoyannis and Schwartz 1977), we synthesized originally the S-sulphonated A- and B-chains of sheep insulin (Katsoyannis, Tometsko, and Fukuda 1963; Katsoyannis, Fukuda, Tometsko, Suzuki, and Tilak, 1964; Katsoyannis, Tometsko, Zalut, and Fukuda 1966; Katsoyannis, Zalut, Tometsko, Tilak, Johnson, and Trakatellis 1971) and eventually the corresponding chains of human insulin (Katsoyannis, Tometsko, Ginos, and Tilak 1966; Katsoyannis, Tometsko, and Zalut 1966, 1967; Katsoyannis, Ginos, Zalut, Tilak, Johnson, and Trakatellis 1971; Schwartz and Katsoyannis 1973).

The pattern followed in all these syntheses consisted of the construction of heneicosapeptides and triacontapeptides embodying the amino-acid sequences found in the A- and B-chains, respectively, with all the functional groups protected. Removal of the protecting groups from the final product and oxidative sulphitolysis of the resulting reduced chain led to the formation of the respective S-sulphonated derivative (Fig. 37.6). In the original synthesis, protecting groups were chosen so that they could be removed in one step, namely, by sodium in liquid ammonia (Sifferd and du Vigneaud 1935), a universally employed reagent in the synthesis of large polypeptides (Katsoyannis and Schwartz 1977). Such treatment of the protected B-chain, however, not only removes its protecting groups, but also causes a cleavage of the chain between the threonine and proline residues at positions 27 and 28 (Katsoyannis, Ginos, Zalut, Tilak, Johnson, and Trakatellis 1971; Katsoyannis, Zalut, Tometsko, Tilak, Johnson, and Trakatellis 1971). The cleavage is quantitative and was indeed used to prepare a truncated insulin, the destripeptide B$^{28-30}$-insulin, which has similar biological behaviour to the natural hormone (Katsoyannis, Zalut, Harris, and Meyer 1971). We overcame this complication and we were able to obtain intact B-chains by the simple expedient of carrying out the sodium in liquid ammonia deblocking step in the presence of sodium amide (Katsoyannis, Ginos, Zalut, Tilak, Johnson, and Trakatellis 1971; Katsoyannis, Zalut, Tometsko, Tilak, Johnson, and Trakatellis 1971). In subsequent synthesis of the B-chain of insulin (Schwartz and Katsoyannis 1973) and of insulin analogues we chose protecting groups for the construction of the protected derivatives of these chains, which could be removed efficiently by treatment with liquid hydrogen fluoride (Sakakibara 1971). The sodium in liquid ammonia treatment did not cause such complications when it was employed to deprotect the synthetic A-chains. Figure 37.7 illustrates one of the early approaches employed for the synthesis of the S-sulphonated A-chain of sheep insulin (Katsoyannis, Tometsko, Zalut, and Fukuda 1966). The C-terminal dodecapeptide (sequence 10–21) was condensed with the adjacent pentapeptide derivative (sequence 5–9) to give the C-terminal heptadecapeptide (sequence 5–21). The latter compound was converted to the amino-free derivative which in turn

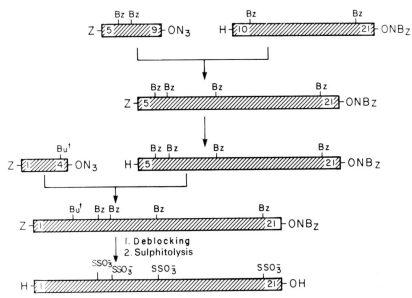

FIG. 37.7. Schematic presentation of the synthesis of the S-sulphonated A-chain by fragment condensation. Abbreviations used: Z, benzyloxycarbonyl; Bz, benzyl; NBz, *p*-nitrobenzyl; Bu^t, tert-butyl.

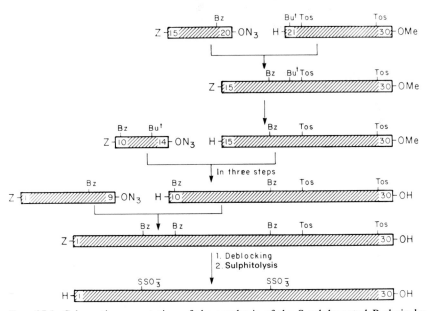

FIG. 37.8. Schematic presentation of the synthesis of the S-sulphonated B-chain by fragment condensation. Abbreviations used: Z, benzyloxycarbonyl; Bz, benzyl; Tos, tosyl; Me, methyl; Bu^t, tert-butyl.

was condensed with the N-terminal tetrapeptide (sequence 1–4) to produce the protected A-chain (sequence 1–21). This compound was deblocked and sulphitolysed to yield the S-sulphonated A-chain. Figure 37.8 shows one of the early approaches employed for the construction of the S-sulphonated B-chain of human insulin (Katsoyannis, Tometsko, Ginos, and Tilak 1966). In this scheme the C-terminal decapeptide fragment (sequence 21–30) was coupled with the adjacent hexapeptide derivative (sequence 15–20) to produce the C-terminal hexadecapeptide (sequence 15–30). The latter peptide derivative was deblocked at the amino end and then condensed with the adjacent pentapeptide fragment (sequence 10–14) to give the C-terminal heneicosapeptide (sequence 10–30). Coupling of the latter compound with the N-terminal nonapeptide derivative (sequence 1–9) afforded the protected triacontapeptide (sequence 1–30). This polypeptide, which is the protected B-chain, was deblocked and sulphitolysed to produce the S-sulphonated derivative. Purification of the synthetic S-sulphonated A- and B-chains was accomplished by chromatography on Sephadex or CM-cellulose using the procedures devised for the purification of their natural counterparts outlined previously (p. 458). The purified synthetic chains exhibited the identical behaviour as the respective natural chains when compared with respect to amino-acid composition after acid and enzymatic hydrolysis, electrophoretic mobility on paper and thin-layer electrophoresis at two pH values, chromatographic behaviour on ion exchange columns, and infrared patterns.

Synthetic insulins

Once we had accomplished the chemical synthesis of the A- and B-chains of sheep and human insulins, we were able to readily combine these chains and produce the respective insulins (Katsoyannis *et al.* 1963, 1964; Katsoyannis, Tometsko, Ginos, and Tilak 1966; Katsoyannis 1966). In these early studies the chain combination was carried on by the method which was employed by Dixon and Wardlaw (1960) to regenerate insulin from natural chains (for a review, see Katsoyannis 1966). Subsequently, implementing our own highly efficient procedures for chain combination and insulin isolation described above (pp. 455 and 458), and using synthetic sheep and human insulin chains and natural bovine and porcine insulin chains, we were able to produce and isolate in pure form several regenerated natural, all-synthetic and half-synthetic (one chain synthetic, the other chain natural) insulins (Katsoyannis, Trakatellis, Zalut, Johnson, Tometsko, Schwartz, and Ginos 1967). These insulins included (Table 37.1) the all-synthetic human, all-synthetic sheep, half-synthetic sheep, bovine and porcine insulins, regenerated bovine and porcine insulins, and an insulin which has not been found in nature and which we tentatively designated as insulin $B_A H_B$. This insulin was produced by combining synthetic human B- and natural bovine A-chains. The $B_A H_B$ insulin, like all the other synthetic and half-synthetic in-

TABLE 37.1

Overall recoveries and specific activities of insulins synthesized by combination of their respective A- and B-chains and isolated by CM-cellulose chromatography. (From Katsoyannis, Trakatellis, Zalut, Johnson, Tometsko, Schwartz, and Ginos (1967), with permission)

Type of chains used for combination	Insulin produced	Recovery as insulin hydrochloride† (%)	Specific activity of isolated insulin‡ (i.u./mg)
Synthetic human A + synthetic human B	Human (all-synthetic)	41	24
Synthetic sheep A + synthetic sheep B	Sheep (all-synthetic)	43	25 (crystalline)§
Natural bovine A + natural bovine B	Bovine (regenerated)	50	25 (crystalline)
Natural porcine A + natural bovine B	Porcine (regenerated)	42	25
Natural bovine A + natural porcine B	Bovine (regenerated)	39	23 (crystalline)
Synthetic sheep A + natural bovine B	Sheep (half-synthetic)	39	25 (crystalline)
Natural bovine A + synthetic sheep B	Bovine (half-synthetic)	37	22 (crystalline)
Synthetic human A + natural bovine B	Porcine (half-synthetic)	52	22
Natural bovine A + synthetic human B	Insulin B$_A$H$_B$	35	22 (crystalline)

† The insulin activity in the combination mixture taken as 100 per cent.
‡ Calculated from determination of protein content (Folin method) and of biological activity (mouse convulsion assay).
§ Crystalline insulins were isolated as zinc complexes.

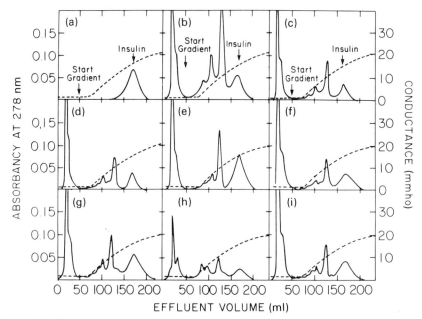

FIG. 37.9. Chromatography of natural bovine insulin and of combination mixtures of various sulphydryl A- and S-sulphonated B-chains on a 0.9×23 cm CM-cellulose column with acetate buffer (pH 3.3) and an exponential NaCl gradient. (a) Natural bovine insulin; (b) combination mixture of synthetic sheep A- and B-chains (all synthetic insulin); (c) combination mixtures of synthetic sheep A- and natural bovine (sheep) B-chains (half-synthetic sheep insulin); (d) combination mixture of natural bovine A- and synthetic bovine B-chains (half-synthetic bovine insulin); (e) combination mixture of natural porcine (human) A- and natural bovine (porcine) B-chains (regenerated porcine insulin); (f) combination mixture of natural bovine A- and natural porcine (bovine) B-chains (regenerated bovine insulin); (g) combination mixture of synthetic human (porcine) A-chain and natural bovine (porcine) B-chains (half-synthetic porcine insulin); (h) combination mixture of natural bovine A- and synthetic human B-chains (half-synthetic $B_A H_B$ insulin); (i) combination mixture of synthetic human A- and B-chains (all synthetic human insulin). (Katsoyannis, Trakatellis, Zalut, Johnson, Tometsko, Schwartz, and Ginos (1967), by permission.)

sulins we produced, possesses the full biological activity of insulin and was obtained in crystalline form.

The chromatographic patterns of the combination mixtures of the A- and B-chains of these insulins on CM-cellulose columns are illustrated in Fig. 37.9. As can be seen, all insulins are eluted with application of the gradient, have the same mobility as the natural hormone, and are the slowest moving components in this chromatographic system. The overall recoveries of these insulins, based on the insulin formed by chain combination, range from 35 to 52 per cent (Table 37.1). The majority of the synthetic insulins were

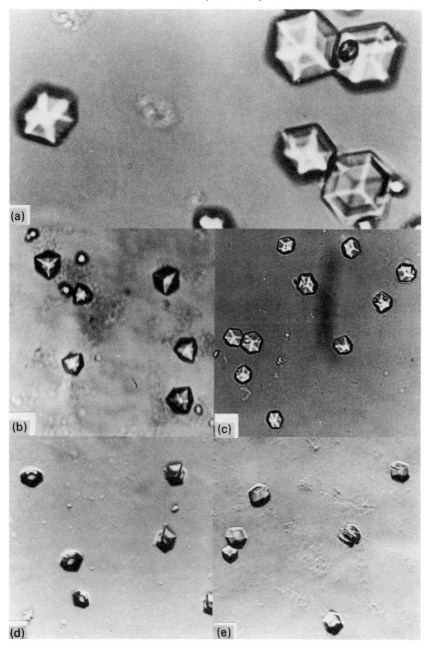

FIG. 37.10. Crystalline zinc insulins: (a) regenerated bovine: (b) half-synthetic sheep (synthetic A- and natural B-chains were used); (c) all-synthetic sheep; (d) half-synthetic bovine (natural A- and synthetic B-chains were used); (d) half-synthetic $B_A H_B$ (natural bovine A- and synthetic human B-chains were used). (Katsoyannis, Trakatellis, Zalut, Johnson, Tometsko, Schwartz, and Ginos (1967), by permission.)

TABLE 37.2

Amino-acid composition† of acid hydrolysates of synthetic and half-synthetic insulins. (From Katsoyannis, Trakatellis, Zalut, Johnson, Tometsko, Schwartz, and Ginos (1967), with permission)

Amino acid	Sheep			Human		Porcine‡		Bovine		$B_A H_B$§	
	Theory	All-synthetic	Half-synthetic	Theory	All-synthetic	Theory	Half-synthetic	Theory	Half-synthetic	Theory	Half-synthetic
Lys	1	1.0	1.0	1	0.8	1	0.9	1	1.1	1	0.9
His	2	1.9	2.0	2	1.9	2	1.9	2	2.0	2	1.9
Arg	1	1.0	1.0	1	1.0	1	1.0	1	0.9	1	0.9
Asp	3	2.6	3.0	3	2.8	3	2.9	3	3.1	3	2.9
Thr	1	0.9	1.0	3	2.5	2	1.9	1	0.9	2	1.8
Ser	2	1.8	1.8	3	2.6	3	2.5	3	2.5	3	2.8
Glu	7	7.2	7.5	7	7.1	7	7.2	7	6.9	7	6.9
Pro	1	1.1	1.0	1	0.9	1	1.1	1	0.9	1	0.8
Gly	5	5.1	5.0	4	4.1	4	4.0	4	4.0	4	4.1
Ala	3	3.1	3.0	1	1.1	2	2.0	3	3.2	2	2.0
Val	5	4.7	4.6	4	3.9	4	3.9	5	4.6	5	4.9
Ile	1	0.8	0.8	2	1.9	2	2.1	1	0.8	1	0.8
Leu	6	6.2	6.0	6	6.1	6	6.2	6	5.8	6	6.0
Tyr	4	ND	ND	4	ND	4	ND	4	ND	4	ND
Phe	3	3.1	3.0	3	3.0	3	3.1	3	3.1	3	3.1
Cys	6	ND	ND	6	ND	6	ND	6	ND	6	ND

† Number of amino-acid residues per molecule.
‡ The A-chains of human and porcine insulins are identical.
§ $B_A H_B$: bovine A- and human B-chains.
ND Not determined.

obtained in crystalline form. Figure 37.10 illustrates (a) the crystals of regenerated bovine, (b) half-synthetic sheep, (c) all-synthetic sheep, (d) half-synthetic bovine and (e) half-synthetic $B_A H_B$ insulins. The synthetic and half-synthetic insulins were compared with their natural counterparts with respect to biological activity, amino-acid composition, and electrophoretic mobility on thin-layer electrophoresis (Katsoyannis, Trakatellis, Zalut, Johnson, Tometsko, Schwartz, and Ginos 1967). The biological activity of the synthetic and half-synthetic insulins is identical to that of the natural hormone (Table 37.1); amino-acid analysis of the synthetic hormones after acid hydrolysis are in excellent agreement with the theoretically expected values (Table 37.2); on thin-layer electrophoresis the synthetic proteins behaved as homogeneous compounds and exhibited identical mobilities with their natural counterparts. All these comparisons of the physical, chemical, and biological properties of the all-synthetic and half-synthetic insulins with those of the natural counterparts justified, in our estimation, the conclusion that the synthetic compounds were identical with the natural hormones, and that the structures proposed for these proteins were correct.

The synthesis of insulin analogues and the pursuit of structure–activity relationships

Alteration of the insulin molecule may modulate its biological profile either by altering its binding ability to the receptor or by altering the 'message region' of the molecule so that the insulin analogue–receptor complex cannot initiate the biological response in the same fashion as the natural hormone–receptor complex. A rational approach to the understanding of the functional areas of the insulin molecule, therefore, requires the availability of a broad spectrum of analogues and the biochemical evaluation of these analogues in terms of biological and immunological activity and receptor binding ability. The availability of efficient methods for the synthesis of the insulin chains, their combination, and isolation of the insulin thus produced, makes the synthesis of a variety of insulin analogues a readily attainable goal. Some forty analogues have been synthesized in our laboratory to date and the synthesis of new analogues is only a matter of time. A brief outline of the procedures currently employed for the synthesis of these analogues follows.

For the synthesis of the S-sulphonated chains of the analogues we follow essentially the scheme outlined in Figs. 37.7 and 37.8. In order to improve the yields during the various synthetic steps, we often vary the length of the intermediate peptide fragments and employ a variety of blocking groups for the protection of their functions. Purification of S-sulphonated B-chain analogues is carried out by the procedure illustrated in Fig. 37.3. We found that the S-sulphonated A-chain analogues can best be purified by chromatography on Ecteola–cellulose columns with Tris·HCl buffer (Ferderigos,

Cosmatos, Ferderigos, and Katsoyannis 1979). Combination of the chains to produce the modified insulin is carried out by our original procedure (Katsoyannis and Tometsko 1966; Katsoyannis, Trakatellis, Johnson, Zalut, and Schwartz 1967; Katsoyannis, Trakatellis, Zalut, Johnson, Tometsko, Schwartz, and Ginos 1967) or a modified version of that procedure (Schwartz and Katsoyannis 1976). Isolation of the insulin analogues from the combination mixture is effected by the procedure described on p. 458 and Figs. 37.4 and 37.5 The last step in this procedure is chromatography on CM-cellulose with acetate buffer and an exponential sodium chloride gradient.

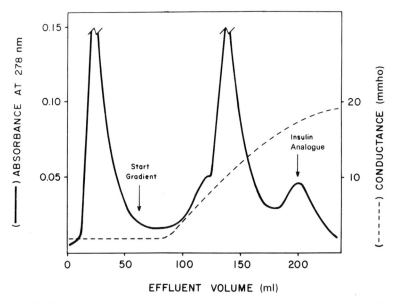

F IG. 37.11. Chromatography of a combination mixture of the sulphydryl form of sheep [sarcosine¹]A-chain and the S-sulphonated sheep (bovine) B-chain (synthetic [sarcosine¹-A] insulin) on a 0.9×23 cm CM-cellulose column with acetate buffer (pH 3.3) and an exponential NaCl gradient. (Okada and Katsoyannis (1975), by permission.)

The chromatographic patterns of all the analogues are almost identical with those obtained when combination mixtures of natural or synthetic A- and B-chains of insulin were chromatographed in this system (Figs. 37.4 and 37.9). A typical chromatogram of an analogue is shown in Fig. 37.11 and depicts chromatography of a combination mixture of the sulphydryl form of [sarcosine¹]A-chain with the S-sulphonated B-chain of sheep insulin (Okada and Katsoyannis 1975). The synthetic analogue, [sarcosine¹-A] insulin is eluted with application of the gradient and is the slowest moving component. It is interesting to point out that, as in the case of the natural chains (p. 459; Fig. 37.4 and 37.9), the chromatogram of the combination mixture of the A- and B-chains of the analogue does not indicate the presence of other components, except the analogue itself, that consists of A- and B-chains. Even in cases when the modified insulin chains combined in low yields to produce an insulin analogue (Schwartz and Katsoyannis 1978), no other component consisting of A- and B-chains was ever detected. A similar situation exists with all the analogues synthesized to date in our laboratory.

Analogues with modifications at the amino- and carboxyl-terminal regions of the A-chain

Our synthetic analogues, which differ from the parent molecule in that the amino and/or carboxyl terminal regions of the A-chain have been modified, are shown in Table 37.3. Elimination of the amino-terminal tetrapeptide sequence from the A-chain (des[tetrapeptide A^{1-4}] insulins) leads to total inactivation, whereas replacement of the α-amino group of the A^1 glycine with hydrogen [deamino A^1-insulin] results in a molecule that possesses 35 per cent of the activity of insulin. Substitution of the A^1 glycine with sarcosine ([sarcosine^1-A] insulin), a modification that increases the basic character of that residue (pK_2 of glycine, 9.6, and of sarcosine, 10.0), results in but a modest decrease of the biological activity and a more pronounced decrease of the immunoreactivity of insulin. Data from other laboratories (for a review, see Blundell, Dodson, Hodgkin, and Mercola 1972) also indicate that modifications at the A^1 position affect considerably the biological activity of insulin and, furthermore, that a decrease in biological activity is concomitant with changes of the circular dichroism spectra of the analogues. It is apparent that in insulin the A^1-glycine is involved in the stabilization of a three-dimensional structure commensurate with high biological activity. This idea is in accord with the X-ray model (Blundell *et al.* 1972) which indicates that the A^1-glycine is involved in interactions with the C-terminal region of the A-chain that are important in the maintenance of the structure of this protein.

Replacement of the A^1-clycine with L- and D-alanine ([L-alanine^1-A] and [D-alanine^1-A] insulins) results in analogues which, in the *in vitro* assays, have relative potencies, as compared to insulin, of approximately 10 and 100 per cent, respectively. It is interesting to note that, for both analogues, their relative binding affinities to isolated fat cells are the same as their relative *in vitro* biological potencies. Our data are consistent with the suggestion that the A^1-glycine is one of the amino-acid residues involved in receptor binding (Pullen, Lindsay, Wood, Tickle, Blundell, Wollmer, Krail, Brandenburg, Zahn, Gliemann, and Gammeltoft 1976). The difference in the relative binding affinity and relative biological activity of the L- and D-analogues reveals a biologically significant stereochemical discrimination of the two hydrogens of the α-carbon of the A^1-residue. Substitution of one of these hydrogens with a methyl group, as is the case with the L-alanine analogue, interferes with

TABLE 37.3

Insulin analogues with modifications at the N- and C-terminal amino-acid residues of the A-chain†

Insulin analogue	(S=Sheep, P=Porcine)	Potency (i.u./mg)				Receptor binding (% of insulin)	References
		MC	GO	DGT	RIA		
Des(tetrapeptide A$^{1-4}$)	(S)	0	—	—	—	0.03 (LM)	Katsoyannis and Zalut (1972b)
Des(tetrapeptide A$^{1-4}$) [Deamino-A^1]	(P)	0	—	—	—	—	Katsoyannis and Zalut (1972b)
[Sarcosine^1-A]	(S)	7–10	—	—	—	—	Katsoyannis and Zalut (1972a)
	(S)	20	—	—	9	—	Okada and Katsoyannis (1975)
[L-Alanine^1-A]	(S)	7.5–9	2.3	3	—	10 (FC)	Cosmatos, Cheng, Okada, and Katsoyannis (1978)
[D-Alanine^1-A]	(S)	10.5–12	23	24	—	100 (FC)	Cosmatos et al. (1978)
[Isoasparagine^{21}-A]	(S)	21	—	—	16	—	Cosmatos, Okada, and Katsoyannis (1976)
[D-Asparagine^{21}-A]	(S)	8	—	—	4	—	Cosmatos et al. (1975)
[Arginine^{21}-A]	(S)	10.5–12	9.5	—	8.6	40 (FC); 18 (LM)	Ferderigos, Cosmatos, Ferderigos, and Katsoyannis (1979)
[Asparaginamide^{21}-A]	(S)	17	3.5	—	2.6–4	50 (FC); 64 (LM)	Unpublished data
[Sarcosine^1-, isoasparagine^{21}-A]	(S)	15	—	—	7	—	Cosmatos et al. (1976)

† MC, mouse convulsion assay; GO, glucose oxidation in isolated fat cells; DGT, 2-deoxy-D-glucose transport assay; RIA, radioimmunoassay; LM, liver membranes; FC, fat cells. The potency of natural insulin is 24–25 i.u./mg.

receptor binding and hence leads to a less biologically active analogue. On the other hand, substitution of the other hydrogen with a methyl group, to produce the D-alanine analogue, does not interfere with binding and does not alter the biological activity of the molecule. It should be pointed out that the differences in potency observed when these analogues were assayed by the *in vivo* and *in vitro* methods, has also been observed with many other analogues, and may be a reflection of the multiplicity of processes involved in the *in vivo* assays which include, but are not confined to, different rates of absorption, distribution, and degradation of the analogues as compared to the natural hormone.

The parity of the relative binding affinity and the relative *in vitro* biological potencies observed in the L- and D-alanine analogues is not unique. Similar results are found with several of our analogues which we have tested as well as with insulin analogues prepared in other laboratories (Freychet, Brandenburg, and Wollmer 1974; Gliemann and Gammeltoft 1974). This prompted the speculation (Freychet *et al.* 1974; Gliemann and Gammeltoft 1974) that the insulin binding site and the site responsible for activating cellular processes may reside on the same region of the molecule. Recent findings in our laboratory, however, to be discussed later in this section, and the biological behaviour of hagfish insulin (Emdin, Gammeltoft, and Gliemann 1977), indicate that this speculation might not be valid.

The X-ray model of insulin reveals that the B-chain folds itself within the insulin molecule so that the A^{21}-asparagine and the B^{22}-arginine are brought into such juxtaposition as to ensure salt-bridge formation between the α-carboxyl group of A^{21}-asparagine and the guanidinium group of the B^{22}-arginine. This structural arrangement appears to stabilize the tertiary structure of insulin (Blundell, Dodson, Dodson, Hodgkin, and Vijayan 1971; Blundell *et al.* 1972). Courtauld atomic models of the natural hormone for the region involved in the salt-bridge formation reveal (Cosmatos, Johnson, Breier, and Katsoyannis 1975) that in the conformation most favourable for salt-bridge formation, the α- and β-carboxyl groups of A^{21}-asparagine are sterically nearly equivalent. On the other hand, Courtauld models of the same region of the [D-asparagine^{21}-A] insulin indicate that the α-carboxyl group, as compared with the natural hormone, is less favourably disposed for salt-bridge formation with the guanidinium group of the B^{22}-arginine. Consistent with these structural features are the findings that [isoasparagine^{21}-A] insulin, which has a free β-carboxyl group and an amidated α-carboxyl group, has similar biological activity while [D-asparagine^{21}-A] insulin has significantly less biological activity than the natural hormone.

Substitution of the A^{21}-asparagine with arginine results in an analogue ([arginine^{21}-A] insulin) that possesses approximately 50 per cent of the biological activity of the natural hormone. Yet in this analogue the A^{21}-arginine still retains the A^{21}–B^{22} salt-bridge-forming capability. This implies that factors other than the above interaction, but involving the side-chain of the A^{21}-amino-acid residue, may come into play. The suggestion was made (Pullen *et al.* 1976) that the A^{21} residue is involved in the binding of the insulin monomer to its receptor on the surface of the target cells. If this is the case, one would expect that the replacement of the A^{21}-asparagine with the bulkier and positively charged arginine might distort the conformation of the binding region of insulin and thus impede its interaction with the receptor. This speculation is strengthened by our finding that the [arginine^{21}-A] insulin has a considerably decreased binding affinity to fat cell and liver membranes as compared to the natural hormone.

The biochemical evaluation of the [asparaginamide^{21}-A] insulin is of considerable interest. This analogue, which differs from the natural hormone in that the carboxyl group of the A^{21}-asparagine is amidated, possesses *in vivo* and *in vitro* activities approximately 70 and 15 per cent, respectively, of that of the natural hormone. The high *in vivo* activity may indicate partial enzymatic deamidation and generation of the native hormone in such systems. On the other hand, the low *in vitro* biological activity indicates the importance for biological activity of a free carboxyl group in the A^{21}-position, either for salt-bridge formation with the B^{22}-arginine or for as yet unknown reasons. The most interesting aspect of this analogue is that, contrary to the observations with all other of our analogues or with modified natural insulin (Freychet *et al.* 1974; Gliemann and Gammeltoft 1974), there is a wide divergence between its relative binding affinity and relative *in vitro* biological activity. A similar situation was encountered with only one naturally occurring insulin, the hagfish insulin, which was found to possess approximately 23 per cent receptor binding affinity but only 4.6 per cent ability to activate lipogenesis as compared to insulin (Emdin *et al.* 1977). This may indicate that the same general area of the insulin molecule is functionally involved in the hormone's biological activities. However, regions in this area which are associated with particular biological activities are not identical but overlapping.

As was the case with the [sarcosine^{1}-A] insulin, the [sarcosine^{1}–isoasparagine^{21}-A] insulin possesses a modestly lower biological activity but a significantly lower immunoreactivity than the natural hormone. This implies that the change in the relative positive charge at the N-terminal residue of the A-chain may be responsible for the considerable decrease in the immunoreactivity of these analogues.

Analogues with modifications of the intra- and interchain disulphide bridges

The relative importance of the intra- and interchain cyclic systems to the biological activity of insulin has been probed by the synthesis of the analogues shown in Table 37.4. Replacement of the cysteine residues at positions A^{6} and A^{11}, which are involved in the formation of the 20-membered intrachain disulphide system of insulin, by alanine results in an analogue ([alanine[6, 11]-A] insulin) that lacks the intrachain cyclic system. This analogue possesses approximately 10 per cent of the activity of the natural hormone. Implicit in this finding is that the intrachain cyclic system is not involved directly in the expression of the biological activity of insulin, but its presence may impose constraints on the molecule leading to topochemical features commensurate with high biological activity. The rather limited role of the 20-membered cyclic system is further manifested by the synthesis of [homocysteine[6, 11]-A] insulin. This analogue, which has a 22-membered intrachain cyclic system, was found to possess 30 per cent of the biological activity of insulin. These data are in agreement with the X-ray model which indicates that the intrachain disulphide system is completely buried and is part of the hydrophobic core of the insulin molecule. Hence, it is not ex-

TABLE 37.4

Insulin analogues with alteration in the intra- and interchain disulphide bridges†

Insulin analogue (sheep)	Potency (i.u./mg)		References
	MC	*RIA*	
[Alanine[6,11]-A] (no intrachain —S—S—)	2–2.5	2–2.5	Katsoyannis, Okada, and Zalut (1973)
[Homocysteine[6,11]-A] (enlarged intrachain —S—S—)	7–8	7–8	Cosmatos and Katsoyannis (1973)
[Homocysteine[7,20]-A] (enlarged interchain —S—S—)	0	2	Cosmatos and Katsoyannis (1975)

† MC, mouse convulsion assay; RIA, radioimmunoassay. The potency of natural insulin is 24–25 i.u./mg.

pected to be involved directly in any physicochemical interplay between the hormone and the responsible cell. On the other hand, substitution of the cysteine residues at positions A^7 and A^{20}, which are involved in the formation of the interchain disulphide system, with homocysteine results in an analogue ([homocysteine[7, 20]-A] insulin) with an enlarged interchain cyclic system and is devoid of biological activity. Apparently this small change in the length of the interchain cross links, less than 1.53 Å, leads to a displacement of the chains that cause a dramatic alteration of the biological profile of the hormone. This, of course, indicates that the biological activity of insulin depends critically on a particular geometry conferred on the molecule by the proper disposition of its individual chains.

Analogues with modifications in the interior of the A-chain

Table 37.5 records our synthetic analogues which differ from insulin in that amino-acid residues located in the interior of the A-chain have been

TABLE 37.5

Insulin analogues with alterations in the interior of the A-chain†

Insulin analogue (sheep)	Potency (i.u./mg)			Receptor binding (% of insulin)	References
	MC	GO	RIA		
[Norleucine²-A]	1	0.22	3	0.6 (LM)	Unpublished data
[D-Tyrosine¹⁹-A]	—	(0.38)	0.22	1.4 (LM)	Unpublished data
[Tyrosine¹⁸-A, asparagine¹⁹-, arginine²¹-A]	0	—	0.15–0.36	<0.1 (LM)	Unpublished data
[Diaminobutyric acid²-, glutamic acid¹⁹,-A]	0	—	0	—	Ferderigos and Katsoyannis (1977)
[Phenylalanine¹⁹-A]	11–12	1.7	7–10	10 (LM)	Unpublished data
[Threonine⁵-A]	6.5–7.5	—	5.3	—	Unpublished data
[Leucine⁵-A]	6–9	(4)	7–10	30 (LM)	Unpublished data

† MC, mouse convulsion assay; GO, glucose oxidation in isolated fat cells; RIA, radioimmunoassay; values in parentheses indicate preliminary data; LM, liver membrane. The potency of natural insulin is 24–25 i.u./mg.

substituted by other residues. The three-dimensional structure indicates that in the insulin monomer the A-chain is folded upon itself so that the amino- and carboxyl-terminal residues are brought together. In this conformation the A^2-isoleucine, which is buried and is part of the invariant hydrophobic core of the molecule, is in van der Waals contact with the phenyl ring of the A^{19}-tyrosine. The latter residue is on the surface of the molecule with one side exposed to the solvent environment and the other side facing the hydrophobic core. This arrangement, the X-ray model predicts, generates interactions which contribute to the stabilization of the molecule, and their deletion will adversely affect its tertiary structure. These predictions are consistent with the data obtained from synthetic analogues with modifications at the A^2- and A^{19}-positions. Substitution of the A^2-isoleucine with norleucine ([nor-leucine^2-A] insulin) results in a dramatic decrease of the biological activity and of the binding affinity as compared to insulin. Sedimentation studies demonstrate that this analogue is predominantly monomeric whereas the natural hormone under similar conditions exists as a dimer. The circular dichroism measurements further suggest that this analogue exhibits a some-what diminished helical content. In view of the monomeric character of the [norleucine^2-A] insulin, one might speculate that the disruption of its secon-dary structure is at the monomer–monomer contact region. Equally interest-ing are the data obtained with analogues produced by modification of the A^{19}-position of the insulin molecule. Replacement of the L-tyrosine with the D-isomer([D-tyrosine^{19}-A]insulin) leads to a drastic decrease of the bio-logical activity and the binding affinity, whereas inversion of the location of the A^{19}-tyrosine and A^{18}-asparagine, as is the case with [tyrosine^{18}-, asparagine^{19}-, arginine^{21}] insulin, results in total inactivation. Even replace-ment of the A^{19}-tyrosine with phenylalanine ([phenylalanine^{19}-A] insulin) affords an analogue with considerably decreased *in vitro* biological activity and receptor-binding affinity, as compared to the natural hormone. The bio-logical evaluation of [α,γ-diaminobutyric acid^2-, glutamic acid^{19}-A] insulin, in which the A^2-isoleucine and A^{19}-tyrosine have been substituted with α,γ-diaminobutyric acid and glutamic acid, respectively, demonstrates the special topochemical relationship of the A^2- and A^{19}-residues which secures in the insulin molecule a conformation commensurate with high biological activity. This analogue, in which the hydrophobic A^2- and A^{19}-residues have been substituted with strongly interacting polar amino-acid residues, is totally in-active. Undoubtedly the potential interaction between α,γ-diaminobutyric acid and glutamic acid in the analogue is stronger than the van der Waals interaction between isoleucine and tyrosine in the natural hormone. How-ever, the structural changes induced in the molecule by the insertion of α,γ-diaminobutyric acid and glutamic acid at positions 2 and 19 have deleterious effects on its biological behaviour.

Glutamine at A^5-position, an invariant amino-acid residue in insulin sequences from different species, is on the surface of the molecule. Substitution of this hydrophilic residue, either with the hydrophilic threonine or with the hydrophobic leucine, affords the [threonine^5-A] and [leucine^5-A] insulins, respectively, which exhibit *in vivo* only approximately 30 per cent of the biological activity of insulin. The relative binding affinity and *in vitro* biological activity of the [leucine^5-A] insulin are approximately 30 and 17 per cent, respectively, as compared to the natural hormone. It is suggested that the A^5-glutamine is in the receptor-binding region of insulin.

Analogues with modifications at the amino- and carboxyl-terminal regions of the B-chain

The X-ray analysis of the three-dimensional structure of insulin has shown that in the B-chain the amino-acid residues B^{20}–B^{23} form a U-turn within the insulin molecule so that the carboxyl terminal segment of the B-chain is folded back and lies in antiparallel fashion against the helical, B^9–B^{19}, segment of that chain (Blundell *et al.* 1971, 1972). This conformation generates interactions which appear to be important to the maintenance of the tertiary structure of the hormone. The B^{27}–B^{30} segment, however, does not appear to be involved in stabilizing interactions which might affect the structure and hence the activity of insulin. The X-ray model also indicates that the amino-terminal segment of the B-chain is on the surface of the monomeric and dimeric forms of insulin and from position B^7 is folded across and forms an extended chain lying antiparallel and against the central portion of the A-chain moiety of the molecule. In this conformation the B^1–B^3 sequence does not appear to be involved in stabilizing interactions that might be of consequence to the tertiary structure of insulin and hence to its biological properties. Hydrogen-bond contacts appear, however, to be established between the B^4-glutamine and the A^{11}-cysteine residues, and between the imidazole nitrogen of the B^5-histidine and the carbonyl oxygen of the A^7-cysteine residues. The latter interaction appears to be of significant structural importance.

These predictions of the X-ray model are generally consistent with the data obtained by the synthesis and biological evaluation of our analogues recorded in Table 37.6. The des[tripeptideB$^{28-30}$] insulin, which differs from the parent molecule in that the tripeptide sequences from the C-terminus of the B-chain has been eliminated and the newly exposed residue, threonine, at position B^{27} has been converted to an amino-alcohol or amino-aldehyde derivative, is fully active. This indicates that the C-terminal tripeptide sequence of the B-chain is not important to the biological activity of insulin. We have found, however, that elimination of the C-terminal tetrapeptide sequence and termination of the B-chain with the B^{26}-tyrosine, bearing a free carboxyl group, results in an analogue (des[tetrapeptide B$^{27-30}$] insulin) possessing approximately 54 per cent of the biological activity of insulin.

TABLE 37.6

Insulin analogues with alterations at the N- and C-terminal regions of the B-chain†

Insulin analogue	(H = Human, B = Bovine)	Potency (i.u./mg) MC	GO	RIA	Receptor binding (% of insulin)	References
Des[tripeptide B$^{28-30}$]	(B)	25	—	—	—	Katsoyannis et al. (1971)
Des[tripeptide B$^{28-30}$]	(H)	17–20	—	—	80 (LM)	Katsoyannis et al. (1971)
Des[tetrapeptide B$^{27-30}$]	(H)	11–15	—	7	—	Katsoyannis, Ginos, Cosmatos, and Schwartz (1973)
Des[pentapeptide B$^{26-30}$]	(H)	8.5–9	—	11	—	Katsoyannis, Ginos, Schwartz, and Cosmatos (1973)
[Des(tetrapeptide B$^{27-30}$), tyrosinamide^{26}-B]	(H)	17	—	10.8	—	Cosmatos, Ferderigos, and Katsoyannis (1979)
[Des(pentapeptide B$^{26-30}$), phenylalaninamide^{25}-B]	(H)	10.5	—	14	—	Cosmatos et al. (1979)
Des(tetrapeptide B$^{1-4}$)	(H)	13	—	7.6	—	Schwartz and Katsoyannis (1978)
Des(pentapeptide B$^{1-5}$)	(H)	—	1.2	3.7	—	Schwartz and Katsoyannis (1978)

† MC, mouse convulsion assay; GO, glucose oxidation in isolated fat cells; RIA, radioimmunoassay; LM, liver membranes. The potency of natural insulin is 24–25 i.u./mg.

Furthermore, elimination of the C-terminal pentapeptide sequence and termination of the B-chain with the B^{25}-phenylalanine, bearing a free carboxyl group, results in an analogue (des[pentapeptide B^{26-30}] insulin) possessing approximately 36 per cent of the biological activity of insulin. These data show that the C-terminal pentapeptide sequence of the B-chain is not functionally involved in the expression of the biological profile of insulin. It is apparent, however, that a gradual decline of the biological activity of insulin occurs as amino-acid residues from the C-terminus of the B-chain are eliminated.

A modest enhancement of the biological activity of these truncated insulins can be effected by amidation of the carboxyl group of the newly exposed C-terminus of the B-chain. Thus [des(tetrapeptide B^{27-30}), tyr$(NH_2)^{26}$-B] insulin and [des(pentapeptide B^{26-30}), phe$(NH_2)^{25}$-B] insulin possess 70 per cent and 44 per cent, respectively, of the biological activity of the natural hormone compared to 54 and 36 per cent, respectively, for the non-amidated parent molecules. An explanation of this observation may be provided from conclusions derived from the X-ray model. Indeed, as was mentioned previously, the folding of the C-terminal segment of the B-chain brings into contact the non-polar B^{24}-phenylalanine, the B^{25}-phenylalanine, and the B^{26}-tyrosine with the hydrophobic residues B^{12}-valine and B^{16}-tyrosine of the B^9–B^{19} helical segment of that chain. Consequently, a largely invariant non-polar region on the surface of the insulin monomer is formed, which contributes to the stability of its tertiary structure and appears to be the contact area between the insulin monomers in the formation of the insulin dimer (Blundell *et al.* 1971, 1972). It is suggested (Insulin Research Group 1974; Pullen *et al.* 1976) that this non-polar region is involved in the binding of the insulin monomer to its receptor in a fashion comparable to its involvement in the binding of the insulin monomers during the formation of the insulin dimer. In view of these considerations, it is tempting to speculate that the gradual decline in the biological activity of insulin observed as amino-acid residues are eliminated from the C-terminus of the B-chain may be due to the proximity of the strongly polar carboxylate ion (of the newly formed C-terminal residue) to the hydrophobic core of the molecule with consequent deleterious conformational changes. It is then reasonable to expect that conversion of the carboxylate ion to the less polar carboxamide group will have a less disruptive influence on the structure of the molecule and hence on its biological activity.

The considerable sequence variation observed in the B^1–B^4 segment of the B-chain of insulin in several species implies that this segment may not be essential to the biological activity of insulin. A similar implication arises also, as was mentioned previously, from the X-ray model of the hormone. The predicted non-involvement of these N-terminal residues of the B-chain in the biological activity of insulin was substantiated by the biological evaluation of analogues prepared in other, as well as in our own, laboratories. Selective Edman degradation of insulin led to the preparation of des(B^1) (Brandenburg 1969), des(dipeptide B$^{1-2}$) (Geiger and Langner 1973), and des(tripeptide B$^{1-3}$) (Geiger and Langner 1975) insulins. The former two analogues were found to be fully active, and the latter analogue was approximately 70 per cent as active as the natural hormone. The synthetic des(tetrapeptide B$^{1-4}$) insulin (Table 37.6) was found to possess approximately 54 per cent of the biological activity of insulin, demonstrating that the B^4-glutamine, as is the case for the B^3-asparagine, B^2-valine, and B^1-phenylalanine, is not involved directly in the expression of the biological activity of insulin. There is a dramatic decrease, however, in the biological activity of the hormone when the B^5-histidine residue is also eliminated. The des(pentapeptide B$^{1-5}$) insulin (Table 37.6) is only approximately 5 per cent as active as the natural hormone. This demonstrates that the B^5-histidine, an invariant amino-acid residue in all insulin species, plays a crucial role in the expression of the biological activity of the hormone and supports the prediction of the X-ray model that this residue is involved in intramolecular interactions of structural importance. We do not exclude the possibility that substitution of the B^5-histidine with other amino-acid residues capable of establishing hydrogen-bond contact with the A^7-carbonyl oxygen might not result in insulin analogues as active as the natural hormone. It is interesting to note that the combination of des(pentapeptide B$^{1-5}$) B-chain with the A-chain to produce the des(pentapeptide B$^{1-5}$) insulin proceeds in a very low yield (Schwartz and Katsoyannis 1978). Similarly, recent work in our laboratory (Schwartz and Katsoyannis, unpublished data) indicates that the [asparagine^5] B-chain, a synthetic chain in which the B^5-histidine is replaced with asparagine, also combines in a very low yield with the A-chain to produce the respective insulin analogue. We are tempted to speculate that the histidine residue at position B^5 is uniquely involved in interchain interactions so as to dispose the A- and B-chains for a more efficient combination.

Analogues with modifications in the interior of the B-chain

Analogues with modifications in the interior of the B-chain are shown in Table 37.7. Substitution of the B^{22}-arginine (pK of the guanidinium group, 12.48) with lysine (pK of the ε-amino group, 10.53) results in the [lysine^{22}-B] insulin which possesses approximately 56 per cent of the biological activity of the natural human hormone. The natural guinea-pig insulin, which has an aspartic acid residue (pK of the β-carboxyl group, 3.65) at B^{22}, possesses only aproximately 9 per cent of the activity of human insulin (Zimmerman, Kells, and Yip 1972). It is interesting to note that the plot of log (biological activity) v. pK of the functional group of the B^{22}-amino-acid residue of these insulins is a straight line (Fig. 37.12) (Katsoyannis, Ginos, Cosmatos, and Schwartz 1975). The apparent dependence of the magnitude of the biological

TABLE 37.7

Insulin analogues with alterations in the interior of the B chain†

Insulin analogue (human)	Potency (i.u./mg)				Receptor binding (% of insulin)	References
	MC	GO	LIP	RIA		
[Lysine[22]-B]	13–14	—	—	8.1	—	Katsoyannis et al. (1975)
[Leucine[9]-B]	13–14	—	—	11–12	30 (LM)	Schwartz and Katsoyannis (1976)
[Leucine[10]-B]	10–11	—	—	9	—	Schwartz and Katsoyannis (1977)
[Lysine[10]-B]	—	—	(5)	—	(19 (LM)	Unpublished data
[Asparagine[10]-B]	—	—	(9.5)	—	(60) (LM),(FC)	Unpublished data
[Asparagine[12]-B]	—	0.01–0.04	—	—	0.03–0.07 (LM)	Unpublished data

† MC, mouse convulsion assay; GO, glucose oxidation in isolated fat cells; LIP, lipogenesis in isolated fat cells; RIA, radioimmunoassay; LM, liver membranes; FC, fat cells; values in parentheses indicate preliminary data. The potency of natural insulin is 24–25 i.u./mg.

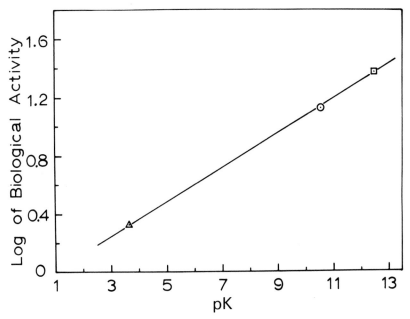

FIG. 37.12. Plot of the log of biological activity v. the pK of the functional group of the amino acid residue at B^{22} of insulin: □ bovine, human (B^{22}Arg); ○ synthetic human [lysine^{22}-B] (B^{22}Lys); △ guinea pig (B^{22}Asp).

activity on the pK value of the functional group of the B^{22}-residue of insulin is intriguing. Whether this dependence is related to the involvement of the B^{22}-residue and the A^{21}-carboxylate ion in the establishment of a salt bridge (B^{22}: Arg or Lys), as the X-ray model predicts, or of a hydrogen bond (B^{22}: Asp), as might be the case with guinea-pig insulin, or to their participation in some other as yet unknown interactions, remains to be proven. Since, however, [asparaginamide^{21}-A] insulin, which cannot form an A^{21}–B^{22} salt bridge, is still biologically active, the importance of this particular putative interaction is at present in doubt. (See also pp. 495–498.)

Replacement of the hydrophilic residue serine at position B^9, which is on the surface of the insulin molecule, with the hydrophobic leucine results in [leucine^9-B] insulin which possesses approximately 55 per cent of the biological activity of the natural hormone. The relative binding affinity in liver membranes of this analogue is approximately 30 per cent of that of insulin. Modification of the B^9-amino-acid residue, as the first member of the B^9–B^{19} helical segment of insulin, is not expected to distort the helix and hence alter the biological activity of the hormone. We are therefore tempted to attribute the lower biological activity of [leucine^9-B] insulin, as compared to the natural hormone, to the perturbing influence that the bulky hydrophobic side-chain of B^9-leucine may have on the interactions of that region with the aqueous environment or to steric effects that come into play during hormone–receptor interactions.

The X-ray model suggests that the histidine at position B^{10} is involved in the formation of zinc insulin hexamers. Guinea-pig (Smith 1966), coypu (Smith 1972), and hagfish (Peterson, Steiner, Emdin, and Falkmer 1975) insulins, which do not contain a histidine at B^{10}, do not form stable zinc hexamers. In addition, these insulins have been found to be considerably less active (< 9 per cent) than mammalian insulins (Smith 1966, 1972; Zimmerman *et al.* 1972). The decreased biological activity, however, cannot be attributed solely to the lack of B^{10}-histidine since these insulins differ substantially in their primary structure from the more active mammalian insulins. Indeed, the fact that the human [leucine^{10}-B] insulin (Table 37.7) was found to possess 45 per cent of the potency of the natural human insulin demonstrates that the B^{10}-histidine, in contrast to the B^5-histidine, is not involved directly in the biological activity of the hormone. The lower potency of the [leucine^{10}-B] insulin, as compared to the natural hormone, however, cannot be attributed to the replacement of the polar B^{10}-histidine with the non-polar leucine, since the [lysine^{10}-B] insulin (Table 37.7) was also found to possess approximately 20 per cent of the biological activity of the natural hormone and approximately 18 per cent of its binding ability in liver membranes. It appears now likely that the magnitude of the biological activity of insulin analogues modified at the B^{10}-position depends on steric effects that the substitute amino-acid residue at position B^{10} exerts on the conformation of the protein molecule. This speculation is based on data obtained recently in our laboratory from the synthesis of human [asparagine^{10}-B] insulin (Schwartz and Katsoyannis, unpublished data). Preliminary assays indicate that this analogue, which contains asparagine instead of histidine at B^{10}, possesses at least 38 per cent of the biological activity of insulin and approximately 60 per cent of its binding ability in liver membranes.

The amino-acid residue valine at position B^{12} is a member of the hydrophobic core of the insulin molecule which is the contact region in the formation of the insulin dimer from monomers, and presumably is part of the contact area of the insulin monomer to its receptor in the target cell. One would then anticipate that alterations in this contact region would affect the binding ability of insulin to its receptor and consequently the biological activity of the hormone. Our data are in agreement with these postulates. The [asparagine^{12}-B] insulin, an analogue in which the hydrophobic valine at B^{12} is substituted with the hydrophilic asparagine, exhibits only traces of *in vitro* biological activity and receptor-binding affinity in liver membranes. We believe this constitutes the first evidence in support of the involvement of the hydrophobic core of insulin in receptor binding and hence in the expression of the biological activity of this hormone.

Conclusions

Advances in synthetic methodology and in the refinement of purification and analytical techniques made possible the chemical synthesis of a complicated protein, insulin. This accomplishment opened the road to the synthesis of insulin analogues and the pursuit of structure–activity relationships in an unlimited way. Already considerable information has been accumulated regarding the contributions of various structural features of insulin to the expression of the biological profile of this hormone and to its receptor-binding affinity. The majority of the available data support the idea that the binding site and the site responsible for activating cellular processes may reside in the same region of the insulin molecule. On the other hand, there is some evidence indicating that the structural features involved in the expression of the biological activity of insulin and in the binding to its receptor, apparently overlapping, may be dissociated. The synthesis of a wide spectrum of analogues might solve this problem. The synthesis of tailor-made analogues that possess more desirable properties than the natural hormone might then become a reality.

Acknowledgements

I am indebted to my co-workers, Drs G. Schwartz, A. Cosmatos, N. Ferderigos, Y. Okada, C. Zalut, J. Ginos, A. C. Trakatellis, A. Tometsko, K. Suzuki, M. Tilak, K. Cheng, and G. T. Burke, who have contributed at various time intervals to this work. This research was supported by the National Institute for Arthritis, Metabolism and Digestive Diseases, US Public Health Service (AM 12925).

References

BAILEY, J. L. and COLE, R. D. (1959). *J. biol. Chem.* **234**, 1733.

BLUNDELL, T., DODSON, G., HODGKIN, D., and MERCOLA, D. (1972). *Adv. Protein Chem.* **26**, 279.

—— DODSON, E. J., HODGKIN, D., and VIJAYAN, M. (1971). *Recent Prog. Horm. Res.* **27**, 1.

BRANDENBURG, D. (1969). *Hoppe-Seyler's Z. physiol. Chem.* **350**, 741.

COSMATOS, A. and KATSOYANNIS, P. G. (1973). *J. biol. Chem.* **248**, 7304.

—— (1975). *J. biol. Chem.* **250**, 5315.

—— FERDERIGOS, N., and KATSOYANNIS, P. G. (1979). *Int. J. Peptide Protein Res.* **14**, 457.

—— OKADA, Y., and KATSOYANNIS, P. G. (1976). *Biochemistry* **15**, 4076.

—— CHENG, K., OKADA, Y., and KATSOYANNIS, P. G. (1978). *J. biol. Chem.* **253**, 6586.

—— JOHNSON, S., BREIER, B., and KATSOYANNIS, P. G. (1975). *J. chem. Soc. Perkin I* 2157.

DIXON, G. H. and WARDLAW, A. C. (1960). *Nature, Lond.* **188**, 721.

DU, Y.-C., JIANG, R.-Q., and TSOU, C.-L. (1965). *Scientia sin.* **14**, 229.

—— ZHANG, Y.-S., LU, Z.-X., and TSOU, C.-L. (1961). *Scientia sin.* **10**, 84.

EMIDIN, S. O., GAMMELTOFT, S., and GLIEMANN, J. (1977). *J. biol. Chem.* **252**, 602.

FERDERIGOS, N. and KATSOYANNIS, P. G. (1977). *J. chem. Soc. Perkin I* 1299.

—— COSMATOS, A., FERDERIGOS, A., and KATSOYANNIS, P. G. (1979). *Int. J. Peptide Protein Res.* **13**, 43.

FREYCHET, P., BRANDENBURG, D., and WOLLMER, A. (1974). *Diabetologia* **10**, 1.
GEIGER, R. and LANGNER, D. (1973). *Hoppe-Seyler's Z. physiol. Chem.* **354**, 1285.
——— (1975). In *Peptides 1974* (ed. Y. Wolman) p. 159. J. Wiley, New York.
GLIEMANN, J. and GAMMELTOFT, S. (1974). *Diabetologia* **10**, 105.
Insulin Research Group, Academia Sinica (1974). *Scientia sin.* **17**, 779.
KATSOYANNIS, P. G. (1966). *Science, N.Y.* **154**, 1509.
——and SCHWARTZ, G. (1977). *Meth. Enzym.* **47**, 501.
——and TOMETSKO, A. (1966). *Proc. natn. Acad. Sci. U.S.A.* **55**, 1554.
——and ZALUT, C. (1972a). *Biochemistry* **11**, 1128.
———(1972b). *Biochemistry* **11**, 3065.
——OKADA, Y., and ZALUT, C. (1973). *Biochemistry* **12**, 2516.
——TOMETSKO, A., and FUKUDA, K. (1963). *J. Am. chem. Soc.* **85**, 2863.
———and ZALUT, C. (1966). *J. Am. chem. Soc.* **88**, 166.
———(1967). *J. Am. chem. Soc.* **89**, 4505.
——GINOS, J., COSMATOS, A., and SCHWARTZ, G. (1973). *J. Am. chem. Soc.* **95**, 6427.
—————(1975). *J. chem. Soc. Perkin I* 464.
———SCHWARTZ, G., and COSMATOS, A. (1973). *J. chem. Soc. Perkin I*, 1311.
——TOMETSKO, A., GINOS, J., and TILAK, M. (1966). *J. Am. chem. Soc.* **88**, 164.
———ZALUT, C., and FUKUDA, K. (1966). *J. Am. chem. Soc.* **88**, 5625.
——ZALUT, C., HARRIS, A., and MEYER, R. J. (1971) *Biochemistry* **10**, 3884.
——FUKUDA, K., TOMETSKO, A., SUZUKI, K., and TILAK, M. (1964). *J. Am. chem. Soc.* **86**, 930.
——TOMETSKO, A., ZALUT, C., JOHNSON, S., and TRAKATELLIS, A. C. (1967). *Biochemistry* **6**, 2635.
——TRAKATELLIS, A. C., JOHNSON, S., ZALUT, C., and SCHWARTZ, G. (1967). *Biochemistry* **6**, 2642.
——GINOS, J., ZALUT, C., TILAK, M., JOHNSON, S., and TRAKATELLIS, A. C. (1971). *J. Am. chem. Soc.* **93**, 5877.
——ZALUT, C., TOMETSKO, A., TILAK, M., JOHNSON, S., and TRAKATELLIS, A. C. (1971). *J. Am. chem. Soc.* **93**, 5871.
——TRAKATELLIS, A. C., ZALUT, C., JOHNSON, S., TOMETSKO, A., SCHWARTZ, G., and GINOS, J. (1967). *Biochemistry* **6**, 2656.
OKADA, Y. and KATSOYANNIS, P. G. (1975). *J. Am. chem. Soc.* **97**, 4366.
PETERSON, J. D., STEINER, D. F., EMDIN, S. D., and FALKMER, S. (1975). *J. biol. Chem.* **250**, 5183.
PULLEN, R. A., LINDSAY, D. G., WOOD, S. P., TICKLE, I. J., BLUNDELL, T., WOLLMER, A., KRAIL, G., BRANDENBURG, D., ZAHN, H., GLIEMANN, J., and GAMMELTOFT, S. (1976). *Nature, Lond.* **259**, 369.
SAKAKIBARA, S. (1971). In *Chemistry and biochemistry of amino acids, peptides and proteins* (ed. B. Weinstein) p. 51. Marcel Dekker, New York.
SANGER, F. (1959). *Science, N.Y.* **129**, 1340.
SCHWARTZ, G. and KATSOYANNIS, P. G. (1973). *J. chem. Soc. Perkin I* 2894.
———(1976). *Biochemistry* **15**, 4071.
———(1977). *J. chem. Res.* (S) 220; *J. chem. Res.* (M) 2453.
———(1978). *Biochemistry* **17**, 4550.
SIFFERD, R. H. and DU VIGNEAUD, V. (1935). *J. biol. Chem.* **108**, 753.
SMITH, L. F. (1966). *Am. J. Med.* **40**, 662.
——(1972). *Diabetes* **21**, Suppl. 2, 457.
ZIMMERMAN, A. E., KELLS, D. I. C., and YIP, C. C. (1972). *Biochem. biophys. Res. Commun.* **46**, 2127.

38. Pharmacokinetic studies with semisynthetic tritiated insulin

M. BERGER, H. J. CÜPPERS, J. G. DAVIES, P. A.
HALBAN, S. M. HOARE, AND R. E. OFFORD

THE preparation of semisynthetic [³H]-insulin, which has been specifically labelled at the N-terminal phenylalanine of the B chain, has been described previously (Halban and Offord 1975). The route is shown in Fig. 38.1. This labelling represents a minimal alteration to the protein and its authenticity as a tracer for insulin has been established (Halban, Karakash, Davies, and Offord 1976; Halban et al. 1978a, b; Davies 1978; Berger et al. 1978; Berger et al. 1979; Halban, Berger, and Offord 1979). Therefore this material should be useful in the study of the pharmacokinetics of subcutaneously injected insulin where the presence of circulating antibodies and naturally occurring insulin makes radioimmunoassay difficult to interpret. We believe that tritiation is preferable to iodination of insulin because, however carefully the iodo-insulin is prepared, it still represents a chemically altered form of the hormone. The size and apolarity of the iodine atom are great enough for it to be quite possible for the substitution to make a significant difference to the properties of the iodinated insulin derivative.

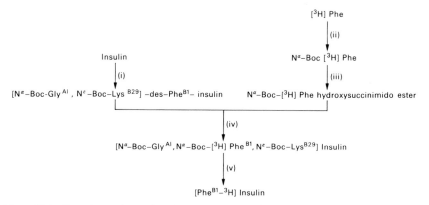

FIG. 38.1. Reaction scheme for the preparation of semisynthetic [³H]-insulin from native insulin and [³H]-phenylalanine.

When tritiated insulin is injected subcutaneously into rats it might seem that the appearance of insulin in the plasma and its distribution to target organs can be followed simply by measurement of the radioactivity. However, once introduced into the animals, the tritiated insulin will undergo degradation and metabolic processing, which will give rise to new tritiated species. A simple count of total radioactivity, therefore, will not be sufficient to determine the levels of labelled insulin the samples studied.

Insulin can be separated from tritiated products of different molecular weight by gel filtration on Sephadex G-50 (Halban *et al.* 1978*b*). For example, the plasma from a rat that had been subcutaneously injected with tritiated insulin gives, when applied to the column, three peaks of radioactivity. The first has a molecular weight substantially higher than that of insulin, the second has a molecular weight indistinguishable from that of insulin, and the third has a molecular weight substantially lower than that of insulin. Similar profiles are obtained when organ extracts are solubilized in sodium dodecyl sulphate and gel-filtered on Sephadex G-75 in detergent buffers. Complete sequence analysis could establish the nature of the material in the three peaks, but when the [^{3}H]-insulin is administered in physiological amounts, there is far too little material for this to be done. However, it is possible to learn a great deal about the nature of the peaks by investigating their behaviour in chromatographic and electrophoretic separating systems. Also, since the semisynthesis has inserted the label in a defined position within the covalent structure of the molecule, sequence analysis by radiochemical techniques can be quite informative.

The high-molecular-weight material is heterogeneous on gel-filtration in Sephadex G-100, G-75, or Sephacryl S-200, and produces no lower molecular-weight radioactivity under dissociating conditions (e.g. in detergent buffers). Performic acid oxidation followed by gel-filtration released no material of lower molecular weight. This observation indicates that the material is not, as had been suggested by other workers (Antoniades, Stathakos, and Simon 1974), a polymeric form of insulin linked by disulphide bonds. Immunoprecipitation failed to reveal any insulin immunoreactivity. Isoelectric focusing showed that the radioactivity is distributed throughout the serum protein bands rather than forming one or a few discrete zones.

If any of the insulin sequence remains in this high-molecular-weight material, the fact should be detectable by comparing the labelled peptides produced from it by enzymic digestion with those produced from a similar digest of the [^{3}H]-insulin. Paper electrophoresis at pH 1.9 and paper chromatography were used to separate the peptides resulting from subtilisin digestion of (a) [^{3}H]-insulin; (b) the high-molecular-weight material from plasma; and, as a further control, (c) high-molecular-weight material obtained after subcutaneous injection not of [^{3}H]-insulin but of [^{3}H]-phenylalanine. Only two tritiated peptides can be isolated from the digest of [^{3}H]-insulin, while the digests of both the other types of sample show a distribution of label over very many peptides. Quantitative comparison of the distribution of label in the three cases (to be reported elsewhere) led us to conclude that the high-molecular-weight fraction obtained after tritiated insulin injection arises by random re-incorporation into plasma proteins of label liberated on the break-down of [^{3}H]-insulin. We find that less than 1 per cent of this fraction can still be present in the original insulin sequence.

The intermediate-molecular-weight peak is fully precipitable by anti-in-sulin. Cellulose acetate electrophoresis of this material at pH 6.5 shows that the tritium runs principally in the position of insulin. The loss of more than a very few amino acids from the ends of either insulin chain would result in the loss of a charged side-chain and thus a change in mobility: the material under investigation therefore appears to contain substantially intact insulin molecules.

When the low-molecular-weight fraction is gel-filtered on Sephadex G-10, 80 per cent of the radioactivity co-elutes with added unlabelled phenylalanine in a characteristic position just after the salt. This fraction subsequently behaves like phenylalanine on electrophoresis at pH 3.5 and pH 1.9 and on paper chromatography. The remaining 20 per cent appears to represent di- and tri-peptides resulting from partial cleavage of the insulin B-chain.

Previous work (Halban *et al.* 1978*a, b*; Berger *et al.* 1979) has suggested that in addition to degradation at the site of action and in such organs as the liver, insulin is also degraded at the site of subcutaneous injection; this observation is contrary to the results of other workers (Binder 1969).

Although the existence of the degradation could be observed, quantitative estimation of the extent to which it occurs is difficult. Insulin and the products of its degradation will continuously diffuse from the injection site into the rest of the body. Furthermore, the different materials will diffuse at different rates. Therefore an indirect approach had to be adopted to see if the pheno-menon took place to an extent that would constitute a significant factor in determining the patient's insulin requirement. The approach, which exploited the discovery of an inhibitor of the degradation, will now be described.

Tritiated insulin was incubated *in vitro* with pieces of rat perirenal fat and the extent of degradation analysed by the Sephadex G-50 system (Offord, Philippe, Davies, Halban, and Berger 1979). This *in vitro* system was in-tended only for screening possible inhibitors of insulin degradation by damaged fatty tissue. Two potential inhibitors were selected. The first was bovine pancreatic trypsin inhibitor (Trasylol, Bayer AG), which was thought likely to inhibit proteolytic degradation of insulin. The second was ophthal-mic acid (γ-glutamyl-2-amino-*n*-butanoylglycine), which, since it is an almost isosteric analogue of glutathione, was thought likely to inhibit any reduction of disulphide bridges that might be catalysed by glutathione-requiring enzymes. Both were found to act as inhibitors in the *in vitro* system, as did performic acid-oxidized glutathione (γ-glutamyl-cysteinesulphonic acid glycine).

Trasylol alone (1100 kallikrein units/ml) reduced degradation to 60 per cent of control values, while ophthalmic acid (30 mM) reduced it to 42 per cent. If both substances were used together degradation was only 33 per cent: the effect of these may therefore be multiplicative.

Trasylol (10 000 kallikrein units) was mixed with a pharmacologically active dose of insulin (Actrapid, 10 international units) and administered to normal fasting human volunteers who had not received insulin or Trasylol before. The experiments (full details of which are to be published elsewhere (Berger, Cüppers, Halban, and Offord 1980)) were carried out in conformity with the Helsinki Declaration of 1975. The addition of Trasylol caused the levels of circulating insulin to rise at a significantly faster rate. Also, the average value for the peak insulin concentration when Trasylol was present was 127 per cent of the average of the values obtained in the same individuals after injection of insulin alone. This was shown to be a statistically significant difference ($p < 0.001$, $n = 4$). Similarly, blood-sugar levels were found to be significantly lower when Trasylol was present; the difference was most marked at 45 minutes after injection (76 per cent of insulin control, $p < 0.001$, $n = 4$). Trasylol alone did not affect blood levels of endogenous insulin or of sugar.

The effect of the inhibitor leads us to conclude that degradation at site may take place to a significant extent. The possibility therefore exists that a noticeable proportion of the insulin used in the treatment of diabetes may be degraded before it can have any effect.

Acknowledgements

We thank the German Research Association for contributions towards the cost of this study and for grants to Professor Berger. J. G. Davies and S. M. Hoare thank the Medical Research Council of the United Kingdom for grants. We thank Ms M. Schütte for skilled technical assistance.

References

ANTONIADES, H. N., STATHAKOS, D., and SIMON, J. D. (1974). *Endocrinology* **95**, 1543–53.

BINDER, C. (1969). *Acta pharmacol. toxicol.* (Suppl. 2), **27**, 1–87.

BERGER, M., CÜPPERS, H. J., HALBAN, P. A., and OFFORD, R. E. (1980). *Diabetes* **29**, 81–3.

——— HALBAN, P. A., GIRARDIER, L., SEYDOUX, J., OFFORD, R. E., and RENOLD, A. E. (1979). *Diabetologia* **17**, 97–9.

——————— MÜLLER, W. A., OFFORD, R. E., RENOLD, A. E., and VRANIC, M. (1978). *Diabetologia* **15**, 133–40.

DAVIES, J. G. (1978). In *Semisynthetic peptides and proteins* (ed. R. E. Offord and C. Di Bello) pp. 249–52. Academic Press, London.

HALBAN, P. A. and OFFORD, R. E. (1975). *Biochem. J.* **151,** 219–25.

——BERGER, M., and OFFORD, R. E. (1979). *Metabolism* **28,** 1097–104.

——KARAKASH, C., DAVIES, J. G., and OFFORD, R. E. (1976). *Biochem. J.* **160,** 409–12.

——BERGER, M., GIRARDIER, L., SEYDOUX, J., OFFORD, R. E., and RENOLD, A. E. (1978a). *Diabetes* **27,** Suppl. 2, 439.

——BERGER, M., GJINOVCI, A., RENOLD, A., VRANIC, M., and OFFORD, R. E. (1978b). In *Semisynthetic peptides and proteins* (ed. R. E. Offord and C. Di Bello) pp. 237–47. Academic Press, London.

OFFORD, R. E., PHILIPPE, J., DAVIES, J. G., HALBAN, P. A., and BERGER, M. (1979). *Biochem. J.* **182,** 249–51.

39. The study of the insulin molecule

ZHANG YOU-SHANG†

An old Chinese proverb says: 'A sparrow has a whole set of viscera, tiny as it is.' In modern Chinese, 'to dissect a sparrow' means to study a typical case with the aim of gaining some general knowledge from it. Insulin is such a sparrow in protein research. Although a small protein with a molecular weight no more than 6000, its structure has all the characteristics typical of a globular protein. Owing to this structural complexity and its important physiological function, the study of insulin since its discovery in 1921 (Banting and Best 1921) has played a leading role in the development of protein chemistry and protein hormone research. The achievements of insulin research represent important milestones in many scientific fields. Insulin is one of the earliest crystallized proteins with important biological functions (Abel 1926), X-ray-diffraction photographs of it were obtained by Dorothy Hodgkin as early as 1935 (Crowfoot 1935). Though the X-ray analysis of insulin crystals was not accomplished until 1969, it is still the only protein hormone to have had its three-dimensional structure completely determined (Adams *et al.* 1969; Peking Insulin Structure Research Group 1973). In 1955, its primary structure was elucidated for the first time through the pioneer work of Sanger and his co-workers (Ryle, Sanger, Smith, and Kitai 1955). Ten years later, insulin became the first protein to be chemically synthesized and crystallized in the laboratory (Kung *et al.* 1965; Katsoyannis *et al.* 1967; Zahn, Danho, and Gutte 1966). In the hospital, radioimmunoassay—with far-reaching influence on hormone research—was first worked out by Berson and Yalow (Berson, Yalow, Baumann, Rothschild, and Newerly 1956; Yalow and Berson 1960) through their investigation of insulin-treated patients. Although insulin has been so extensively studied, the mechanism of insulin action is still not clear and its elucidation remains a challenge to us. In other words, the rich store of information to be gained from insulin

† The work on insulin structure and function in the Shanghai Institure of Biochemistry cited in this article has been done in collaboration with Chu Shang-chuan, Lu Zi-xian, Tsao Chiu-ping, and other members of the Insulin Research Group, and has been constantly encouraged by Professors Wang Ying-lai, Tsao Tien-chin, and Niu Ching-i.

has not yet been exhausted and more rewarding achievements can be expected in the near future.

The total synthesis of insulin

The project of insulin synthesis was initiated two decades ago—at a time when conditions seemed favourable. However, in two important respects there were uncertainties which had to be fully considered in the strategy of this project. The largest peptide which had been synthesized at the time was alpha-melanostimulating hormone with only 13 amino-acid residues. Would the methods of peptide synthesis be powerful enough to deal with a protein molecule with 51 amino-acid residues? In addition, the three-dimensional structure of a native protein molecule was known to be essential for its biological activity. Would the synthetic insulin finally obtained turn out to be a denatured and biologically inactive material? In 1959, the A- and B-chains of native insulin were successfully recombined and the resynthesized insulin was obtained in crystalline form, and so both these problems were solved at one blow (Du, Zhang, Lu, and Tsou 1961). We used to describe the resynthesis of native insulin as the green light, which showed that the total synthesis of insulin would be done by making two separate chains with 21 and 30 amino-acid residues. In Canada, the regeneration of insulin activity from the A- and B-chains was done independently by Dixon and Wardlaw (1960). Later, semisynthetic insulins from synthetic and natural chains were made for the first time, and then the synthetic chains which proved satisfactory in the semisynthesis were used to achieve the total synthesis of insulin. Insulin synthesis as a whole molecule instead of via its two chains was achieved much later, by the work of Sieber *et al.* (1974). It is also interesting to note that the high yield of the A- and B-chains in the DNA recombinant experiment apparently favoured the production of insulin from these two chains rather than from its natural precursor—proinsulin.

In the preparation of the two chains from native insulin, the reaction was carried out in 8 M urea and all three disulphide bridges were split, so the crystallization of recombined insulin demonstrated the basic principle that all the information of higher-ordered structure of a protein was contained in its primary structure. Therefore we did not have to worry about the three-dimensional structure of our synthetic insulin. Finally, the crystallization of synthetic insulin as typical 2Zn insulin rhombohedra proved this principle unequivocally.

Structure and function relationships of insulin

The semisynthesis of insulin analogues

Our work on the structure and function of insulin is a natural development of the chemical synthesis and structural analysis of crystalline insulin. From the strategic point of view, there are also two major problems. First, for a protein molecule, even one as small as insulin, the number of random substitutions of amino-acid residues, carried out simply by a process of trial and error, would be far too great to cope with. Therefore we need some indication of which part of the molecule is likely to be interesting. In this respect, the information provided by the three-dimensional structure of the insulin molecule proves to be very helpful. Secondly, although insulin and its analogues could be made entirely by chemical synthesis, yet the time and labour needed are still too much in spite of recent improvements in the synthetic methods. So a semisynthetic approach would be preferable if such a method could indeed be worked out. The concept of semisynthesis involving peptide-bond formation had already been in our minds in the early days of the chemical synthesis of insulin, when we had set up a peptide bank to collect the natural peptides of insulin and tried to join them up through papain catalysed transpeptidation. However, the experimental problems were not fully solved and the semisynthesis involving peptide-bond formation was not actually used in the intermediate stages of insulin synthesis. In recent years, a lot of work has been done by Offord on the preparation of protected natural peptides and their potential use in the semisynthesis (Offord 1969, 1972). In our laboratory, we have elaborated a semisynthetic method and have employed it for studying the structure and function of insulin (Shanghai Insulin Research Group 1973†).

In the three-dimensional structure of the insulin dimer, the peptide chains B23–B27 of the two molecules run closely antiparallel to form a beta-pleated sheet and the non-polar side-chains of B24 Phe and B26 Tyr are involved in the formation of a hydrophobic region. In addition, the insulin molecule with the last eight amino-acid residues of the B-chain removed by tryptic hydrolysis is no longer active—as shown by previous studies (Bromer and Chance 1967). Therefore, our attention has been focused on this C-terminal part of the B-chain where modification of the amino-acid sequence could be done by our semisynthetic method, as shown in the scheme on p. 489.

In this scheme, there are certain points worth mentioning. The tryptic hydrolysis of insulin hexamethyl ester is carried out in 50 per cent DMF at pH 8.0 where the substrate has good solubility and the enzyme has high activity. As the reaction proceeds, the insoluble desoctapeptide insulin (DOI) pentamethyl ester comes out of solution and can be readily separated from the reaction mixture. This solubility difference has also been utilized to purify the hexamethyl esters of insulin and its analogues, because they can be extracted with the same solvent, leaving behind the unreacted DOI pentamethyl ester. Before DOI is coupled with synthetic peptides, its alpha-amino groups and carboxyl groups in the side-chain and A21 must be protected.

† In developing this method, Xü Geng-jun of the Shanghai Institute of Biochemistry has played a major role.

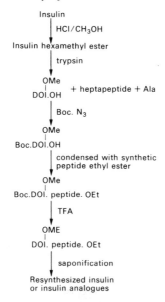

The protection of amino groups with Boc presents no problem. Insulin reversibly treated with Boc is fully active and can be crystallized. However, esterification followed by saponification gives rise to side-reaction which had already been recognized in the chemical synthesis of the A-chain. This difficulty was avoided afterwards by leaving the C-terminal carboxyl group free. We have found that insulin after esterification and saponification loses an appreciable amount of activity and cannot be crystallized. Recently, the nature of this side-reaction was identified by Gattner and Schmitt, who showed that A21 Asn was transformed to a cyclic imide (Gattner and Schmitt 1977). It has long been known that A21 desamidoinsulin with full activity can be obtained by acid treatment, so we suggest that A21 desamidoinsulin might be used in place of insulin in order to avoid the above complication. In our opinion, the methyl esterification of natural peptides should have a wide application. The side-reaction which occurred in insulin owing to the presence of the C-terminal Asn is exceptional and can be dealt with specially.

Recently, we have further extended the range of our semisynthetic approach by removing B22 Arg with carboxypeptidase B from the DOI pentamethyl ester so that we are able to prepare insulin analogues with B22 Arg replaced (Chu, Li and Tsao 1978). When B22 Arg is removed, B20 and B21 residues could be removed subsequently by carboxypeptidase A, thus they should also become accessible to substitutions.

Using our semisynthetic method, we have prepared a series of insulin analogues with modified sequence in the C-terminal part of the B-chain. Before doing this, we have prepared resynthesized insulin, the activity of which after partial purification is 50 per cent *in vivo* as compared to native insulin. This result shows that our semisynthetic approach is feasible. When the three aromatic amino acids B24, B25, and B26 are replaced by alanines, the product is no longer active, showing that at least one of these hydrophobic amino-acid residues is indispensable for insulin activity. In later experiments we have found that B24 Phe is the most interesting because deshexapeptide insulin (DHI) obtained from DOI and glycylphenylalanine has an activity *in vivo* after partial purification of 40 per cent. Furthermore, an active molecule can only be obtained when this phenylalanine is replaced by aromatic amino acids, e.g. Tyr and Tryp, though there is with these residues some reduction

TABLE 39.1

Activities of insulin and its analogues

Crude product	Relative activity (%)† (mouse-convulsion test)
DOI	< 0.15
DOI–Gly.Phe.Phe.Tyr.Thr.Pro.Lys.Ala (resynthesized insulin)	10 (50)‡
DOI–Gly.Ala.Ala.Ala.Thr.Pro.Lys.Ala	< 0.4
DOI–Gly.Phe (DHI)	10 (40)‡
DOI–D-Ala.Phe	10 (40)‡
DOI–D-Val.Phe	10 (40)‡
DHI analogues with B22 and B23 replaced	
Desnonapeptide insulin (DNI)	< 0.6
DNI–Arg.D-Ala.Phe	20
DNI–Lys.D-Ala.Phe	20
DNI–Asp.D-Ala.Phe	20

† Compared to crystalline porcine insulin with full activity (100 per cent).

‡ After partial purification.

in activity. The B23 Gly is structurally important because it is involved in the B20–B23 reverse turn structure. We have found that it can only be replaced by certain D-amino acids, e.g. D-Ala and D-Val, but not by any L-amino acids (Chu, Li, Tsao, Chang, and Lu 1973). This interesting result can be nicely explained by the Ramachandran plot because the dihedral angles of B23 Gly lie in the allowed region for D-configuration (Peking Insulin Structure Research Group 1974). This implies that replacement of B23 Gly by L-amino acids will result in severe steric hindrance which destroys the native conformation of the polypeptide chain. In the case of D-amino acids, the orientation of the side-chain would not be incompatible with the proper conformation. We have also noticed that the dihedral angles of B8 and B20 glycines are also in the allowed region for D-configuration. Based on this information from the conformational map, we are going to see whether these two glycine residues can be substituted in a similar way. The replacement of B20 seems more practical if our extended semisynthetic method could be used.

All insulins, with the exception of guinea-pig insulin and porcupine insulin (Horuk, private communication), have an Arg at B22. In the three-dimensional structure of insulin, between B22 Arg and A21 Asn, there is a salt linkage which might be important in stabilizing the native conformation of the insulin molecule. Previous reports emphasized that B22 Arg could only be replaced by basic amino-acid residues (Weitzel, Oertel, Rager, and Kemmler 1971). However, we have found that B22 Arg of DHI analogue can well be replaced by Asp, an amino acid with opposite charge. This unexpected finding shows that this salt linkage is not so important as formerly suggested. In addition, the replacement of B22 Arg by Asp might be advantageous in resisting enzymatic degradation.

The preparation of active fragments by limited enzymatic hydrolysis

This is another approach to the study of the structure and function of insulin. For instance, despentapeptide insulin (DPI) is readily obtained from insulin by limited peptic digestion as reported by Shvachkin (Shvachkin, Shmeleva, Krivtsov, Fedotov, and Ivanova 1972) and Gattner (Gattner 1975). We have obtained crystalline DPI suitable for X-ray analysis (Peking Insulin Structure Research Group 1976). The activity of DPI is 20 i.u./mg in the mouse-convulsion test. Compared to insulin, its *in vitro* activity in the rat diaphragm assay is 80 per cent and its receptor-binding activity is 88 per cent (Insulin Research Group *et al.* 1976). The K_D of DPI determined by the gel-filtration method is 0.58, corresponding to a molecular weight of 4900 (Insulin Research Group 1976). In contrast to insulin, the K_D of DPI is concentration-independent and the elution peak is symmetrical. Therefore, in solution DPI does not associate and exists as monomers. Considering the essentially full biological activity of DPI, it is reasonable to say that the functional unit of insulin is a monomer which binds with insulin receptor.

The comparative study of insulins from different species

This approach has also proved very useful in structure and function studies. In this respect, both unity and diversity in structure among different insulins may give us valuable information. The invariant 21 of the 51 amino-acid residues are considered to be essential for insulin activity. Thus the invariance of B23 Gly and B24 Phe is consistent with our finding that the addition of Gly and Phe to the inactive DOI has the remarkable effect of restoring activity.

Insulins with unusual properties, such as the hagfish, guinea-pig, chicken, and snake insulins, are especially interesting. Hagfish insulin (Peterson, Coulter, Steiner, Emdin, and Falkmer 1974) is an ancient insulin the study of which is significant from the evolutionary point of view. Guinea-pig insulin is unique among mammalian species. For instance, its B22 residue is Asp instead of Arg. Its exceptionally low activity was considered to be attributable to this difference. However, our results have shown that B22 Arg of DHI analogue can be replaced by Asp without any influence on its activity. Therefore, this difference is not important and substitutions in other positions must be responsible for the low activity of guinea-pig insulin.

We have worked out a new adsorption method for the large-scale production of porcine insulin, without the need for vacuum distillation (Chang, Y. S. *et al.*, unpublished work). In the laboratory, this method can be conveniently used for the preparation of insulin from different species (Tsao, Li, Peng, and Zhang 1979). We have prepared chicken insulin in typical rhombohedral crystalline form (Fig. 39.1). In all insulins so far studied, only

FIG. 39.1. Crystalline chicken insulin (× 96).

turkey insulin (later found to be identical to chicken insulin) was reported to have much higher activity (Weitzel *et al.* 1969). However, results in different laboratories and from different methods in the same laboratory are in conflict. From our results, at least, the *in vivo* activity of chicken insulin (mouse-convulsion test) was not appreciably different from that of porcine insulin. The crystalline shape of chicken insulin is similar to that of porcine insulin, so their three-dimensional structures should also be very similar. The amino-acid residues of chicken insulin which are different from those of porcine insulin are B1 Ala, B2 Ala, A8 His, A9 Asn, and A10 Thr. From previous structure and function studies of insulin, the variation in such positions should not have much influence on insulin activity. Therefore, further studies must be done in order to see if the activity of chicken insulin is really much higher. Crystalline snake insulin has also been obtained by the same method from the pancreas of a non-venomous species *Zaocys dhumnades* (*Cantor*) near Shanghai. Snake insulins have been little studied, the only case reported being rattlesnake insulin (Kimmel, Mahler, Pollock, and Vensel 1976). The snake insulin we have studied is different from rattlesnake insulin because the former is more acidic. The crystals of *Zaocys dhumnades* snake insulin are dodecahedral (Fig. 39.2), which are quite different from porcine zinc insulin crystals but similar to cubic zinc-free insulin crystals. The sequence of snake insulin has been determined and the B10 residue is still histidine. The snake insulin might crystallize as zinc hexamers in a cubic form as proposed by Pitts *et al.* (1980). However, because of the limited supply of snake pancreas, there have been difficulties in accumulating sufficient amounts of crystalline snake insulin for the determination of its primary and three-dimensional structures.

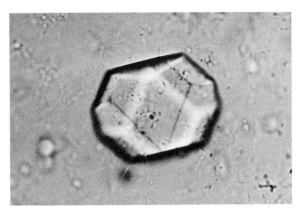

FIG. 39.2. Crystalline snake insulin (×96).

In summary, evidence accumulated through studies by semisynthesis, limited enzymatic hydrolysis, and comparison of insulins from different species has led us to the conclusion that the C-terminal part of the B-chain, especially B23 Gly, and B24 Phe, is important for insulin activity. Our results in the structure and function studies of insulin are consistent with the hypothesis that the hydrophobic surface involving B12 Val, B16 Tyr, B24 Phe, and B26 Tyr might be a part of the binding site of the insulin molecule with its receptor.

References

ABEL, J. J. (1926). *Proc. natn. Acad. Sci. U.S.A.* **12**, 132.
ADAMS, M. J., BLUNDELL, T. L., DODSON, E. J., DODSON, G. G., VIJAYAN, M., BAKER, E. N., HARDING, M. M., HODGKIN, D. C., RIMMER, B., and SHEAT, S. (1969). *Nature, Lond.* **224**, 491.

BANTING, F. G. and BEST, C. H. (1921). *J. lab. clin. Med.* **7**, 464.

BERSON, S. A., YALOW, R. S., BAUMANN, A., ROTHSCHILD, M. A., and NEWERLY, K. (1956). *J. clin. Invest.* **35**, 170.

BROMER, W. W. and CHANCE, R. E. (1967). *Biochim. biophys. Acta* **133**, 219.

CHU, S. C., LI, T. F., and TSAO, C. P. (1978). *Acta biochim. biophys. sinica* **10**, 199.

——— CHANG, Y. S., and LU, T. H. (1973). *Scientia sin.* **16**, 71.

CROWFOOT, D. (1935). *Nature, Lond.* **135**, 591.

DIXON, G. H. and WARDLAW, A. C. (1960). *Nature, Lond.* **188**, 721.

DU, Y. C., ZHANG, Y. S., LU, Z. X., and TSOU, C. L. (1961). *Scientia sin.* **10**, 84.

GATTNER, H.-G. (1975). *Z. physiol. Chem.* **356**. 1397.

——— and SCHMITT, E. W. (1977). *Z. physiol. Chem.* **358**, 105.

Insulin Research Group, Shanghai Institute of Biochemistry (1976). *Scientia sin.* **19**, 475.

———(1976). *Scientia sin.* **19**, 351.

KATSOYANNIS, P. G., TRAKATELLIS, A. C., ZALUT, C. JOHNSON, S., TOMETSKO, A., SCHWARTZ, G., and GINOS, J. (1967). *Biochemistry* **6**, 2656.

KIMMEL, J. R., MAHLER, M. J., POLLOCK, H. G., and VENSEL, W. H. (1976). *Gen. Comp. Endocrinol.* **28**, 320.

KUNG, Y. T., DU, Y. C., HUANG, W. T., CHEN, C. C., KE, L. T., HU, S. C., JIANG, R. Q., CHU, S. Q., NIU, C. I., HSU, J. Z., CHANG, W. C., CHENG, L. L., LI, H. S., WANG, Y., LOH, T. P., CHI, A. H., LI, C. H., SHI, P. T., YIEH, Y. H., TANG, K. L., and HSING, C. Y. (1965). *Scientia sin.* **14**, 1710.

OFFORD, R. E. (1969). *Nature, Lond.* **221**, 37.

———(1972). *Biochem. J.* **129**, 499.

Peking Insulin Structure Research Group (1973). *Scientia sin.* **16**, 136.

———(1974). *Scientia sin.* **17**, 752.

———(1976). *Scientia sin.* **19**, 358.

PETERSON, J. D., COULTER, C. L., STEINER, D. F., EMDIN, S. O., and FALKMER, S. (1974). *Nature, Lond.* **251**, 239.

PITTS, J. E., WOOD, S. P., HORUK, R., BEDARKAR, S., and BLUNDELL, T. L. (1980). *Insulin: chemistry structure, and function of insulin and related hormones* (ed. D. Brandenburg and A. Wollmer), p. 678. De Gruyter, Berlin.

RYLE, A. P., SANGER, F., SMITH, L. F., and KITAI, R. (1955). *Biochem. J.* **60**, 541.

Shanghai Insulin Research Group (1973). *Scientia sin.* **16**, 61.

SHVACHKIN, YU, P., SHMELEVA, G. A., KRIVTSOV, V. F., FEDOTOV, V. P., and IVANOVA, A. I. (1972). *Biokhimiya* **37**, 966.

SIEBER, P., KAMBER, B., HARTMANN, A., JÖHL, A., RINIKER, B., and RITTEL, W. (1974). *Helv. chim. Acta* **57**, 2617.

TSAO, C. P., LI, T. F., PENG, X. H., and ZHANG, Y. S. (1979). *Proceedings of the 3rd Chinese Biochemical Symposium*, p. 2.

WEITZEL, G., OERTEL, W., RAGER, K., and KEMMLER, W. (1969). *Z. physiol. Chem.* **350**, 57.

——— WEBER, U., MARTIN, J., and EISELE, K. (1971). *Z. physiol. Chem.* **352**, 1005.

YALOW, R. S. and BERSON, S. A. (1960). *J. clin. Invest.* **39**, 1157.

ZAHN, H., DANHO, W., and GUTTE, B. (1966). *Z. Naturf.* **21b**, 763.

40. Research on the insulin crystal structure in China

THE BEIJING INSULIN STRUCTURE
RESEARCH GROUP†

It is a special pleasure for us to have the opportunity of writing a paper for this volume in celebrating the seventieth birthday of Dorothy Hodgkin. Professor Hodgkin is a close friend of Chinese protein crystallographers and a number of friendly and interesting exchange visits have taken place between us. Professor Hodgkin started on her pioneering studies on insulin crystal structure in the early thirties when protein crystallography had barely come into being. Inspired by the achievement of the total synthesis of crystalline bovine insulin with chemical methods in our country, we began to investigate the crystal structure of insulin in the mid-sixties and benefited from the experience of Dorothy Hodgkin's laboratory in crystal growth and the preparation of some of the derivatives. The work on X-ray analysis of the same protein gave the two insulin groups an excellent opportunity of becoming acquainted with each other. In the summer of 1972 Professor Hodgkin visited us in Beijing and we were able to compare two sets of electron-density maps (resolution at 1.9 Å and 2.5 Å respectively) which we had achieved independently in our two countries. This occasion was valuable for both groups. To our great satisfaction, most peaks in the corresponding sections of maps of these two sets matched well. After many discussions, we all agreed that the X-ray analysis was worth carrying out by the two separate groups simultaneously. Then, during the Ninth International Congress of Crystallography in Japan, Dorothy Hodgkin spoke about the work Chinese crystallographers had been doing on insulin structure, and gave the results in detail. Since we had little chance to exchange our views with foreign scientists at that time, Professor Hodgkin's report played an important role in making our work known better abroad.

After our paper on the insulin crystal structure at 1.8 Å resolution was published (Peking Insulin Structure Research Group 1974), Professor Hodgkin wrote an article in *Nature* (1975) and expressed the idea that it would

† The convention in this chapter describing the two molecules in the 2Zn insulin crystal asymmetric unit is opposite to that used in Oxford (p. 527).

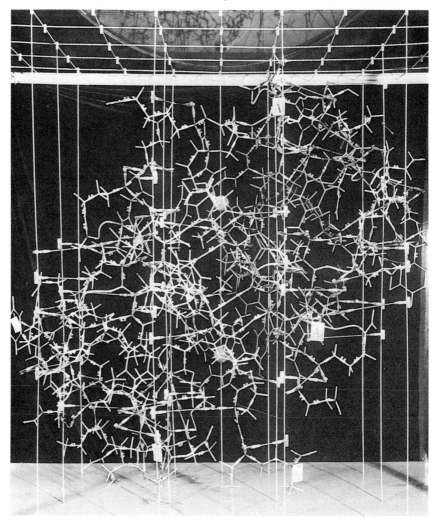

Fig. 40.1 Wire skeletal model of insulin dimer (photograph taken along the non-crystallographic axis).

be splendid if we could some day soon all meet and talk over the very interesting observations that were accumulating, in the East and in the West, on the structure and function of insulin. Two years later, by the end of 1977, Dorothy Hodgkin's desire was realized. She came to China again, making her fourth visit. Again, we compared the electron-density maps (Dorothy Hodgkin's results by refinement at 1.5 Å resolution and ours obtained from isomorphous replacement at 1.8 Å resolution) and had a wide-ranging discussion.

The most important problem we discussed at that time was to what extent the refinement results were reliable. Dorothy Hodgkin appreciated our experimental results at 1.8 Å resolution and wanted to get a reliable assessment of the refinement error from the comparison of the two sets of the maps. We observed that Dorothy Hodgkin's refinement at high resolution had made some ambiguous parts of our electron-density map clearer and more definite. This made it easier to locate atoms and identify water molecules in the solvent system as well. By that time refinement of our map was delayed owing to lack of computing facilities. Now we are refining the structure so as to improve the precision, and at the same time we are attempting to raise the resolution further.

In this paper we would like to republish the photograph of the skeletal

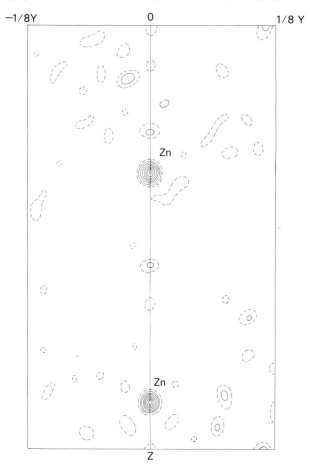

FIG. 40.2 The map of electron density on the $x=0$ section (in the 2.5–1.8 Å range), showing two zinc atoms.

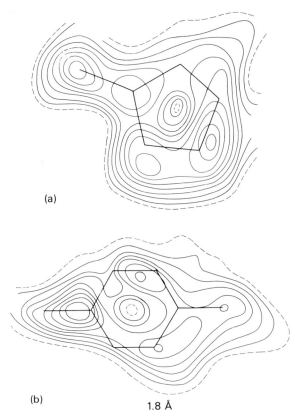

(a)

(b)

1.8 Å

Fig. 40.3 A tyrosine (a) and histidine (b) in the electron-density map.

model of insulin dimer (see Fig. 40.1) and some maps of electron-density phased by isomorphous replacement method at 1.8 Å resolution, showing zinc atoms, some amino-acid residues, a stretch of α-helix, and several water molecules (see Figs. 40.2–40.7). About 60 water molecules could be identified from the electron-density map. Through careful examination of the distribution of these water and solvent molecule positions, we have come to some preliminary conclusions as follows. (1) Water molecules are mainly distributed in three regions: the water molecules located in the large regions between the hexamers are predominantly in semi-flowing state; large patches of solvent system exist in the cavity around the threefold axis; the water molecules on the surface of the dimer are bound individually to the polar groups of the protein molecule. (2) The distribution of water molecules suggests an approximate twofold symmetry. About 85 per cent of water molecules are bound to the same groups of the two individual insulin molecules in a dimer. (3) The insulin molecule tends to bind as many water mole-

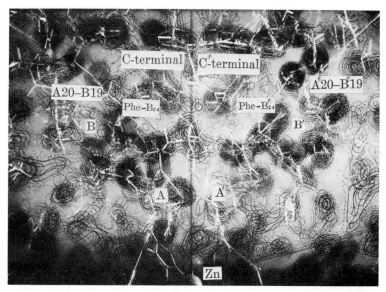

FIG. 40.4. The electron density distributed in the vicinity of non-crystallographic two-fold axis (composite of sections 26/60–36/60). The line in the middle indicates the position of the non-crystallographic twofold axis. In general on its left side is Insulin Molecule II and on its right is Insulin Molecule I. A–B is a stretch of α-helix (B9–B19), the axis of which is approximately situated in the line connecting A and B.

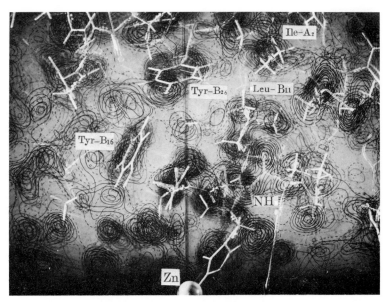

FIG. 40.5. Some amino-acid residues in the electron-density map (composite of sections 35/60–44/60).

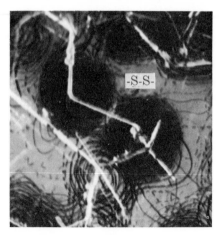

FIG. 40.6 A disulphide bridge of Molecule I (Cys-A20—Cys-B19) in the electron-density map.

cules as possible. Almost all the polar groups bind with water molecules except where polar groups lie embedded within the molecule and are therefore inaccessible, and where hydrogen bonds have already been formed either within or between polypeptide chains. Of the 77 polar groups in the B-chain we find there are only eight not bound with water, while among 56 in A-chain there are only three. The water-binding tendency of CO and NH groups in the main chain and the polar groups of the side-chain are not significantly different. The fine quality of our electron-density map is due to the fine quality of the native crystal and its derivatives. The statistical analysis of intensity

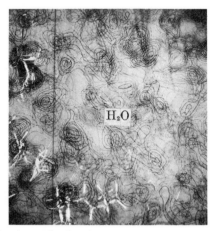

FIG. 40.7. Electron-density peaks showing some water molecules and solvent system on the surface of the molecule.

data for the native insulin crystals shows that in the range of 2.2–1.8 Å 90 per cent of the reflections have an intensity greater than twice the value of the estimated standard deviation. Among the three heavy-atom derivatives used, mercuric ethyl chloride insulin proved to be the best one. It made a substantial contribution to phase determination up to 1.8 Å resolution. Recently an attempt has been made to calculate an isomorphous difference Patterson projection with the $hk0$ data within the 1.9–1.5 Å range. This shows that the heavy-atom vector peaks are still the highest of all. Therefore a very interesting question is raised: Could the experimental resolution be further improved?

Asymmetry of three helical stretches of the main chain between two molecules in a dimer was observed on careful examination of our electron-density maps at 1.8 Å resolution. We were very interested in such remarkable structural asymmetry in a protein in the crystalline state. Many similar structural features, including those found by Dorothy Hodgkin and her colleagues in the rhombohedral 4-zinc insulin structure, have been unravelled by protein crystallography. All of these make us further realize what an active protein demands for its conformation. Not long before, with the aid of our computer programs for automatic stereochemical analysis of proteins, we made a detailed study of the stereochemistry of insulin based on the preliminary atomic co-ordinates obtained by the 1.8 Å analysis. This gives us a more or less quantitative concept of the difference between the conformations of two insulin molecules in an asymmetric unit, and provides us with a more accurate basis for further discussions and studies on the flexibility of protein molecular conformation. In the meantime, 1300 atomic contact distances within the 4 Å range, 3000 within the 5 Å range, and as many as 6000 within the 6 Å range have been identified within an insulin molecule. This means that a great number of complicated weak interactions are involved in the molecule. These results may be of significance for the understanding of the stabilization of the three-dimensional conformation.

Penetrating work is being carried on by research workers in different fields in an attempt to clarify the primary process of insulin's biological effect. We are convinced that crystallography will also make its due contribution to this end. Studies on insulin analogues and insulins of various species and crystal forms generally exhibit parallel binding abilities and biological activities, in spite of the remarkable differences in their hormonal activities. Therefore we cannot say with any certainty whether it is possible to differentiate a binding site from an active site in the molecule. Many results, however, show that the surface of contact of the two molecules in a dimer still plays an important biological role. Studies on crystal structure and conformation in solution of B-chain C-terminal despentapeptide insulin (DPI) are in progress in our laboratory, and these will probably throw some light on this problem (Peking Insulin Structure Research Group 1976).

We tried to study the C2 DPI crystal structure by the molecular-replacement method. According to the preliminary results obtained, it seems that, compared with the insulin dimer, certain changes have occurred at the surface of contact between the two DPI molecules related by the crystallographic twofold axis. However, some of the surface features found in rhombohedral insulin crystal may still be maintained. For instance, a not very large hydrophobic surface involving residues B24 and B25 may still be formed between these two molecules, and the contact distance between the two C_x atoms of residues B25 of the two molecules is the shortest of all. The expected peaks in rotation and translation functions are not very prominent; perhaps it is due to the fact that somewhat large differences exist between the structures of insulin and DPI. The reliability of these results awaits confirmation by further work.

Attempts are also being made in our laboratory to determine the structure of DPI with the isomorphous replacement method. Three or four heavy-atom sites have been found from the difference Patterson analysis with K_2HgI_4-DPI crystal at 4 Å resolution. The parameters have been refined by Fourier and full matrix least-squares method and the phase angle has been determined by the probability method. The interpretation of the electron-density map is in progress.

Recently, at the Institute of Biochemistry in Shanghai, chicken and snake insulins and B-chain C-terminal desheptapeptide insulin have been prepared and purified and microcrystals have been obtained. At the same time, the B1-modified (replaced by tryptophan) insulin has been prepared in our laboratory in collaboration with Dr Dietrich Brandenburg from Aachen, and growth of large crystals suitable for X-ray diffraction analysis has been achieved. Crystal structure analysis of this modified insulin is under way. We look forward to further exchanges of our results with those of Dorothy Hodgkin and her co-workers in the near future.

Before ending this paper, we would like to express our cordial wishes for Dorothy Hodgkin's health and happiness and for the flourishing of the friendship between the crystallographers and peoples of our two countries.

References

HODGKIN, D. C. (1975). *Nature, Lond.* **255**, 103.
Peking Insulin Structure Research Group (1974). *Scientia Sin.* **17**, 752.
——(1976). *Scientia Sin.* **19**, 358.

41. Hydrogen atoms and hydrogen bonding in rhombohedral 2Zn insulin crystals by X-ray analysis at 1.2 Å resolution[†]

NORIYOSHI SAKABE, KIWAKO SAKABE, AND
KYOYU SASAKI

To understand hydrogen bonding in protein crystals, non-bonded inter-action, and the protonation of ionizable groups, which are all very important not only for the stabilization of conformation but also for reactivity, it is necessary to locate hydrogen atoms accurately. Schoenborn (1971) found the co-ordinates of hydrogen atoms in myoglobin at 2.0 Å resolution by neutron diffraction, the first example of such an analysis. However, the resolution in these studies was not high enough to analyse hydrogen bonded and non-bonded interactions precisely.

It is well known that protein structure analysis by neutron diffraction is the technique most suited to the positioning of hydrogen atoms. However, there are several difficulties; access to high-flux neutron equipment is diffi-cult, and the crystals must be very large. These problems can be solved, at least in principle, by using X-ray diffraction. Positioning hydrogen atoms by X-ray diffraction is difficult as the magnitude of X-ray scattering for a hydrogen atom is only one-sixth of that for a carbon atom, so that it is neces-sary to obtain accurate high-resolution data for the refinement of crystal structures. The refinement of crystal structure of proteins is difficult, laborious, expensive, and time-consuming, and no one has yet refined a pro-tein structure in such a way that hydrogen atoms can be observed.

In this paper we describe the refinement of the crystal structure of rhombo-hedral 2Zn porcine insulin with X-ray diffraction data to 1.2 Å resolution which has allowed the positioning of many hydrogen atoms in the insulin molecule.

[†] In this chapter the Oxford convention for molecules I and II is used.

Crystallographic refinement of the structure

Insulin is one of the most widely studied of all hormone molecules. Porcine insulin consists of two chains: the A-chain containing 21 amino-acid residues and the B-chain of 30 residues. The two chains are linked covalently by two disulphide bridges. The complete sequence of porcine insulin was determined by Sanger and his colleagues (Ryle, Sanger, Smith, and Ktai 1955).

2Zn porcine insulin crystals were prepared at pH 6.2 from a mixture containing 1.0 per cent porcine insulin, 12.5 per cent acetone, 0.042 M sodium citrate, and 0.01 M HCl. They are of the rhombohedral space group, R3. It is convenient to refer the structure to a hexagonal unit cell with the cell dimensions $a = 82.46$ Å and $c = 33.94$ Å. An asymmetric unit of these crystals contains two equivalent insulin monomers related by a local twofold axis (Dodson, Harding, Hodgkin, and Rossmann 1966). The three-dimensional structure has been solved by the multiple isomorphous replacement (m.i.r.) method in Oxford and Peking (Adams *et al.* 1969; Blundell, Dodson, Hodgkin, and Mercola 1972; Peking Insulin Research Group 1971, 1974). We have analysed it by using single isomorphous method with an anomalous dispersion effect (Sakabe, Sakabe, and Katayama 1972).

Our intensity data up to 1.2 Å resolution were collected with Ni-filtered CuKα radiation on a Hilger & Watts four-circle diffractometer placed in a room maintained at 4 °C. The absorption errors were corrected by the method of Katayama, Sakabe, and Sakabe (1972).

The starting atomic co-ordinates of the insulin molecules and two zinc ions obtained from 1.9 Å m.i.r. phased maps were provided by Dorothy Hodgkin. We have refined the co-ordinates with 1.3 Å resolution data in the initial stages of the calculations and later with 1.2 Å resolution data. With 1.3 Å resolution data, positional parameters, x, y, and z were refined by a differential-difference Fourier synthesis, and individual thermal parameters (B) were calculated by the equation

$$B_{new} = (1 - 2\Delta\rho/\rho_{obs}) \times B_{old}$$

where ρ is the electron density (Lenhert 1968). At 1.2 Å resolution, both positional and thermal parameters were refined by a diagonal matrix least-squares method (Isaacs and Agarwal 1978; see also Chapter 27) with about 21 000 reflections. The structure model thus obtained was standardized every few cycles of the refinement by using the energy refinement program (Levitt 1974) in 1.3 Å resolution and by using the 'model fit' program (Dodson, Isaacs, and Rollet 1976) at 1.2 Å resolution. Both F_0-synthesis (electron-density map) and $(F_0 - F_c)$-synthesis (difference map) were used for correcting the positional parameters of ill-defined atoms of the insulin molecule and for finding disordered side-chains, water molecules, and a citrate molecule. All their parameters were then refined. The positional and thermal parameters of hydrogen atoms were estimated on the basis of the assumptions that hydrogen atoms keep a staggered conformation and that the thermal parameter of a hydrogen atom is equal to that of the bonded atom. Hydrogen atoms were included in the calculation of the structure amplitudes, although they have not been refined yet. The value of

$$R = \frac{\sum ||F_\mathrm{o}| - |F_\mathrm{c}||}{\sum |F_\mathrm{o}|}$$

reduced to 0.14 with 1.2 Å resolution data. Difference maps were calculated with phases from which hydrogen atoms had been excluded, and the hydrogen atoms were divided into four classes according to the value of the electron density at their estimated positions. The left column in Table 41.1.shows the number of hydrogen atoms in each class against the thermal parameter. The hydrogen atom bonded to the atom with low thermal parameter appears clearer in the difference map than that with higher thermal parameter, and about 80 per cent of the hydrogen atoms which can be estimated in the insulin structure were resolved in positive regions of the electron-density map as shown in Table 41.1

After this step, the refinement was switched from a diagonal-matrix least-squares method to a block diagonal-matrix least-squares method. The ordered atoms whose isotropic thermal parameters were less than 30 Å² were refined with anisotropic thermal parameters. Disordered atoms and atoms with thermal parameters greater than 30 Å² were refined with isotropic thermal parameters. The R value reduced to 0.11, and about 90 per cent of hydrogen atoms bonded to insulin were confirmed at the expected positions on the difference maps. From Table 41.1, it is clear that this refinement with anisotropic thermal parameters is the preferable method for the detection of hydrogen atoms on the difference map.

The accuracy of the phases

The resolution of atoms on the electron-density maps depends upon the number of reflections and the quality of phases. The electron density of ring side-chains is illustrated to show the evidence for the increased resolution and the quality of phases.

Figure 41.1 shows the electron-density map in the plane of tyrosyl and imidazole rings at 1.2 Å resolution where the skeletal model of the ring is superimposed.

Another proof of the high accuracy of the phases is demonstrated by the appearance of hydrogen atoms on the difference maps. Figure 41.2 shows the composite difference map calculated with phases from which hydrogen atoms have been excluded.

TABLE 41.1

The appearance of hydrogen atoms at the estimated positions on the difference map†

Thermal parameter‡ (Å²)	Isotropic thermal parameters (R=0.14)				Anisotropic thermal parameters (R=0.11)				Sum
	Electron density				Electron density				
	Positive			Negative	Positive			Negative	
	very strong	strong	weak		very strong	strong	weak		
$B \leqslant 10$	77	45	43	4	141	26	2	0	169
$10 < B \leqslant 20$	75	90	117	28	158	105	41	6	310
$20 < B \leqslant 30$	18	27	57	36	24	48	53	13	138
$30 < B \leqslant 40$	1	12	17	13	8	12	15	8	43
$40 < B \leqslant 50$	0	2	8	6	1	3	10	2	16
$50 < B$	1	2	8	8	0	5	6	8	19
Total number	172	178	250	95	332	199	127	37	695§
Percentage	24.7	25.6	36.0	13.7	47.8	28.6	18.3	5.3	100.0

† The difference map calculated with phases without regard for hydrogen atoms.
‡ The thermal parameter of the atom bonded to hydrogen atom.
§ The total number of hydrogen atoms bonded to insulin is 770. The hydrogen atoms bonded to disordered atom, COOH, $CONH_2$ and tyrosyl OH were not included.

The thermal parameter (B) is related to the root mean square displacement (Δu) of atom by following equation $B = 8\pi^2(\Delta u)^2$.

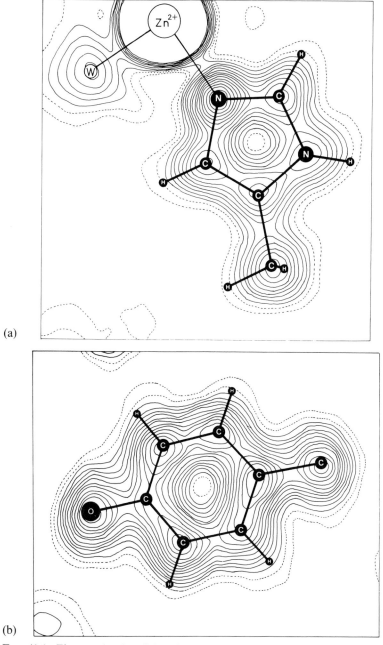

FIG. 41.1. Electron density of ring planes (a) B10-His, (b) B16-Tyr side-chains in molecule II. The atomic positions including assumed hydrogen atoms are superimposed.
Contours are at 0.4 (dashed), 0.6 (dashed), 0.8, 1.0, 1.2, ...eÅ^{-3}.

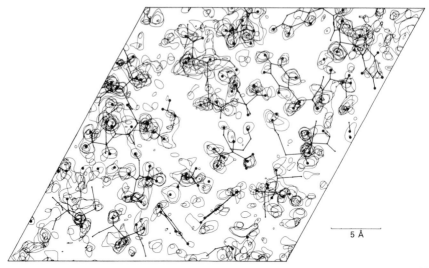

FIG. 41.2. Difference composite maps calculated with the phases from which hydrogen atoms have been excluded. Nine sections along the z axis from $-4/72$ to $4/72$ are superimposed. Black circles show the assumed position of hydrogen atoms.

The protonation of the imidazole ring

Zinc ions are essential for the formation of rhombohedral 2Zn insulin crystals. They bond six ligands: three crystallographically equivalent Nε atoms from B10-histidines and three water molecules in a distorted octahedral coordination around the threefold axis (Sakabe, Sakabe, and Sasaki 1980; see also Chapter 42). It is clear that these B10-histidines are located on the surface of the insulin molecule in the crystal.

Tanford and Epstein (1954*a*, *b*) have compared the complete hydrogen-ion titration curve for zinc-free insulin with the curve for crystalline zinc insulin and concluded that the two imidazole groups which are titrated between pH 6.5 and pH 7 in zinc-free insulin appeared instead in the direct titration curve of crystalline zinc insulin near pH 4. Figure 41.3 shows the difference map of the imidazole ring of B10-histidine. In the map, the electron density peak of the hydrogen atom bonded to the Nδ atom is of nearly equal height to those of hydrogen atoms bonded to the Cβ, Cδ, and Cε atoms. However, we could not observe the electron density on the map which corresponds to the hydrogen atom of Nε. These observations are consistent with the conclusion of Tanford and Epstein (1954).

It is very difficult to know how the charges of the zinc ions are neutralized. One possible explanation is that the charges may spread over to the carboxyl group of B13-glutamic acid via the imidazole ring and a water molecule which are connected by a hydrogen-bond network as shown in Fig. 41.4. However,

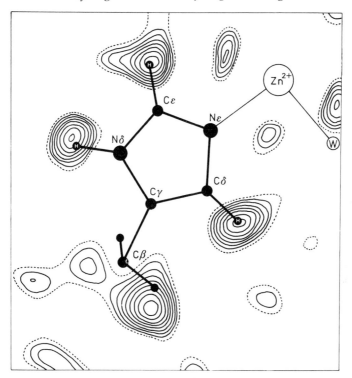

FIG. 41.3. The difference map of the imidazole ring plane on B10-His in molecule I with the phases from which hydrogen atoms have been excluded. Contours are at 0.05 (dashed), 0.07, 0.09, 0.11, ... e Å^{-3}.

it is possible that the charges may be neutralized by OH$^-$ which is assigned as water molecules around the zinc cation and/or negative charges caused by the electric dipole moment developed in a right-handed α-helix (Wada 1976).

The conformation of the terminal methyl group of side-chains

The simplifying assumption usually made for the conformation of side-chains is that a terminal methyl group is staggered with respect to a connected methylene group. In order to justify the assumption we need to check hydrogen atoms attached to the terminal methyl group of alanine, valine, leucine, and isoleucine residues on difference maps since many of these residues are important members of the non-polar core (see p. 538). For example the three small peaks corresponding to three hydrogen atoms on the Cγ carbon atom of B18-valine are shown in Fig. 41.5.

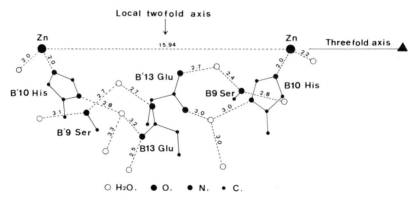

FIG. 41.4. The hydrogen bond network around threefold axis. Positive charges on the zinc ions may distribute to glutamate via an imidazole ring and a water molecule. Interatomic distances are shown by Å units. B′ indicates the B-chain in molecule II.

Generally speaking, most of the terminal methyl groups differ in conformation by less than 10° from a strictly staggered conformation. On the other hand, the rotation from a staggered conformation of about 25° was observed in one terminal methyl group of the B11-leucine residue in molecule I. If we assume that this methyl group is a strictly staggered conformation, the distance between the hydrogen atom of Cα and one of the hydrogen atoms

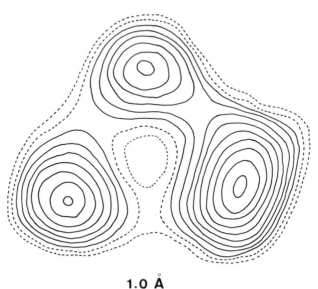

1.0 Å

FIG. 41.5. The difference map for the hydrogen atoms of the methyl group of B18-Val in molecule II with the phases from which hydrogen atoms have been excluded. Contours are 0.05 (dashed), 0.07 (dashed), 0.09, 0.11, ... $e\text{Å}^{-3}$.

of Cδ will be 2.0 Å, which is the same as the van der Waals contact distance between two hydrogen atoms (Baur 1972). Therefore, the small rotation from a staggered conformation helps to remove this short contact and gives the resultant conformation lower energy.

Hydrogen bonding in the insulin molecule

Definition of parameters

Parameters are defined in Fig. 41.6 where l is hydrogen bond length, θ is the angle between the standard line X–X′ and hydrogen bond, ζ is the angle between the direction of hydrogen bond and the projection of it on the peptide plane measured anticlockwise looking from C or N to Cα in the residue containing C or N, and ξ is the angle between the projection of the hydrogen bond on the peptide plane and the standard line X–X′, measured from the standard line X–O or X–N to the Cα in the residue containing O or N.

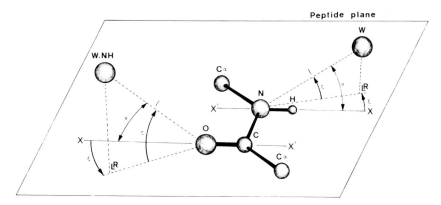

FIG. 41.6. The definition of parameters (l, θ, ζ, ξ) of the hydrogen bond between peptides or between a peptide and a water molecule.

Hydrogen bonding in the secondary structure

Helix

There is a central piece of typical α-helix in the B-chain and the geometry of hydrogen bonds (Table 41.2) shows that this α-helix has very similar conformation in molecules I and II (see Fig. 42.6(b), p. 536).

The hydrogen bond lengths vary from 2.8 to 3.2 Å, and the mean hydrogen bond length is 2.99 ± 0.09 Å, which is nearly equal to the hydrogen bond length (3.01 Å) for amino acid and simple peptide bonds (Fuller 1959). The mean values of angles θ, ζ, and ξ are 27°, 22°, and 13° respectively. The long hydrogen bonds tend to have a larger value of θ.

TABLE 41.2

The geometry of hydrogen bonds in α-helix from B9 to B19

Residue no.			Molecule I				Molecule II			
Donor	Acceptor									
#	NH	...O=C	$l(\text{Å})$	$\theta(°)$	$\zeta(°)$	$\xi(°)$	$l(\text{Å})$	$\theta(°)$	$\zeta(°)$	$\xi(°)$
1	B13	B 9	2.99	23	16	17	3.03	26	21	14
2	B14	B10	3.16	34	29	19	3.11	35	31	18
3	B15	B11	2.94	19	15	13	2.97	22	14	18
4	B16	B12	2.93	27	25	11	2.99	33	31	10
5	B17	B13	2.98	27	23	15	3.11	31	24	22
6	B18	B14	2.97	25	21	13	3.04	20	16	12
7	B19	B15	2.86	24	24	−4	2.78	22	22	−2

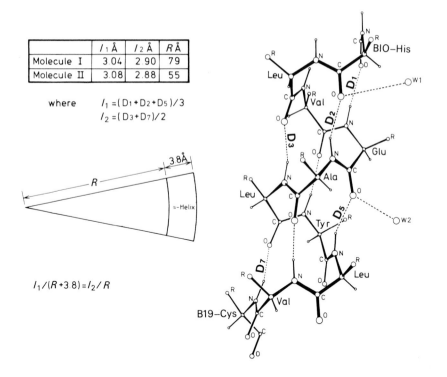

	l_1 Å	l_2 Å	R Å
Molecule I	3.04	2.90	79
Molecule II	3.08	2.88	55

where
$$l_1 = (D_1 + D_2 + D_5)/3$$
$$l_2 = (D_3 + D_7)/2$$

$$l_1/(R+3.8) = l_2/R$$

α-Helix in molecule I

FIG. 41.7. The curve of the α-helix in the B-chain. The α-helix is on the surface of the molecule and is curving towards hydrophobic side. The water molecule (W2) which is hydrogen bonded to B13 in molecule I is not found in molecule II.

The accuracy of these parameters is reasonably high since the root mean square difference of parameters between two molecules is 0.04 Å in l and less than 2° in angles θ, ζ, and ξ on an arithmetic average. If the difference between the mean value and an individual value is larger than the root mean square difference, such differences are likely to have some meaning. The hydrogen bonds # 1, # 2, and # 5 (Table 41.2) lie on the surface of the insulin molecule and those # 3 and # 7 face the inner side of the molecule. The mean values of l and θ for the former hydrogen bonds are slightly larger than those of the latter ones (Fig. 41.7). This means that the helix is not straight and tends to curve against the inner side of the molecule; this curvature is 80 Å for molecule I and 60 Å for molecule II (Fig. 41.7).

Comparing the corresponding hydrogen bonds between two molecules, one can still find a small difference in the hydrogen bond length # 5 in

TABLE 41.3

The geometry of hydrogen bonds in helix from A1 to A8

	Residue no.		Molecule I				Molecule II			
	Donor	Acceptor								
#	NH	O=C	l(Å)	θ(°)	ζ(°)	ξ(°)	l(Å)	θ(°)	ζ(°)	ξ(°)
1	A5	A1	2.98	21	5	21	3.09	21	−6	20
2	A6	A2	2.95	19	19	−2	2.77	23	22	−4
3	A7	A3	†3.60	39	32	24	2.83	47	42	−25
4	A8	A4	2.64	11	7	9	*3.82	78	62	−64

† No hydrogen bonding.

Table 41.2. The deviation of the length between two molecules is 3.3 times larger than the root mean squares difference of the hydrogen bond length mentioned above. One of the reasons may be the fact that the side-chain of B17-leucine takes two alternative conformations in molecule II; it is a disordered structure.

On the other hand, the conformation of the segment from A1 to A8 is nearly an α-helix and it is different in the two molecules (see Fig. 42.4, p. 534). This segment contains three hydrogen bonds in molecule I and II with different geometries (Table 41.3). This difference comes from the inter-molecular packing between hexamers; the carbonyl oxygens at the final turn of these two helices being in unsymmetrical contact with the two B5-histidines from adjacent hexamers (Chapter 42). Such a difference may express the flexibility of local conformation (Sakabe, Sakabe, and Sasaki 1979).

β-Sheet

One antiparallel β-sheet with four hydrogen bonds can be located on the dimer-forming surface (see Fig. 42.3(b), p. 533), and their hydrogen bond

TABLE 41.4

Hydrogen bonds in β-sheet

#	NH	...O=C	$l(\mathring{A})$	$\theta\,(°)$	$\zeta\,(°)$	$\xi\,(°)$
	Donor	*Acceptor*				
1	B 24	B′26	2.93	26	−11	−23
2	B′26	B 24	2.84	6	5	−4
3	B 26	B′24	2.84	13	13	−4
4	B′24	B 26	2.98	28	−1	−29
5	B 25	A 19	3.19	12	−9	−9
6	B′25	A′19	3.09	5	1	−6

B′ indicates the B-chain in molecule II.

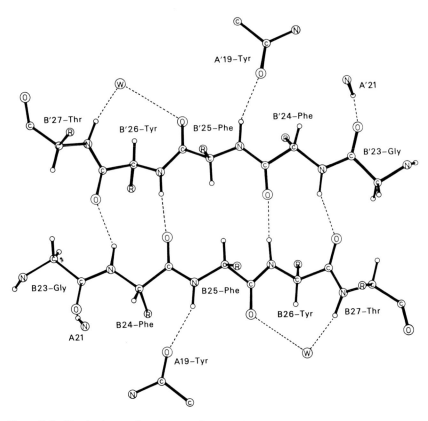

FIG. 41.8. The hydrogen bond network around the anti-parallel β-sheet. W: water molecule.

Conformation of β-turn

Mol.	β-turn	Tetrapeptide 1 2 3 4	$C_1^\alpha \ldots C_4^\alpha$ (Å)	$\phi 2$ (°)	$\psi 2$ (°)	$\phi 3$ (°)	$\psi 3$ (°)	l (Å)	θ (°) $C{=}O_1 \ldots HN_4$	ζ (°)	ξ (°)	Type‡
I	A16-A19	Leu—Glu—Asn—Tyr	5.6	−70	−13	−102	3	2.86	50	47	20	I
II			5.7	−66	−18	−99	7	3.00	60	55	29	I
I	B7-B10	Cys—Gly—Ser—His	5.5	55	−135	−60	−26	3.01	40	−31	28	II′
II			5.5	60	−137	−60	−29	3.00	43	−34	29	II′
I	B20-B23	Gly—Glu—Arg—Gly	5.4	−50	−25	−79	−22	2.74	51	42	31	III
II			5.7	−49	−32	−69	−30	2.92	48	43	24	III
I	B27-B30	Thr—Pro—Lys—Ala	†	−81	168	−88	164					
II			5.7	−57	−30	−75	−22	3.11	60	56	27	III

† This segment is not a β-turn.
‡ Venkatachalam (1968).

geometry is shown in Table 41.4. The mean value of l in the middle of the
β-sheet is 2.84 Å and that at the ends is 2.96 Å; the former is shorter than
the latter by 0.12 Å. Similar differences were observed in the hydrogen bond
angles; the mean values of angles θ, ζ, and ξ are 10°, 9°, and $-4°$, respectively,
while those in the ends are 27°, $-6°$, and $-26°$, respectively. These values
suggest that the hydrogen bonds in the middle are stronger than those in
the ends owing to the co-operative effects of the hydrogen bonds.

There is another hydrogen bond between the B25-NH and A19-CO
groups in both molecules the geometry of which (Table 41.4) is similar to
that of the hydrogen bond in the middle of the β-sheet just mentioned. Thus
there are three β-sheet structures in an insulin dimer as shown in Fig. 41.8.
One water molecule is attached to both B27-NH and B25-CO. B23-
CO is hydrogen bonded to A21-NH in two molecules, and these geometries are similar to that of the β-turn.

β-Turn

A β-turn involves four consecutive residues where the polypeptide chain folds
back on itself by nearly 180°. Venkatachalam (1968) has classified β-turns
into three types, and all these types are contained in the insulin structure.
The dihedral angles and the geometry of the hydrogen bonds for the β-turns
are shown in Table 41.5. The conformation of type III is found in the segment
from B27 to B30 in molecule II, whereas the same segment in molecule I

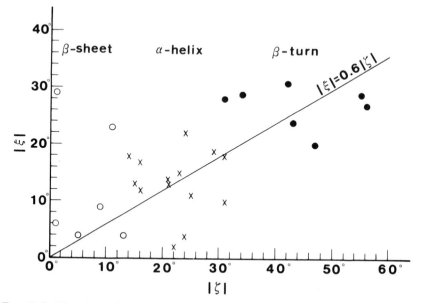

FIG. 41.9. The relation between hydrogen bond parameters ζ and ξ found in the secondary structure. $\bigcirc$: anti-parallel β-sheet; $\times$: α-helix; $\bullet$: β-turn.

takes an expanded conformation. This difference between the two molecules is probably caused by the difference of the intramolecular contact with B29 and B30; B29-NH is hydrogen bonded to the carboxyl side-chain of A4-glutamic acid and the side-chain of B29-lysine contacts with both the terminal carboxyl group of B30 and the carboxyl side-chain of A4-glutamic acid in molecule II. No such interactions are observed in molecule II.

In order to characterize the geometry of the hydrogen bonds in the secondary structure, $|\zeta|$ and $|\xi|$ are plotted in two dimensions (Fig. 41.9) where one can find the linear relation $|\zeta| \simeq 0.6|\xi|$.

Another finding is that there are consistent relations between the value of the angle and the type of the secondary structure; $|\zeta|$ for β-sheet being from $0°$ to $10°$, that in the α-helix being $15°$ to $30°$, and that in β-turn being $30°$ to $50°$.

Hydrogen bonding between water molecules and the peptide backbone

The crystal contains about 280 water molecules in an asymmetric unit. If the first hydration shell can be considered to extend to 3.5 A, it contains about 150 water sites. If we take a value of 3.2 Å as the maximum distance for a hydrogen bond, 73 per cent of water molecules in the first hydration shell are hydrogen bonded to polar atoms of insulin molecules, and their mobility is higher than that of an insulin atom bonded to the water molecule (Sakabe, Sakabe, and Sasaki 1980). Almost all of the other water molecules in the first hydration shell are hydrogen bonded to water molecules which are hydrogen bonded to the polar atoms in insulin, and only a few water molecules are located near non-polar atoms of insulin at a distance of 3.2–3.5 Å.

About 60 water molecules are hydrogen bonding to NH or CO in the peptide backbone and these water molecules are divided into six types as shown in Fig. 41.10. Types I and II water molecules have one hydrogen bond to the peptide backbone and the other types of water molecules have two or three hydrogen bonds.

It is likely that the geometry of these hydrogen bonds are restricted by the protein structure. One characteristic example is that types V and VI water

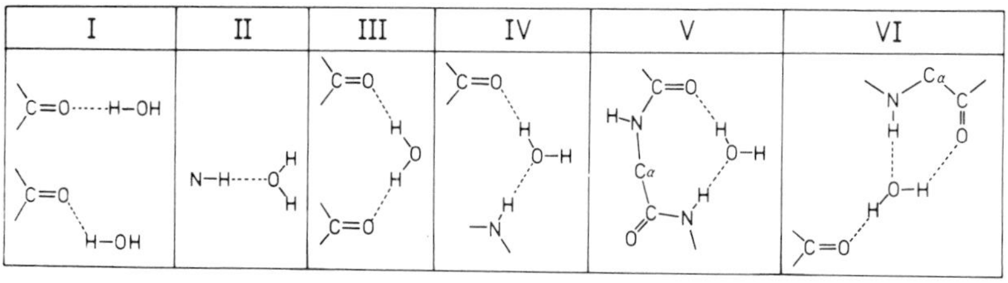

FIG. 41.10. Classification of the hydrogen bond between the peptide backbone and a water molecule.

molecules are found at limited conformations in the insulin structure (Fig. 41.10). Type V water molecule is found when the dihedral angles ϕ and ψ are about $-130°$ and $140°$ to $170°$, and type VI water molecules are found when the dihedral angles ϕ and ψ are $-100°$ to $-140°$ and about $110°$ respectively.

We have analysed the geometry of hydrogen bonds according to the definition shown in Fig. 41.6. The hydrogen bond length is in the range 2.5 Å to 3.2 Å and the pronounced maximum is observed in the range 2.9 Å to 3.1 Å, which is longer than the hydrogen bond length of OH --- O 2.7 Å (Brown 1976). To characterize the geometry of these hydrogen bonds, ζ and ξ are plotted on a stereonet (Fig. 41.11). In this figure, the distribution of points

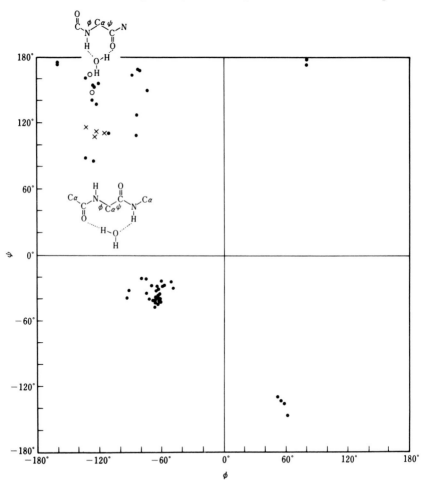

FIG. 41.11. Ramachandran plot of the B-chain where some residues which are hydrogen bonded to water molecules are shown as ×: type V; ●: type VI.

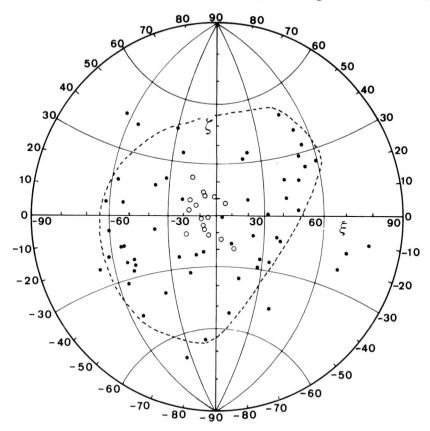

• OC, ○ HN

FIG. 41.12. Distribution of the geometry of the water molecules hydrogen bonded to the peptide backbone. Most of water molecules are inside the dashed line. ○: hydrogen bonded to NH; ●: hydrogen-bonded to C=O.

is not uniform and local. In the hydrogen bond NH---OH, ζ tends to lie within $\pm 20°$ and ξ tends to lie between $0°$ to $20°$. Whereas in the hydrogen bond C=O---OH, ζ is distributed within $\pm 50°$ and ξ between $-70°$ and $55°$. These results suggest that the direction of a hydrogen bond between a water molecule and an amide group of the peptide backbone is effectively restricted by the hydrogen donor group since NH is a hydrogen donor and CO is a hydrogen acceptor in this case.

The chemically expected number of hydrogen bonding sites for water molecules is one for each NH group and one or two for each CO group.

However, CO groups near the terminal of the chain are attached to more than two water molecules. In many cases of the direction of CO to HOH tends to be close to the direction of sp^2 lone-pair orbital, and in a few cases it tends to be close to the direction between the orbitals.

Acknowledgement

We are grateful to Dorothy Hodgkin for helpful advice at the early stage of the refinement. We are thankful to Dr G. G. Dodson and Professor T. L. Blundell for commenting on our manuscript and correcting the language. This work was partially supported by the Yamada Science Foundation. The computation was carried out at the Computer Center of Institute for Molecular Science and at the Computer Center of Nagoya University.

References

ADAMS, M. J., BLUNDELL, T. L., DODSON, E. J., DODSON, G. G., VIJAYAN, M., BAKER, E. N., HARDING, M. M., HODGKIN, D. C., RIMMER, B., and SHEAT, S. (1969). *Nature, Lond.* **224**, 491.

BAUR, W. (1972). *Acta crystallogr.* **B28**, 1456.

BLUNDELL, T., DODSON, G., HODGKIN, D., and MERCOLA, D. (1972). *Adv. Protein Chem.* **26**, 279.

BROWN, I. D., (1976). *Acta crystallogr.* **A23**, 24.

DODSON, E. J., ISAACS, N. W., and ROLLET, J. S. (1976). *Acta crystallogr.* **A32**, 311.

—— HARDING, M. M., HODGKIN, D. C., and ROSSMANN, M. G. (1966). *J. molec. Biol.* **16**, 227.

FULLER, W. (1959). *J. chem. Phys.* **63**, 1705.

ISAACS, N. and AGARWAL, R. C. (1978). *Acta crystallogr.* **A34**, 782.

KATAYAMA, C., SAKABE, N., and SAKABE, K. (1972). *Acta crystallogr.* **A28**, 293.

LENHERT, P. G. (1968). *Proc. R. Soc.* **A303**, 45.

LEVITT, M. (1974). *J. molec. Biol.* **82**, 393.

Peking Insulin Structure Research Group (1971). *Peking Rev.* **40**, 11.

—— (1974). *Scientia sin.* **17**, 752.

RYLE, A. P., SANGER, F., SMITH, L. F., and KITAI, R. (1955). *Biochem. J.* **60**, 541.

SAKABE, K., SAKABE, N., and SASAKI, K. (1980). In *Water and metal cations in biological systems* (ed. B. Pullman and K. Yagi) p. 117. Japan Scientific Societies Press, Tokyo.

SAKABE, N., SAKABE, K., and KATAYAMA, C. (1972). *Acta crystallogr.* **S34**.

—————— and SASAKI, K. (1978). *Proc. Symp. Proinsulin, Insulin and C-Peptide*, p. 73. (Baba, T. Kaneko, and N. Yanaihara) p. 73. Excerpta Medica. Amsterdam.

SCHOENBORN, B. P. (1971). *Cold Spring Harb. Symp. quant. Biol.* **XXXVI**, 569.

TANFORD, C. and EPSTEIN, J. (1954a). *J. Am. chem. Soc.* **76**, 2163.

—————— (1954b). *J. Am. chem. Soc.* **76**, 2170.

VENKATACHALAM, C. M. (1968). *Biopolymers* **6**, 1425.

WADA, A. (1976). *Adv. Biophys.* **9**, 1.

42. Similarities and differences in the crystal structures of insulin†

J. F. CUTFIELD, S. M. CUTFIELD,
E. J. DODSON, G. G. DODSON
C. D. REYNOLDS, AND D. VALLELY

THE motives for determining the crystal structure of insulin are obvious enough; the interactions and folding of the two chains we hoped would add to the picture of protein design and organization; the character of the molecule's surfaces we hoped would explain the hormone's solution properties and chemical behaviour. And from considerations of sequence and the biological potency of modified insulins some picture of the region which binds to the receptor and is responsible for the expression of biological activity might emerge. All these hopes which had sustained the research for 34 years were realized when, in 1969, the crystal structure of 2Zn insulin was determined at 2.8 Å resolution in Dorothy Hodgkin's laboratory (Adams *et al.* 1969; Blundell *et al.* 1971).

The molecule in the crystal was arranged as three equivalent dimers each co-ordinating through the B10 histidines to two zinc ions positioned on the threefold axis and separated by about 16 Å (Fig. 42.1). Earlier studies with the rotation and translation function had demonstrated the two molecules in the dimer were related by a local twofold axis. This axis, within close limits, intersected and was perpendicular to the threefold crystal axis (Dodson,

† The Chinese and Oxford conventions for naming the two independent molecules in the 2Zn insulin asymmetric unit are reversed—evidence of independence in the two studies! The two molecules were distinguished structurally by the position of the B25 phenylalanine side-chain. In the Oxford convention molecule I, the side-chain lies across the local twofold axis, while in molecule II it turns back and lies against the A19 tyrosine of the same molecule. Viewed along the local axis towards the hexamer centre, molecule I appears on the right. In the Chinese analysis the alternative indexing was chosen which has the effect of rotating (by 180°) the unit cell about an axis perpendicular to the threefold axis. The molecules on the right of the local twofold axis viewed towards the origin is the Chinese molecule I but the Oxford molecule II.

Because it is molecule II in the Oxford convention that does not change in 4Zn insulin, the related rhombohedral form, and because hagfish insulin has the molecule II (Oxford) structure, the Chinese convention (which calls it molecule I) has proved to be the more appropriate in some ways. The Oxford convention is also used by Sakabe *et al.* (Chapter 41).

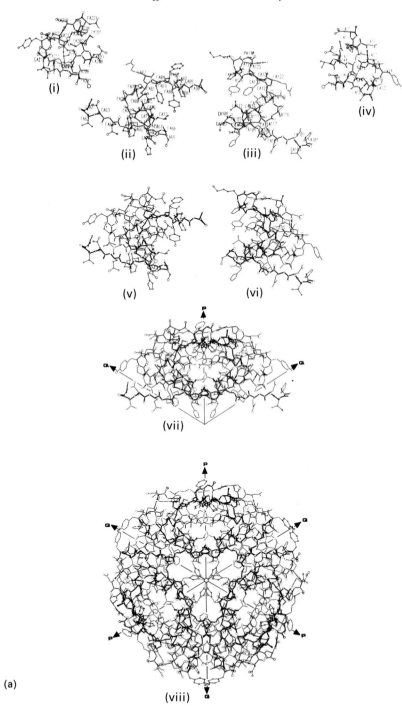

(i)

(ii)

(iii)

(iv)

(v)

(vi)

(vii)

(viii)

(a)

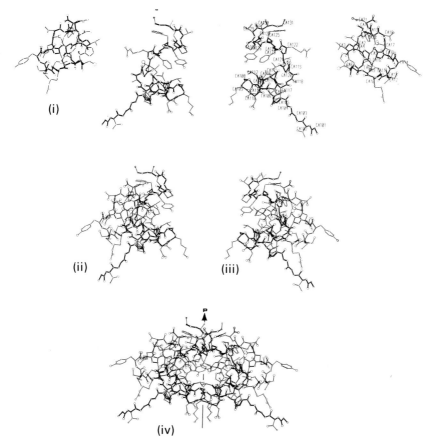

FIG. 42.1. (a) The assembly of the 2Zn pig insulin hexamer from its components, viewed down the threefold axis. (i) and (ii) A-chain and B-chain, molecule I, (iii) and (iv) A-chain and B-chain, molecule II, (v) and (vi) the monomers, molecules I and II, (vii) the dimer, (viii) the hexamer.

(b) The assembly of the hagfish insulin dimer viewed in an equivalent direction to 2Zn insulin in (a). (i) A-chain, (ii) B-chain, (iii) the monomer, and (iv) the dimer.

Harding, Hodgkin, and Rossmann 1966). The electron density, however, showed that while the two independent molecules in the asymmetric unit were closely related by the local axis of twofold symmetry they were distinctly different in their conformation at the A-chain N-terminus, at the B5 histidine and at B25 phenylalanine. We presumed that this was caused by either crystal contacts which were not twofold related or the packing requirements for the subunit organization in the hexamer (or dimer). However, the limited detail in the 2.8 Å map prevented us from tracing the contacts which were responsible for the structural differences.

To resolve these and other features of the structures the phases were extended to 1.9 Å, giving a much improved electron-density map (Hodgkin 1974). In this map, too, there were considerable defects and consequently we decided to extend the X-ray-diffraction data to 1.5 Å spacing and refine the structure crystallographically. These studies have provided us with co-ordinates whose accuracy ranges between 0.03 and 0.50 Å and have explained, too, how the structural differences at the A-chain N-terminus were produced (Dodson, Dodson, Hodgkin, and Reynolds 1980).

The Japanese 2Zn insulin study on data measured out to 1.2 Å spacing and with crystals maintained at 4 °C has extended the accuracy and defects, such as hydrogen positions on the structural picture. This work is discussed in Chapter 41.

While the 2Zn insulin crystal analysis was being continued other insulin crystal forms were studied. The most complete determinations so far are those on the related rhombohedral crystal, 4Zn insulin and hagfish insulin. Like 2Zn insulin the 4Zn insulin crystal unit cell contains a hexamer composed of three symmetry-related dimers. Its unit cell dimensions are a little different to that of 2Zn insulin, but its X-ray diffraction is markedly so. From the preliminary work with the rotation and translation function it was evident that a local axis related the molecules in the dimer, but that this local axis in contrast to 2Zn insulin was not perpendicular to and did not intersect the crystal threefold axis (Dodson *et al.* 1966). An isomorphously phased electron-density map was calculated in 1976 (Bentley, Dodson, Dodson, Hodgkin, and Mercola 1976) on terms extending to 2.8 Å spacing. This gave a clear description of the insulin molecules' folding and showed that one of the insulin molecules (molecule I) had undergone extensive rearrangement although the general organization of the hexamer was preserved (Fig. 42.2). In addition the zinc co-ordination by molecule I was now quite different. Both B_5 and B_{10} histidine were now involved in an off axial tetrahedral arrangement.

We were surprised by the changes in this insulin molecule, particularly since within each dimer the other molecule was not significantly different to its equivalent in 2Zn insulin. This seemed to throw a complicating light on the idea that sequence defines structure; for some proteins apparently the sequence can define more than one structure. The structural differences in the two molecules within the dimer reduced the local symmetry. In addition the direction of the local axis had moved (just as had been predicted). In order to describe the new insulin structure in more detail X-ray data were extended to 1.5 Å spacing and the atomic positions refined by fast Fourier

FIG. 42.2. The assembly of the 4Zn pig insulin hexamer from its constituent parts projected down the threefold axis. (i) and (ii) A-chain and B-chain, molecule I, (iii) and (iv) A-chain and B-chain, molecule II, (v) and (vi) the monomers, molecules I and II, (vii) the dimer, (viii) the hexamer.

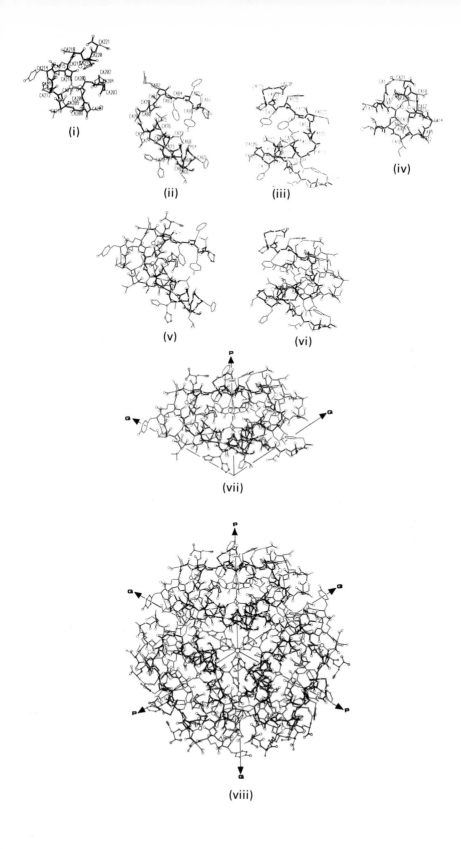

least squares and difference Fourier calculations. This refinement is now practically completed, only two or three residues remain poorly defined, and most of the better-order water molecules have been located.

A quite different picture comes from hagfish insulin, the most primitive insulin that has so far been sequenced. This insulin forms dimers but not hexamers and does not bind zinc. The dimers in this crystal contain the crystallographic twofold axis of symmetry and the asymmetric unit contains only one molecule (Cutfield, Cutfield, Dodson, Dodson, and Sabesan 1974; Peterson, Steiner, Emdin, and Falkmer 1975). The initial isomorphously phased map was calculated on data extending to 3.1 Å spacing only (Cutfield, Cutfield, Dodson, Dodson, Emdin, and Reynolds 1979). In spite of the more limited resolution the molecules' folding could be easily determined. Following this analysis the data were extended to 1.8 Å and refinement calculations begun. The data were limited by the small crystal size—beyond 1.9 Å they are very weak and this has made the refinement more difficult. While the accuracy of the atomic positions in this insulin is less than for the pig 2Zn and 4Zn rhombohedral forms it is sufficient to allow useful comparisons. These have been carried out between the structure in the two pig insulin rhombohedral crystals and the tetragonal hagfish insulin crystals.

The comparison of the insulin structures

The crystal data and refinement details for the hagfish, 2Zn pig insulin, and 4Zn pig insulin are given in Table 42.1 from where it can be seen there are five independent insulin molecules whose structures can be compared. In this paper we shall examine first the folding and conformation within the individual chains. Secondly we shall compare some interior non-polar contacts

TABLE 42.1

Crystal and refinement details

Insulin crystal	Space group	Molecules in asymmetric unit	Subunit organization	Resolution in Å	No. of terms	Agreement† factor	Range of‡ in atomic positions e.s.d.s
Hagfish	$P4_12_12$	1	Dimer	1.9	4093	25.2	0.1 –1.0 Å
2Zn pig	R_3	2	Hexamer	1.5	13 668	17.9	0.03–1.0 Å
4Zn pig	R_3	2	Hexamer	1.5	14 755	19.0	0.03–1.0 Å

† $R = \dfrac{\Sigma |F_o| - |F_c|}{\Sigma |F_o|}$ · all observed terms.

‡ The lower limit applies to the well-defined atoms, the upper to the least well defined, mostly solvent molecules.

and finally the structural relations between the surfaces involved in subunit contacts and interactions.

An overall impression of the dimers that occur in each of the three crystal forms is given in Fig. 42.3. The view is in the direction of the twofold axis. There is a gradation in the twofold symmetry within the dimer from the hagfish where it is exact, to the 2Zn pig insulin where it is mostly well obeyed to 4Zn pig insulin where it is obeyed only in the envelope about the twofold local axis.

Insulin Dimers viewed down the 2 fold axis direction

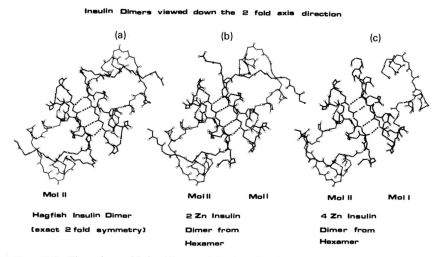

(a) (b) (c)

Mol II   Mol II   Mol I   Mol II   Mol I

Hagfish Insulin Dimer   2 Zn Insulin   4 Zn Insulin

(exact 2 fold symmetry)   Dimer from   Dimer from

Hexamer   Hexamer

FIG. 42.3. The polypeptide backbones of the three insulin dimers viewed down the local twofold axis: (a) hagfish insulin, (b) 2Zn pig insulin, (c) 4Zn pig insulin.

All the comparisons illustrated are, unless otherwise stated, based on overlapping the main chain atoms from the α-helical region B_9-B_{19} for the B-chain and from A_7-A_{21} for the A-chain. The overlap was optimized by least squares using a programme devised by Dr David Smith.

The A-chain

The A-chain, of 21 amino acids, consists in all five structures of two roughly antiparallel helical structures, one at the N-terminus and the other at the C-terminus. Figure 42.4 illustrates the overlapped A main chain structures viewed down the threefold crystal axis. This organization brings the N and C terminal residues together to form a contiguous surface, stabilized in part by the van der Waals contacts between A_2 isoleucine and A_{19} tyrosine sidechains. The helix extending from A_1 to A_8 contains mainly α-helical interactions. In 2Zn insulin this structure is different in the two molecules (Fig.

42.4(b)). The changes appear to arise from the crystal packing forces; constraining the two B_5 histidines to lie along the local axis between hexamers and thus violating the symmetry in their contacts with carbonyl oxygens at the final turn of the N-terminal helix. This apparently generates in molecule

Overlap of A Chains

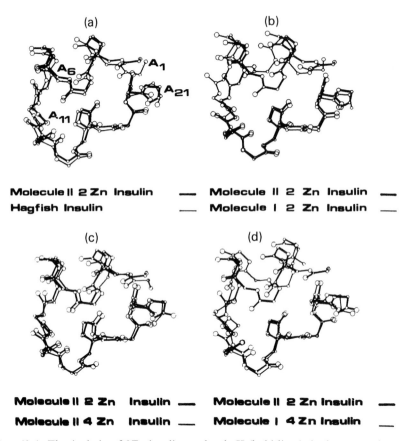

FIG. 42.4. The A-chain of 2Zn insulin, molecule II (bold lines), is shown overlapped on to the A-chains of (a) hagfish insulin, (b) 2Zn insulin, molecule I, (c) 4Zn insulin, molecule II, and (d) 4Zn insulin, molecule I.

I another set of interactions produced by 30° rotation about the A5—A6 peptide bond (Dodson *et al.* 1980).

There is a general similarity in the conformation of the A-chain N-terminal helix in the two molecules in the 4Zn insulin dimer, even though they have undergone a relative displacement of 1–2Å, and the two molecules are so considerably different in other regions (Fig. 42.4). The similarity in conformation at the A-chain N-terminus can be explained by the rearrangement at B_1–B_5 (molecule I) removing the B_5 histidine molecule I from close contact in the crystal with B_5 histidine molecule II. This relieves 4Zn insulin of the interactions made in 2Zn insulin molecule I at the A-chain N-terminal helix (Fig. 42.5). The N-terminal A_1 glycine in molecule I 4Zn insulin is poorly defined and its atomic positions are not yet satisfactorily determined.

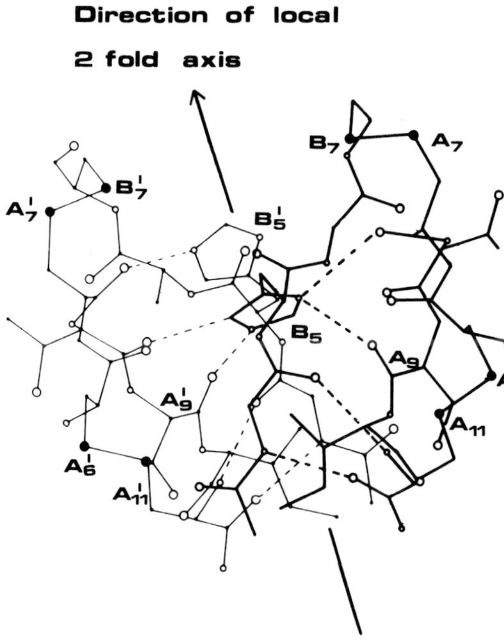

FIG. 42.5. Contacts between the A- and B-chains of molecules I (thin lines) and II (thick lines) for 2Zn insulin, viewed along the threefold axis.

The closest agreements shown by the A-chain N-terminal structures are clearly those between hagfish insulin and molecule II 2Zn and molecule II 4Zn insulin.

The B-chain

The B-chain main chain structure for hagfish insulin and 2 and 4Zn insulin is shown in Fig. 42.6. All five of the insulin molecules contain a well-defined and structurally constant α-helix from B_9 to B_{20}. In the case of 4Zn insulin molecule I, the helix is seen to begin at B_2 and continues to B_{20}. The N-terminal phenylalanine B_1 appears to be disordered and no reliable atomic

Overlap of B Chains

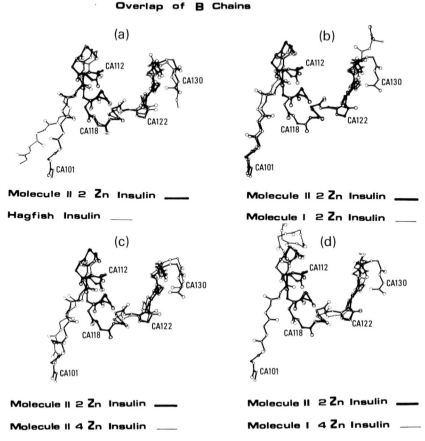

FIG. 42.6. The B-chain of 2Zn insulin, molecule II (bold lines), is shown overlapped on to the B-chains of (a) hagfish insulin, (b) 2Zn insulin, molecule I, (c) 4Zn insulin, molecule II, and (d) 4Zn insulin, molecule I.

positions have been determined for this residue yet. In the other molecule the six B-chain N-terminal residues are in an extended conformation, where, in spite of markedly different sub-unit and crystal contacts they retain a rather similar structure to that seen in 2Zn insulin.

The C-terminal residues are extended in all five molecules. Their direction is defined by a β-turn involving residues B_{20}–B_{23}. Like the α-helical structure this is a well-preserved interaction. It helps to direct the spatial arrangement of the structurally important helix at B_9–B_{19} and the C-terminal residues B_{24}–B_{26}.

There are considerable differences at the two C-terminal B-chain residues B_{29} and B_{30} in molecule I of both 2 and 4Zn insulin. These residues are poorly defined and B_{30} in both 2 and 4Zn insulin appears to be disordered. In the

2Zn insulin crystal the B_{29} and B_{30} (molecule I) make contacts with A_4 glutamic acid (molecule II) and with B_{29} and B_{30} (molecule II). Probably these interactions affect the conformations B_{29} and B_{30} assume in molecule I; the equivalent B_{29} and B_{30} in molecule II appear less affected since their conformation is close to that found in the hagfish insulin crystal where there are quite different contacts.

The non-polar core

The packing of the residues buried in the molecules' interior play an important part in defining the overall chain folding. Comparison of the buried residues, illustrated in Fig. 42.7, shows that the B-chain buried residues have similar conformation in all five molecules, reflecting the general stability of the region between B_9–B_{26}. The A-chain buried residues in the pig insulin are more variable. The changes at the A-chain N-terminal helix seem to affect A_2 isoleucine and A_{19} tyrosine in 2Zn insulin. In 4Zn insulin where the A-chain N-terminal helix is much more similar, the A_2 isoleucine has the same conformation in both molecules and A_{19} tyrosine is much less altered in position. There are in 4Zn insulin, as in 2Zn insulin, substantial changes between molecules I and II at the disulphide A_6–A_{11} associated in 2Zn insulin with the rotation about A_5–A_6 peptide and in 4Zn insulin with the rearrangement of B_7–A_7 following the repositioning of the B-chain N-terminal residues. The leucines B_{11}, A_{13}, and A_{16} show considerable changes between 2Zn insulin and 4Zn insulin and hagfish insulin. These reflect the altered positions and subunit contacts made by A_{13} leucine and A_{14} tyrosine in these two crystals, already discussed.

There are no significant discrepancies between hagfish insulin and pig insulin (molecule II). In particular, the cystines and A_{19} tyrosine are closely similar. The largest differences occur at leucines A_{13} and A_{16}, and these as in 2 and 4Zn insulin are presumably caused by the different sequence and interactions made by the surface residues (A_{13} leucine and A_{14} tyrosine) in those two molecules. The side-chain A_2 isoleucine is rather poorly defined in hagfish insulin molecules and some atomic positions are not shown. Interestingly this side-chain in 2Zn insulin molecule I proved difficult to position and its thermal parameters still remain a little high (E. J. Dodson, G. G. Dodson, D. C. Hodgkin, and N. W. Isaacs, unpublished work).

The dimer-forming residues

The residues involved in dimer formation in hagfish insulin, 2Zn and 4Zn pig insulin are shown in Figs. 42.1 and 42.2. In Fig. 42.8 these residues are each overlapped on to the 2Zn insulin molecule II dimer forming surface. Apart from B_{25} phenylalanine side-chain these structures are closely similar.

Overlap of Internal Residues

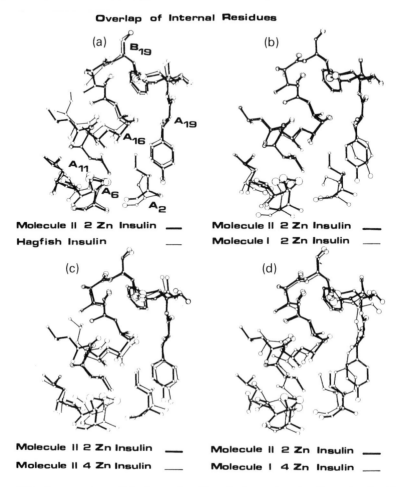

FIG. 42.7. A comparison of the internal residues in the various insulin molecules. The residues from 2Zn insulin, molecule II (bold lines), are shown overlapped on to the corresponding residues from (a) hagfish insulin, (b) 2Zn insulin, molecule I, (c) 4Zn insulin, molecule II, and (d) 4Zn insulin, molecule I.

The r.m.s. differences between these various structures range from 0.11 to 0.25 Å for the backbone atoms and 0.2 to 0.4 Å for the side-chain atoms. Hence these residues and the contacts they make within the dimer are a remarkably constant feature of the insulin molecule, preserved even when very considerable changes occur elsewhere in the molecule. Whether in the free monomer these residues are unaffected by the changes elsewhere in insulin structure, or whether they are restored to this conformation on dimer formation, cannot be decided directly. The similar values reported for the dimeriza-

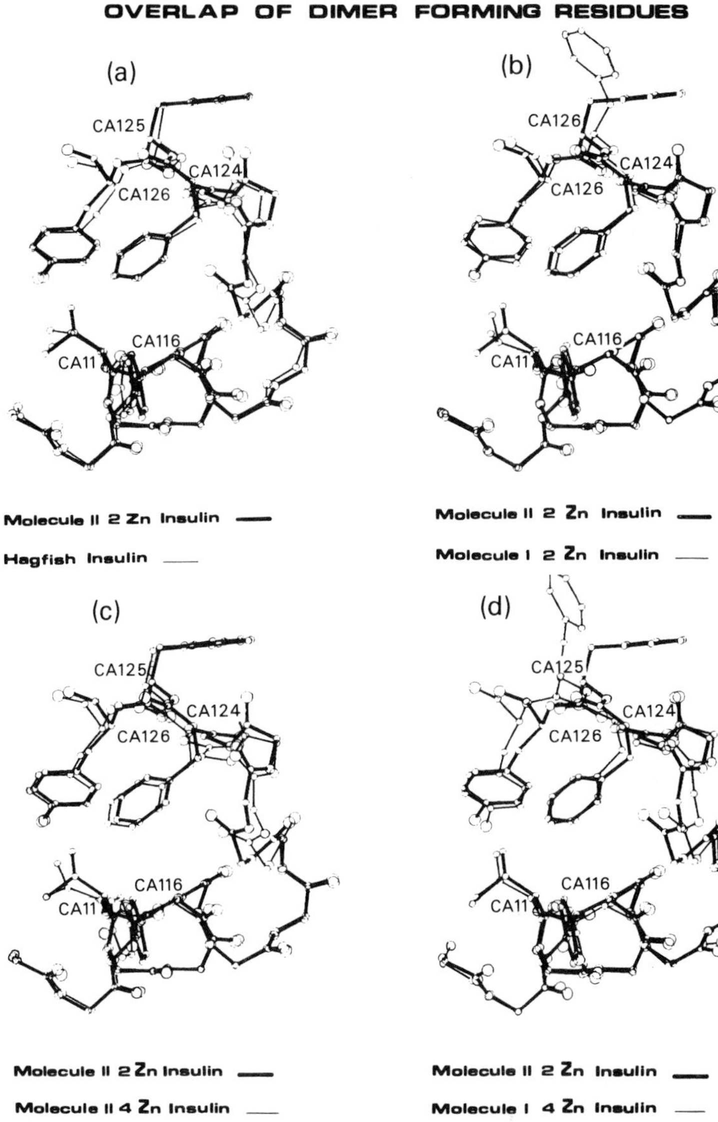

OVERLAP OF DIMER FORMING RESIDUES

(a)

CA125
CA124
CA126

CA116
CA11

Molecule II 2 Zn Insulin ——

Hagfish Insulin ——

(b)

CA126
CA124
CA126

CA116
CA11

Molecule II 2 Zn Insulin ——

Molecule I 2 Zn Insulin ——

(c)

CA125
CA124
CA126

CA116
CA11

Molecule II 2 Zn Insulin ——

Molecule II 4 Zn Insulin ——

(d)

CA125
CA124
CA126

CA116
CA11

Molecule II 2 Zn Insulin ——

Molecule I 4 Zn Insulin ——

FIG. 42.8. Overlaps of the side-chains involved in dimer formation. The positions of B_{12}, B_{16}, B_{24}, B_{25}, and B_{26} in 2Zn insulin, molecule II (bold lines) are compared with those in (a) hagfish insulin, (b) 2Zn insulin, molecule I, (c) 4Zn insulin, molecule II, and (d) 4Zn insulin, molecule I.

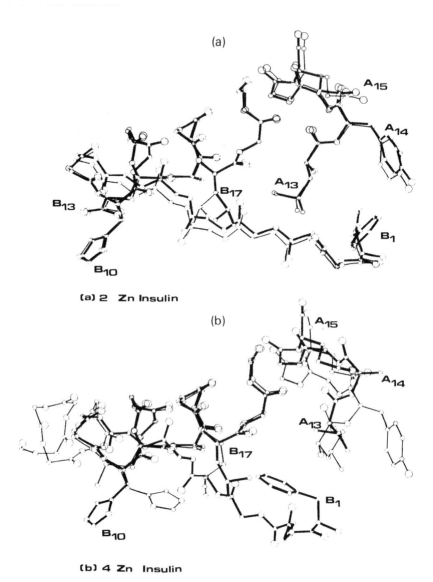

(a)

A15

A14

B17

A13

B13

B1

B10

(a) 2 Zn Insulin

(b)

A15

A14

A13

B17

B1

B10

(b) 4 Zn Insulin

FIG. 42.9. Comparison of the side-chain positions of the residues in molecule I (thin lines) and II (bold lines) involved in contacts between dimers within the hexamer for (a) 2Zn and (b) 4Zn insulin.

tion constant with hagfish, 2 and 4Zn insulin suggest, however, these surfaces are little altered in these different conditions.

The hexamer-forming residues

A comparison of the residues which come into contact upon the hexamer's assembly can only include the 2 and 4Zn insulin structures. These are illustrated in Fig. 42.9. In 2Zn insulin these residues also obey the local axis well—except that the side-chain B_{17} leucine finds an alternative arrangement by rotating 180° about the CA—CB bond (Fig. 42.9). The local symmetry in 4Zn insulin applies only to the B-chain helical residues where (as in 2Zn insulin) the B_{17} leucine side-chain appears in two alternative arrangements. The N-terminal residues B_1 phenylalanine molecule II have no equivalent owing to the repositioning of B_1–B_8 in molecule I. The A-chain structures in this region correspond poorly. There is a general movement in the C-terminal helix of 1–2 Å as well as other conformational changes, already noted. These movements can be seen as adjustments by the more flexible A-chain to the removal of the molecule I B-chain N-terminal residues from the dimer–dimer contacts (Fig. 42.2).

The zinc co-ordination

The zinc co-ordination by the B_{10} histidines is the driving force for the assembly of the hexamer. In 2Zn insulin the two zincs are axial and octahedrally co-ordinated by the three symmetry related B_{10} histidines and three waters in a geometrically similar scheme. In the 4Zn insulin crystals the zinc co-ordination by molecule I proves to be more complicated than we first thought (Bentley *et al.* 1976). We now find the B_{10} histidine in molecule I is involved in two co-ordination arrangements—as is illustrated in Fig. 42.10. There is co-ordination to an axial zinc ion (as in 2Zn insulin molecule I) but with the difference that in 4Zn insulin the co-ordination is tetrahedral, not octahedral. This is because there is not space for off-axial water molecules owing to the nearby B_6 leucine side-chain. There is also the tetrahedral co-ordination to a second zinc ion in a general position as was originally described from the 2.8 Å isomorphously phased map. The zinc ion here is co-ordinated to B_5 and B_{10} histidines and to two water molecules. The other axial zinc ion which is linked to the three symmetry-related molecules II has octahedral co-ordination geometry.

Although there are two zinc co-ordination schemes in molecule I, about equally present, there is no evidence that the protein-folding in this molecule is disordered. The changes in protein structure and subunit contacts are therefore not directed by the zinc co-ordination.

2 Zn Insulin – Zn Coordination Molecule I

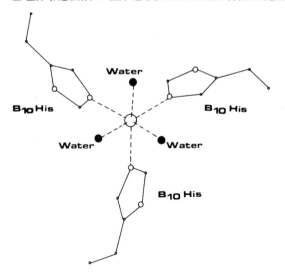

4 Zn Insulin – Zn Coordination Molecule I

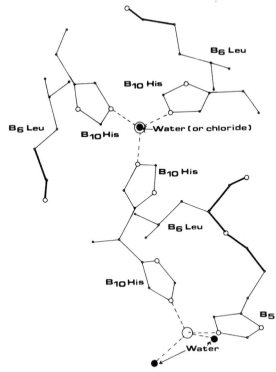

FIG. 42.10. Zinc co-ordination for molecule I in (a) 2Zn and (b) 4Zn insulin, viewed down the threefold crystal axis.

Discussion

The comparison between these three insulin crystals has revealed that, within each of the dimers that constitute the asymmetric units in 2 and 4Zn insulin, one molecule (molecule II) has essentially the same structure. And that structure is essentially the same as that found in the symmetrical hagfish insulin dimer. The similarity is most clearly defined by the conformations at the A-chain N-terminus, the B_5 histidine and the B_{25} phenylalanine side-chain, illustrated in Fig. 42.11. Furthermore the similarity between those three molecules extends to the other structural features; we find less relative movement between chains, closer preservation of their secondary structure, and less movement among the buried residues and the residues involved in assembly. Only at the B-chain N-terminus are there significant differences which are produced by their very different subunit interactions.

The presence of the hagfish insulin structure in molecule II in 2 and 4Zn pig insulin is a striking example of how precisely a protein's internal interactions can determine a three-dimensional structure, notwithstanding the considerable differences in sequence (19 in 51) and aggregation state. It is interesting that this structure has persisted while its companion in the dimer has changed. As yet we do not know whether those varying structures have any particular significance for insulin's structure as a monomer in solution. We have noted that hagfish insulin and the molecule II in 2 and 4Zn insulin make more intramolecular contacts than the molecule I structure. Thus B_{25} phenylalanine is folded back to make van der Waals contacts to A_{19} tyrosine and the B-chain C-terminal carboxylic acid is involved in a network of balancing charges and H bonds with the A-chain N-terminus within the same molecule. It seems more likely therefore that molecule II will be closer to the monomeric structure than molecule I.

The most constant feature of the insulin molecules we have studied is the structure and contacts of the dimer-forming residues. This is associated with the preservation of the structure between B_9 and B_{28} on which the dimer-forming residues are embedded, as illustrated in Figs. 42.1, 42.2, 42.6, and 42.8. The insulin molecule therefore evidently contains structurally stable regions and less stable regions which following certain interactions will assume distinctly different conformations. The transformation from 2 to 4Zn insulin, and vice versa, can be brought about within the crystal simply by raising or lowering the chloride-ion concentration (Bentley *et al.* 1978). Such an extensive rearrangement implies a complex pathway between the two structures and therefore a considerable flexibility in the molecule.

Although many proteins appear to have a constant structure with only small changes occurring between crystal forms or during their function the changes observed in insulin's folding are not unique and differential movement on this scale is clearly an important functional property in some

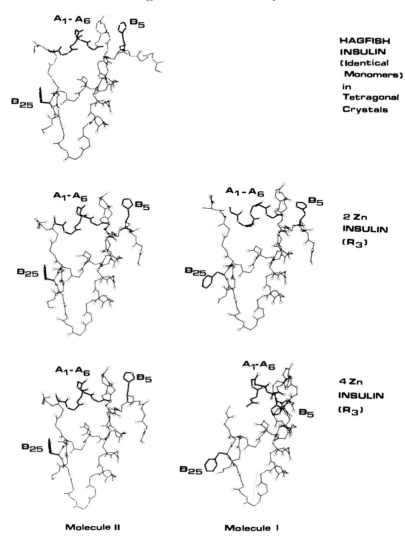

Molecule II   Molecule I

FIG. 42.11. Comparison of the five independent insulin molecules viewed perpendicular to the twofold axis showing the arrangement of the N-terminus of the A-chain (A_1–A_6), B_5 histidine and B_{25} phenylalanine.

proteins. Large structural changes have been observed both in monomeric proteins and in oligomeric proteins where the subunit structure itself can also be affected.

Haemoglobin, which is a tetrameric molecule, is probably the best described example of a protein whose function depends on substantial conformational changes (Perutz 1970; Baldwin 1975). There is also convincing crystallographic evidence of substantial movement in lactic acid dehydrogenase where the repositioning of the residues 100–120 seems to be associated with the enzyme's catalytic mechanism (Rossmann *et al.* 1971). Hexokinase, which is a large enzyme, undergoes extensive structural changes when, as a dimer, it binds its substrate glucose (Fletterick, Bates, and Steitz 1975). Another kinase, adenylate kinase, has been crystallized in three crystal forms, two of which can be interconverted in the crystalline state (as we find is also possible for 2 and 4Zn insulin). The differences between the two adenylate kinase structures are rather intricate and range between 2 and 6 Å (Sachsenheimer and Schulz 1977).

Insulin differs from these, and other proteins (Banks *et al.* 1979) where structured changes have been detected in two ways. First the structural changes involve a considerable proportion of the molecule and are extensive—more than 20 Å at the B-chain N-terminus. Secondly the altered molecules co-exist in the 2 and 4Zn insulin hexamer with unaltered partners. We have often wondered whether a symmetrical hexamer exists; there is some evidence from nuclear magnetic resonance studies (R. J. P. Williams and K. Williamson, private communication) that they can in solution. Recently a new variety of rhombohedral crystals has been obtained at elevated temperatures (see Chapter 43). The structure of insulin in these crystals is clearly our next step.

Insulin's structural behaviour has given us some surprises and it is likely we shall have more as the studies continue. The value that detailed structural knowledge has for chemistry and biochemistry is a major lesson from Dorothy Hodgkin's own research on sterols, penicillin, vitamin B_{12}, and, of course, insulin. We like to think that in studying insulin's structural behaviour we are keeping to these traditions.

References

ADAMS, M. J., BLUNDELL, T. L., DODSON, E. J., DODSON, G. G., VIJAYAN, M., BAKER, E. N., HARDING, M. M., HODGKIN, D. C., RIMMER, B., and SHEAT, S. (1969). *Nature, Lond.* **224**, 491–5.

BALDWIN, J. M. (1975). *Prog. Biophys. molec. Biol.* **29**, 225–320.

BANKS, R. D., BLAKE, C. C. F., EVANS, P. R., HASER, R., RICE, D. W., HARDY, G. W., MERRETT, M., and PHILIPS, A. W. (1979). *Nature, Lond.* **279**, 773–7.

BENTLEY, G. A., DODSON, E. J., DODSON, G. G., HODGKIN, D. C., and MERCOLA, D. A. (1976). *Nature, Lond.* **261**, 166–8.

BLUNDELL, T. L., CUTFIELD, J. F., CUTFIELD, S. M., DODSON, E. J., DODSON, G. G., HODGKIN, D. C., MERCOLA, D. A., and VIJAYAN, M. (1971). *Nature, Lond.* **231**, 506–511.

CUTFIELD, J. F., CUTFIELD, S. M., DODSON, E. J., DODSON, G. G., and SABESAN, M. (1974). *J. molec. Biol.* **87**, 23–30.

——————EMDIN, S. O., and REYNOLDS, C. D. (1979). *J. molec. Biol.* **132**, 85–100.

DODSON, E. J., DODSON, G. G., and HODGKIN, D. C. (1980). In *Frontiers of bioinorganic chemistry and molecular biology* (ed. A. N. Ananchenko) pp. 145–50. Pergamon Press, Oxford.

—————— and REYNOLDS, C. D. (1979). *Can. J. Biochem.* **57,** 469–79.

—— HARDING, M. M., HODGKIN, D. C., and ROSSMANN, M. G. (1966). *J. molec. Biol.* **16,** 227–34.

FLETTERICK, R. J., BATES, D. J., and STEITZ, T. A. (1975). *Proc. natn. Acad. Sci. U.S.A.* **72,** 38–42.

HODGKIN, D. C. (1974). *Proc. R. Soc.* **338,** 251–75.

PETERSON, J. D., STEINER, D. F., EMDIN, S. O., and FALKMER, S. (1975). *J. biol. Chem.* **250,** 5183–91.

PERUTZ, M. F. (1970). *Nature, Lond.* **228,** 726–39.

ROSSMANN, M. G., ADAMS, M. J., BUCHNER, M., FORD, G. C., HACKERT, M. L., LEUTZ, P. J., MCPHERSON, A., SCHEVITZ, R. W., and SMILEY, I. E. (1971). *Cold Spring Harb. Symp. quant. Biol.* **XXXVI,** 179–91.

SACHSENHEIMER, W. and SCHULTZ, G. E. (1977). *J. molec. Biol.* **114,** 23–36.

43. Effects of destabilizing agents on the insulin hexamer structure

R. A. G. DE GRAAFF, A. LEWIT-BENTLEY, AND S. P. TOLLEY

THE capacity of insulin to assume different structures in different crystal forms has been studied in order to gain more insight into the conformational properties of insulin. The study in hand is concerned with the role of metal ions and the effects of various destabilizing agents on the insulin structure in the crystalline state. As the crystals are usually not destroyed during the transformations and continue to diffract quite well, it has been possible to characterize the end states.

The co-ordination of the metal ions in 2Zn insulin is octahedral in both the molecules of the asymmetric unit (molecules I and II) (Dodson, Dodson, Hodgkin, and Reynolds 1979). In 4Zn insulin the co-ordination in molecule II is octahedral but that in molecule I proves to be tetrahedral. The observation that 4Cu, 4Ni, and 4Cd insulin crystals have not been obtained indicates that the less specific co-ordination geometry of the zinc cation is important. This study has therefore included experiments to test the effects of these other transition metal cations on the transformation from a hexamer containing two metal ions to a hexamer containing four metal ions. The similar stabilities of zinc in the tetrahedral and octahedral states with oxygen and nitrogen ligands is possibly decisive in allowing transformations to occur between the different subunit structures.

The transformation from 2Zn to 4Zn insulin depends on the concentration and type of salt used in crystallizations. We examined the limits of NaCl concentration which produce 4Zn crystals, as well as the limiting concentrations of other ions of the Hofmeister lyotropic series.

Another object of these studies was to explore conditions under which new crystal forms with other folding patterns in the insulin molecule might possibly be produced. In these experiments crystallizations were carried out in the presence of phenol at various temperatures.

Experimental details and results

Crystallization experiments

Crystallization of Ni, Cu, and Cd insulins

Crystals of 2Ni, 2Cu, and 2Cd insulin were grown using the usual 2Zn insulin method (Schlichtkrull 1958; Harding, Hodgkin, Kennedy, O'Connor, and Weitzmann 1966), substituting nickel, copper, or cadmium acetate (0.15 M solution) for zinc acetate, as appropriate. Monocomponent pig zinc insulin (0.41 per cent zinc; NOVO Terapeutisk, Copenhagen) was used in all crystallizations of zinc insulin, while amorphous pig insulin (0.0086 per cent zinc; NOVO Terapeutisk, Copenhagen) was used in other metal insulin crystallizations. Crystals were grown in a pH range pH 6.2–6.4. Despite many attempts to grow 4Ni, 4Cu, and 4Cd insulin crystals by the 4Zn insulin method, crystals were never obtained.

Crystallization with salts of the Hofmeister series of ions

Crystallizations of pig zinc insulin with salts other than sodium chloride show that different ions can induce the 2Zn to 4Zn transformation. To study the effect quantitatively it was decided first to determine the minimum sodium chloride concentration required for 4Zn formation by crystallization.

TABLE 43.1

The concentration of sodium chloride required for 2Zn and 4Zn insulin crystals

Insulin concentration (%)	Maximum concentration of NaCl for 2Zn	Minimum concentration of NaCl for 4Zn
0.25	0.25 M	0.35 M
0.50	0.20 M	0.30 M
1.00	0.35 M	0.40 M

Knowing the 2→4Zn transition point with sodium chloride other salts were then examined. Destabilizing effects by these salts seem to follow the Hofmeister lyotropic series of ions.

Salts below sodium chloride in the series, for example sodium sulphate and sodium acetate, could not be used in crystallization experiments because they precipitate with zinc.

Sodium nitrate shows an anomaly here. According to its position in the Hofmeister series it should cause the 4Zn conformation at a lower concentration than that of sodium chloride. Repeated attempts to obtain 4Zn insulin crystals with a sodium nitrate concentration above 0.5 M have been unsuccessful.

TABLE 43.2

The concentrations of salts of the Hofmeister lyotropic series required for 2Zn and 4Zn insulin crystal formation

Salt	Maximum concentration for 2Zn	Minimum concentration for 4Zn
CsCl	0.1 M	Incomplete
CaCl$_2$	0.01 M	Precipitates at higher concentration
KCl	0.45 M	0.5 M
NaCl	0.35 M	0.4 M
NaNO$_3$	0.5 M	No crystals obtained above 0.5 M
NaI	0.055 M	0.1 M
K$_2$Pt(CN)$_4$	Incomplete	0.05 M
KSCN	0.02 M	0.03 M

All crystallizations with 1 per cent insulin.

Crystallization with phenol

(*a*) *Crystals grown at room temperature.* Crystallizations in the presence of other destabilizing agents, notably phenol, have been carried out under varied conditions. The crystal forms obtained in the presence of phenol include a monoclinic and at least two rhombohedral forms. All these modifications were grown using the same method varying only pH and temperature. The basic recipe is as follows:

50 mg monocomponent pig zinc insulin (NOVO Terapeutisk)
0.02 M hydrochloric acid (5.0 ml)
0.15 M zinc acetate (0.5 ml)
Sodium chloride (600 mg)
0.2 M trisodium citrate (2.5 ml)
5 per cent aq. phenol (2.0 ml)
pH 7.8–8.0

Monoclinic crystals were also obtained at a much lower pH (pH 6.7) when sodium chloride was omitted from the preparation. These crystals were basically very similar to the 'high chloride' type, though they show some intensity differences and are generally more susceptible to radiation damage.

The usual procedure followed in these crystallizations was to adjust to the desired pH (HCl/NaOH), warm the solution to 50 °C, filter using a millipore filter (Millipore Corporation, Bedford, Mass., USA) whilst still warm, and store in a dewar flask containing water at 50 °C packed with polystyrene and straw to allow slow cooling to room temperature. Although actual room temperature varied throughout the year, from 10 °C in winter to 25 °C in summer, it was found that the dewar minimized these differences and the usual temperature inside the dewar after cooling was 18 °C ($\pm$ 2 °C).

(b) *Crystallization of a new rhombohedral form by dialysis.* The room temperature (RT) rhombohedral form was grown by dialysis of an insulin solution against a buffer at lower pH. The solutions were made up as follows:

Solution A: 7.5 mg pig zinc insulin
90 mg NaCl
15 mg phenol
0.375 ml 0.04 M HCl
0.15 ml 0.72 M $ZnSO_4$
0.375 ml 0.2 M sodium citrate
0.6 ml H_2O
pH adjusted to 9.2, placed in dialysis sac

Solution B: 0.36 g NaCl
60 mg phenol
0.6 ml 0.72 M $ZnSO_4$
1.5 ml 0.2 M sodium citrate
3.9 ml H_2O
pH adjusted to 7.43

Crystals were obtained after dialysing solution A against solution B at room temperature for 4–10 days.

(c) *Crystals grown at elevated temperatures.* On some occasions the dewar was allowed to cool from 50 °C in an oven set at 30 °C. Some of these crystals, although grown from the same solutions that had previously produced the monoclinic form, proved to be rhombohedral and grew in the presence of the monoclinic crystals (Fig. 43.1).

Similarly crystals were also grown in an oven or water bath at 37 °C (corresponding to body temperature). Some of the crystals were monoclinic while others were rhombohedral; the pH was between 7.9 and 8.0. These were well-formed crystals but deteriorated rapidly when kept at a temperature below 37 °C, even when stored at 30 °C. After a few hours the tubes became full of precipitate and the crystals began to decompose or else became so fragile that they were impossible to use for X-ray diffraction studies. Provided that these crystals were mounted in capillaries and sealed immediately on removing them from the 37 °C oven they remained more stable. Some preliminary photographs have been taken showing that these crystals possess a threefold symmetry.

MP

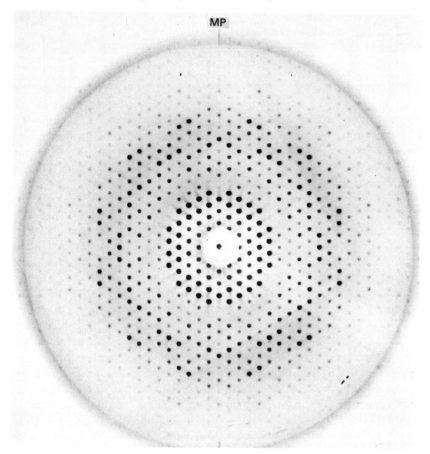

FIG. 43.1. The X-ray diffraction pattern (*hki*0) of a crystal of insulin grown using the monoclinic recipe (see pp. 549–50) but at a temperature of 30 °C. The crystal is rhombohedral, space group R32.
Note the presence of the mirror plane (MP).

Transformation experiments

$2 M^{2+} \rightarrow 4 M^{2+}$ transformations

As all our attempts to grow 4Cu, 4Ni, and 4Cd crystals had failed, experiments were carried out to try to obtain these $4 M^{2+}$ crystals by transformation from the $2 M^{2+}$ form as is possible with zinc insulin crystals.

Good crystals of 2Cu, 2Ni, and 2Cd insulin were taken and soaked in a 4Zn-like mother liquor containing:

0.02 M hydrochloric acid (0.5 ml)
0.15 M copper acetate (or nickel acetate or cadmium acetate (0.75 ml))
Sodium chloride (600 mg)

Acetone (1.5 ml)
0.2 M trisodium citrate (2.5 ml)
Water (4 ml)
pH 6.2–6.4

The crystals were left in the soaking solution for up to three days, and when removed and examined under the microscope all the crystals showed extensive cracking, the nickel and copper insulin crystals were opaque. The diffraction patterns showed the 2Ni and 2Cd crystals to be unchanged, the soaked 2Cu crystals did not diffract. On one occasion only, a 2Cu crystal transformed to give an $hk0$ pattern close to the 4Zn type but with a double c axial length in the $0kl$ projection.

Transformations involving phenol

Soaking experiments were carried out on high-temperature rhombohedral crystals (grown at 30 °C), to examine the effects of various mother liquor solutions. The solutions used were a 4Zn mother liquor, a 2Zn mother liquor, and a monoclinic mother liquor without phenol (equivalent to a 4Zn mother liquor without acetone). The crystals shattered when placed in the two former solutions, but interesting results were obtained on soaking in the phenol-free monoclinic solution. The crystals began to crack after 30 seconds in the soaking solution. After a further one or two minutes the cracks became extensive but the crystal retained its shape. On leaving the crystals in the soaking solution for five to ten minutes more, the cracks were seen to disappear and the overall appearance of the crystal was much improved, now showing only minor cracking. The crystals at this stage were still usable and diffracted reasonably well. One soaked crystal was photographed for 12 hours and gave a good diffraction pattern, which on close inspection seemed to be more like 2Zn than 4Zn (Fig. 43.2).

Discussion

The detailed knowledge now available of the structure of 2Zn insulin (Dodson *et al.* 1979) and 4Zn insulin (Bentley, Dodson, Dodson, Hodgkin, and Mercola 1976; Vallely, to be published) has given us some insight into the flexibility of the insulin structure in the crystal. The extensive rearrangements necessary for the transformation of 2→4Zn insulin (Bentley, Dodson, and Lewitova 1978) can be accomplished more or less reversibly within the crystal.

Similar transformations for 2Cu, 2Ni, and 2Cd could not be induced, an observation which is consistent with the failure to obtain the 4Cu, 4Ni, and 4Cd forms by crystallization. A possible explanation for the failure of the Cu, Ni, and Cd insulins to assume the 4Zn insulin folding is the reluctance of these ions to co-ordinate in a tetrahedral fashion. Zinc, on the other

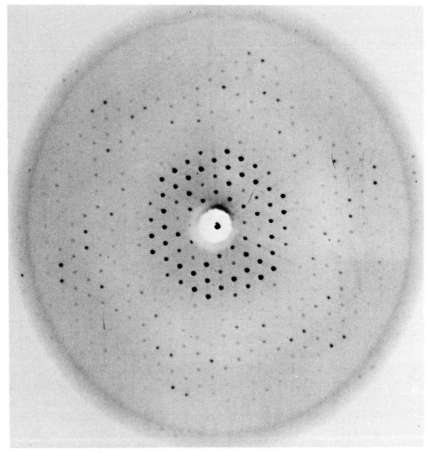

FIG. 43.2. The X-ray diffraction pattern (*hki*0) of the crystal in Fig. 43.1 after soaking in a monoclinic mother liquor without phenol, for 10 minutes. Note that the mirror plane is absent.

hand, with the appropriate ligands (oxygen and nitrogen), can assume a tetrahedral co-ordination and hence can be accommodated easily into the 4Zn rearrangement.

Study of Table 43.2 suggests that the 2→4Zn transformation follows the Hofmeister series of ions. This transformation is not confined to crystals, it has been shown to take place in solution by n.m.r. studies, and still follows the Hofmeister series (Williamson and Williams, in 1979).

An interesting thermodynamical model, based on the hypothesis that the effect is due to action of the ions on the empty volume of the solvent, has been suggested by Hey, Clough, and Taylor (1976). This action would lead to a modified solvent structure favouring a looser conformation of the macromolecule.

The addition of a neutral salt to proteins in solution has many effects (Von Hippel and Schleich 1969) including: helix to coil transformations (for various macro-molecules including synthetics and DNA); salting in–salting out (i.e. aggregation phenomena); modification of the rate of enzyme reactions where they require a conformational change of the enzyme (changing the hydration of groups involved in the conformational change); and affecting the 'melting temperature', T_m, of reversible conformational changes.

The effect goes:

$$SO_4^{2-} < CH_3COO^- < Cl^- < Br^- < NO_3^- < ClO_4^- < I^- < CNS^-$$

$$Cs^+ < K^+ < Na^+ < Ca^{2+}$$

Helix, salting out	⟷	Coil, salting in
higher T_m		lower T_m
2Zn		4Zn

It appears that 4Zn is a looser conformation than 2Zn and should have a larger surface area, usually meaning that it would be a less stable conformation. The 2→4Zn equilibrium may therefore be influenced by the thermodynamic state of the solvent, i.e. the ions act through the solvent structure and not solely by direct interaction with the protein. However, heavy ions (e.g. I^- and $Pt(CN)_4^{2-}$) do form isomorphous derivatives of 4Zn insulin, thus pointing to some specific ion interaction as well.

It is interesting that another destabilizing agent, phenol, leads to different crystal forms of insulin, a monoclinic form first studied by Harding *et al.* (1966) and two rhombohedral forms. The first of the rhombohedral crystal forms displays a doubling of the cell axes (RT rhombohedral form, see Table 43.3 and Figs. 43.3 and 43.4). The second is a new rhombohedral form with space group R32 suggesting that the asymmetric unit contains a monomer.

TABLE 43.3

Cell constants of some insulin crystals

	a	b	c	α	β	γ
Monoclinic	62.3 Å	61.8 Å	47.8 Å	90°	110°	90°
4Zn	80.7 Å	80.7 Å	37.6 Å	90°	90°	120°
2Zn	82.5 Å	82.5 Å	34.0 Å	90°	90°	120°
RT form	159.6 Å	159.5 Å	40.6 Å	90°	90°	120°
30 °C form	80.1 Å	80.1 Å	41.0 °A	90°	90°	120°
30 °C soaked	80.1 Å	80.1 Å	40.3 Å	90°	90°	120°

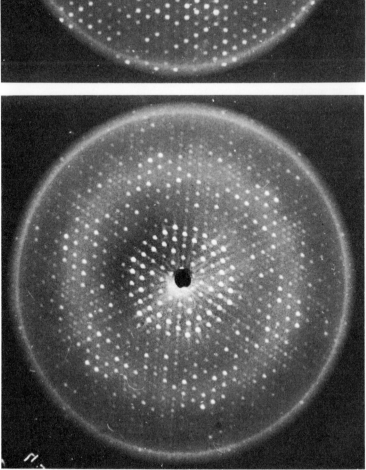

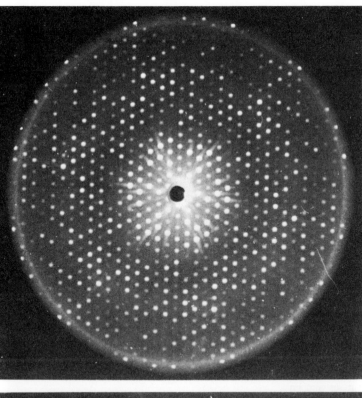

FIG. 43.3. The X-ray diffraction pattern (*hk*0) of a new rhombohedral form of insulin. This was grown by dialysis in the presence of phenol. Note the doubling of the cell axes.

FIG. 43.4. The X-ray diffraction pattern (*hk*0) of a 4-zinc insulin crystal.

We hope to study this form in more detail in the future especially since preliminary soaking experiments in phenol-free monoclinic solution have given a pattern more akin to 2Zn than 4Zn although the solution contained a high concentration of sodium chloride which would normally induce the 4Zn conformation. If we do find the 2Zn conformation it would support the theory that 2Zn is the more stable form of insulin in the crystalline state and the appearance of exact symmetry in a modification of insulin crystallizing under such similar conditions as the monoclinic form suggests that the symmetry of the hexamer in the monoclinic form is more exact than was hitherto supposed.

Acknowledgements

We are grateful for the financial help provided by the British Diabetic Association, Kroc Foundation, Medical Research Council, and the Science Research Council.

References

BENTLEY, G., DODSON, E. J., DODSON, G. G., HODGKIN, D. C., and MERCOLA, D. A. (1976). *Nature, Lond.* **261**, 166–8.
—— DODSON, G., and LEWITOVA, A. (1978). *J. molec. Biol.* **126**, 871–5.
DODSON, E. J., DODSON, G. G., HODGKIN, D. C., and REYNOLDS, C. D. (1979). *Can. J. Biochem.* **57**, 469–79.
HARDING, M. M., HODGKIN, D. C., KENNEDY, A. F., O'CONNOR, A., and WEITZMANN, P. (1966). *J. molec. Biol.* **16**, 212.
HEY, M. J., CLOUGH, J. M., and TAYLOR, D. J. (1976). *Nature, London.* **262**, 807–9.
HIPPEL, P. VON and SCHLEICH, T. (1969). In *Structure and stability of biological macromolecules* (ed. S. N. Timasheff and G. D. Fasman). Dekker, New York.
SCHLICHTKRULL, J. (1958). In *Insulin crystals.* Munskgaard, Copenhagen.
WILLIAMSON, K. L. and WILLIAMS, R. J. P. (1979). *Biochemistry* **18**, 5966–73.

44. The crystal structure of insulin and solution phenomena: use of the high-resolution structure in the calculation of the optical activity of the tyrosyl residues

DAN MERCOLA AND AXEL WOLLMER

F R O M the earliest investigations by Pasteur (1848) progress in understanding optical activity of molecules in solution has been linked to knowledge of their crystalline state. A property of proteins that may be clarified by this association is the intense optical activity of the aromatic side-chains of many proteins in solution including insulin (Grosjean and Tari 1964). From the first observations of these effects by Simmons and Blout (1960), Meyer and Edsall (1965) and others it was realized that the phenomenon could be rationalized if it were grounded in the existence of a rigid structure, although the nature of the structure for many proteins was still unclear. For insulin the justification was well foreshadowed in the first X-ray studies of insulin by Dorothy Hodgkin:

> The measurements obtained ... fix quite definitely the arrangement of the molecules with respect to one another and their approximate size and shape since this follows directly from the crystal lattice, while the variation of the intensities of the spectra strongly suggests that the arrangement of atoms within the molecules is also of a perfectly definite kind [Crowfoot 1935].

and later,

> wet crystals examined in their mother liquor show reflections with spacings of 2.4 Å or less, which indicate that here there is regularity in the structure to atomic dimensions [Crowfoot and Riley 1939].

The connection was reinforced by the observation that the tyrosyl optical activity of insulin is sensitive to various manipulations and chemical modifications that alter the structure (Beychok 1965; Carpenter and Hayes 1966);

Morris, Mercola, and Arquilla 1968; Brugman and Arquilla 1974) and is very sensitive to the aggregation state of the molecule (Goldman and Carpenter 1967, 1974; Morris *et al.* 1968). It has been difficult to exploit these results since it was not possible to define anything about the details of the origin of these changes in optical activity and rarely possible to identify which of the four tyrosyl residues of insulin were involved in a given change (Morris *et al.* 1968). Today the situation is very different. The structure of a 'perfectly definite kind' is perfectly definitely known at near atomic resolution (see Chapters 39–42 of this volume), and this provides us with a very secure reference point. At the same time there have been advances in the methods for calculating optical activity. These methods have been collected together in a form that is very convenient for the calculation of tyrosyl optical activity of proteins by E. H. Strickland (1972). He first showed that such methods have some validity by applying them to the crystal structure co-ordinates of ribonuclease (then 2.0 Å resolution) and correctly calculated the sign and about 70 per cent of the intensity of the tyrosyl optical activity. A more critical assessment, however, requires a knowledge of the effects of inaccuracies in the model and the ability of the method to account for the optical activity of various structures. The various aggregation states and high resolution of the insulin structure provide us with an excellent test case for both of the requirements (Strickland and Mercola 1976).

Here we describe the comparison of the observed and calculated tyrosyl circular dichroism of insulin using the high-resolution structure co-ordinates of 2Zn insulin refined at 1.5 Å resolution to an overall residual value of 17.9 per cent (Chapter 42). Experiments in progress on the application of the method to the identification of structural variations of insulin in solution are discussed.

How is the circular dichroism of tyrosine calculated?

The origin of optical activity was first derived on quantum mechanical grounds by Rosenfeld (1928):

$$[\alpha]_\lambda = A\sum_j \text{Im} \, (\boldsymbol{\mu}_1 \cdot \mathbf{m}_1) \frac{\lambda_j^2}{\lambda^2 - \lambda_j^2}, \qquad (44.1)$$

where $[\alpha]_\lambda$ is specific rotation at wavelength λ, A is a characteristic constant of the solution,† and Im indicates the imaginary part of the following dot product. $\text{Im}(\boldsymbol{\mu}_1 \cdot \mathbf{m}_1)$ is the rotatory strength of the jth absorption transition, R_j. The principal conclusion from the Rosenfeld equation is that, while absorption of light depends on the existence of either an electric, $\boldsymbol{\mu}$, or magnetic,

† $A = (n + \frac{2}{3})(9600\pi N/hcM)$ where n, N, h, c, and M are index of refraction, Avogadro's number, Planck's constant, speed of light, and molecular weight, respectively.

m, transition, the rotation of light depends on the presence of both moments oriented in such a way that their dot product is non-zero, i.e. it is dependent on three-dimensional structure. A corresponding absorption manifestation of rotation is that there is an unequal absorption of left and right circularly polarized light. The difference, $\varepsilon_1 - \varepsilon_r = \Delta\varepsilon$, is called circular dichroism (CD) and is a convenient measure of optical activity. The area, or for simple shapes, the maximum, $\Delta\varepsilon$, of such a difference absorption band is proportional to R_j: $\Delta\varepsilon \cong \frac{1}{2} \times R_j \times 10^{40}$.

The problem of understanding the origin of a given CD band is the problem of understanding the origin of simultaneous electric and magnetic dipole moments. Four mechanisms have been proposed (Condon, Alter, and Erying 1937; Tinoco 1962; Schellman 1966) and the total CD at any one transition is the sum of effects. One mechanism called electric dipole–electric dipole coupling is proving particularly important in explaining the CD bands of proteins (Hsu and Woody 1971; Fleischhauer and Wollmer 1970; Strickland 1972; Strickland and Mercola 1976). This mechanism illustrates the general features and assumptions of the calculation. In the formulation of Hsu and Woody (1971) the near-UV absorption band of any phenolic group with transition dipole moment $\mathbf{\mu}_1$ that is placed in suitable proximity and orientation to any other transition dipole $\mathbf{\mu}_2$ will become optically active with a contribution to R_j given by:

$$R_{j_{12}} = \frac{2\pi V_{12}\lambda_1\lambda_2}{hc(\lambda_2^2 - \lambda_1^2)} (\mathbf{R}_{12} \cdot \mathbf{\mu}_2 \times \mathbf{\mu}_1), \qquad (44.2)$$

where λ_1 is the wavelength of the near-UV tyrosyl absorption band, λ_2 is the wavelength of the second transition, R_{12} is the separation between $\mathbf{\mu}_1$ and $\mathbf{\mu}_2$ which are located in the centre of the phenolic ring and second chromophoric group respectively; V_{12}, the interaction potential, is discussed below. The pairwise contributions given by eqn 44.2 are summed for the interaction of each tyrosine residue with the transitions of all side-chains of Phe, His, other Tyr, Gln, Asn, Arg as well as carboxylate groups and all amide groups of the peptide backbone. Equation 44.2 shows that the importance of $\mathbf{\mu}_2$ decreases with spectral separation $(\lambda_2^2 - \lambda_1^2)$ especially for short wavelength transitions and, after accounting for V_{12} (see below), decreases with distance, R_{12}. The requisite magnetic term is implicit in the vector product because charge displacement along one dipole can be considered to induce displacement along the other dipole and, for all but coplanar arrangements of the dipoles, together they define an element of arc current with its associated magnetic moment. Any parallel components between the electric and magnetic moments produce a non-vanishing dot product.

An important advance was the introduction of the monopole distribution for the calculation of V_{12} and the dipoles (Woody 1968; Hsu and Woody

1971; Strickland 1972). In the case of V the previous generalization of Coulomb's law is replaced by a multipole-type expression:

$$V_{12} = \sum_j \sum_k \frac{q_{2j} q_{1k}}{\varepsilon_{2jlk} R_{2jlk}}, \qquad (44.3)$$

where the q_{1k} and q_{2j} are the monopoles approximating the charge configuration of the phenolic group transition near 275 nm and the second transition respectively. R_{2jlk} is the monopole separation and ε_{2jlk} is the dielectric constant. In the case of tyrosine, for example, monopoles are placed 1.08 Å above and below the nuclear plane at the carbon atom positions. Thus the monopoles are fixed properties of all groups and their separation and locations are simple functions of the crystal structure co-ordinates.

The monopole approximations have the advantage that they are intrinsically much more accurate at short separations of the order of the dipole length than calculations using a generalized Coulomb's law. They also illustrate one of the major assumptions of the method, the attempt to approximate excited state charge distribution in the absence of exact knowledge of the wave functions. This is undoubtedly the most important assumption of the method.

Other difficulties include the selective treatment of $\pi \rightarrow \pi^*$ transitions. For example $\sigma \rightarrow \sigma^*$ transitions were not considered. These vacuum ultraviolet transitions are spectrally remote from the tyrosyl band at 275 nm and poorly characterized. Further, since every bond of the structure has a $\sigma \rightarrow \sigma^*$ transition, extensive cancelling of positive and negative contributions may be expected (*see below*). Finally, there are uncertainties in the use of certain quantities. For another example the appropriate value for the dielectric constant (eqn 44.3) cannot be determined precisely. For groups that are nearly in van der Waals contact a value of 1 appears to be appropriate. However, for larger separations where the intervening medium is relevant, values up to 2.5 may be appropriate (Strickland 1972). In this case contributions may be overestimated by an uncertain factor up to 2.5 when groups have large separations, e.g. 10 Å. It will be seen that contributions of this kind are usually quite small.

These assumptions are mentioned to provide an idea of the limitations to be expected. It is of course difficult to know how accurate such calculations are, apart from tests such as those described here.

Two other mechanisms may be mentioned. A special case of eqn 44.1 is the interaction between identical phenolic transitions (degenerate dipole–dipole coupling). This is mechanism 2 and does not contribute to net CD at 275 nm although it may alter the shape of the nearby spectrum. Mechanism 3, the 'one-electron mechanism', describes a contribution of nearby static charges such as carboxylate and imidazole groups and the treatment of Schellman (1966) is followed here.

In comparing calculated and observed CD a major assumption involves the relationship of the structure of insulin in the crystals and in solution. It is only the crystal structure of insulin that is known. There is then the interesting but problematical quandary of which monomer structure of the two variants in the rhombohedral structures, if either, occurs in solutions of monomers, dimers, and hexamers. The basic pattern of monomer II has

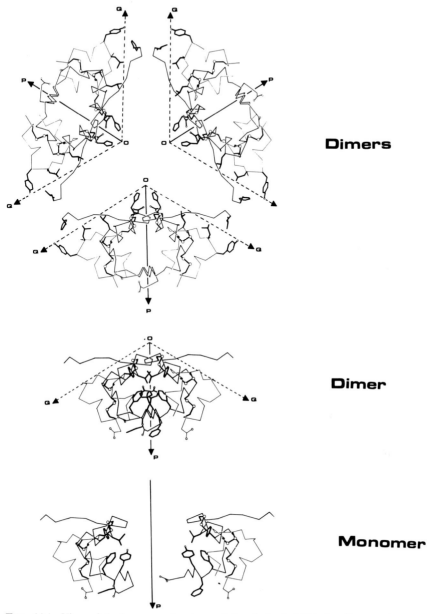

Dimers

Dimer

Monomer

FIG. 44.1. View of the hexamer structure of rhombohedral 2Zn insulin as refined at 1.5 Å resolution ($R = 17.9$ per cent) showing the relationship of monomer I and II to each other along the dimer axis, OP, and along the axes relating dimers in the hexamer, OQ. The Oxford convention for labelling the two molecules is used.

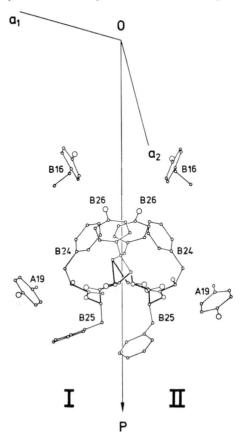

FIG. 44.2. The relationships among the aromatic groups of insulin that are brought about by dimer formation as viewed looking down the threefold axis of rhombohedral insulin (cf. Fig. 44.1).

appeared as a common feature in several other crystal structures of insulin (Cutfield, Cutfield, Dodson, Dodson, and Sabesan 1974; Bentley, Dodson, Dodson, Hodgkin and Mercola 1976; Dodson, Dodson, Hodgkin, and Reynolds 1979) and this structure may be closely related to a stable form in solution (Figs. 44.1–3). As more details become available it should be possible to extend the calculations to dimers and hexamers of a single form and this may be helpful in settling this question.

What are the predicted causes of the tyrosyl circular dichroism of insulin?

Of the dozens of neighbouring atoms that make up the environment about the tyrosyl residues the major sources of circular dichroism (CD) are calcu-

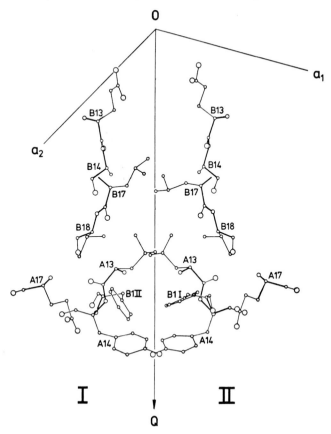

FIG. 44.3. The relationships among the aromatic groups and nearby peptide bonds that are brought about by hexamer formation (note that the B1 phenyl rings are projected on to the opposite side of the local twofold axis OQ, from that of the respective monomer structure). Viewed looking down the threefold axis (cf. Fig. 44.1).

lated to be largely due to two effects: close contacts with other aromatic groups or, in some cases, nearby stretches of peptide bonds. Two properties of proteins appear to be primarily responsible for this result. One is the non-polar nature of the groups around most phenolic side-chains and the second is a consequence of the relatively large number of peptide bonds. To see how these two properties may influence the CD of the tyrosyl residues it is helpful to briefly consider the nature of the contribution of various classes of neigh-bouring groups (Table 44.1).

The most important contributions to CD are summarized for the three aggregation states of insulin in Table 44.1. First, looking down the totals column an overall picture emerges. For each monomer structure the tyrosyl CD from all sources is negative and this negative value increases a further threefold when additional interactions of the dimer and hexamer are taken into account. This enhancement occurs even though the monomer structure used in the calculations is assumed to be unchanged. The enhancement is related to the increased number of nearby groups for those tyrosyl rings near surfaces involved in aggregation of the molecules (Figs. 44.2 and 3). Table 44.1 shows that, for any level of aggregation, only two kinds of groups, peptide bonds and aromatic groups, strongly influence tyrosyl CD. The effects of these groups as contributors to tyrosyl CD at various separations in the hexamer are compared in Fig. 44.4. Here each group has been subdivided according to whether a given interaction contributes to positive or negative tyrosyl CD. First it is evident that, for either the peptide bonds or

TABLE 44.1

Calculated tyrosyl circular dichroism of 2Zn insulin contributions of different classes of perturbing groups

Environment	Perturbing groups	$\Delta\varepsilon$ [275 nm; $M^{-1} \cdot cm^{-2} \cdot 10^{-3}$]					
		Tyr, Phe	*Peptide bonds*	*Asn, Gln amides*	*Asp, terminal carboxyls*	*His*	*Totals*
In isolation	Monomer I	−0.95	−0.58	+0.23	−0.08	−0.04	−1.42
	Monomer II	−0.04	−1.13	+0.27	+0.18	+0.16	−0.57
	Average	−0.50	−0.85	+0.25	+0.05	+0.06	−1.00
As part of the dimer	Monomer I	−1.79	−1.23	+0.13	−0.14	+0.05	−2.84
	Monomer II	−0.43	−1.79	+0.29	+0.08	+0.30	−1.56
	Average	−1.11	−1.51	+0.21	−0.03	+0.18	−2.20
As part of the hexamer	Monomer I	−2.94	−1.10	+0.04	−0.11	−0.40	−4.49
	Monomer II	−0.33	−1.60	+0.14	−0.01	+0.28	−1.52
	Average	−1.64	−1.35	+0.09	−0.06	−0.06	−3.01

aromatic groups, tyrosyl CD falls off with separation somewhat faster than $1/r$ as is expected from eqns 44.2 and 44.3. Second, and more interesting from the present point of view, is that in the case of the peptide bond contributions (Fig. 44.4 (A)) at any separation the positive and negative values are nearly equal in magnitude and so tend to cancel each other (Fig. 44.4 (A), solid line). This kind of effect is expected for large numbers of identical perturbing contributors that are distributed randomly about a given tyrosyl ring. Although a random distribution of peptide bonds almost certainly does not hold for any one tyrosyl residue, the sum for all tyrosine residues in the hexamer appears to approximate this situation.

We might anticipate, then, that for very large proteins where tyrosyl groups are more consistently 'buried' positive and negative contributions would tend to balance even more precisely. For insulin the *relative* contribution of pep-

tide bonds to the total CD decreases by about half as the size of the aggregation state increases to the hexamer (Table 44.1). Thus this effect of large
numbers appears to be more important than the *absolute* increase of the peptide bond contribution which accompanies aggregation (Table 44.1). As a
result usually only local short stretches of peptide backbone have significant
influence (examples are given in the Appendix). We will return to this effect
in discussing the experimental uses of CD, especially the question of which,
if either, monomer structure persists in dilute solutions of insulin.

There are many fewer aromatic groups than peptide bonds. As a result

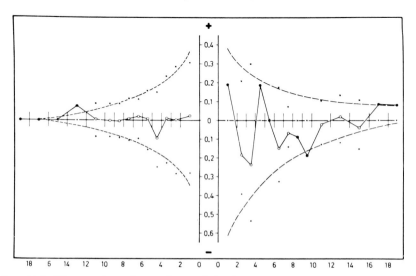

FIG. 44.4. Distance-dependence of tyrosyl circular dichroism. For successive shells of
separation (2.0 Å each) from the tyrosyl groups the positive and negative contributions
to tyrosyl CD (dashed lines) due to peptide bonds, A, and aromatic groups, B, decrease
with separation from tyrosyl groups as $1/r$. The actual positive and negative sums for
a given shell are indicated by '+' and 'x' respectively.

the cancelling amongst contributions to tyrosyl CD from all other aromatic
groups is much less precise (Fig. 44.4 (B)), solid curve. In this case the sign
and magnitude of the resultant CD at any separation is much more dependent
upon the structural details of the aromatic neighbours about tyrosyl groups.
For example, the largest CD value for any separation (Fig. 44.4 (B), solid
curve) is largely due to a single interaction between the B-1 phenylalanyl side-
chain of molecule II at 2.7 Å (minimum monopole separation) from the A-
14 phenolic group of molecule I (this effect is the basis of one test of the
accuracy of the calculations since the calculated results for desphe-B1-insulin
can be compared to the experimental values, *see below*). The calculations
with insulin suggest that aromatic interactions are the greatest single source

of tyrosyl CD and are largely responsible for the 'fingerprint'-like character of the near-UV CD spectra, especially for non-tryptophan-containing proteins.

A third and quantitively less important class of contributors are polar side-chains such as carboxyl and amide groups. On average they are not close to tyrosyl residues (average minimum separation is 18 Å compared to 10 Å for ring groups) and so produce little CD. There are numerous exceptions though (Appendix I). These tend to occur at intermediate separations, and for all aggregation states there is a great deal of cancelling among them (Table 44.1). These polar groups are not expected to be an important source of tyrosyl CD of insulin.

For the remaining amino acids with chromophoric or charged side-chains, histidine and arginine, there is less certainty about their consequences on tyrosyl CD because there is less known about the spectral properties of these groups experimentally and the calculations are of unknown reliability. Fortunately the two guanido groups of the dimer are not close to any tyrosyl side-chain (minimum separation is 9.2 Å) which precludes either guanido group on the dimer from making a significant contribution to tyrosyl CD (cf. Fig. 44.1). Imidazole groups are also not immediate neighbours of phenolic groups in insulin. However, several moderately close approaches do occur (Appendix I). Even at these separations contributions are expected to be small (e.g. Fig. 44.1) and this is consistent with the estimates for these groups (Appendix I).

The disulphide bond chromophore presents a special case as they themselves may be optically active in the wavelength of tyrosyl CD and therefore need to be calculated if comparisons with observations are to be made. The intensity of disulphide bond CD is dependent upon the dihedral angle. Theoretical and experimental determinations suggest that there is no CD at values near $\pm 90°$ (for a review see D. B. Boyd 1974). This is very nearly the value for the dihedral angle of the three disulphide bridges of insulin therefore no disulphide CD is expected (Dodson and Hodgkin, private communication).

In summary we find that nearby peptide bonds and aromatic groups are the major predicted causes of tyrosyl CD. The effect of peptide bonds is dampened somewhat as a consequence of their ubiquitous nature. Each group can only produce a positive or negative contribution and so for large numbers much cancelling occurs. This effect becomes more apparent as the size of the aggregate increases (decreasing surface-to-volume ratio). On the other hand tyrosyl ring interaction with other aromatic groups cancels less exactly and so individual interaction may dominate in a particular aggregation state. These effects tend to be constant with increasing aggregation state. There are relatively fewer close contacts between polar and phenolic groups (Appendix I) and this general property of aromatic groups in proteins may limit the role of polar groups in generating significant tyrosyl CD.

How do calculated and observed values compare?

The calculated and observed values for CD and 275 nm are compared in Table 44.2. A very striking feature of insulin in solution is the large increase in negative CD observed upon aggregation, approximately fourfold. The

TABLE 44.2

Comparison between experimental CD of insulin and tyrosyl CD calculated for 2Zn insulin

Aggregation state	$\Delta\varepsilon$ [275 nm; $M^{-1} cm^{-2} 10^{-3}$][a]				
	Observed for insulin	Ref.	Corrected to single species[b]	Calculated for tyrosine	Difference[c] obs./calc.
Monomer		g, i	−1.2	I: −1.42	−0.22, −0.72
		j	−0.7	II: −0.57	+0.63, +0.13
				Av: −1.00	−0.2, +0.3
Dimer	−2.9	g	−3.0		−0.8
	−3.1	l	−3.3[j]	−2.2	−1.1
Hexamer	−4.2	d–i	−4.2	−3.0	−1.2

[a] Per monomer molecular weight (∼ 5.750).
[b] Details of correction are given in Strickland and Mercola (1976).
[c] The difference in observed and calculated CD per monomer molecular weight. Value may represent disulphide CD intensity.
[d] Morris *et al.* (1968).
[e] Menendez and Herskovitts (1970).
[f] Ettinger and Timasheff (1971).
[g] Goldman and Carpenter (1974).
[h] Wollmer, personal communication.
[i] Wood *et al.* (1975).
[j] Value given by Wood *et al.* (1975).

effect is especially noticeable for dimer formation where there is a threefold increase. Hexamer formation leads to an additional factor of 1.3.

Unfortunately there are no direct observations of 100 per cent monomer population at neutral pH (Table 44.2)—most published values refer to acidic or alkaline conditions—and large corrections have been applied by using the association equilibrium constants of Goldman and Carpenter (1974) for insulin (e.g. Wood *et al.* 1975; Strickland and Mercola 1976). Much uncertainty remains about the true value for the monomer. Although it is clear that large enhancement of CD does occur upon dimer formation, the amount given here (Table 44.2) can only be regarded as approximate. There is more agreement about the dimer values and essentially identical measurements have been made for the hexamer of insulin in seven laboratories (Strickland and Mercola 1976).

The calculated values of the aggregation-dependent increase in CD intensity predict an approximate doubling for dimer formation and a further increase by a factor of 1.3 for hexamer formation. The final results for the hexamer account for about 70 per cent of the observed CD at 275 nm.

Several considerations may be helpful when judging these results. For one, the calculated CD values are based on point atom co-ordinates and so strictly should be compared to observation at 0 K! Both theoretical and experimental results show that temperature changes have only a modest influence on CD over the range 0–300 K. The extent of thermally induced changes can be estimated by the use of the isotropic atomic temperature factors (*B*-values) which have also been determined during the course of the crystal structure refinement. *B*-values provide an estimate of the r.m.s. amplitude of vibration of the respective atoms and, by assumption, the associated monopoles. Therefore these values provide an estimate of error of position of the monopoles: $SD_{position} = \sqrt{(B/8\pi^2)}$. This value can then be used to define a Gaussian probability distribution of 'monopole density' at any position. Very close monopoles with high *B*-values may now have overlapping charge distributions and so reduced electric potential (eqn 44.3) and therefore reduced CD (eqn 44.2). So far these calculations are preliminary and have only been applied to the

TABLE 44.3

Calculated tyrosyl circular dichroism of insulin dependence on X-ray crystallographic resolution and refinement

Environment		$\Delta\varepsilon$ [275 nm; M^{-1} cm^{-2} 10^{-3}]†					
		1971	1975	1976	1977	1978	1978‡
In isolation	Monomer I	−0.20	−0.62	−0.59	−1.37	−1.42	−1.49
	Monomer II	−1.55	−0.15	−1.17	−0.71	−0.57	−0.56
	Average	−0.88	−0.39	−0.88	−1.04	−1.00	−1.02
As part of the dimer	Monomer I	−1.91	−2.20	−2.38	−2.85	−2.84	−2.91
	Monomer II	−2.57	−1.72	−2.41	−1.90	−1.56	−1.58
	Average	−2.24	−1.96	−2.40	−2.38	−2.20	−2.24
As part of the hexamer	Monomer I	−3.31	−2.90	−4.60	−4.76	−4.49	−4.56
	Monomer II	−4.08	−3.71	−3.10	−2.22	−1.52	−1.53
	Average	−3.70	−3.31	−3.85	−3.49	−3.01	−3.05
Resolution (Å)		2.8	1.9	1.5	1.5	1.5	1.5
R-value (%)		42.0	37.6	19.1	18.7	17.1	17.1

† $\Delta\varepsilon$ is given per monomer molecular weight (~ 5750).
‡ Corrected for temperature factors *B*.

interaction potential (eqn 44.3) and must be extended to the dipole quantities as well (eqn 44.2). However, CD is linear with potential and this should provide a good estimate of the likely extent of the effect. The results are given in terms of CD for comparison in Table 44.3. Comparison of the last two columns shows that very little change is expected in interaction potential owing to the vibrational properties found in 2Zn insulin.

These considerations are also experimentally justified. If thermal effects in fact do make very little difference it would be expected that the observed near-UV CD spectrum of insulin would not be substantially altered at liquid nitrogen temperatures (77 K). Such an experiment is shown for the hexameric state in Fig. 44.5. Some influence of low temperature is apparent, especially in the enhanced resolution of the phenylalanyl fine structure (low-wavelength side of the band at 275 nm). However, the intensity change at 275 nm is less than 10 per cent. Other experiments using zinc-free insulin (presumably dimers) show virtually no change.

These results illustrate the well-defined and stable nature of the insulin structure and show that the tyrosyl CD is largely but not completely independent of temperature.† Accounting for the observed temperature dependence of insulin, the calculated CD for the hexamer is 60 per cent of the observed value at 77 K.

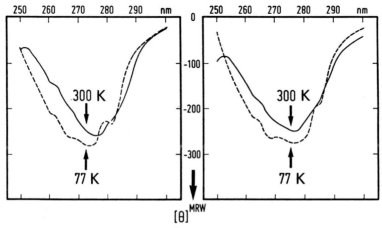

FIG. 44.5. Near-UV CD spectra of pig insulin and desphenylalanyl-B1-insulin at 10^{-5} M in 0.1 M NaCl, 0.05 M Na_2HPO_4, 0.1 mM $ZnCl_2$ (pH = 7.4), and 50 per cent v/v glycerol at ambient temperature and 77 K.

A final consideration is that not all of the CD at 275 nm is due to tyrosyl groups. One way that this can be seen is in the complete spectra of the various aggregation states (e.g. Fig. 44.5) which show that the spectrum is broadened in regions where tyrosine itself has no absorption (> 300 nm) and this implies the presence of other sources of negative CD that overlap the tyrosyl position at 275 nm.‡ These effects may be partially eliminated by comparing differences between aggregation states rather than absolute values. An added

† B-values must be expected to diminish with decreasing temperature and therefore the definition of the X-ray structure may improve at lower temperatures; however, our calculations show that CD values are largely insensitive to B-values less than 40 Å².

‡ Some broadening may be due to the phenolate character of some tyrosyl residues on insulin, e.g. where hydroxyl functions may be involved in hydrogen bonding with the solvent or neighbouring side-chains (Morris *et al.* 1968). The phenolate ion absorption maximum is 295 nm. Optically active disulphide bond chromophores may also produce this effect although the basis of this effect if present cannot be easily understood in terms of the crystal structure and present theory of the optical activity of the disulphide chromophore (Boyd 1974).

advantage is that all other influences on the observed CD that are common to all aggregation states, such as the effects of temperature, tend to be eliminated as well. The most reliable difference is that for the dimer-to-hexamer step. The calculated difference for this step, $-0.8\,M^{-1}\,cm^{-1}$, compares well with the observed range of -0.9 to $-1.2\,M^{-1}\,cm^{-1}$. Although the observed difference is not well defined, the comparison with calculated values shows better agreement than comparisons on an absolute scale. Further, three of the four tyrosyl residues are close to the associating surfaces in this aggregation step and the calculations show that they make significant contributions to the final hexamer value (Appendix I). Therefore this single comparison has some generality. In terms of the absolute scale these results agree with previous estimates (Morris *et al.* 1968; Strickland and Mercola 1976) that no more than 85 per cent of the CD observed at 275 nm can be due to tyrosyl groups. Our best final yardstick for the tyrosyl CD intensity for hexameric insulin on an absolute scale is $4.2\,M^{-1}\,cm^{-1} \times 1.10 \times 0.85 = 3.9\,M^{-1}\,cm^{-1}$. This is to be compared to a calculated value of $3.0\,M^{-1}\,cm^{-1}$ (Table 44.2).

These results are encouraging. They predict a major spectral property of insulin, the existence of aggregation-dependent CD values, and account for about 75 per cent of the tyrosyl CD. Further comparison will require refined CD measurements. There is a particular need for measurements of solutions of precisely defined populations of monomers and dimers, and experiments along these lines are in progress. In addition it is of interest to ask if more rigorous tests of some of the larger predicted effects can be devised. The analogue desphenylalanyl-B1-insulin provides such a test.

Do the calculations account for the circular dichroism of desphe-insulin?

The desphenylalanyl-B1-insulin (desphe-insulin (Brandenburg 1969)) derivative is attractive as a test case because the B1 aromatic groups contribute about a quarter of the entire hexamer CD largely owing to their proximity to the A14 phenolic groups (Fig. 44.3). On this ground alone removal of the B1 residue may be expected to lead to a large observable decrease in CD. In addition desphe-insulin is perhaps the best characterized derivative of insulin known (for a review see Wollmer *et al.* 1977). It retains full biological activity and, for all properties that have been examined, behaves very like insulin. A medium-resolution crystal structure has been determined in York and most of the structure is similar to insulin (de Graaff *et al.*, manuscript in preparation). However, a very interesting situation has developed in the region of the A14 tyrosyl residues. The A14 phenolic group of monomer II appears to be disordered and is not visible in the electron-density map beyond C_β (cf. Fig. 44.3). The other A14 phenolic group (monomer I), although somewhat less well defined than in insulin, remains in the same position as in 2Zn insulin.

In 2Zn insulin the two A14 phenolic groups are linked to each other by hydrogen bonding through their hydroxyl functions and their positions are related by the non-crystallographic twofold axis OQ (Fig. 44.3). The reason for the unequal disorder in this pair of rings in desphe-insulin is uncertain. In 4Zn insulin, for example, the phenylalanine residues are rearranged and not near the pair of A14 residues (Bentley *et al.* 1976; see also Chapter 42). Here too the phenolic side-chains behave as in desphe-insulin crystals. The unequal disorder may be related to the presence of a hydrogen bonded solvent structure in monomer I linking the hydroxyl function of the A14 to the A17 amide groups and A18 carboxyl groups. Also there is a larger number of nearby groups in the adjacent hexamer which may inhibit rotation of the A14 ring group of monomer I. The additional disorder of desphe-insulin also effects the predictions of the tyrosyl CD since disordered groups cannot contribute to CD.

In the absence of the B1 aromatic groups, the most important remaining contribution to the CD of *either A14 phenolic group is the neighbouring A14 phenolic group* (see Appendix I). This reciprocal effect is simply a consequence of proximity and the symmetry relating these two phenolic groups. As a result disorder in either group largely abolishes the remaining CD of both groups. Therefore when considering how the desphe hexamer may exist free in solution it is not necessary to know whether the unequal pattern of disorder seen in the crystal structure is retained or if both groups are disordered. A large loss of *positive* CD is predicted for the A14 tyrosines of desphe-insulin in either case, i.e. opposite that expected for loss of the B1 Phe alone.

The calculated and observed results for desphe-insulin are compared in Table 44.4. Somewhat surprisingly, for identical concentrations and aggregation states (within experimental error) the *observed* CD values are very similar to those for insulin. The explanation is provided by the calculations and the unusual results found in the crystal structure of desphe-insulin. The effects of the loss of the B1 phenylalanyl residues ($-0.62 \text{M}^{-1} \text{cm}^{-1}$) tends to be equal and opposite to the effects of any disorder in the pair of A14 tyrosyl residues in the hexamer. For example, a change of $+0.64 \text{M}^{-1} \text{cm}^{-1}$ is expected if both A14 groups are disordered (e.g. Table 44.4) and $+0.42 \text{M}^{-1} \text{cm}^{-1}$ if only the A14 ring of monomer II is disordered. Each of these changes is a relatively large effect, comparable to the average CD of the monomer. Thus when the effects of disorder are taken into account it is seen that very little change is expected in the tyrosyl CD of hexameric desphe-insulin in good agreement with observation (Table 44.4).

For monomers and dimers of desphe-insulin there is much less change predicted from the values for insulin and the desphe derivative does not provide a sensitive test.

All of these calculations have been carried out with the refined co-ordinates of 2Zn insulin and it will be necessary to confirm these conclusions when refined co-ordinates of desphe-insulin become available. However, preliminary calculations for the hexamer co-ordinates of desphe-insulin as

TABLE 44.4

Comparison of observed and calculated tyrosyl CD ($\Delta\varepsilon_{275}$) for insulin and desphe insulin†

Concentration (M)	Observed				Calculated		
	Aggregation state $(M_W/M_i)_{obs.}$		$(\Delta\varepsilon_{275})_{obs.}$			$(\Delta\varepsilon_{275})_{calc.}$	
	Insulin	Desphe insulin	Insulin	Desphe insulin	Aggregation state	Insulin	Desphe insulin
3.5×10^{-4}	6.2 ± 0.4	5.8 ± 0.4	-4.22	-4.22	Hexamer	-3.00	-2.99
3.5×10^{-5}	3.5 ± 0.3	3.2 ± 0.3	-4.06	-3.80	Dimer	-2.20	-2.01
3.5×10^{-6}	1.4 ± 0.2	1.4 ± 0.2	-2.84	-2.12	Monomer	-1.00	-1.3

† In all cases $\Delta\varepsilon_{275}$ is given per mole of insulin ($M_i = 6000$). The CD of insulin and desphe-insulin were measured in 0.025 M Tris.HCl, pH 7.8, using a Cary 61 CD spectrometer. M_W for insulin was measured in the same buffer as for CD measurements while M_W for desphe-insulin was measured in 0.169 M Tris.HCl, pH 8.0. The highest values of M_W/M_i are extrapolated values for a linear portion of the concentration-dependence of M_W (observed values of M_W/M_i at 1.8×10^{-4} M are 5.5 and 6.6 for insulin and desphe-insulin respectively). Calculated CD values for desphe-insulin are based on the co-ordinates of 2Zn insulin ($d = 1.5$ Å; $\bar{R} = 17.9$ per cent) and the results given here are based on the assumption that both crystallographically independent A14 phenolic groups are completely disordered in solution.

determined at an intermediate resolution (2.5 Å) yield a value of -3.15 M^{-1} cm^{-1}, very similar to the results described here (Table 44.3).

The analysis of desphe-insulin suggests that the calculation method accurately represents the CD properties of the B1 and A14 residues of insulin and encourages us to think that the method may be of practical value in examining the structural properties of insulin in solution and some of these applications are considered below.

How sensitive are the calculated CD values to variation in the atomic co-ordinates?

Two questions about sensitivity are considered here. Are the particular results seen with insulin inevitable, given a crude model of the structure, or are the calculations sensitive to the accuracy of the co-ordinates? Conversely, can precisely defined point atom models of insulin valid for the crystal structure environment be appropriate to solutions at finite temperatures?

A convenient way of examining the first point is to compare the calculated CD results for successive stages of the crystallographic refinement from the first set of co-ordinates obtained in 1971 (Blundell *et al.* 1971) to the present (Table 44.3). Several trends are apparent. For all levels of aggregation, as the refinement progressed, monomer I became progressively more levorotatory and monomer II became progressively more dextrorotatory, so much so that their respective CD intensities nearly reverse in the process. The totals

Interaction partners	Resolution	$\Delta\varepsilon_{calc}$ [275 nm; M^{-1} cm^{-2} 10^{-3}]	
		1.9 Å	1.5 Å
	R-value	37.6 %	17.1 %
Tyr A14 I	120 Phe B1 II	+0.08	−0.95
	120 Tyr A1 4II	−0.59	+0.33
	0 Phe B1 I	±0	−0.33
Tyr A14 II	120 Phe B1 I	−1.65	+0.19
	0 Phe B1 II	−1.43	+0.10

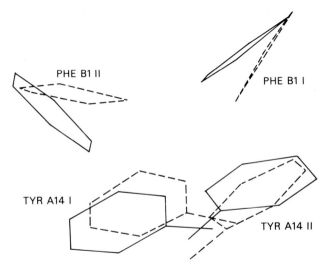

FIG. 44.6. Dependence of selected pairwise interactions on the quality of the X-ray structure. ----$d=1.9$ Å, $R=37.6$ per cent; —— $d=1.5$ Å, $R=17.9$ per cent.

for each aggregation state, although more constant, also change steadily, increasing in intensity for the monomer and decreasing for the hexamer. Thus the aggregation-dependent enhancement of CD is always apparent, but less so at the end of the refinement.

There is considerable variation between sets of calculations, of the order of 20 per cent. This is also true for the last two sets of calculations even though the crystallographic co-ordinates were approaching convergence (ΔR—1.6 per cent). Much of this effect is due to small changes affecting a few large interactions and this is illustrated in Fig. 44.6 for the familiar case of the pair of A14 tyrosyl residues. The positions of these phenolic groups and nearby B1 aromatic side-chains for two different stages of the crystallographic refinement are shown. The distance separating each phenolic group from its three perturbing neighbours has changed very little; however, small changes in geometry have occurred. The calculated CD contributions to each A14 phenolic group (Fig. 44.6, box) have changed considerably. There are even sign reversals. This illustrates the need for highly accurate co-ordinates. The point has already been made that tyrosyl CD is dominated by relatively few large contributions of this kind (Appendix I). It is then not surprising that small geometric changes involving some of these groups can have large consequences on the overall value. The actual sensitivities are the derivatives of eqn 44.2. For a dominating interaction CD would change approximately as $R\mu_1\mu_2\sin\theta_1\cos\theta_2$,† which of course implies continuous variation, sign reversals and, in limiting cases, changes in CD as the square of these trigometric functions.

The sensitivity of the calculated results to small changes in co-ordinates alerts us again to the point that the calculated results strictly apply to a crystalline environment at 0 K and may be irrelevant to solutions at ambient temperatures where greater flexibility may alter the geometry.

We have seen that purely thermally induced effects such as vibrations appear to play a minor role in the insulin structure. Another possible source of energy upon transfer of insulin from the crystal to the solution environment is crystal lattice energy. This is almost certainly an insignificant effect. Although reliable estimates of lattice energies for proteins are lacking, it is well known that protein crystals are delicate compared to inorganic crystals. What is more, whatever energy that is derived from this source must be partitioned amongst all atoms of the structure by some means (perhaps temperature factor weighting?) but in any case this will necessitate division by a large number approximately the number of non-hydrogen atoms in the structure. Even considering inorganic crystal values this results in very little energy per bond.

Undoubtedly the major structural reasons for error between calculated and observed CD are specific structural rearrangements between states of nearly equal energy such as may occur between monomers I and II when crystal contacts are lost. Indeed defined transformations can occur entirely within the crystalline state (Bentley, Dodson, and Lewitova 1978). Many of the alternative states are known from crystal structure analyses and these together with the sensitivity of CD to structural change, provide a means of identifying which of these forms exist in solution. Possible examples are considered now.

What are the practical applications?

The positions of the tyrosyl residues in many proteins allow them to serve as probes of local structural features or conformational changes (e.g. Strick-

---

† Where θ_1 and θ_2 are the respective angles between μ_1 and μ_2 and the z-axis of local Cartesian co-ordinates with origins at the R–μ intersections.

land 1972). Practically speaking tyrosyl groups may serve as a probe for environmental changes within a shell of about 5–10 Å in most directions from the nuclear plane† (cf. Fig. 44.2). The possible ways in which this effect may be expected to distinguish structural alternatives are well illustrated with insulin.

The two monomers of the 2Zn crystal structure are predicted to have very different tyrosyl CD values, yet much of their structures are very similar (Chapters 40–42). This may be understood by considering the approximate twofold symmetry that relates the two monomers. The B16 tyrosine, for example, is on the hydrophobic face of the molecule that is involved in dimer formation and lies between the C-terminal B-chain (peptide bonds B25 through B27 and B29) and the peptide backbone of residues A17 and A18. The structure in these regions is very similar for both monomers and the contributions of all nearby peptide bonds (Appendix I) to the CD of the B16 tyrosine residues in monomers I and II are similar, $-0.64\,\mathrm{M^{-1}\,cm^{-1}}$ and $-0.44\,\mathrm{M^{-1}\,cm^{-1}}$ respectively. The addition of all other side-chains and contributions does not substantially alter this situation, the total CD for the B16 residues of monomer I and II are $-0.84\,\mathrm{M^{-1}\,cm^{-1}}$ and $-0.73\,\mathrm{M^{-1}\,cm^{-1}}$ respectively. Therefore the optical properties of the B16 tyrosyl groups then largely reflect the twofold symmetry relating the monomers.

The B26 tyrosyl groups are also located on this face of the molecule. The peptide bond contributions to B26 tyrosyl CD are again comparable, $-0.14\,\mathrm{M^{-1}\,cm^{-1}}$ and $-0.21\,\mathrm{M^{-1}\,cm^{-1}}$ for monomers I and II respectively. However, these phenolic groups are moderately close to the B25 phenylalanyl rings (Appendix I). One of the most striking differences between the two monomers is the positions of these B25 ring groups (Fig. 44.1). The B25 ring of monomer I lies directly on the twofold axis while that of monomer II is displaced towards the surface of the molecule and, importantly, towards the A19 tyrosine. The B26 phenolic groups act as probes sensitive to this asymmetry. The difference in the contributions of the B25 rings to the CD of the B26 phenolic groups in monomers I and II is $0.42\,\mathrm{M^{-1}\,cm^{-1}}$ and this largely accounts for the very large differences in the final CD values for these tyrosyl groups $(-1.34\,\mathrm{M^{-1}\,cm^{-1}}$ and $-0.20\,\mathrm{M^{-1}\,cm^{-1}}$ for monomers I and II respectively).

As a final example we consider the largest backbone difference between the two monomers. This involves differences in the helix of the N-terminal A-chain (A1 to A6) and is described in detail in Dodson, Dodson, Hodgkin, and Reynolds (1979) and Chapters 41 and 42. The A19 tyrosyl groups are very close to this region (Appendix I) and act as probes for this structural difference. The tyrosyl CD of A19 of monomer II is $-0.48\,\mathrm{M^{-1}\,cm^{-1}}$ more negative than that of monomer I owing to this stretch of backbone. Further,

† Perturbing groups lying on planes of symmetry of the phenolate ring and other limiting cases (cf. eqn 44.2) produce no CD.

as noted above, the B25 ring is displaced toward the A19 phenolic group and contributes a further difference of $-0.33 \,\mathrm{M}^{-1}\,\mathrm{cm}^{-1}$. As a result the total CD for the A19 tyrosyl residues are rather different: $-0.23 \,\mathrm{M}^{-1}\,\mathrm{cm}^{-1}$ v. $-0.80 \,\mathrm{M}^{-1}\,\mathrm{cm}^{-1}$ for monomers I and II respectively. These examples illustrate that the calculated difference between the two monomers, especially as they influence the B16 and B26 tyrosyl groups (Appendix I), may be understood in terms of the major structural differences between the two molecules.

The calculated CD values for the two monomers may be a useful criterion for distinguishing which, if either, of the two monomer structures of crystalline insulin exists in solution. It is expected that monomer II is more closely related to the stable form of the molecule in solution. The large predicted CD differences between these two forms should be easily detected and experiments are now in progress to measure the tyrosyl CD of 100 per cent monomer populations in dilute solutions at neutral pH.

Dimer and hexamers of insulin *in solution* may only contain one kind of monomer structure, such as is the case for dimers of hagfish insulin which only contain monomer II in the crystal structure (Cutfield *et al.* 1974; Chapters 35 and 42). Calculations based on such 'hypothetical' aggregation states may be useful in improving the agreement described here and confirming the nature of the species found in solution.

Several other variants of the insulin structure are known from crystallographic studies (see Chapter 42). The most dramatic of these is the large rearrangements of monomer I in 4Zn insulin crystals (Bentley *et al.* 1976). In this structure the N-terminal B-chain (B1 to B7) is rearranged from an extended chain to a continuation of the B-chain helix and numerous other smaller adjustments take place. Little is known of the independent existence or importance of this or other forms in solution. The possible importance of the ability of insulin to adopt different conformations as a factor of its biological activity has been considered (Dodson 1976; Dodson *et al.* 1979). Calculations of the tyrosyl CD from refined co-ordinates may provide a criterion for identifying and studying the conditions of transformation among some of these forms in solution.

Recently several other hormones such as relaxin (Isaacs *et al.* 1978; Bedarkar, Turnell, Blundell, and Schwabe 1978) and insulin-like growth factor (Blundell *et al.* 1978) have been suggested to be homologous to the structure of insulin. Comparison of the observed and calculated CD values based on models of these structures may be helpful in confirming these conclusions.

In general any model structure for which accurate co-ordinates are available may be used to test the model or theories of changes that may take place near tyrosyl residues. In practice, structures free of tryptophan, based on highly defined co-ordinates and which have large expected CD values seem likely to lead to the best results.

Acknowledgements

We are grateful to a number of collaborators who have contributed substantially to the work summarized here. All the calculations described here were carried out by Dr Wolfgang Strassburger using a computer program based on the original program kindly provided by Dr E. H. Strickland. We are also indebted to Drs Guy Dodson, Colin Reynolds, and Eleanor Dodson for help in using the crystallographic results, and to Professor Jorg Fleischhauer for his continued guidance and support.

References

BEDARKAR, S., TURNELL, W. G., BLUNDELL, T. L., and SCHWABE, C. (1977). *Nature, Lond.* **271**, 278–81.

BENTLEY, G., DODSON, G. G., and LEWITOVA, A. (1978). *J. molec. Biol.* **126**, 871–5.

—— DODSON, E. J., DODSON, G. G., HODGKIN, D. C., and MERCOLA, D. A. (1976). *Nature, Lond.* **261**, 166–8.

BEYCHOLK, S. (1965). *Science, N.Y.* **154**, 1288–99.

BLUNDELL, T. L., BEDARKAR, S., RINDERKNECHT, E., and HUMBEL, R. E. (1978). *Proc. natn. Acad. Sci. U.S.A.*, **75**, 180–4.

—— CUTFIELD, J. F., CUTFIELD, S. M., DODSON, E. J., DODSON, G. G., HODGKIN, D. C., MERCOLA, D. A., and VIJAYAN, M. (1971). *Nature, Lond.* **231**, 506–11.

BOYD, D. (1974). *Int. J. quantum Chem.: Quantum Biology Symposium, No. 1*, 13–19.

BRANDENBERG, D. (1969). *Hoppe-Seyler's Z. physiol. Chem.* **350**, 741–50.

BRUGMAN, T. and ARQUILLA, E. R. (1974). *Biochemistry* **12**, 727–32.

CARPENTER, F. C. and HAYES, S. L. (1966). *Pacific Slope Biochemical Conference abstracts*, p. 102. University of Oregon, Eugene, Oreg.

CONDON, E. V., ALTER, W., and ERYING, H. (1973). *J. chem. Phys.* **5**, 753–75.

CROWFOOT, D. (1935). *Nature, Lond.* **135**, 591–5.

—— and RILEY, D. (1939). *Nature, Lond.* **144**, 1011–14.

CUTFIELD, J. F., CUTFIELD, S. M., DODSON, E.J., DODSON, G. G., and SABESAN, M. N. (1974). *J. molec. Biol.* **87**, 23–30.

DODSON, E. J., DODSON, G. G., HODGKIN, D. C., and REYNOLDS, C. D. (1979). *Can. J. Biochem.* **57**, 469–79.

DODSON, G. G. (1976). Life Science Research Report 3, Berlin, pp. 17–28.

ETTINGER, H. J. and TIMASHEFF, S. N. (1971). *Biochemistry* **10**, 824–40.

FLEISCHHAUER, J. and WOLLMER, A. (1970). *Z. Naturf.* Teil B. Anorg. Chem. Biophys. Biol. **27B.1**, 530–2.

GOLDMAN, J. and CARPENTER, F. C. (1967). *Pacific Slope Biochemical Conference abstracts*, p. 17. University of Washington, Seattle, Wash.

———— (1974). *Biochemistry* **13**, 4566–74.

GROSJEAN, M. and TARI, M. (1964). *C. r. hebd. Séanc. Sci. Paris*, **258**, 2034–7.

HSU, M.-C. and WOODY, R. W. (1971). *J. Am. chem. Soc.* **93**, 3515–25.

ISAACS, N., JAMES, R., NIALL, H., BRYANT-GREENWOOD, G., DODSON, G., EVANS, A., and NORTH, A. C. T. (1978). *Nature, Lond.* **271**, 278–81.

MENENDEZ, C. and HERSKOVITTS, T. (1970). *Archs Biochem. Biophys. Acta* **160**, 145–50.

MEYER, E. and EDSALL, J. T. (1965). *Proc. natn. Acad. Sci. U.S.A.* **53**, 169–77.

MORRIS, J. W. S., MERCOLA, D. A., and ARQUILLA, E. R. (1968). *Biochim. biophys. Acta* **160**, 145–50.

PASTEUR, L. (1848). *Annls Chim. Phys.* **24**, 442–60.

ROSENFELD, V. L. (1928). *Z. Physik.* **52**, 161–70.

SCHELLMAN, J. A. (1966). *J. chem. Phys.* **44**, 55–63.
——(1968). *Accts. Chem. Res.* **1**, 144–51.
SIMMONS, N. S. and BLOUT, E. (1960). *Biophys. J.* **1**, 55–62.
STRICKLAND, E. H. (1972). *Biochemistry* **11**, 3465–74.
——and MERCOLA, D. A. (1976). *Biochemistry* **15**, 3875–83.
TINOCO, I., JR (1962). *Adv. chem. Phys.* **4**, 113–60.
WOLLMER, A., FLEISCHHAUER, J., STRASSBURGER, W., THIELE, H., BRANDENBURG, D., DODSON, G., and MERCOLA, D. A. (1977). *Biophys. J.* **20**, 233–43.
WOOD, S. BLUNDELL, T. L., WOLLMER, A., LAZARUS, N., and NEVILLE, R. (1975). *Eur. J. Biochem.* **55**, 531–43.
WOODY, R. W. (1968). *J. chem. Phys.* **49**, 4797–806.

Appendix I

Calculated major contributions (groups providing either strong μ–μ coupling or short separations from the Tyr side-chain) to the CD of individual Tyr residues are listed according to aggregation state. R_{min} is the shortest separation in Å between monopoles of interacting groups. These monopoles are located in the orbitals above and below the plane of each group (Strickland 1972). The peptide bond is represented as Pb.

Tyr A14 I

	Perturbing group	R_{min} (Å)	$\Delta\varepsilon_{275}$
Monomer	Tyr A19 I	11.1	−0.15
	Phe B1 I	2.7	−0.32
	Pb (A10–A17) I	(1.9–9.9)	0.08
	Pb (B1–B3) I	(5.0–8.8)	0.18
	Pb B14 I	9.9	0.04
	Pb B17 I	9.7	−0.05
	Pb B18 I	8.4	−0.10
	Glu A17 I	2.9	0.15
	Gln A15 I	7.3	0.08
	Asn A18 I	7.2	−0.07
	Asn B3 I	6.9	0.04
Dimer		—	—
Hexamer	Tyr A14 II	3.0	0.33
	Phe B1 II	2.7	−0.95
	Pb A12 II	9.2	0.04
	Pb A13 II	7.8	−0.10
	Pb B1 II	8.5	−0.04
	Pb B17 II	8.8	−0.08
	Pb B18 II	8.5	0.05
	Glu A17 II	6.9	−0.04

Tyr A14 II

	Perturbing group	R_{min} (Å)	$\Delta\varepsilon_{275}$
Monomers	Tyr A19 II	10.9	−0.21
	Phe B1 II	2.3	0.09
	Pb (A10–A17) II	(2.1–9.9)	0.12
	Pb (B1–B3) II	(5.5–9.5)	0.16
	Pb B17 II	9.4	−0.03
	Pb B18 II	7.9	−0.11
	Glu A17 II	7.9	0.16
	Gln A15 II	2.9	0.06
	Asn A18 II	7.7	−0.03
	Asn B3 II	7.0	0.05
Dimer		—	—
Hexamer	Tyr A14 I	3.2	0.27
	Tyr A19 I	15.8	−0.12
	Phe B1 I	1.9	0.19
	Pb (A12–A14) I	(7.6–9.9)	0.00
	Pb B1 I	7.6	−0.10
	Pb B2 I	9.9	0.02
	Pb B17 I	8.7	−0.08
	Pb B18 I	8.4	0.02
	Glu A17 I	6.3	−0.02

Tyr A19 I

	Perturbing group	R_{min} (Å)	$\Delta\varepsilon_{275}$
Monomer	Tyr B16 I	13.2	−0.11
	Tyr B26 I	5.9	−0.53
	Phe B24 I	6.5	0.13
	Phe B25 I	6.1	0.18
	Pb (A1–A6) I	(4.5–9.9)	0.03
	Pb (A10–A20) I	(1.2–9.7)	0.08
	Pb (B11–B15) I	(7.7–9.8)	0.02
	Pb (B18–B28) I	(3.4–8.8)	0.40
	Glu A4 I	7.3	−0.01
	Glu A17 I	8.6	0.02
	Gln A5 I	6.3	−0.03
	Gln A15 I	6.9	0.00
	Asn A18 I	4.4	0.19
	Asn A21 I	8.2	0.01
	COO A21 I	8.6	−0.05
Dimer	Phe B24 II	9.1	0.00
	Asn A21 II	9.3	0.03
Hexamer	Tyr A14 II	14.8	0.12
	Pb B23 II	9.6	−0.05
	Pb B24 II	8.3	0.09

Tyr A19 II

	Perturbing group	R_{min} (Å)	$\Delta\varepsilon_{275}$
Monomer	Tyr A14 II	10.5	0.17
	Tyr B26 II	6.0	−0.39
	Phe B24 II	6.8	0.13
	Phe B25 II	2.9	−0.15
	Pb (A1–A6) II	(2.8–7.8)	−0.45
	Pb (A9–A20) II	(1.5–9.9)	−0.18
	Pb (B11–B15) II	(6.9–9.3)	0.02
	Pb B18 II	8.7	0.00
	Pb (B23–B29) II	(3.9–9.2)	0.22
	Glu A4 II	7.4	0.08
	Glu A17 II	8.8	0.05
	Gln A5 II	3.9	−0.03
	Gln A15 II	6.5	0.00
	Asn A18 II	4.4	0.27
	Asn A21 II	7.9	−0.04
	COO A21 II	9.4	−0.01
	COO B30 II	5.9	−0.04
Dimer	Tyr B26 I	12.1	0.13
	Phe B24 I	9.7	0.01
	Phe B25 I	8.3	−0.09
	Pb B23 I	9.7	−0.03
	Pb B24 I	8.7	0.07
	Asn A21 I	9.5	0.00
Hexamer	—	—	—

Tyr B16 I

	Perturbing group	$R_{\min}$ (Å)	$\Delta\varepsilon_{275}$
Monomer	Phe B24 II	4.6	0.15
	Pb A19 I	9.8	−0.05
	Pb (B9–B24) I	(1.7–8.7)	−0.64
	Glu B13 I	4.0	−0.13
	Glu B21 I	6.6	0.01
Dimer	Tyr B26 II	3.0	−0.30
	Phe B24 II	7.6	0.07
	Pb A2 II	8.9	−0.10
	Pb A6 II	9.9	−0.11
	Pb (B6–B15) II	(1.8–9.3)	−0.48
	Pb (B25–B28) II	(7.1–9.1)	0.08
	His B10 II	6.4	0.16
	Glu B13 II	7.9	−0.04
Hexamer	Tyr A14 II	14.2	−0.16
	Pb A11 II	9.9	−0.09
	Pb (B1–B7) II	(5.9–9.7)	−0.16
	Pb B9 II	9.1	0.05
	Pb B10 II	7.9	−0.01
	Pb B13 II	8.5	−0.04
	His B5 II	9.8	−0.04
	His B10 II	5.5	−0.13
	Glu B13 II	9.1	−0.03
	Gln B4 II	2.4	−0.27

Tyr B16 II

	Perturbing group	$R_{\min}$ (Å)	$\Delta\varepsilon_{275}$
Monomer	Phe B24 II	4.7	0.12
	Pb A19 II	9.9	−0.05
	Pb (B9–B24) II	(1.5–8.5)	−0.44
	Glu B13 II	3.9	−0.12
	Glu B21 II	8.9	0.04
	Asn A21 II	9.8	0.05
Dimer	Tyr B26 I	2.7	−0.14
	Phe B24 I	7.8	0.07
	Pb A2 I	9.4	−0.04
	Pb A6 I	9.7	−0.11
	Pb (B6–B15) I	(1.8–9.6)	−0.70
	Pb (B25–B29) I	(9.3–7.2)	0.12
	His B10 II	6.3	0.15
	Glu A4 I	9.4	−0.04
	Glu B13 I	7.9	−0.03
Hexamer	Tyr A14 I	14.2	−0.17
	Pb (B1–B7) I	(5.8–8.7)	−0.23
	Pb B9 I	9.0	0.05
	Pb B10 I	7.7	−0.03
	Pb B13 I	8.4	−0.03
	His B10 I	9.9	−0.04
	His B5 I	8.6	0.14
	His B10 I	5.3	−0.13
	Glu B13 I	8.9	−0.03
	Glu B4 I	2.9	−0.28

Tyr B26 I

	Perturbing group	R_{min} (Å)	$\Delta\varepsilon_{275}$
Monomer	Tyr A19 I	5.2	−0.29
	Tyr B16 I	10.2	−0.14
	Phe B24 I	4.6	0.16
	Phe B25 I	7.8	−0.14
	Pb (A1–A6) I	(4.9–9.0)	−0.07
	Pb A16 I	9.7	0.00
	Pb A18 I	9.3	−0.08
	Pb A19 I	8.3	0.02
	Pb (B6–B15) I	(3.4–7.4)	−0.63
	Pb (B23–B29)	(1.9–8.9)	0.56
	His B10 I	9.3	0.00
	Glu A4 I	5.9	−0.04
	Glu B13 I	9.8	0.04
	COO B30 I	9.8	−0.03
Dimer	Tyr B16 II	3.2	−0.74
	Tyr B26 II	9.7	0.00
	Phe B24 II	1.9	0.29
	Phe B25 II	9.2	−0.19
	Pb A19 II	8.2	0.00
	Pb A20 II	7.9	−0.08
	Pb (B11–B25) II	(3.4–3.4)	−0.13
	Glu B13 II	8.5	0.03
	Glu B21 II	9.7	0.00
	Asn A21 II	5.6	−0.11
	COO A21 II	8.4	0.03
Hexamer	Glu B4 I	7.4	0.02

Tyr B26 II

	Perturbing group	R_{min} (Å)	$\Delta\varepsilon_{275}$
Monomer	Tyr A19 II	6.0	−0.09
	Tyr B16 II	10.8	−0.15
	Phe B24 II	4.9	0.31
	Phe B25 II	6.4	0.28
	Pb (A1–A6) II	(4.6–9.3)	−0.25
	Pb A16 II	9.4	0.01
	Pb A18 II	9.3	−0.09
	Pb A19 II	8.2	0.04
	Pb (B5–B15) II	(3.6–9.7)	−0.58
	Pb (B23–B29) II	(2.1–9.4)	0.62
	His B10 II	9.5	0.02
	Glu A4 II	8.0	−0.006
	COO B30 II	7.9	0.07
Dimer	Tyr B16 I	3.06	−0.57
	Tyr B26 I	8.9	−0.02
	Phe B24 I	2.4	0.23
	Phe B25 I	7.7	0.00
	Pb A19 I	8.4	−0.01
	Pb A20 I	7.8	−0.08
	Pb (B11–B25) I	(3.4–9.8)	−0.08
	Glu B13 I	8.7	0.03
	Glu B21 I	7.5	−0.06
	Asn A21 I	6.1	0.04
	COO A21 I	9.0	−0.02
Hexamer	Glu B4 II	7.0	−0.02

45. Chemistry, structure, and function of insulin†

DIETRICH BRANDENBURG

Introduction

'No attempts have been made to speculate about the effects of modification on the insulin structure, especially in view of the fact that Professor Dorothy Hodgkin has elucidated the crystal structure, and we have to become acquainted with it first.' This sentence concluded a communication (Brandenburg, Biela, Herbetz, and Zahn 1970) on the effect of A1-modification and intramolecular crosslinking on the biological activity of insulin, presented on 18 September 1969 at the 5th Annual Meeting of the European Association for the Study of Diabetes at Montpellier.

At that time, structure–function studies could be not much more than attempts to search for relationships between the hormone's *primary* structure and its biological properties. The description of the tertiary structure, published in the 1 November issue of *Nature* (Adams *et al.* 1969), and later the availability of three-dimensional models created an exciting turning point, and provided the basis for systematic studies linking primary to tertiary structure and ultimately to hormonal function (for reviews see Blundell, Dodson, Hodgkin, and Mercola 1972; Geiger 1976; Brandenburg 1978; Brandenburg and Saunders 1979; Brandenburg, Saunders, and Rudolph 1980).

Ten years later, not less than one-third of the communications presented at the 2nd International Insulin Symposium in Aachen were directly related to the three-dimensional structure of insulin. Therefore, it appears more than appropriate to dedicate the Proceedings of this Symposium (Brandenburg and Wollmer 1980*a*) to Dorothy Hodgkin. In this paper a short summary is given of those communications which describe recent results from the Wool Research Institute and from studies carried out in co-operation with other groups. Further reports on the Meeting have appeared (Sönsken 1979; Blundell 1980; Brandenburg and Wollmer 1980*b*).

† Selected topics from the 2nd International Insulin Symposium, Aachen, Federal Republic of Germany, 4–7 September 1979.

Currently our chemical work comprises the preparation of three different groups of analogues and derivatives:

  (i) variants of the monomeric hormone modified at either a single site or in several positions by substitution of functional groups, addition, removal or replacement of amino acids;
  (ii) cross-linked derivatives in which two reactive sites are covalently linked within one monomer, or intermolecularly between different monomers;
  (iii) derivatives with photo-activatable groups as ligands for subsequent *in situ* cross-linking to receptors and other macromolecules.

Homogeneous derivatives of groups (ii) and (iii) are prepared by chemical modification and semisynthetic alteration of the native hormone. For modified monomers (group (i)), total synthesis is now extending these possibilities.

Variations of the monomer structure

The N-terminus of the A-chain is, on the basis of the three-dimensional structure (Adams *et al.* 1969; Blundell *et al.* 1972) and previous structure function data (Blundell *et al.* 1972; Geiger 1976; Brandenburg 1978; Brandenburg and Saunders 1979; Brandenburg *et al.* 1980; Pullen *et al.* 1976), important with respect to stabilization of the monomer and to receptor binding. Transformation of each of the three amino groups into dimethyl amino groups, a modification which preserves the positive charge, has now been achieved by reductive methylation of suitably protected intermediates. Replacement of the two A1-hydrogens has almost no effect on bioactivity *in vitro*, but may interfere with crystallization (Uschkoreit, Brandenburg, and Friesen 1980).

Elongation of the A-chain by a glutamic acid residue shifts the amino group by three atoms, abolishes the net charge in position A0, and reduces the *in vitro* potency to 22 per cent. This analogue crystallizes well, and X-ray data have been collected to 1.9 Å resolution (Bedarkar *et al.* 1979). While A1-modified insulins usually show a direct correlation between receptor binding and activity, analogues in which the A-chain is extended by one to three basic amino acids show a discrepancy: the lipogenic potency in fat cells was consistently between 35 and 40 per cent, but affinities for receptors in beef liver plasma membranes ranged between 80 (Lys^{A-1}—Arg^{A0}—) and 125 per cent (Arg_3-insulin) (Rösen *et al.* 1980).

A-chains in which the tyrosine positions A14 and A19 are varied have been assembled by total synthesis in solution and combined with natural B-chains to give, after purification, two homogeneous analogues (Wieneke, Danho, Büllesbach, Gattner, and Zahn 1979; Danho, Sasaki, Büllesbach, Gattner, and Wollmer 1980). The presence of one iodine atom at the invariant A19–tyrosine led to a marked reduction of activity (20 per cent) (Wieneke *et al.* 1979). The *in vitro* potency of insulin is unaffected if the OH group is absent, as in [Phe^{A14}]-insulin (Danho *et al.* 1980). The A14-tyrosine side-chain is in close contact with the N-terminus of the B-chain (Adams *et al.* 1969; Blundell *et al.* 1972). This region proved to be important for dynamic aspects of insulin structure (Wollmer *et al.* 1980). The X-ray analysis of des-Phe^{B1}-insulin and energy calculations point towards high thermal mobility not only of B1-phenylalanine but also of A14-tyrosine. This explains the discrepancy between calculated and observed tyrosine circular dichroism (Chapter 44). Mobility of A14-tyrosine would also be in line with an undefined biological role.

Replacement of the phenyl residue by the 3-monoiodo-4-hydroxyphenyl group and simultaneous omission of the amino group (i.e. incorporation of 3-iodo-desaminotyrosine in position B1) has no effect on activity (Bahrami and Brandenburg 1980). Even the 3,5-diiodo derivative, in which the size of the side-chain is tripled, has retained almost 80 per cent potency *in vitro* (Bahrami and Brandenburg 1980). Obviously, the larger side-chain is also mobile and can assume a position which does not interfere with the monomer structure. Replacement of the primary by a tertiary amino group does not reduce the activity (Uschkoreit *et al.* 1980). While these observations appear to be in agreement with the concept that the N-terminus of the B-chain is not directly related to interactions with insulin receptors (Blundell *et al.* 1972; Pullen *et al.* 1976), photochemical labelling (see below) clearly points to a different direction. In order to obtain a semisynthetic insulin with selectively iodinated B16-tyrosine, the corresponding tryptic fragment B(1–22)-*bis*-*S*-sulphonate containing only one tyrosine has been subjected to iodination (Zewail, Gattner, and Danho 1979).

The C-terminal part of the B-chain plays a central role in current considerations regarding the receptor binding region of insulin. Parallels have been drawn between the phenomena of binding and dimerization (Pullen *et al.* 1976). Systematic variations in this region are hampered by the fact that solid-phase synthesis has not proved to be a fast and easy route to homogeneous insulin analogues, and total synthesis in solution is still a demanding task. Unfortunately semisynthesis using des-octapeptide (B23-30)-insulin as intermediate also creates many problems which are mainly related to carboxyl protection. Enzymatic peptide coupling by means of trypsin in aqueous organic solvent mixtures (Inouye *et al.* 1979) may help to overcome difficulties. The method is currently being investigated in detail but seems to be very sensitive with respect to sequence (Gattner, Danho, and Naithani 1980). On the other hand, transformation of porcine to human insulin by enzymatic replacement of B30 proceeded smoothly with high yields (Gattner, Danho, and Naithani 1980).

For studies concerning position B25 a [Tyr^{25}]-B-chain has been synthesized by fragment condensation in solution (Knorr, Danho, Büllesbach, Gattner, and Zahn 1980). A synthetic octapeptide B(23–30) in which 25-phenylalanine is replaced by L-4-nitrophenylalanine has been prepared for incorporation into B-chain (Casaretto, Danho, Gattner, and Zahn 1980). This residue is, like azido compounds (see below), light activatable and will serve as a photo-label of this important site. Neither activity nor the ability of insulin to crystallize is affected by replacing B29-lysine with dimethyllysine (Uschkoreit *et al.* 1980).

The three amino groups of insulin are the major sites for specific chemical modification and often the starting point for sequential alterations of the primary structure. Relative reactivities have been discussed as an indicator of structural accessibility and, with respect to differential protection, an essential strategy in semisynthesis (Friesen 1980).

Cross-linked insulins

The crystal structure of A1-B29-LL-diaminosuberoyl (Dsu) insulin, the only intermolecularly cross-linked derivative which has so far been obtained in rhombohedral 2Zn crystals, has been determined at 3.1 Å spacing (Dodson, Cutfield, Hoenjet, Wollmer, and Brandenburg 1980). The X-ray analysis and CD-data, which indicate an intact association behaviour in solution, show that the three-dimensional structure of this derivative is very similar to insulin and, in particular, that the monomer–monomer interface is unchanged. If there are parallels between the phenomena of monomer–monomer association and receptor binding, which for Dsu-insulin is reduced to a value as low as 5 per cent, then it appears that the correct spatial arrangement of the interacting residues is insufficient for proper binding, and that other features are involved. The cross-link may, for instance, reduce the flexibility of the hormone from assuming a particular conformation before or during the binding process (Dodson *et al.* 1980).

Covalent insulin dimers are a special group of cross-linked derivatives since each half can be considered both as carrier and substituent. All the six possible isomers, three symmetrical and three unsymmetrical ones, have been prepared by specific cross-linking of amino groups, in suitably protected intermediates by means of suberic acid *bis-p*-nitrophenyl ester and similar reagents (Schüttler and Brandenburg 1980).

As shown in three subsequent reports (Schüttler and Brandenburg 1980; Piron *et al.* 1980; Willey, Tatnell, Jones, Schüttler, and Brandenburg 1980), these dimers exhibit very interesting and often unexpected biological properties, such as slower association to, and dissociation from, receptors, differences between *in vivo* and *in vitro* bioactivity, tissue-dependent binding (liver cells or membranes < adipocytes), and divergence between affinity and *in vitro* potency. Receptor-binding studies with ^{125}iodo-labelled dimers gave three different types of Scatchard plots indicating lost negative co-operativity and, in some cases, bivalent binding (Piron *et al.* 1980).

A synoptical interpretation of the biological data so far obtained is at present very difficult, since our knowledge is limited to the spatial structure of the constituing monomer. Computer drawings of the six space-filled models (R. B. Feldmann, in Piron *et al.* 1980) give a first idea of the possible arrangement of the dimers. This new class of modified insulins appears to become a useful tool for studying monomer–monomer interactions and receptor topography.

Photo-affinity labelling

While information about the receptor-binding site of insulins from groups (i) and (ii) is indirect, affinity labelling of the receptor was expected to provide direct evidence. Photochemical labelling (Levy 1973) appeared to be a promising approach and is presently being extensively studied (Thamm, Saunders, and Brandenburg 1980; Saunders, Thamm, and Brandenburg 1980; Wisher, Thamm, Saunders, Sönksen, and Brandenburg 1980; Diaconescu, Thamm, Saunders, and Brandenburg 1980).

About 15 photo-activatable insulins ('P-insulins') have been prepared in homogeneous form and thoroughly characterized. They contain a light-sensitive group such as 4-azidophenyl acetyl (Apa), 4-azido-benzimidyl (Ab), 2-nitro, 4-azido-phenyl acetyl (Napa), 2-nitro, 4-azido phenyl (Nap) and Nap—Gly bound to the amino groups of A1-glycine, B1-phenylalanine, or B29-lysine of insulin, to B2-valine of des-Phe^{B1}-insulin, or the side-chain of [A1-D-α,γ diamino butyric acid]-insulin (Thamm *et al.* 1980; Saunders *et al.* 1980). A1,B29-di-substituted derivatives and a doubly labelled dimer have also been obtained.

Specific photo-induced binding of radio-iodinated P-insulins to rat liver membranes (Wisher *et al.* 1980) could be demonstrated. It is remarkable that specific attachment did not only occur at A1 or B29 but also at position B1 which previously was thought to be remote from the receptor-binding region of insulin. The molecular weight of covalent conjugates isolated from liver-cell membranes were about 300 K and, after reduction, of 130 K. The true molecular weight is probably lower and in the order of 90 K. When B2-Napa-des-Phe^{B1}-insulin was coupled to viable adipocytes a permanent activation was observed as monitored by lypogenesis (Diaconescu *et al.* 1980). These experiments show that photochemically labelled insulins are valuable for studying hormone receptor interactions and for obtaining information on the events following binding.

Conclusions

The elucidation of the three-dimensional structure of insulin has greatly influenced the direction and extent of insulin research in our Institute. But in spite of extensive work in this and other laboratories over the last ten years, it appears difficult to establish simple correlations between structural features and functional aspects, and to draw meaningful conclusions. This may reflect the fact that this hormone is a protein of minimal size and that the monomer has a less well-defined topography when released from the stabilizing environment of the hexamer. Thus the insulin model appears to be a constant stimulus of yet undetermined dynamics, rather than a rigid dogma.

Acknowledgements

Sincere thanks are due to Professor H. Zahn for his constant interest and support, and all members of the Insulin Division for their enthusiasm over the course of the years. Particular gratitude is expressed to our friends from other laboratories with whom the co-operative studies have been carried out; to Axel Wollmer for his tireless efforts as co-editor of the Proceedings, and to Pamela Kaplin for assistance in the preparation of this manuscript.

References

ADAMS, M. J., BLUNDELL, T. L., DODSON, E. J., DODSON, G. G., VIJAYAN, M., BAKER, E. N., HARDING, M. M., HODGKIN, D. C., RIMMER, B., and SHEAT, S. (1969). *Nature, Lond.* **224,** 491.

BAHRAMI, S. and BRANDENBURG, D. (1980). In *Chemistry, structure and function of insulin and related hormones*. Proc. 2nd Int. Insulin Symp., Aachen (ed. D. Brandenburg and A. Wollmer) p. 177. Walter de Gruyter, Berlin.

BEDARKAR, S., BLUNDELL, T. L., WOOD, S. P., TICKLE, I. J., FRIESEN, H.-J., GATTNER, H.-G., BRANDENBURG, D., and ZAHN, H. (1979). Abstract, 2nd Int. Insulin Symp., Aachen, p. 7.

BLUNDELL, T. L. (1980). *FEBS Lett.* **109,** 167–70.

—— DODSON, G. G., HODGKIN, D. C., and MERCOLA, D. A. (1972). *Adv. protein Chem.* **26,** 279–402.

BRANDENBURG, D. (1978). In *Amino-acids, peptides, and proteins*, Vol. 9 (Senior reporter R. C. Sheppard). The Chemical Society, London.

—— and SAUNDERS, D. (1979). In *Amino-acids, peptides, and proteins*, Vol. 10 (Senior reporter R. C. Sheppard). The Chemical Society, London.

—— and WOLLMER, A. (1980*a*). *Chemistry, structure and function of insulin and related hormones*. Proc. 2nd Int. Insulin Symp., Aachen. Walter de Gruyter, Berlin.

———— (1980*b*). *Nachr. Chem. Tech.* **28,** 11–15.

—— SAUNDERS, D., and RUDOLPH, B.-E. (1980). In *Amino-acids, peptides, and proteins*, Vol. 11 (Senior reporter R. C. Sheppard). The Chemical Society, London.

—— BIELA, M., HERBERTZ, L., and ZAHN, H. (1970). *Diabetologia* **6,** 38.

CASARETTO, M., DANHO, W., GATTNER, H.-G., and ZAHN, H. (1980). In *Chemistry, structure and function of insulin and related hormones*. Proc. 2nd Int. Insulin Symp., Aachen (ed. D. Brandenburg and A. Wollmer) p. 73. Walter de Gruyter, Berlin.

DANHO, W., SASAKI, A., BÜLLESBACH, E., GATTNER, H.-G., and WOLLMER, A. (1980). In *Chemistry, structure and function of insulin and related hormones*. Proc. 2nd Int. Insulin Symp., Aachen (ed. D. Brandenburg and A. Wollmer) p. 59. Walter de Gruyter, Berlin.

DIACONESCU, C., THAMM, P., SAUNDERS, D., and BRANDENBURG, D. (1980). In *Chemistry, structure and function of insulin and related hormones*. Proc. 2nd Int. Insulin Symp., Aachen (ed. D. Brandenburg and A. Wollmer) p. 351. Walter de Gruyter, Berlin.

DODSON, G. G., CUTFIELD, S., HOENJET, E., WOLLMER, A., and BRANDENBURG, D. (1980). In *Chemistry, structure and function of insulin and related hormones*. Proc. 2nd Int. Insulin Symp., Aachen (ed. D. Brandenburg and A. Wollmer) p. 17. Walter de Gruyter, Berlin.

FRIESEN, H.-J. (1980). In *Chemistry, structure and function of insulin and related hormones*. Proc. 2nd Int. Insulin Symp., Aachen (ed. D. Brandenburg and A. Wollmer) p. 125. Walter de Gruyter, Berlin.

GATTNER, H.-G., DANHO, W., and NAITHANI, V. K. (1980). In *Chemistry, structure and function of insulin and related hormones*. Proc. 2nd Int. Insulin Symp., Aachen (ed. D. Brandenburg and A. Wollmer) p. 117. Walter de Gruyter, Berlin.

GEIGER, R. (1976). *Chemikerzeitung* **100**, 111.

INOUYE, K., WATANABE, K., MORIHARA, K., TOCHINO, Y., KANAYA, T., EMURA, J., and SAKAKIBARA, S. (1979). *J. Am. Chem. Soc.* **101**, 751–2.

KNORR, R., DANHO, W., BÜLLESBACH, E., GATTNER, H.-G., and ZAHN, H. (1980). In *Chemistry, structure and function of insulin and related hormones*. Proc. 2nd Int. Insulin Symp., Aachen (ed. D. Brandenburg and A. Wollmer) p. 67. Walter de Gruyter, Berlin.

LEVY, D. (1973). *Biochim. biophys. Acta* **322**, 329.

PIRON, M., MICHIELS-PLACE, M., WAELBROECK, M., DE MEYTS, P., SCHÜTTLER, A., and BRANDENBURG, D. (1980). In *Chemistry, structure and function of insulin and related hormones*. Proc. 2nd Int. Insulin Symp., Aachen (ed. D. Brandenburg and A. Wollmer) p. 371. Walter de Gruyter, Berlin.

PULLEN, R., LINDSAY, D., WOOD, S., TICKLE, I., BLUNDELL, T., WOLLMER, A., KRAIL, G., BRANDENBURG, D., ZAHN, H., GLIEMANN, J., and GAMMELTOFT, S. (1976). *Nature, Lond.* **259**, 369–73.

RÖSEN, P., SIMON, M., REINAUER, H., BRANDENBURG, D., FRIESEN, H.-J., and DIACONESCU, C. (1980). In *Chemistry, structure and function of insulin and related hormones*. Proc. 2nd Int. Insulin Symp., Aachen (ed. D. Brandenburg and A. Wollmer) p. 403. Walter de Gruyter, Berlin.

SAUNDERS, D., THAMM, P., and BRANDENBURG, D. (1980). In *Chemistry, stucture and function of insulin and related hormones*. Proc. 2nd Int. Insulin Symp., Aachen (ed. D. Brandenburg and A. Wollmer) p. 317. Walter de Gruyter, Berlin.

SCHLÜTER, K., PETERSEN, K.-G., SCHÜTTLER, A., BRANDENBURG, D., and KERP, L. (1980). In *Chemistry, structure and function of insulin and related hormones*. Proc. 2nd Int. Insulin Symp., Aachen (ed. D. Brandenburg and A. Wollmer) p. 433. Walter de Gruyter, Berlin.

SCHÜTTLER, A. and BRANDENBURG, D. (1980). In *Chemistry, structure and function of insulin and related hormones*. Proc. 2nd Int. Insulin Symp., Aachen (ed. D. Brandenburg and A. Wollmer) p. 143. Walter de Gruyer, Berlin.

SÖNKSEN, P. H. (1979). *Nature, Lond.* **282**, 11–12.

THAMM, P., SAUNDERS, D., and BRANDENBURG, D. (1980). In *Chemistry, structure and function of insulin and related hormones*. Proc. 2nd Int. Insulin Symp., Aachen (ed. D. Brandenburg and A. Wollmer) p. 309. Walter de Gruyter, Berlin.

USCHKOREIT, J., BRANDENBURG, D., and FRIESEN, H.-J. (1980). In *Chemistry, structure and function of insulin and related hormones*. Proc. 2nd Int. Insulin Symp., Aachen (ed. D. Brandenburg and A. Wollmer) p. 191. Walter de Gruyter, Berlin.

WIENEKE, H.-J., DANHO, W., BULLESBACH, E., GATTNER, H.-G., and ZAHN, H. (1979). Abstract, 2nd Int. Insulin Symp., Aachen, p. 167.

WILLEY, K. P., TATNELL, M. A., JONES, R. H., SCHÜTTLER, A., and BRANDENBURG, D. (1980). In *Chemistry, structure and function of insulin and related hormones*. Proc. 2nd Int. Insulin Symp., Aachen (ed. D. Brandenburg and A. Wollmer) p. 425. Walter de Gruyter, Berlin.

WISHER, M. H., THAMM, P., SAUNDERS, D., SÖNKSEN, P. H., and BRANDENBURG, D. (1980). In *Chemistry, structure and function of insulin and related hormones*. Proc. 2nd Int. Insulin Symp., Aachen (ed. D. Brandenburg and A. Wollmer) p. 345. Walter de Gruyter, Berlin.

WOLLMER, A., STRASSBURGER, W., HOENJET, E., GLATTER, U., FLEISCHHAUER, J., MERCOLA, D. A., DE GRAAF, R. A. G., DODSON, E. J., DODSON, G. G., SMITH, D. G., BRANDENBURG, D., and DANHO, W. (1980). In *Chemistry, structure and function of insulin and related hormones*. Proc. 2nd Int. Insulin Symp., Aachen (ed. D. Brandenburg and A. Wollmer) p. 27. Walter de Gruyter, Berlin.

ZEWAIL, M., GATTNER, H.-G., and DANHO, W. (1979). Abstract, 2nd Int. Insulin Symp., Aachen, p. 183.

Publications of
Dorothy Mary Crowfoot Hodgkin

Steroids

BERNAL, J. D. and CROWFOOT, D. (1934). X-ray crystallographic measurements of some derivatives of cardiac aglucones. *Chemy Ind.* 953–6.

BERNAL, J. D. and CROWFOOT, D. (1935). Structure of some hydrocarbons related to the sterols. *J. chem. Soc.* 93–100.

CROWFOOT, DOROTHY. (1935). X-ray crystallography of the toad poisons bufagin and cinobufagin and of strophanthidin. *Chemy Ind.* 568–9.

BERNAL, J. D. and CROWFOOT, D. (1935). Molecular shape of calciferol and related substances. *Chemy Ind.* 701–2.

CROWFOOT, D. M., RAPSON, WM. SAGE, and ROBINSON, ROBERT (1936). Synthesis of substances related to sterols. XI. The constitution of the condensation products from acetylcyclopentene or acetylcyclohexene and methoxytetralone. *J. chem. Soc.* 757–9.

BERNAL, J. D. and CROWFOOT, D. M. (1936). X-ray crystallographic data on the sex hormones, estrone, androsterone, testosterone and progesterone and related substances. *Z. Kristallogr. Kristallgeom.* **93**, 464–80.

CROWFOOT, DOROTHY, and JENSEN, H. (1936). Molecular weight of cinobufagin. *J. Am. chem. Soc.* **58**, 2018–19.

CROWFOOT, D. and BERNAL, J. D. (1937). X-ray crystallography and the chemistry of sterols and sex hormones. *Chem. Weekbl.* **34**, 19–22.

BERNAL, J. D., CROWFOOT, DOROTHY, and FANKUCHEN, I. (1940). X-ray crystallography and the chemistry of the steroids. I. *Trans. R. Soc.* **A239**, 135–82.

CROWFOOT, D. M. and LOW, B. W. (1943). A note on the crystallography of helvolic acid and the methyl ester of helvolic acid. *Br. J. exp. Path.* **24**, 120.

CROWFOOT, DOROTHY (1944). X-ray crystallography and sterol structure. *Vitams Hor.* **2**, 409–61.

CARLISLE, C. H. and CROWFOOT, D. (1945). The crystal structure of cholesteryl iodide. *Proc. R. Soc.* **A184**, 64–83.

CROWFOOT, DOROTHY and DUNITZ, J. D. (1948). Structure of calciferol. *Nature, Lond.* **162**, 608–9.

HODGKIN, DOROTHY CROWFOOT, and SAYRE, D. (1952). Crystallographic examination of the structure of lumisterol. *J. chem. Soc.* 4561–6.

HODGKIN, DOROTHY C., WEBSTER, MONICA S., and DUNITZ, J. D. (1957). Structure of calciferol. *Chemy Ind.* 1148–9.

TAYLOR, NOEL, E., HODGKIN, DOROTHY CROWFOOT, and ROLLETT, J. S. (1960). The X-ray crystallographic determination of the structure of bromomirestrol. *J. chem. Soc.* 3685–95.

SAUNDERSON, CAROL P. and HODGKIN, DOROTHY CROWFOOT (1961). Crystal structure of suprasterol II. *Tetrahedron Lett.* 573–8.

592 *Publications of Dorothy Mary Crowfoot Hodgkin*

HODGKIN, DOROTHY CROWFOOT, RIMMER, BERYL M., DUNITZ, J. D., and TRUEBLOOD, K. N. (1963). The crystal structure of a calciferol derivative. *J. chem. Soc.* 4945–56.
HODGKIN, DOROTHY CROWFOOT (1963). X-ray crystallography in hormone research. *Tech. Endocr. Res., Proc. Stratford-upon-Avon, Eng.* 7–22.
COOPER, A. and HODGKIN, D. C. (1967). The crystal structure and absolute configuration of fusidic acid methyl ester 3-*p*-bromobenzoate. *Tetrahedron* **24**, 909–22.
HARRISON, H. R., HODGKIN, DOROTHY C., MASLEN, E. N., MOTHERWELL, W. D. S. (1971). Crystal and molecular structures of 6β-bromoacetoxy- and 6β-chloroacetoxy-3,5α-cholestane (1-cholesteryl bromoacetate and chloroacetate). *J. chem. Soc.C* 1275–81.

Penicillins and related antibiotics

CROWFOOT, D., BUNN, C. W., ROGERS-LOW, B. W., and TURNER-JONES, A. (1949). X-ray crystallographic investigation of the structure of penicillin. In *Chemistry of penicillin* (ed. H. T. Clarke *et al.*), pp. 310–66. Princeton University Press.
HODGKIN, DOROTHY CROWFOOT (1949). X-ray analysis of the structure of penicillin. *Advmt Sci.* **6**, 85–9.
HODGKIN, DOROTHY CROWFOOT and MASLEN, E. N. (1961). The X-ray analysis of the structure of cephalosporin C. *Biochem. J.* **79**, 393–402.
ABRAHAMSSON, S., HODGKIN, DOROTHY CROWFOOT, and MASLEN, E. N. (1963). Crystal structure of phenoxymethylpenicillin. *Biochem. J.* **86**, 514–35.
JAMES, M. N. G., HALL, D., and HODGKIN, D. C. (1968). Crystalline modifications of ampicillin I: the trihydrate. *Nature, Lond.* **220**, 168–70.
VIJAYAN, KALYANI, ANDERSON, B. F., and HODGKIN, D. C. (1973). Crystal and molecular structure of a synthetic compound related to the penicillins and cephalosporins, 3-benzyl 7-tert-butyl 2,2-dimethyl-8-oxo-4-thia-1-aza-6αH-bicyclo 4,2,0-octane-3β,5α-dicarboxylate. *J. chem. Soc., Perkin Trans.* **1**, 484–8.

Vitamin B₁₂ and corrins

HODGKIN, DOROTHY, C., PORTER, M. W., and SPILLER, R. C. (1950). Crystallographic measurements of the antipernicious anemia factor. *Proc. R. Soc.* **B136**, 609–13.
BRINK, CLARA, HODGKIN, DOROTHY CROWFOOT, LINDSEY, JUNE, PICKWORTH, JENNY, ROBERTSON, JOHN H., and WHITE, JOHN G. (1954). X-ray crystallographic evidence on the structure of vitamin B₁₂. *Nature, Lond.* **174**, 1169–70.
HODGKIN, DOROTHY, C. (1955). X-ray crystallographic study of the structure of vitamin B₁₂. *Bull. Soc. fr. Minér. crystallogr.* **78**, 106–15.
HODGKIN, DOROTHY CROWFOOT, PICKWORTH, JENNY., ROBERTSON, JOHN H., TRUEBLOOD, KENNETH, N., PROSEN, RICHARD J., and WHITE, JOHN G. (1955). The crystal structure of the hexacarboxylic acid derived from B₁₂ and the molecular structure of the vitamin. *Nature, Lond.* **176**, 325–8.
KAMPER, M. J. and HODGKIN, DOROTHY CROWFOOT (1955). The crystal structure of a chlorine-substituted vitamin B₁₂. *Nature, Lond.* **176**, 551–3.
HODGKIN, DOROTHY CROWFOOT, JOHNSON, A. W., and TODD, ALEXANDER R. (1955). Structure of vitamin B₁₂. *Chem. Soc. (London) Spec. Publ.* **3**, 109–123.
HODGKIN, DOROTHY C., KAMPER, JENNIFER, MACKAY, MAUREEN, PICKWORTH, JENNY, TRUEBLOOD, KENNETH N., and WHITE, JOHN G. (1956). Structure of vitamin B₁₂. *Nature, Lond.* **173**, 64–6.
HODGKIN, DOROTHY C., KAMPER, JENNIFER, LINDSEY, JUNE, MACKAY, MAUREEN,

PICKWORTH, JENNY, ROBERTSON, J. H., SHOEMAKER, CLARA B., WHITE, J. G., PROSEN, R. J., and TRUEBLOOD, K. N. (1957). The structure of vitamin B_{12}. I. An outline of the crystallographic investigation of vitamin B_{12}. *Proc. R. Soc.* **A424**, 228–63.

HODGKIN, DOROTHY CROWFOOT (1958). X-ray analysis and the structure of vitamin B_{12}. *Fortschr. Chem. org. NatStoffe.* **15**, 167–220.

HODGKIN, DOROTHY C., PICKWORTH, JENNY, ROBERTSON, J. H., PROSEN, R. J., SPARKS, R. A., and TRUEBLOOD, K. N. (1959). Structure of vitamin B_{12}. II. Crystal structure of a hexacarboxylic acid obtained by the degradation of vitamin B_{12}. *Proc. R. Soc.* **A251**, 306–52.

HODGKIN, DOROTHY C., KAMPER, M. J., TRUEBLOOD, K. N., and WHITE, J. G. (1960). The structures of wet and air-dried crystals of vitamin B_{12}. *Z. Kristallogr. Kristallgeom.* **113**, 30–43.

LENHERT, P. GALEN and HODGKIN, DOROTHY CROWFOOT (1961). Structure of the 5,6-dimethylbenzimidazoylcobamide coenzyme. *Nature, Lond.* **192**, 937–8.

HODGKIN, DOROTHY CROWFOOT, LINDSEY, JUNE, MACKAY, MAUREEN, and TRUEBLOOD, K. N. (1962). The structure of vitamin B_{12}. IV. The X-ray analysis of air-dried crystals of B_{12}. *Proc. R. Soc.* **A266**, 475–93.

HODGKIN, DOROTHY CROWFOOT, LINDSEY, JUNE, SPARKS, R. A., TRUEBLOOD, K. N., and WHITE, J. G. (1962). Structure of the air-dried crystals of vitamin B_{12}. *Proc. R Soc.* **A266**, 494–517.

DALE, DAVID, HODGKIN, DOROTHY CROWFOOT, and VENKATESAN, K. (1963). Determination of the crystal structure of factor V_{1a}. *Cryst. Crystal Perfect., Proc. Symp., Madras* 237–42.

HODGKIN, DOROTHY CROWFOOT (1964). Vitamin B_{12} and the porphyrins. *Fedn Proc. Fedn Am. Socs. exp. Biol.* **23**, 592–8.

BRINK-SHOEMAKER, CLARA, CRUICKSHANK, D. W. J., HODGKIN, DOROTHY CROWFOOT, KAMPER, M. JENNIFER, and PILLING, DIANA (1964). Structure of vitamin B_{12}. VI. Structure of crystals of vitamin B_{12} grown from and immersed in water. *Proc. R. Soc.* **A278**, 1–26.

HODGKIN, DOROTHY CROWFOOT (1965). The structure of the corrin nucleus from X-ray analysis. *Proc. R. Soc.* **A288**, 294–305.

HODGKIN, DOROTHY C. (1965). A discussion on recent experiments on the chemistry of corrins. Postscript. *Proc. R. Soc.* **A288**, 359–60.

HODGKIN, DOROTHY CROWFOOT (1966). Vitamin B_{12}. *Revta colomb. Quim.* **25**, 6–10.

MOORE, F. M., WILLIS, B. T. M. and HODGKIN, DOROTHY CROWFOOT (1967). Structure of a monocarboxylic acid derivative of vitamin B_{12}. Crystal and molecular structure from neutron diffraction analysis. *Nature, Lond.* **214**, 130–3.

NOCKOLDS, C., RAMASESHAN, K. S., HODGKIN, DOROTHY CROWFOOT, WATERS, T. N. M., and WATERS, J. M. (1967). Structure of a monocarboxylic acid derivative of vitamin B_{12}. Crystal and molecular structure from X-ray analysis. *Nature, Lond.* **214**, 129–30.

HODGKIN, DOROTHY C. (1969). Vitamin B_{12}. *Proc. R. Instn Gt Br.* **42**, 377–96.

HODGKIN, DOROTHY C., HARRISON, H. R., and HODDER, O. J. R. (1971). Crystal and molecular structure of 8,12-diethyl-2,3,7,13,17,18-hexamethylcorrole. *J. chem. Soc. B* 640–5.

VENKATESAN, K., DALE, D., HODGKIN, DOROTHY C., NOCKOLDS, C. E., MOORE, F. H. and O'CONNOR, B. H. (1971). Structure of vitamin B_{12}. IX. Crystal structure of cobyric acid, factor V_{1a}. *Proc. R. Soc.* **A323**, 455–87.

STOECKLI-EVANS, HELEN, EDMOND, E., and HODGKIN, DOROTHY, C. (1972). Further refinement of the crystal structure of neo-vitamin B_{12}. *J. chem. Soc., Perkin Trans.* **2**, 605–14.

594   *Publications of Dorothy Mary Crowfoot Hodgkin*

ANDERSON, BRIAN F., BARTCZAK, TADEUSZ J., and HODGKIN, DOROTHY C. (1974). Crystal and molecular structure of 8,12-diethyl-2,3,7,13,17,18-hexamethylcorrole hydrobromide. *J. chem. Soc. Perkin Trans.* **2**, 977–80.

EDMOND, ERIC D. and HODGKIN, DOROTHY C. (1975). Crystal and molecular structure of rac-15-cyano-1,2,2,7,7,12,12-heptamethylcorrin hydrochloride, metal free corrin. *Helv. chim. Acta* **58**, 641–54.

HODGKIN, D. C. (1979). New and old problems in the structure analysis of vitamin B_{12}. *Vitamin B_{12}* (Ed. B. Zagalak and W. Friedrich), pp. 19–36. De Gruyter, Berlin (1979).

Insulin

CROWFOOT, DOROTHY (1935). X-ray single-crystal photographs of insulin. *Nature, Lond.* **135**, 591–2.

CROWFOOT, D. (1937). Two crystalline modifications of insulin. *Nature, Lond.* **140**, 149–50.

CROWFOOT, DOROTHY (1938). Crystal structure of insulin. I. Investigation of air-dried insulin crystals. *Proc. R. Soc.* **A164**, 580–602.

CROWFOOT, DOROTHY and RILEY, DENNIS (1939). X-ray measurements on wet insulin crystals. *Nature, Lond.* **144**, 1011–12.

HARDING, MARJORIE M., HODGKIN, DOROTHY CROWFOOT, KENNEDY, ANN F., O'CONNOR, A., and WEITZMANN, P. D. J. (1966). The crystal structure of insulin. II. An investigation of rhombohedral zinc insulin crystals and a report of other crystalline forms. *J. molec. Biol.* **16**, 212–26.

DODSON, ELEANOR, HARDING, MARJORIE M., HODGKIN, DOROTHY CROWFOOT, and ROSSMANN, MICHAEL G. (1966). The crystal structure of insulin III. Evidence for a 2-fold axis in rhombohedral zinc insulin. *J. molec. Biol.* **16**, 227–41.

ADAMS, M. G., COLLER, E., HODGKIN, DOROTHY CROWFOOT, and DODSON, G. G. (1966). X-ray crystallographic studies on zinc insulin crystals. *Am. J. Med.* **40**, 667–71.

ADAMS, M., COLLER, E., DODSON, G., HODGKIN, D. C., and RAMASESHAN, S. (1966). X-ray crystallographic studies on zinc insulin crystals. *Acta crystallogr.* Suppl. **21**, A156.

ADAMS, M. J., DODSON, G., DODSON, E., and HODGKIN, DOROTHY CROWFOOT (1967). A report on recent calculations on rhombohedral insulin crystals containing lead. *Conform. Biopolym., Pap. Inst. Symp. Madras* **1**, 9–16.

ADAMS, MARGARET JOAN, BLUNDELL, T. L., DODSON, E. J., DODSON, G. G., VIJAYAN, M., BAKER, E. N., HARDING, M. M., HODGKIN, D., RIMMER, B., and SHEAT, S. (1969). Structure of rhombohedral 2-zinc insulin crystals. *Nature, Lond.* **224**, 491–5.

HODGKIN, D. C. (1970). Structure of insulin. *Chemikerzeitung* **94**, 647.

HODGKIN, DOROTHY C. (1970). Crystal structure of insulin. *Verh. schweiz. naturf. Ges.* **150**, 93–101.

HODGKIN, DOROTHY C. (1970). Demonstration. *Mem. Soc. Endocr.* **19**, 519–22.

HODGKIN, D. C. (1971). X-rays and structure of insulin. *Br. med. J.* iv, 447–51.

BLUNDELL, T. L., CUTFIELD, J. F., CUTFIELD, S. M., DODSON, E. J., DODSON, G. G., HODGKIN, D. C., MERCOLA, D. A., and VIJAYAN, M. (1971). Atomic positions in rhombohedral 2-zinc insulin crystals. *Nature, Lond.* **231**, 506–11.

HODGKIN, D. C. and BAKER, E. N. (1971). Crystal structure of insulin and its relation to biological activity. *Chem. N.Z.* **35**, 130.

HODGKIN, DOROTHY C. (1971). Insulin molecules. Extent of our knowledge. *Pure appl. Chem.* **26**, 375–84.

BLUNDELL, T. L., CUTFIELD, J. F., DODSON, G. G., DODSON, E., HODGKIN, D. C., and MERCOLA, D. (1971). Structure and biology of insulin. *Biochem. J.* **125**, 50–1P.

BLUNDELL, T. L., DODSON, G. G., DODSON, E. J., HODGKIN, D. C., and VIJAYAN, M. (1971). X-ray analysis and the structure of insulin. *Recent Prog. Horm. Res.* **27**, 1–40.

BLUNDELL, T. L., CUTFIELD, J. F., DODSON, E. J., DODSON, G. G., HODGKIN, D. C., and MERCOLA, D. A. (1971). Crystal structure of rhombohedral 2 zinc insulin. *Cold Spring Harb. Symp. quant. Biol.* **36**, 233–41.

HODGKIN, DOROTHY C. (1972). Structure of insulin. *Dansk. Tidsskr. Farm* **46**, 1–28.

BLUNDELL, T. L., CUTFIELD, J. F., CUTFIELD, S. M., DODSON, E. J., DODSON, G. G., HODGKIN, D. C., and MERCOLA, D. A. (1972). Three-dimensional atomic structure of insulin and its relation to activity. *Diabetes* **21**, Suppl. 2, 492–505.

HODGKIN, DOROTHY CROWFOOT (1972). Structure of insulin. *Diabetes* **21**, 1131–50.

HODGKIN, DOROTHY CROWFOOT and MERCOLA, D. (1972). Secondary and tertiary structure of insulin. *Handbk. Physiol., Sect. 7, Endocrinol.* **1**, 139–57.

HODGKIN, DOROTHY CROWFOOT, BLUNDELL, T. L., CUTFIELD, J. F., CUTFIELD, S. M., DODSON, G. G., DODSON, E. J., MERCOLA, D. A., and VIJAYAN (1972). Arrangement in three dimensions of the atoms in insulin molecules and crystals. *Insulin Action, Proc. Symp. 1971* 1028 (ed. I. B. Fritz). Academic Press, New York.

HODGKIN, D. C. (1972). Crystal structure of insulin. *Soviet Phys. Crystallogr.* **16**, 1054–60. Translated from the Russian in *Kristallografiya* **16**, 1203–9 (1971).

BLUNDELL, TOM, DODSON, GUY, HODGKIN, DOROTHY, and MERCOLA, DAN (1972). Insulin: the structure in the crystal and its reflection in chemistry and biology. *Adv. Protein Chem.* **36**, 279–402.

HODGKIN, D. and BLUNDELL, T. L. (1972). Features of the conformation of insulin molecules observed in 2 zinc rhombohedral insulin crystals. *Struct. Act. Relat. Protein Polypeptide Horm., Proc. 2nd Int. Symp., 1971* (ed. M. Margoulies *et al.*) pp. 161–6. Exerpta Medica, Amsterdam.

BLUNDELL, T. L., CUTFIELD, J., CUTFIELD, S., DODSON, G., DODSON, E., HODGKIN, D. C., and MERCOLA, D. (1973). Spatial structure of insulin and its relation to activity. *Pept., Proc. 12th Eur. Pept. Symp., 1972* (ed. H. Hanson *et al.*) pp. 255–69. North-Holland, Amsterdam.

BENTLEY, G. A., BLUNDELL, T. L., DODSON, E. J., DODSON, G. G., CUTFIELD, J. F., CUTFIELD, S. M., HODGKIN, D. C., MERCOLA, D., and VIJAYAN, M. (1973). Crystal structure of insulin. *Symp. Pap.—4th Int. Biophys. Congr., 1972* 150–68.

HODGKIN, DOROTHY CROWFOOT (1974). The Bakerian Lecture, 1972. Insulin, its structure and biochemistry. *Proc. R. Soc.* **A338**, 251–75.

HODGKIN, DOROTHY C. (1974). Insulin, its chemistry and biochemistry. *Proc. R. Soc.* **B186**, 191–215.

HODGKIN, D. C. (1974). Varieties of insulin (Sir H. Dale Lecture for 1974). *J. Endocr.* **63**, 3–14.

HODGKIN, DOROTHY CROWFOOT (1975). Chinese work on insulin. *Nature, Lond.* **255**, 103.

CUTFIELD, J. F., DODSON, E. J., DODSON, G. G., HODGKIN, D. C., ISAACS, N. W., SAKABE, K., and SAKABE, N. (1975). A high resolution structure of insulin: a comparison of results obtained from least-squares phase refinement and difference Fourier refinement. Abstract 2.3-2. *Acta crystallogr.* **A31**, s21.

BENTLEY, G. A., CUTFIELD, J. C., CUTFIELD, S. M., DARGAY, J., DODSON, E. J., DODSON, G. G., HODGKIN, D. C., MERCOLA, D., and SABESAN, M. (1975). Some insulin structures—a comparative study. Abstract 03.1-1. *Acta crystallogr.* **A31**, s24.

BENTLEY, GRAHAM, DODSON, ELEANOR, DODSON, GUY, HODGKIN, DOROTHY, and MERCOLA, DAN (1976). Structure of insulin in 4-zinc insulin. *Nature, Lond.* **261**, 166–8.

HODGKIN, D. C. (1976). The structure of insulin. *Proc. 5th Int. Wolltextil-Forschungskonf., 1975*.

HODGKIN, D. C. (1976). X-rays and the structure of insulin. *J. Natn. Sci. Counc. Sri Lanka* **4** (2), 87–98.

HODGKIN, D. C. (1977). The molecular basis of insulin action: the structure of insulin. *Int. Congr. Ser.-Excerpta Med.* **413**, (Diabetes), 155–62.

DODSON, E. J., DODSON, G. G., HODGKIN, D. C., and REYNOLDS, C. D. (1979). Structural relationships in the two-zinc insulin hexamer. *Can. J. Biochem.* **57**, 469–79.

DODSON, E. J., DODSON, G. G., and HODGKIN, D. C. (1980). The conformations observed of the N terminal A chain residues of insulin. In *Frontiers of bioorganic chemistry and molecular biology* (Ed. Ananchenko), p. 145. Pergamon Press, Oxford.

Protein structure in general

BERNAL, J. D. and CROWFOOT, D. (1934). X-ray photographs of crystalline pepsin. *Nature, Lond.* **133**, 794–5.

CROWFOOT, DOROTHY and RILEY, DENNIS (1938). X-ray study of Palmer's lactoglobulin. *Nature, Lond.* **141**, 521–2.

CROWFOOT, DOROTHY and FANKUCHEN, I. (1938). Molecular weight of a tobacco-seed globulin. *Nature, Lond.* **141**, 522–3.

CROWFOOT, DOROTHY (1939). X-ray studies of protein crystals. *Proc. R. Soc.* **A170**, 74–5; **B127**, 35–6.

CROWFOOT, DOROTHY (1941). A review of some recent X-ray work on protein crystals. *Chem. Rev.* **28**, 215–28.

CROWFOOT, DOROTHY (1941). Recent work on the properties of crystalline proteins and viruses. *Proc. 7th Int. Genetical Congr., Edinburgh, Aug. 23–30, 1939*, pp. 92–3.

CROWFOOT, D. and SCHMIDT, G. M. J. (1945). X-ray crystallographic measurements on a single crystal of a tobacco necrosis virus derivative. *Nature, Lond.* **155**, 504–5.

CROWFOOT, DOROTHY (1946). Белковые кристаллы [Protein crystals]. *Успехи химии*, том **XVB**, 2.

HODGKIN, DOROTHY CROWFOOT (1949). An X-ray crystallographic study of certain peptides. *Int. Congr. Biochem., Abstrs. 1st Congr., Cambridge, Engl.* 159–60.

HODGKIN, DOROTHY CROWFOOT (1950). X-ray analysis and protein structure. *Cold Spring. Symp. quant. Biol.* **14**, 65–78.

COWAN, PAULINE and HODGKIN, DOROTHY C. (1951). Comparison of X-ray measurements on air-dried tobacco necrosis protein crystals with electron-microscope data. *Acta crystallogr.* **4**, 160–1.

COWAN, P. M., and HODGKIN, DOROTHY CROWFOOT (1953). Observations on peptide chain models in relation to crystallographic data in gramicidin B and insulin. *Proc. R. Soc.* **B141**, 89–92.

SCHMIDT, G. M. J., HODGKIN, DOROTHY CROWFOOT, and OUGHTON, BERYL M. (1957). Crystallographic study of some derivatives of gramicidin. *Biochem. J.* **65**, 744–50.

HODGKIN, DOROTHY CROWFOOT and OUGHTON, BERYL M. (1957). Appendix 2. Possible molecular models for gramicidin S and their relationship to present ideas of protein structure. *Biochem. J.* **65**, 752–6.

HODGKIN, DOROTHY CROWFOOT and RILEY, DENNIS PARKER (1968). Ancient history of protein X-ray analysis. In *Struct. Chem. Mol. Biol.* (ed. A. Rich and N. Davidson) pp. 15–28. W. H. Freeman Co., San Francisco.

HODGKIN, DOROTHY M. C. (1970). A discussion on structures and functions of proteolytic enzymes. Introductory remarks. *Phil. Trans. R. Soc.* **B257**, 65.

HODGKIN, DOROTHY C. (1974) Problems in the X-ray analysis of proteins. *Chem. Scr.* **6**, 145–57.

HODGKIN, D. C. (1979). Crystallographic measurements and the structure of protein molecules as they are. *Ann. N.Y. Acad. Sci.* **325**, 121–48.

General

POWELL, H. M. and CROWFOOT, D. M. (1932). Layer-chain structures of thallium dialkyl halides. *Nature, Lond.* **130**, 131–2.

BERNAL, J. D. and CROWFOOT, D. (1933). Crystal structure of vitamin B_1 and of adenine hydrochloride. *Nature, Lond.* **131**, 911–12.

BERNAL, J. D. and CROWFOOT, D. (1933). Crystalline phases of some substances studied as liquid crystals. *Trans. Faraday Soc.* **29**, 1032–49.

BERNAL, J. D. and CROWFOOT, D. (1933). The structure of the Diels' hydrocarbon $C_{18}H_{16}$. *Chemy Ind.* 729–30.

POWELL, HERBERT M. and CROWFOOT, DOROTHY M. (1934). The crystal structures of dimethylthallium halides. *Z. Kristallogr. Kristallgeom.* **87**, 370–8.

BERNAL, J. D., CROWFOOT, D. M., ROBINSON, B. W., and WOOSTER, W. A. (1933). Crystallography 1932–33. *Rep. Prog. Chem.* 360–433.

BERNAL, J. D. and CROWFOOT, D. (1934). Use of the centrifuge in determining the density of small crystals. *Nature, Lond.* **134**, 809–10.

CROWFOOT, D. (1935). The interpretation of Weissenberg photographs in relation to crystal symmetry. *Z. Kristallogr. Kristallgeom.* **90**, 215–36.

MANN, FREDERICK G., CROWFOOT, DOROTHY, GATTIKER, DAVID C., and WOOSTER, NORA (1935). The structure and configuration of certain diammino-palladium compounds. *J. chem. Soc.* 1642–52.

BERNAL, J. D., CROWFOOT, D. M., EVANS, R. C., and WELLS, A. F. (1935). Crystallography 1933–35. *Rep. Prog. Chem.* 181–242.

CROWFOOT, D. M. (1936). X-ray crystallographic measurements on phrenosinic (cerebronic) acid and its oxidation product. *J. chem. Soc.* 716–18.

BLOUNT, B. K. and CROWFOOT, D. (1936). Veratrine alkaloids. III. Preparation of cevanthrol, and the X-ray crystallographic examination of cevanthrol and cevanthridine. *J. chem. Soc.* 414–15.

COX, E. G. and CROWFOOT, D. M. (1936). Crystallography 1936. *Rep. Prog. Chem.* 196–227.

CROWFOOT, DOROTHY (1938). Molecular weight of fichtelite. *J. chem. Soc.* 1241–2.

CARLISLE, C. H. and CROWFOOT, DOROTHY (1941). Determination of molecular symmetry in the α,α'-diethylbenzyl series. *J. chem. Soc.* 6–9.

CHALMERS, J. G. and CROWFOOT, DOROTHY (1941). The elimination of 3,4-benzopyrene from the animal body after subcutaneous injection. II. Changed benzopyrene. *Biochem. J.* **35**, 1270–5.

BERENBLUM, I., CROWFOOT, D., HOLIDAY, F. R., and SCHOENTAL, R. (1943). The metabolism of 3,4-benzopyrene in mice and rats. II. The identification of the isolated products as 8-hydroxy-3,4-benzopyrene and 3,4-benzopyrene-5,8-quinone. *Cancer Res.* **3**, 151–8.

CARLISLE, C. H. and CROWFOOT, D. (1943). X-ray crystallographic examination of fatty acids obtained from Mona wax. *Biochem. J.* **37**, 197–8.

CROWFOOT, D. and ROGERS-LOW, B. W. (1944). X-ray crystallography of gliotoxin. *Nature, Lond.* **153**, 651–2.

ABRAHAM, E. P., CROWFOOT, D. M., JOSEPH, A. E., and OSBORN, E. M. (1946). An antibacterial substance from *Arctium minus* and *Onopordon tauricum*. *Nature, Lond.* **158**, 744–5.

CROWFOOT, DOROTHY (1948). X-ray crystallographic studies of compounds of biochemical interest. *A. Rev. Biochem.* **17**, 115–46.

HODGKIN, D. CROWFOOT (1950). Crystallography, 1947, 1948, and 1949. Introduction. *Rep. Prog. Chem.* **46**, 57–8.

HODGKIN, D. CROWFOOT (1950). Crystal chemistry. *Rep. Prog. Chem.* **46**, 64–85.

DARWIN, CECILY and HODGKIN, DOROTHY C. (1950). Crystal structure of the dimer of *p*-bromonitrosobenzene. *Nature, Lond.* **166**, 827–8.

HODGKIN, D. CROWFOOT (1951). Crystallography. Introduction. *Rep. Prog. Chem.* **47**, 420.

HODGKIN, D. CROWFOOT, and PITT, G. J. (1951). Organic compounds. *Rep. Prog. Chem.* **47**, 432–69.

HODGKIN, D. C. and PERUTZ, M. F. (1951). Crystallography. *Rep. Prog. Chem.* **47**, 308–382.

MACKAY, MAUREEN and HODGKIN, DOROTHY CROWFOOT (1955). A crystallographic examination of the structure of morphine. *J. chem. Soc.* 3261–7.

HODGKIN, DOROTHY C. (1960). Molecules in crystals. *Nature, Lond.* **188**, 441–7.

DIETRICH, HANS and HODGKIN, DOROTHY CROWFOOT (1961). Crystal structure of the trans-dimer of nitrosoisobutane. *J. chem. Soc.* 3686–90.

HODGKIN, DOROTHY CROWFOOT (1965). X-ray analysis of complicated molecules. *Science N.Y.* **150**, 979–88; *Angew. Chem.* **77**, 954–62. (Nobel prize lecture.)

DALE, DAVID and HODGKIN, DOROTHY CROWFOOT (1965). The crystal structure of black nitrosylpentamminecobalt dichloride. *J. chem. Soc.* 1364–71.

CROWFOOT-HODGKIN, D. (1966). *Die Roentgen-Strukturanalyse einiger Biochemisch Interessant Molekule* [The X-ray structure analysis of some biochemically interesting molecules]. Westdeuts, Verlag: Opladen.

HODGKIN, DOROTHY CROWFOOT (1967). Some observations of crystallography, chemistry and medicine. *Harvey Lect.* **61**, 205–29.

HOSKINS, B. F., WHILLANS, F. D., DALE, D. H., and HODGKIN, DOROTHY C. (1969). Structure of the red nitrosylpentaamminecobalt (III) cation. *Chem. Commun.* 69–70.

CANDELORO, SOFIA, GRDENIC, D., TAYLOR, NOEL, THOMPSON, B., VISWAMITRA, M., and HODGKIN, DOROTHY C. (1969) Structure of ferroverdin. *Nature, Lond.* **224**, 589–91.

ANDERSON, B. F. and HODGKIN, DOROTHY C. (1969). The crystal structure of thiostrepton. *Acta crystallogr.* **A25**, S200.

ANDERSON, BRYAN, HODGKIN, DOROTHY C., and VISWAMITRA, M. A. (1970). Structure of thiostrepton. *Nature, Lond.* **225**, 223–5.

ADAMS, MARGARET JOAN, HODGKIN, DOROTHY C., and RAEBURN, URSULA A. (1970). Crystal structure of a complex of mercury (II) chloride and histidine hydrochloride. *J. chem. Soc. A* 2632–5.

HODGKIN, D. C. (1972). Gerhard Schmidt's first researches in X-ray crystallography. *Israel J. Chem.* **10**, 649–53.

CANDELORO DE SANCTIS, SOFIA and HODGKIN, DOROTHY C. (1973). Structure of ferroverdin. I. Monoclinic ferroverdin crystals. *Proc. R. Soc.* **B184**, 121–35.

CANDELORO DE SANCTIS, SOFIA, GRDENIC, D., TAYLOR, N., and HODGKIN, DOROTHY C. (1973). Structure of ferroverdin. II. Rhombohedral ferroverdin crystals. *Proc. R. Soc.* **B184**, 137–48.

HODGKIN, D. (1976). Structure of molecules in crystals. *Abstr. Pap. Am. chem. Soc.* 6.

HODGKIN, D. (1976). Growing points and sticking points in X-ray analysis. *Abstr. Pap. Am. chem. Soc.* 34.

HODGKIN, DOROTHY (1978). Discoveries and their uses. (Presidential Address of the British Association.)

Biographies and book reviews

HODGKIN, DOROTHY (1964). Book review: Globular protein molecules—their structure and dynamic properties, by Jacob Segal *et al. Acta crystallogr.* **17,** 74–5.

HODGKIN, DOROTHY (1964). 50 years of X-ray diffraction. *Nature, Lond.* **201,** 329–30.

HODGKIN, DOROTHY (1971). Kathleen Lonsdale, FRS, 1903–1971. *Chem. Brit.* **7,** 477–478.

HODGKIN, DOROTHY M. C. (1975). Kathleen Lonsdale, 28 January 1903–1 April 1971, elected FRS 1945. *Biogr. Mem. Fellows R. Soc.* **21,** 447–84.

HODGKIN, D. C. (1976). Dorothy Wrinch, an obituary. *Nature, Lond.* **260,** 564.

HODGKIN, D. C. (1976). Fascination for discovery. The development of X-ray analysis by the late Sir Lawrence Bragg (ed. D. C. Phillips and H. Lipson). *Nature, Lond.* **260,** 733.

HODGKIN, D. C. (1980). J. D. Bernal. *Biogr. Mem. Fellows R. Soc.* In press.

List of theses written under the supervision of Dorothy Hodgkin

D. Phil. theses

1942	D. P. Riley	The crystallography of some proteins and related substances.
1943	C.H. Carlisle	A crystallographic investigation of the stereochemistry of the sterols and related compounds.
1948	B. W. Rogers	X-ray crystallographic measurements in the penicillin series.
1951	D. Sayre	The X-ray analysis of complex organic compounds.
1952	P. M. Cowan	X-ray crystallographic studies in some peptides and proteins.
1955	J. Pickworth	Crystallographic studies on the structure of vitamin B_{12} and related compounds.
1956	M.S. Webster	An X-ray analytical study of the structure of some complex organic molecules.
1959	E. N. Maslen	Crystal structure of compounds related to penicillin.
1960	M. J. Kamper	X-ray crystallographic studies connected with vitamin B_{12}.
1961	M. M. Harding	An X-ray analytical study of the structure of insulin and related problems.
1962	D. H. Dale	A crystallographic study of some problems in the structure of co-ordination compounds.
1963	R. D. Diamand	A crystallographic study of the structure of some antibiotics.
1963	C. P. Saunderson	A study of some problems in steroid chemistry by X-ray crystallographic methods.
1965	A. S. Cooper	Experiments in the X-ray analysis of organic structures.
1966	M. N. G. James	X-ray crystallographic studies of some antibiotic peptides.
1967	F. H. Moore	X-ray and neutron-diffraction studies connected with the structure of vitamin B_{12}.
1967	C. E. Nockolds	An X-ray analytical study of structures related to vitamin B_{12}.
1968	M. J. Adams	X-ray analytical studies bearing on the structure of insulin.
1970	S. Candeloro	The study of some stereochemical problems by X-ray analysis.
1970	E. D. Edmond	Crystallographic studies related to the chemistry of corrin.
1970	H. R. Harrison	X-ray crystal structure analysis of some biologically interesting molecules.
1975	S. M. Cutfield	Structural studies of protein crystals.

Part II theses

1943	B. W. Low	An X-ray crystallographic study of some antibacterial compounds and their degradation products.
1948	M. H. Roberts	An X-ray crystallographic study of certain derivatives of 'gramicidin S'.

1948	L. Weissman	Studies in dimerisation (with E. J. Bowen).
1949	C. Darwin	The structure of some aromatic nitroso compounds.
1951	M. S. Curzon	An X-ray crystallographic study of suprasterol II.
1951	C. Gibb	An X-ray crystallographic investigation of some sterol and terpene derivatives.
1955	M. J. Harrison	Crystallographic measurements bearing on the structure of vitamin B_{12}.
1956	D. M. R. Martin	Crystallographic investigations of some nitroso compounds.
1957	T. J. Harper	An X-ray crystallographic investigation of the structure of *cis*-1-4-dichloro-1-4-dinitroso-cyclohexane.
1957	M. M. Aitken	The crystal structure of mercury dithizonate.
1958	P. J. Fisher	Crystallographic investigation into the structure of cuprous iodide methyl isocyanide.
1959	G. I. Hurdman	Crystallographic experiments on antibiotics.
1960	V. S. Armstrong	The crystal structure of 1:3:5:cyclooctatriene chromium tri-carbonyl.
1962	M. J. Adams	X-ray crystallographic studies of dithizone and related compounds.
1967	U. A. Raeburn	An X-ray crystallographic study of some metal–histidine complexes.

B.Sc. theses

1947	A. E. Joseph	Some methods of identification of organic compounds by means of crystallography.
1948	M. H. Roberts	An X-ray crystallographic study of certain derivatives of 'gramicidin S'.
1949	C. Darwin	The structure of some aromatic nitroso-compounds.
1959	G. I. M. Hurdman	Crystallographic experiments on antibiotics.
1960	V. S. Armstrong	The crystal structure of 1:3:5:cyclooctatriene chromium tri-carbonyl.
1963	T. J. Wiseman	The crystal structure of ferrous *bis* nioxime *bis* imidazole.
1968	R. A. Armstrong	X-ray crystallographic investigations of the structures of some metal complexes.

Index